Theory and Applications of Colloidal Suspension Rheology

An essential text on practical application, theory, and simulation, written by an international coalition of experts in the field and edited by the authors of *Colloidal Suspension Rheology*. This up-to-date work builds upon the prior work as a valuable guide to formulation and processing, as well as fundamental rheology of colloidal suspensions. Thematically, theory and simulation are connected to industrial applications by consideration of colloidal interactions, particle properties, and suspension microstructure. Important classes of model suspensions including gels, glasses, and soft particles are covered so as to develop a deeper understanding of industrial systems ranging from carbon black slurries, paints and coatings, asphalt, cement, and mine tailings, to natural suspensions such as biocolloids, protein solutions, and blood. Systematically presenting the established facts in this multidisciplinary field, this book is the perfect aid for academic researchers, graduate students, and industrial practitioners alike.

Norman J. Wagner holds the Unidel Robert L. Pigford Chair in Chemical and Biomolecular Engineering at the University of Delaware. Internationally recognized for his research and applications of colloidal suspensions, he received the Bingham medal and was elected to both the National Academy of Engineering and National Academy of Inventors.

Jan Mewis is Emeritus Professor of the Chemical Engineering Department at the Katholieke Universiteit Leuven. He has lectured worldwide and was chairman of the International Committee on Rheology. He received the Bingham medal in the US, as well as the Gold medal of the British Society of Rheology.

Cambridge Series in Chemical Engineering

Books in the Series

Baldea and Daoutidis, *Dynamics and Nonlinear Control of Integrated Process Systems*
Chamberlin, *Radioactive Aerosols*
Chau, *Process Control: A First Course with Matlab*
Cussler, *Diffusion: Mass Transfer in Fluid Systems, Third Edition*
Cussler and Moggridge, *Chemical Product Design, Second Edition*
De Pablo and Schieber, *Molecular Engineering Thermodynamics*
Deen, *Introduction to Chemical Engineering Fluid Mechanics*
Denn, *Chemical Engineering: An Introduction*
Denn, *Polymer Melt Processing: Foundations in Fluid Mechanics and Heat Transfer*
Dorfman and Daoutidis, *Numerical Methods with Chemical Engineering Applications*
Duncan and Reimer, *Chemical Engineering Design and Analysis: An Introduction 2E*
Fan, *Chemical Looping Partial Oxidation Gasification, Reforming, and Chemical Syntheses*
Fan and Zhu, *Principles of Gas–Solid Flows*
Fox, *Computational Models for Turbulent Reacting Flows*
Franses, *Thermodynamics with Chemical Engineering Applications*
Leal, *Advanced Transport Phenomena: Fluid Mechanics and Convective Transport Processes*
Lim and Shin, *Fed-Batch Cultures: Principles and Applications of Semi-Batch Bioreactors*
Litster, *Design and Processing of Particulate Products*
Marchisio and Fox, *Computational Models for Polydisperse Particulate and Multiphase Systems*
Mewis and Wagner, *Colloidal Suspension Rheology*
Morbidelli, Gavriilidis, and Varma, *Catalyst Design: Optimal Distribution of Catalyst in Pellets, Reactors, and Membranes*
Nicoud, *Chromatographic Processes*
Noble and Terry, *Principles of Chemical Separations with Environmental Applications*
Orbey and Sandler, *Modeling Vapor–Liquid Equilibria: Cubic Equations of State and their Mixing Rules*
Pfister, Nicoud, and Morbidelli, *Continuous Biopharmaceutical Processes: Chromatography, Bioconjugation, and Protein Stability*
Petyluk, *Distillation Theory and its Applications to Optimal Design of Separation Units*
Ramkrishna and Song, *Cybernetic Modeling for Bioreaction Engineering*
Rao and Nott, *An Introduction to Granular Flow*
Russell, Robinson, and Wagner, *Mass and Heat Transfer: Analysis of Mass Contactors and Heat Exchangers*
Schobert, *Chemistry of Fossil Fuels and Biofuels*
Shell, *Thermodynamics and Statistical Mechanics*
Sirkar, *Separation of Molecules, Macromolecules and Particles: Principles, Phenomena and Processes*
Slattery, *Advanced Transport Phenomena*
Varma, Morbidelli, and Wu, *Parametric Sensitivity in Chemical Systems*
Vassiliadis, del Rio Chanona, Yuan, and Kähm, *Optimization for Chemical and Biochemical Engineering*
Weatherley, *Intensification of Liquid–Liquid Processes*
Wolf, Bielser, and Morbidelli, *Perfusion Cell Culture Processes for Biopharmaceuticals*

Theory and Applications of Colloidal Suspension Rheology

Edited by

NORMAN J. WAGNER
University of Delaware

JAN MEWIS
KU Leuven

CAMBRIDGE
UNIVERSITY PRESS

University Printing House, Cambridge CB2 8BS, United Kingdom

One Liberty Plaza, 20th Floor, New York, NY 10006, USA

477 Williamstown Road, Port Melbourne, VIC 3207, Australia

314–321, 3rd Floor, Plot 3, Splendor Forum, Jasola District Centre, New Delhi – 110025, India

79 Anson Road, #06–04/06, Singapore 079906

Cambridge University Press is part of the University of Cambridge.

It furthers the University's mission by disseminating knowledge in the pursuit of education, learning, and research at the highest international levels of excellence.

www.cambridge.org
Information on this title: www.cambridge.org/9781108423038
DOI: 10.1017/9781108394826

First published 2021

A catalogue record for this publication is available from the British Library.

ISBN 978-1-108-42303-8 Hardback

To Sabine and Ria, and all of the students, postdocs, and colleagues who enriched our research and teaching.

Contents

Contributors

Antony N. Beris
University of Delaware, USA

Surita Bhatia
Stony Brook University, USA

Michel Cloitre
ESPCI, CNRS, France

Jan K. G. Dhont
Forschungszentrum Jülich, Germany

Wendy Hom
NYU Tandon School of Engineering, USA

Ronald G. Larson
University of Michigan, USA

Jan Mewis
KU Leuven, Belgium

Gerhard Nägele
Forschungszentrum Jülich, Germany

George Petekidis
University of Crete and FORTH, Greece

Jeffrey Richards
Northwestern University, USA

Nicolas Roussel
Laboratoire Navier CNRS, France

Alex Routh
University of Cambridge, UK

Peter Scales
University of Melbourne, Australia

Dimitris Vlassopoulos
University of Crete and FORTH, Greece

Thomas Voigtmann
Deutsches Zentrum für Luft- und Raumfahrt, Germany

Norman J. Wagner
University of Delaware, USA

Preface

Micron- or nano-sized particles are omnipresent; they occur in nature and have been used by man since antiquity. They are often suspended in water or organic liquids, e.g., when using or producing minerals, composites, paints, inks, etc. In such cases the complex flow properties are of primary importance as they determine the handling and application behavior. The extensive knowledge acquired over the last few decades about their rheology can be used to optimally design these products and the processes they are used in. It also provides a powerful tool in fundamental studies of this class of materials. With *Colloidal Suspension Rheology* (Cambridge University Press, 2012) the basics of this discipline were brought together in a single volume, providing an efficient tool for those starting research, design, or formulation of such materials. The necessary size limitations of the book, however, made it necessary to refer to the literature for some important underlying scientific aspects. Neither was is feasible to demonstrate how these insights into colloidal suspension rheology could be applied to engineer real classes of materials. Therefore, it is warranted to devote an entire, independent book to include these theoretical elements as well as examples of applying colloidal rheology. This required contributions from experts in the various fields, while editing was essential to ensure that the book remained a consistent entity suitable for studying the subject. The goal of the current book is not only to present a discussion of the theory and applications of colloidal suspension rheology, but to do so in a systematic and coherent presentation of the best-known status of the field.

The book starts with a short review of the basic principles of colloid science and rheology, as required for the book. The next chapters deal with three pillars of modern colloidal rheology. Chapter 2 (G. Nägele, J. Dhont, and T. Voigtmann), on theory, explains how the complex hydrodynamics of particles in a flowing medium are approached fundamentally, which requires accounting for the suspension's microstructure as it evolves during flow. Two theoretical routes are presented and important theorems and formulas are developed, some of which become the basis of simulation methods, microrheological techniques, and experimental methods. Analytical solutions to such problems are limited. Consequently, numerical simulation at the particle level becomes an important tool to gain further insight. Therefore, Chapter 3 (R. Larson) surveys the possibilities and restrictions of simulation methods used for research into colloidal suspension rheology. The third element of this part of the book, Chapter 4 (N. Wagner), presents two experimental techniques used to analyze the microstructure of colloidal suspensions during flow.

Taken together, these four chapters provide foundational material for the model system studies and applications to follow.

The focus of Chapters 5 and 6 are important classes of colloidal dispersions that occur widely in practice and where significant, recent advances in our understanding of model system rheology and microstructure have been made. The resulting knowledge can be applied to solve processing and formulation problems involving gels, glasses, and soft colloidal dispersions more broadly. Chapter 5 (G. Petekidis and N. Wagner) provides a guide to the complex microstructure and the resulting rheological properties of colloidal glasses and gels. Key questions addressed include the conditions necessary for gel and glass formation and how differences in colloidal interactions can lead to dramatically different material properties in the gel or glass states. The subject of Chapter 6 (D. Vlassopoulos and M. Cloitre) is the behavior of colloidal suspensions containing soft particles, such as polymer-coated particles, star polymers, and microgels. Here, it is shown how particle softness can fundamentally change the structure and rheological properties of colloids, and therefore can be used to tailor the properties of suspensions. This chapter provides understanding that bridges the model systems presented in Chapter 1 and in *Colloidal Suspension Rheology* with the more complex and applied systems addressed in the later chapters.

Chapters 7–9 provide further illustration of how these advances in colloidal rheology are being applied to real systems. Suspensions of biocolloids, including protein solutions, are presented in this context in Chapter 7 (S. Bhatia and W. Hom). Particle softness and self-assembly of surfactants to form micelles can be treated within the context of soft colloids, while the complexity of particle heterogeneity presents challenges for protein solutions that can be addressed to some level. Blood is a thixotropic suspension, and its rheology and modeling are reviewed in Chapter 8 (A. Beris). Finally, in Chapter 9 the application of colloidal rheology to a number of industrial products is demonstrated. These include paints (A. Routh), carbon black dispersions (J. Richards), asphalts (N. Wagner), cement-based products (N. Roussel), and the large scale processing of minerals (P. Scales). While not encyclopedic, these present some of the most common, and critically important examples of industrially relevant colloidal suspensions where formulation for rheological control is often essential. We sincerely hope and anticipate that the knowledge gained through reviewing the material presented in this book will be of value both to researchers of colloidal suspension rheology as well as those seeking guidance in product formulation and design.

Beyond those whose writing is explicitly contained in this book, we also wish to thank our many colleagues and students who have enthusiastically supported the concept of cogent and concise presentation of theoretical and applied aspects of colloidal suspension rheology. Conversations with William Russel (Princeton), Eric Furst (U. Delaware) and Jan Vermant (ETHZ) on aspects of this text are gratefully acknowledged. While an extensive literature is cited and many key works discussed, space limitations necessitate omissions of many fine examples in addition to those presented here, as well as additional applications and theory not explicitly covered. While endeavoring to provide a concise and coherent presentation of what we view to be established knowledge, some new and still unproven, or controversial, avenues of research in the discipline are included as brief discussions of future possibilities. As authors, we assume responsibility for any omissions or errors throughout, and we welcome suggestions for further areas and topics for scholarly research in colloidal suspension rheology not covered here or in the prior, complementary book.

A number of students also contributed to the preparation of figures and referencing in this manuscript, and we thank Cameron Mertz, Laura Smith, and Rong Song, all graduates of the University of Delaware. More broadly, this book would not be possible but for the creative work of generations of students and postdocs who have worked in our laboratories. We acknowledge that this book benefitted from our associations with the Society of Rheology and the International Fine Particle Research Institute, as well as research funded by Proctor and Gamble, DuPont, and Genentech. Financial support for research in the laboratory of N. Wagner by the National Science Foundation and National Institute of Standards and Technology, as well as general support by the University of Delaware and KU Leuven are gratefully acknowledged.

General List of Symbols

A	Hamaker constant [J]
a	particle radius [m]
a_i	particle radius of species/size i [m]
a_T	shift factor for temperature effects
c	mass concentration [kg/m^3]
$\mathbf{D}$	rate of strain tensor [1/s]
D_{ij}	components of the rate of strain tensor [1/s]
$\boldsymbol{\mathcal{D}}$	diffusivity tensor [m^2/s]
$\mathcal{D}_0$	Stokes–Einstein–Sutherland diffusivity [m^2/s]
$\mathcal{D}_{ij}$	components of the diffusivity tensor [m^2/s]
$\boldsymbol{\mathcal{D}}^s$	self diffusivity tensor [m^2/s]
$\mathcal{D}_{ij}^s$	components of the self diffusivity tensor [m^2/s]
$\mathcal{D}^{ss}$	short-time self diffusion coefficient [m^2/s]
e	charge of an electron [C]
F	force [N]
G	shear modulus [N/m^2]
G'	storage modulus [N/m^2]
G''	loss modulus [N/m^2]
G_{pl}	plateau modulus [N/m^2]
g	gravity or acceleration constant [m/s^2]
$g(r)$	radial distribution function [–]
h	surface-to-surface distance between particles [m]
$\mathbf{I}$	unit tensor [–]
k	coefficient in the power law term of nonlinear viscosity models (N.s^n/m)
k'	coefficient in the Cross equation [s]
k_B	Boltzmann constant [J/K]
k_H	Huggins coefficient [–]
L	length [m]
l_b	Bjerrum Length [1/m]
N	number of particles [–]
N_A	Avogadro's number [1/mol]
N_i	ith normal stress difference [Pa]
n	number density [1/m^3]

P pressure [Pa]
P_y compressional yield stress [Pa]
$\boldsymbol{q}$ scattering vector [1/m]
R radius [m]
R_g radius of gyration [m]
r center-to-center distance between particles [m]
S entropy [J/K]
$S(q)$ static structure factor [–]
$S(q,t)$ dynamic structure factor [–]
t time [s]
T temperature [K]
U depth of square well potential [J]
V volume [m^3]
v local velocity [m/s]
$\boldsymbol{v}$ velocity vector [m/s]
v_i velocity component in the i direction (i = x, y, or z) [m/s]
x Cartesian coordinate, in simple shear flow the flow direction [m]
y Cartesian coordinate, in simple shear flow the velocity gradient direction [m]
z Cartesian coordinate, in simple shear flow the vorticity direction [m]
z_i valence of an electrolyte

Greek Symbols

γ strain [–]
γ_0 peak strain [–]
$\dot{\gamma}$ shear rate [s^{-1}]
Δ half-width of the square well potential [m]
δ phase angle [–]
ε dielectric constant [–]
ε_o permittivity of vacuum [F/m]
η (suspension) viscosity [Pa.s]
η' dynamic viscosity [Pa.s]
$[\eta]$ intrinsic viscosity [cm^3/g]
$[\eta]'$ dimensionless intrinsic viscosity [–]
κ Debye-Hückel constant [1/m]
λ structure parameter in thixotropy models [–]
ν number of particles or molecules per volume [$1/m^3$]
Π osmotic pressure [Pa]
θ polar coordinate [–]
ρ density [kg/m^3]
$\boldsymbol{\sigma}$ extra stress tensor [Pa]
σ shear stress in simple shear flow [Pa]
σ_y yield stress [Pa]

σ_y^B Bingham yield stress [Pa]
σ_y^d dynamic yield stress [Pa]
τ relaxation time [s]
τ_B Baxter stickiness parameter [–]
ϕ particle volume fraction [–]
Φ particle interaction potential [J]
Υ_i ith normal stress difference coefficient for suspensions [–]
ψ_i ith normal stress coefficient [Pa.s^2]
Ψ potential field [V]
Ψ_s surface potential [V]
ζ zeta potential [V]
ω frequency [rad/s]
Ω rotational speed [1/s]

Subscripts

D diffusion
dep depletion
E extensional
eff effective
el elastic contribution
FCC face-centered cubic
g glass
gel gel
IGT ideal glass transition
J jamming
lin linearity limit
m suspending medium/mean value
M Maxwell
MRJ maximally random jammed state
max maximum value
p particle
pl plastic
r relative
rcp random close packing
STS shear thickened state
y yield condition
0 limiting value in the zero shear limit
∞ limiting value at high shear rates or frequencies

Superscripts

B Brownian
C Casson
d dispersion
dep depletion
el electrostatic
g gravity
h hydrodynamic
hcY hard core Yukawa (potential)
hs hard sphere
H Herschel-Bulkley
I interparticle contribution
m power law index in the Cross model
n power law index
s surface
$*$ complex

Dimensionless Numbers

Bi Bingham number (σ_y/σ)
De Deborah number (characteristic relaxation time of the system/characteristic time of the flow-specific expressions vary, depending on the case)
Mn Mason number $(6\pi\eta_m a^2\dot{\gamma}/(d\Phi/dr)|_{\max})$
Pe Péclet number $(6\pi\eta_m a^3\dot{\gamma}/k_B T)$
Pe_{dep} Péclet number for depletion gel $(12\pi\xi a^3\dot{\gamma}/U_{dep(r=2a)})$
Pe_g Gravitational Péclet number $Pe_g = \frac{4\pi a^4(\Delta\rho)g}{3k_B T}\phi^{\left(\frac{d_f-1}{d_f-3}\right)}$
Re Reynolds number $(\rho VD/\eta)$
Re_p Particle Reynolds number $(\rho\dot{\gamma}a^2/\eta_m)$
Wi Weissenberg number (N_1/σ)

Useful Physical Constants and Values

(note that many CODATA internationally recommended values can be found at physics.nist.gov/cuu/Constants/)

Constant		Value
e	elementary charge	$1.602176487 \times 10^{-19}$ C
g	standard acceleration of gravity	9.80665 m s^{-2}
k_B	Boltzmann's constant	$1.3806504 \times 10^{-23}$ J K^{-1}
m_u	atomic mass unit	$1.660538782 \times 10^{-27}$ kg
N_A	Avogadro's number	$6.022214170 \times 10^{23}$ mol^{-1}
R	molar gas constant	8.314472 J mol^{-1} K^{-1}
ε_0	electric permittivity of vacuum	$8.854187817 \times 10^{-12}$ C^2 N^{-1} m^{-2} [F m^{-1}]
μ_0	vacuum permeability	$4\pi \times 10^{-7}$ N A^{-2}

Characteristic Values:

k_BT	4.1×10^{-21} J (at room temperature)
k_BT/e	25.7 mV (at room temperature)
κ^{-1}	3.08 nm for a 10 mM 1:1 electrolyte in water at room temperature
l_b	0.7 nm for water at room temperature
Q	typically of O(1) μC cm^{-2}

Properties of water at 298 K:

ε	relative dielectric constant	80
η	viscosity	8.90×10^{-4} Pa s
ρ	density	997 kg m^{-3}

Useful Hamaker constants in water (units 10^{-20} J)

decane	0.46
fused silica	0.85
gold	30
polystyrene	1.3
poly(methyl methacrylate)	1.05
poly(tetrafluroethylene)	0.33

1 Introduction to Colloidal Suspension Rheology

Norman J. Wagner and Jan Mewis

1.1 Structure of this Chapter and the Book

Dispersions of small particles in liquid media, either natural or manmade, can be found essentially everywhere. One of their major characteristics is that their physical properties can vary over very large ranges. This is illustrated in particular by their flow behavior, which can be that of a simple Newtonian fluid, but can also display strong nonlinearities, including shear thinning, shear thickening, a yield stress, viscoelasticity, thixotropy, and even solid-like behavior. This vast variability in rheological behavior is understood to be due, in large part, to the variation of the *microstructure*. It is a central tenet of colloid science and rheology that these bulk, physical properties arise from microstructure, and that microstructure itself is the result of both the colloidal particle characteristics and the interactions between these particles as mediated by the suspending medium. In *Colloidal Suspension Rheology*, which henceforth will be referred to as *CSR*, the basic rheological phenomena and how these collective, bulk properties arise from ever increasing complexity in particle nature, fluid properties, and particle interactions, have been systematically reviewed [1]. In the present book, we expand on the underlying theoretical, simulation, and experimental approaches to determining such microstructure, as well as illustrate how colloid rheology can be applied to real-world systems and products. This requires some insight into basic colloidal phenomena, which can be found in CSR and many textbooks [2–5], and in the brief review in this chapter.

Figure 1.1 provides a conceptual roadmap to this book. As a guide to the reader, this book is designed to complement and extend the teachings presented in *CSR* by expounding on the relationship between microstructure and bulk rheology as well as demonstrating how powerful the colloid rheology viewpoint can be for understanding some of humanity's most important complex fluids and materials. Importantly, this chapter briefly summarizes our modern understanding of the canonical case of colloidal hard spheres (HS) as a basis for understanding all real-world systems. Hard sphere colloidal suspensions are the conceptually simplest colloidal suspensions as the interparticle potential is simply the excluded volume interactions. However, as the particles are Brownian and immersed in a medium, this leads to a breadth of rich rheological behavior, as will be shown. As colloidal HS suspensions are actually very difficult to realize in the laboratory and often not suited for many practical applications, we next define model colloidal interaction potentials and simply present their

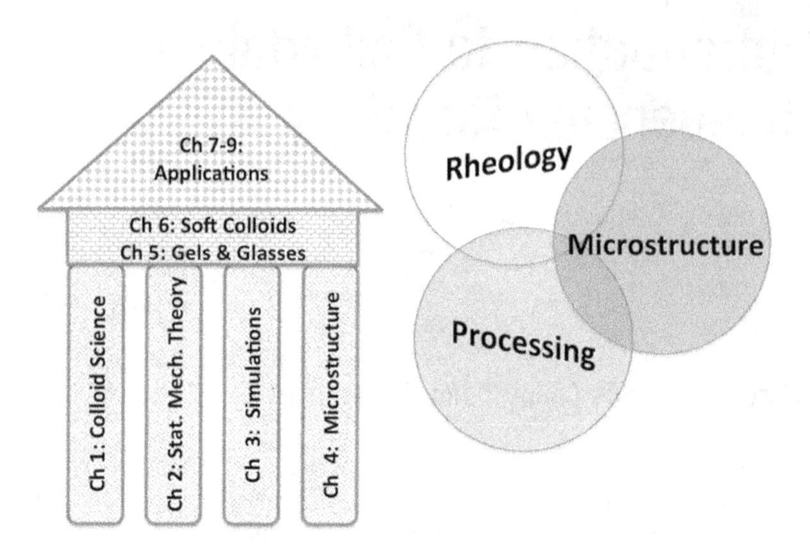

Figure 1.1 Graphical depiction of the structure of this book. The left figure shows the viewpoint of colloidal suspensions, where fundamental colloid science coupled with theory and simulation leads to predictions for microstructure. This enables understanding the rheology of gels, glasses, and soft colloidal materials, which form a foundation for understanding a broad range of applications. It replaces earlier, phenomenological attempts to correlate rheology directly with colloidal particle properties. The right figure depicts the realistic situation where processing and applications affects the microstructure and hence rheology. The structure of this book parallels this understanding.

consequences with respect to state behavior. Colloidal microstructure formation leads to *thixotropy*, which is perhaps one of the most important, practical issues facing formulators and processors working with colloidal dispersions. Thus, we provide a short guide to the salient concepts about thixotropy and its distinction from viscoelasticity in this chapter as well. A very brief introduction to basic rheological terminology is included in the Appendix for reference.

Theory for the microstructure and resulting dynamical properties is presented for both the classical Smoluchowski equation and mode coupling theory in Chapter 2 by Nägele, Dhont, and Voigtmann in some detail so as to provide a foundation for understanding and further research. They also develop the Langevin equation, which is the basis for the simulation methods presented in Chapter 3. Colloidal suspensions are inherently many-body systems, which limits our ability to achieve quantitative predictions from the theory. Larson provides an overview of simulation methods that provide computational routes to solving colloidal-scale models for suspensions in Chapter 3. Chapter 4 provides a brief introduction to two classes of measurement techniques for the microstructure, showing how theory and experiment can be tested at the microscale. These first four chapters cement the relationship between particle properties and interactions, microstructure, and rheological behavior, providing a foundation for understanding practical colloidal suspensions and their applications, which are the subject of Chapters 5–9.

While, in principle, a detailed understanding of the colloidal particles themselves and their interactions in the suspending medium should enable prediction of the rheological behavior, many real systems also depend on more than the chemistry of the system. In particular, how suspensions are formulated and processed is usually of critical importance to their properties. While early attempts in the field to link particle properties directly to product processing and rheology were valuable to product formulation, it has proven much more fruitful to consider not only particle properties and particle interactions, but also the suspension microstructure and how it is influenced by preparation, processing, and use.

Following this theme, the next five chapters examine systems of more complexity and practical interest. Gels and Glasses are the topic of Chapter 5, where Petekidis and Wagner review the rheology of heterogeneous and homogeneous colloidal gels and colloidal glasses. In Chapter 6, Vlassopoulos and Cloitre explore how particle softness can be exploited to produce a wide range of rheological behavior. Particle softness is often evident in biologically relevant colloids, which are reviewed by Bhatia and Hom in Chapter 7. Blood is thixotropic and its rheology of great practical interest. The colloidal viewpoint of blood rheology and associated thixotropic modeling is summarized by Beris in Chapter 8. Finally, in Chapter 9 insight into how powerful the colloidal viewpoint can be to formulate and process practical systems is presented from paints (Routh) and carbon blacks (Richards) to some of mankind's most important complex fluids such as bitumen, asphalt, cement (Roussel), and large scale processing (Scales). This latter half of the text builds upon the foundational concepts from the earlier chapters and *CSR*.

1.2 Introduction and Observations

The reader is almost certainly familiar with the celebrated Einstein relationship for the relative viscosity, η_r, which relates the increase in viscosity over the medium viscosity, η_m, upon the addition of particles (colloidal or otherwise) at a low concentration, characterized by the hard sphere volume fraction ϕ, as:

$$\eta_r = \eta/\eta_m = 1 + 2.5\phi. \tag{1.1}$$

While widely employed, the simplicity of this theoretical result for dilute suspension belies the incredible complexity and richness of colloidal suspension rheology, which we endeavor to provide guidance for learning and understanding in the present book. Essentially, the rich physics of colloidal systems appear in higher order terms that have to be added to the equation above for nondilute systems. Terms of order ϕ^2 and higher order account for interactions between colloidal particles, as well as for the resultant structure that develops in such suspensions. Structure formation can lead to highly nonlinear, viscoelastic, and thixotropic behavior, including a yield stress. Hence, understanding structure-property relationships is essential for effective suspension formulation and product development.

The basic features of the shear rheology of colloidal suspensions are shown in Figure 1.2. This illustrates the increasing complex rheological behavior that emerges

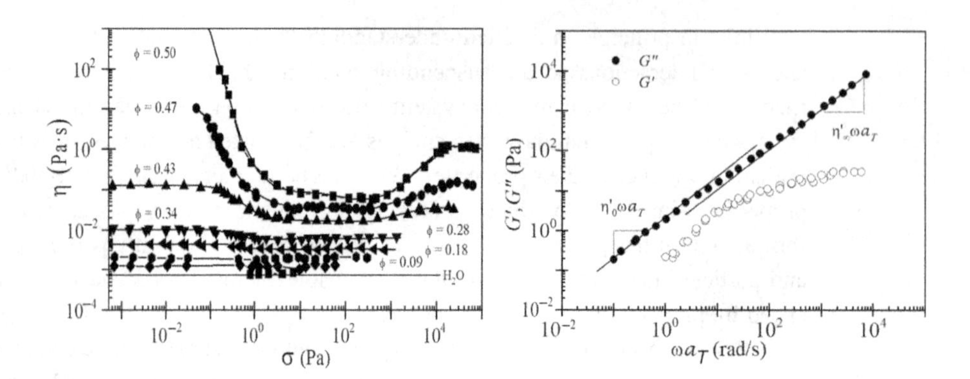

Figure 1.2 Representative rheological behavior of colloidal dispersions. Left: Shear viscosity of aqueous colloidal latex suspensions as a function of the applied shear stress showing the building of viscosity and the appearance of a yield stress, shear thinning, and shear thickening behavior with increasing particle volume fraction. Data from Laun [7], also appearing in *CSR*, figure 3.1; Right: Linear viscoelastic moduli for a concentrated colloidal suspension showing characteristic behavior of a loss modulus being dominated by hydrodynamic dissipation and an elastic modulus arising from Brownian motion. Data from Shikata and Pearson [8], also appearing in *CSR*, figure 3.5

at large volume fractions, including shear thinning and shear thickening as well as viscoelasticity The behavior of various systems can vary drastically as colloidal particles are strongly influenced by surface interactions acting on the nanoscale, with dramatic effects on colloidal stability and phase behavior, A more complete discussion of this background knowledge can be found in *CSR* and related textbooks [2–6]. As a foundation for what follows, we first review the well-developed understanding of hard sphere colloidal dispersions.

1.3 Colloidal Hard Spheres

Colloidal dispersions of hard spheres constitute a canonical system valuable for understanding real-world colloidal dispersions and will be summarized here for this purpose. Hard spheres are *athermal*[1] and assumed to be perfect spheres interacting only through excluded volume and hydrodynamic interactions. The particles are immersed in a suspending medium, which is taken to be an incompressible Newtonian fluid. Hence, the motion of any particle creates forces acting on *all* particles due to the requirement that the suspending medium respond to this motion. Hydrodynamic interactions are ubiquitous and distinguish colloidal dispersions from molecular liquids and gases, and consequently, may not be neglected. Furthermore,

[1] The term *athermal* indicates that the idealized HS interactions have no energy scale, because the interparticle interaction is either zero when separated or infinity upon contact. Therefore, there is no natural energy scale and thus, the phase behavior does not depend on temperature.

hydrodynamic interactions are very long range in nature and this leads to many important phenomena.[2] The particle radius sets a characteristic length scale, along with the separation distance between particles in dispersion. Colloidal hard spheres are also subject to Brownian motion, which sets the timescale governing the rheology as well as the scale for stress, and thus elasticity, in colloidal dispersions.

1.3.1 Characteristic Properties of Brownian Particles

Colloidal dispersions are comprised of discrete particles suspended in a wetting medium, which is often a Newtonian[3] fluid. This physical picture requires a distinct separation in timescales between the motion of the particles and that of the atoms/molecules comprising the suspending medium. A good rule of thumb is a minimum factor of 10 in size scale, which translates to three orders of magnitude difference in mass between the discrete particles and the atoms/molecules comprising the suspending medium. This sets the lower range of the colloidal scale, which includes nanoparticles and proteins of only nanometers in size, for example.

Although the fluid is described as a continuum, the thermal motion of the atoms comprising the fluid leads to stochastic collisions with the colloidal particles. For particles that are not too large, these stochastic collisions lead to a noticeable random motion of the particle that is visible in an optical microscope, such as reported by the botanist Robert Brown.[4] This *Brownian motion* becomes negligible for more massive particles of order tens of microns in characteristic size, thus defining the upper end of the colloidal scale.

Brownian motion is a defining trait of colloidal particles, where the motion is driven by the thermal energy of the solvent. On a per atom, molecule, or particle basis, the thermal energy scales by k_BT, i.e., Boltzmann's constant times the absolute temperature. To derive a characteristic force from the energy one needs a suitable length scale, for which the particle radius a is a logical choice. Hence the characteristic Brownian force is given by:

$$F^B = k_BT/a. \tag{1.2}$$

This force ranges from $\sim 10^{-12}$ to 10^{-15} Newtons for particles from $\sim$10 nm to 10 μm in size at room temperature.

Energy is conserved in a colloidal suspension at rest and so the energy imparted to the colloid by collisions with the atoms or molecules comprising the suspending

[2] The long-range nature creates mathematical challenges for computing the dissipation in the fluid phase during flow, which requires sophisticated mathematical methods and advanced computational methods, as will be discussed in Chapters 2 and 3.

[3] Throughout we refer to the suspending medium which, unless otherwise indicated, will be assumed to behave as an incompressible, Newtonian (i.e., constant viscosity) continuum.

[4] Robert Brown was a botanist who reported the phenomenon in 1827. Jean Perrin was awarded the 1926 Nobel prize for proving a concept central for atomistic theory. Specifically, he calculated Avogadro's number by methods including the quantitative observation of Brownian motion and application of the Stokes–Einstein–Sutherland equation.

medium is returned to the fluid in the form of viscous dissipation between the moving colloid and the surrounding medium. Such viscous drag is given by Stokes' law [9]:

$$F^h = 6\pi\eta_m v_p, \tag{1.3}$$

in which v_p is the velocity of the particle with respect to the surrounding fluid. This characteristic velocity is obtained by balancing the Brownian and hydrodynamic forces, on average. Recognizing that the velocity is the ratio of a characteristic displacement, here given by a, and a characteristic time, t_D, yields:

$$t_\mathrm{D} \sim \frac{6\pi\eta_m a^3}{k_\mathrm{B}T}. \tag{1.4}$$

This characteristic timescale, which ranges from nanoseconds to seconds for colloids in water, is important for understanding relaxation effects, transport, and diffusion in dilute colloidal suspensions. For example, from the theory of stochastic processes, a diffusion coefficient for a single colloidal particle in a suspending medium can be defined as:

$$\mathcal{D}_0 = \frac{a^2}{t_\mathrm{D}} = \frac{k_\mathrm{B}T}{6\pi\eta_m a}, \tag{1.5}$$

known as the Stokes–Einstein–Sutherland equation [10,11]. The mean square displacement $\left\langle \left(\Delta r(t)^2\right)\right\rangle$ a particle has traveled over time t as a result of this process is given by the Einstein–von Smoluchowski equation [12]:

$$\lim_{t\to\infty}\left\langle \left(\Delta r(t)^2\right)\right\rangle = 6\mathcal{D}_0 t. \tag{1.6}$$

The characteristic timescale for Brownian diffusion competes with other processes, such as advection rates due to flow. The ratio of the rate of advection over that of diffusion is known as the Péclet number, *Pe*. For shear flow, the rate of advection is given by the shear rate $\dot{\gamma}$, yielding:

$$Pe = \frac{\text{advection rate}}{\text{diffusion rate}} = \dot{\gamma} t_\mathrm{D} = \frac{6\pi\eta_m a^3\dot{\gamma}}{k_\mathrm{B}T}. \tag{1.7}$$

The characteristic time for diffusion is taken as given in Eq. (1.4). The Péclet number can also be considered a dimensionless shear rate. When this number is much smaller than unity, Brownian motion dominates the advection caused by flow and the suspension structure remains near equilibrium. On the other hand, for $Pe \gg 1$ flow effects dominate, with the possibility of dramatic effects.

Brownian motion also contributes to the suspension stress. Stress is energy per volume and so, a characteristic stress for Brownian motion scales inversely with the *volume* of the particle, which scales with the characteristic size cubed, leading to:

$$\sigma^\mathrm{B} = \frac{k_\mathrm{B}T}{a^3}. \tag{1.8}$$

Typical values range from 10^3 to 10^{-6} Pa across the colloidal size range.

Importantly, the force of gravity F^g acts on a spherical colloidal particle of density ρ_p and volume V_p in a suspending medium of density ρ_m as:

$$F^g = \Delta\rho V_p g = (\rho_p - \rho_m)(4/3)\pi a^3 g. \tag{1.9}$$

A typical gravitational force for a particle with a 100 kg/m^3 larger density than the surrounding medium and a particle size of 1 μm is $F^g \sim 4\times10^{-15}$ N, or ~4 fN. Thus, the gravitational force is often comparable to or less than the Brownian force for colloidal particles. A sedimentation velocity can be calculated from this force balanced against the Stokes drag given in Eq. (1.3). In the rest of this chapter, gravitational effects will be neglected.

Colloidal forces introduce additional phenomena, including rotational Brownian motion and a rotational Péclet number, as well as elasticity that arises from such rotational motion. These effects depend on particle shape. They are covered more completely in Chapter 5 of *CSR*, which details our understanding of particle shape effects, and are not reproduced here.

Finally, colloids are sufficiently small such that inertial effects on the particle scale can be neglected, even for turbulent flows. The particle Reynolds number Re_p gauges the relative importance of inertial to viscous forces on the particle length scale:

$$Re_p = \frac{\rho\dot{\gamma}a^2}{\eta_m}. \tag{1.10}$$

Simulations show significant inertial effects arise in suspension rheology when $Re_p \sim O(0.1)$ [13], which can readily be achieved for micron-sized particles in strong flows. All of these properties and their characteristic properties are summarized in Table 1.1.

1.3.2 Brownian Hard Sphere Phase Behavior and Diffusion

Dispersions of perfect Brownian hard spheres have a relatively (and deceptively!) simple phase behavior (shown in Figure 1.3), with the sole equilibrium phase transition being from a disordered fluid at volume fraction $\phi = 0.494$ to a hexagonal crystal at $\phi = 0.545$ with a limiting volume fraction at hexagonal close packing ($\phi = 0.740$). Realizing this experimentally required microgravity experiments on the International Space Station to avoid the effects of gravitational settling and gravity-induced convection [14]. Images illustrating this phase behavior from pioneering work on hard sphere-like PMMA particles coated with a thin, polymer brush [15] are shown in Figure 5.1.

Crystallization in the laboratory is often slow or frustrated, so phase separation is suppressed in favor of a metastable disordered phase that can persist to random close packing (rcp) or "maximally random jamming" (MRJ) [16]. Calculations suggest that jamming occurs with the maximum degree of randomness at $\phi_{\text{rcp}} = 0.637 \pm 0.0015$ [17]. Theory and simulations suggest the existence of "cages" comprised of neighboring particles in the fluid above $\phi \geq 0.4$. Two types of diffusive motion can be

Table 1.1 Summary of characteristic properties of Brownian suspensions

Term	Meaning	Characteristic Values	Units
a	Particle Radius	10^{-9}–10^{-6}	m
$F^B = k_B T/a$	Brownian Force (Eq. 1.2)	10^{-12}–10^{-15}	N
$\sigma^B = k_B T/a^3$	Brownian Stress (Eq. 1.8)	10^3–10^{-6}	Pa
$t_D \sim 6\pi\eta_m a^3/k_B T$	Brownian Relaxation Time (Eq. 1.4)	10^{-9}–1	s
$\mathcal{D}_0 = a^2/t_D = k_B T/6\pi\eta_m a$	Diffusivity (Eq. 1.5)	10^{-18}–10^{-12}	m^2/s
$F^g = \Delta\rho V_p g$	Gravitational Force (Eq. 1.9)	10^{-24}–10^{-15}	N
$Pe = 6\pi\eta_m a^3\dot{\gamma}/k_B T$	Péclet Number (Eq. 1.7)	$\dot{\gamma} t_D$	–
$Re_p = \rho\dot{\gamma}a^2/\eta_m$	Particle Reynolds Number (Eq. 1.10)	0–1	–

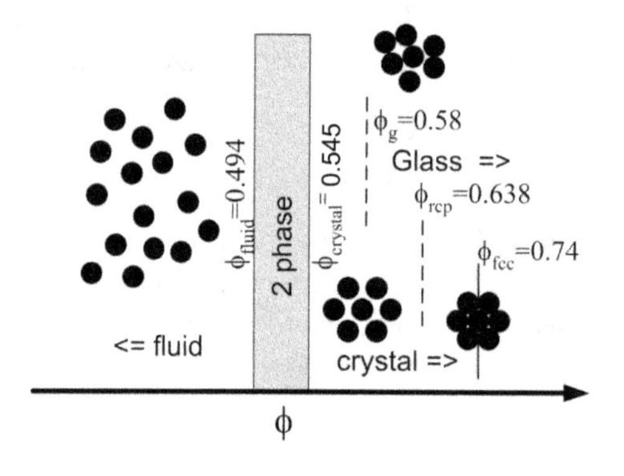

Figure 1.3 Phase diagram of hard spheres. Adapted from Cheng et al. [14], also appearing in CSR, figure 1.11

identified in this regime: short-time, localized diffusion within the cage of neighbors, known as the "β relaxation", while on a longer time frame, known as the "α" relaxation, Fickian diffusion is recovered as fluctuations "melt" the neighboring cages, allowing particles to diffuse, albeit slowly, over macroscopic distances. This is illustrated in Figure 1.4. These cages of neighboring particles restrict Brownian motion, such that theory and experiments suggest the existence of an ideal glass transition at $\phi_{IGT} \sim 0.57$–0.58. A colloidal glass is distinguished from a colloidal liquid in that the long-range particle motion is arrested and it has solid-like mechanical properties, but it is not a crystalline solid, but rather amorphous in microstructure. Above this volume fraction, normal diffusive motion is confined to rattling around inside the cage of neighbors, while bulk relaxation requires activated "hopping" motion outside the cage of nearest neighbors [18]. At and above the static percolation or "jamming" transition, defined as the volume fraction at maximally random

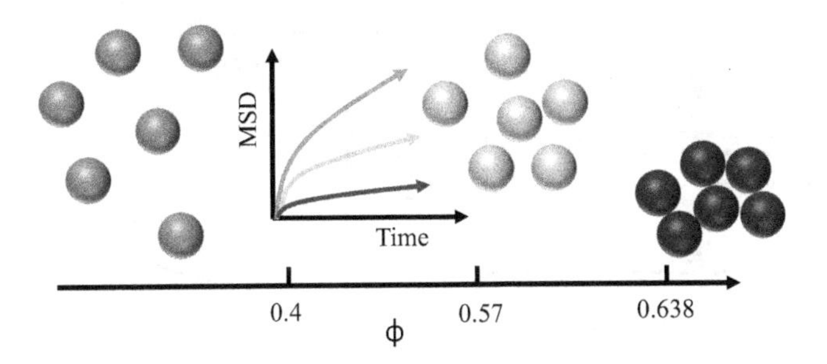

Figure 1.4 Illustration of colloidal dynamics from left to right with increasing concentration: $\phi >\sim 0.4$ normal diffusion; $\sim 0.4 < \phi < 0.57 \sim 0.58$ nearest neighbor cage hinders diffusion; $0.57 \sim 0.58 < \phi < 0.64$ local motion within the cage and hopping diffusion; $\phi > 0.64$ static percolation, all motion ceases.

jamming (0.638), particles are all constrained and, in principle, all motion ceases. Competing with this state is the crystalline state, which can permit local motion and hopping related to defect sites, up to the point of close packing for the FCC lattice as defined by geometry to be 0.74 volume fraction, as shown in Figure 1.3.

Increasing particle concentration leads to an increase in the number of neighboring particles due to crowding. This can be calculated as [19]:

$$g(2a) = \frac{1 - (\phi/2)}{(1 - \phi)^3}, \tag{1.11}$$

where $g(2a)$ is the probability of finding a neighboring particle, which will be discussed in more detail in the following chapters. The function $g(r)$ is known as the *radial distribution function*, and it is evaluated at the distance of minimum approach between particles, i.e., $g(r = 2a)$. This function is described in more detail in Chapter 2. The number of contacts grows nonlinearly such that by 50 vol%, particles have on average six contacts, which is the minimum number of contacts required for jamming of frictionless spheres (i.e., isostatic packing) [20].

The radial distribution function for hard spheres is known and is shown in Figure 1.5, where it is evident that the excluded volume of the reference particle at $r = 0$ creates a local packing of the neighboring particles, which is especially prominent at higher packing fractions. The first neighbors near contact or $r = 2a$ become more prevalent and second and even third and fourth neighboring shells are evident at integer multiples of the particle diameter, $2a$. At long distances the probability of finding another particle is simply given by the average number density. Thus, $g(r)$ is normalized by this number density, such that it goes smoothly to one for large separations. Importantly, this is evidence of *equilibrium microstructure* due to the competing effects of excluded volume and Brownian motion. Interparticle repulsions (such as electrostatic and steric) will cause particles to exclude more volume (larger effective radius) and therefore, such structuring will be evident at much lower packing fractions. Alternatively, attractive interactions (such as due to van der Waals forces, depletion, poor solvent quality) will physically bind particles into networks,

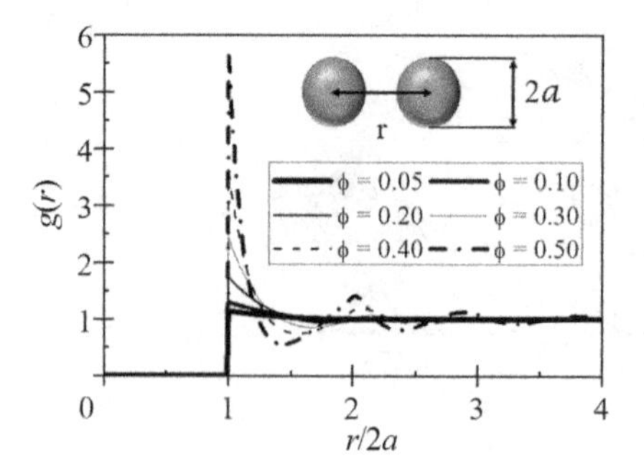

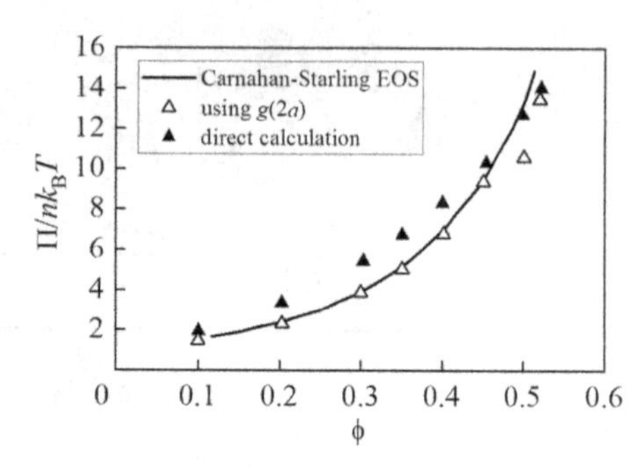

Figure 1.5 Left: Radial distribution function for suspensions of Brownian hard spheres of radius a as a function of dimensionless center-to-center distance $r/2a$, which gives the normalized probability of finding a neighboring particle at a given distance. Note that at low concentrations, the probability is ~ 1, which indicates random arrangement, while at higher concentrations, packing constraints require particles to organize into "shells". Also appearing in *CSR*, figure 3.7. Right: Osmotic pressure for hard sphere suspensions according to Eq. (1.13) as compared to Stokesian Dynamics simulations at low *Pe*, where both Eq. (1.12) and direct summation of forces are used and compared. Data from Yurkovetsky and Morris [21]

sometimes fractal in nature, which can become soft solids at sufficiently high concentrations. The effects of these changes on rheology are the topic of Chapters 5 and 6.

This crowding and resultant microstructure formation leads to a slowing down of the dynamics and an increase in the osmotic pressure and viscosity. The (osmotic) pressure in a Brownian hard sphere suspension is given directly in terms of this contact number $g(2a)$ as:

$$\frac{\Pi}{nk_{\mathrm{B}}T} = 1 + 4\phi g(2a). \tag{1.12}$$

In the above, n is the number of particles per volume. The growing number of neighboring particles gives rise to an isotropic pressure increase in the equilibrium suspension, can be described by the Carnahan–Starling equation:

$$\frac{\Pi}{nk_{\mathrm{B}}T} = \frac{1 + \phi + \phi^2 - \phi^3}{(1 - \phi)^3}, \tag{1.13}$$

which is compared against simulations in Figure 1.5. This increase in the isotropic pressure due to crowding is also matched by a dramatic increase in the zero shear viscosity and high frequency elastic modulus, as shown in Section 1.4.

1.4 Brownian Hard Sphere Rheology

1.4.1 Behavior at Low Shear Rates and Linear Viscoelasticity

The zero shear viscosity and linear viscoelasticity are properties of the equilibrium suspension, which requires that the applied shear rate during measurement be small, or

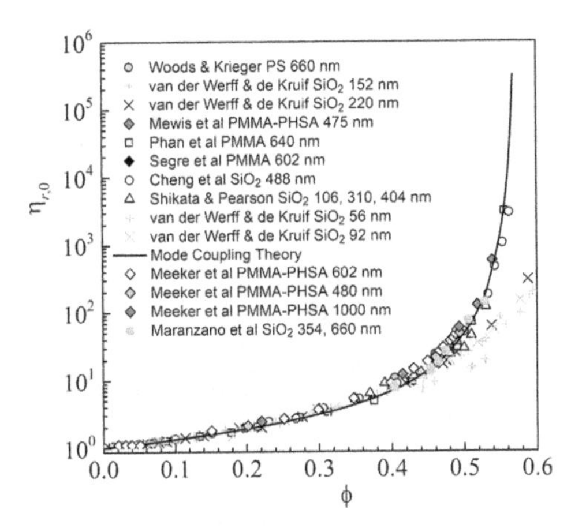

Figure 1.6 Zero shear viscosity of many model, hard sphere-like systems compared to the mode coupling theory. Adapted from Russel et al. [23]

$Pe \ll 1$. As colloidal motion slows down with increasing particle concentration, the low shear viscosity dramatically increases. Liquid-like behavior at rest requires diffusion of the particles inside the sample such that the zero shear viscosity generally tracks the Fickian diffusion of the particles [22]. Consequently, experiments and mode coupling theory, which will be discussed in detail in Chapter 2, predict a dramatic increase in the suspension's zero shear viscosity as the volume fraction is increased in the liquid state toward the ideal glass transition, $\phi_{IGT} \sim 0.57$–0.58 [23].

This transition is shown for many model systems, along with predictions of the mode coupling theory, in Figure 1.6. As observed, the viscosity rises sharply as the particle concentration approaches this ideal glass transition. Diffusion drops dramatically, but is not completely arrested as hopping mechanisms still enable very slow relaxation and creep.

An important feature of the viscous behavior is the asymptotic zero shear viscosity for hard spheres, given by the virial expansion [24]:

$$\eta_{r,0} = 1 + 2.5\phi + 5.9\phi^2 + \dots \quad (1.14)$$

The exact, linear term was calculated by Einstein [25,26] in the early 1900s and accounts for the extra dissipation in the fluid flow by the presence of a small amount of hard particles, while the second term reflects the leading order terms due to particle–particle interactions associated with both Brownian motion and hydrodynamic interactions [24].

Importantly, while the Einstein coefficient of 2.5 is completely due to hydrodynamic dissipation, the coefficient of the ϕ^2 term is comprised of both hydrodynamic dissipation (5.0) and an additional dissipation caused by interparticle interactions due to Brownian motion (0.9). Hence, this latter term depends on the microstructure in the suspension and the result is specific for equilibrium hard sphere Brownian suspensions. With increasing particle concentration this interparticle contribution increases

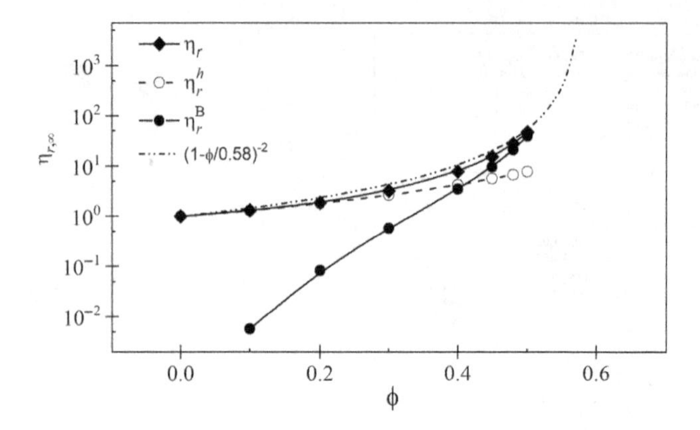

Figure 1.7 Contributions to the zero shear viscosity from purely hydrodynamic and interparticle contributions showing how the Brownian contribution to the stress drives the viscosity divergence. Adapted from Banchio and Brady [27]

much more rapidly than the hydrodynamic contribution. This has been computed for monodisperse hard spheres, as shown in Figure 1.7, using computational methods that will be discussed more completely in Chapter 3.

Experimental measurements of the zero shear viscosity require care at high concentrations, as nonlinear behavior is evident even for very small shear rates and simulations are limited by the propensity to crystallize. Fortunately it is possible to study the rheology of colloidal dispersions with liquid-like microstructure without interference from the underlying crystallization phase transition, which can be fully suppressed by small amounts (approximately 8%) of polydispersity [28]. The model that best describes this divergence in the interparticle contribution to the viscosity is given by MCT and has the following semi-empirical form:

$$\frac{\eta_0 - \eta'_\infty}{\eta_m} \cong 0.9\phi^2 \left(1 - \frac{\phi}{\phi_g}\right)^{-2.46}, \tag{1.15}$$

where the glass transition is taken to be in the range from 0.57 to 0.58. The purely hydrodynamic contribution to the viscosity is equivalent to the viscosity determined by the high frequency measurements, which has been correlated by Lionberger and Russel [29] and shown in Figure 1.8:

$$\eta'_{r,\infty} = \frac{1 + 1.5\phi\left(1 + \phi - 0.189\phi^2\right)}{1 - \phi\left(1 + \phi - 0.189\phi^2\right)}. \tag{1.16}$$

This represents the viscous resistance of the dispersion without Brownian motion, as the frequency of deformation is faster than the particle motion, but with the microstructure inherent to the equilibrium suspension.

Importantly, the hydrodynamic contribution does not diverge at the ideal glass transition, which is a consequence of the fact that the particles are not touching even at

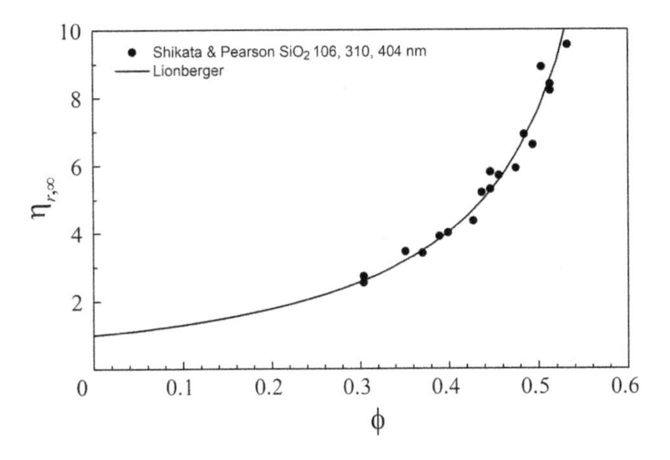

Figure 1.8 High frequency viscosity of Brownian suspensions showing the correlation of Lionberger and Russel [29] and data from Shikata and Pearson [8]. Adapted from Russel et al. [23]

the IGT. Rather, particles are free to move locally within their cage, as depicted in Figure 1.4. Thus, particles can still respond to small amplitude, high frequency oscillations in the medium with a finite dissipation. In contrast, in the limit of random close packing where particles are touching, theory predicts the hydrodynamic viscosity diverges at the maximally random jammed state, $\phi_{MRJ} = 0.638$.

Interparticle forces due to Brownian motion are conservative ones and hence generate both elastic and viscous responses. This is in direct contrast to the hydrodynamic interactions, which only contribute to the viscous forces. Experimental data shown in Figure 1.2 illustrate the overall trend. However, just as for the steady shear viscosity, two distinct contributions to G'' can be identified: the high frequency limit of the loss modulus is purely hydrodynamic and given by $G''_{\omega\to\infty} = \eta'_{\infty}\omega$, while the remainder arises from the viscoelasticity coming from the Brownian motion. The result of subtracting this purely viscous term from the loss modulus reveals the viscoelasticity arising from Brownian motion, as shown in Figure 1.9. This behavior of the moduli is similar to that observed for a simple Maxwell fluid, but with a broad distribution of relaxation times (see section 1.2.3 of *CSR* for a more detailed explanation of viscoelasticity).

The Brownian stress yields a plateau shear modulus, which has been shown to scale with the characteristic stress, Eq. (1.8) as:

$$\frac{G'a^3}{k_B T} \approx 0.78\eta'_{r,\infty}\phi^2 g(2a;\phi) = 0.78\eta'_{r,\infty}\phi^2 \frac{1-(\phi/2)}{(1-\phi)^3}, \tag{1.17}$$

1.4.2 Nonlinear Shear Rheology

The experimental results shown in Figure 1.2 indicate that under an applied stress (or deformation rate) a colloidal suspension can display shear thinning, which is non-Newtonian rheology. Shear thinning can be captured by a simple, empirical equation proposed by de Kruif et al. [30]:

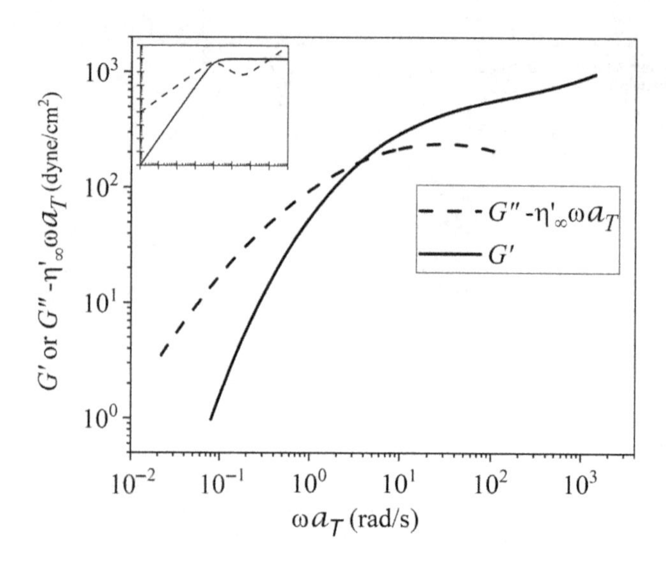

Figure 1.9 Linear viscoelasticity of a colloidal suspension obtained by subtracting off the hydrodynamic contribution from the data shown in Figure 1.2 from reference [8]. The inset shows a single mode, extended Maxwell model for reference, with the dashed line indicating the loss modulus and the solid line the elastic modulus.

$$\frac{\eta_r - \eta_{r,\infty}}{\eta_{r,0} - \eta_{r,\infty}} = \frac{1}{1 + \sigma/\sigma_c} \tag{1.18}$$

The critical stress for shear thinning is a function of the Brownian stress and the volume fraction; an empirical relation is [30]:

$$\frac{\sigma_c a^3}{k_B T} \simeq \begin{matrix} 6\phi & \phi < 0.5 \\ 3 - 21.7(\phi - 0.5) & 0.5 < \phi < 0.638 \end{matrix}. \tag{1.19}$$

The high shear viscosity is not the same as the high frequency viscosity, as it contains contributions from both hydrodynamic interactions as well as interparticle interactions caused by Brownian motion. Furthermore, the suspension microstructure is not the same in these two limits. In Eq. (1.18) the high shear viscosity is also given by a correlation of de Kruif et al. [30] and is shown in Figure 1.10:

$$\eta_{r,\infty} = \left(1 - \frac{\phi}{0.71}\right)^{-2}. \tag{1.20}$$

Note that the high shear viscosity is defined as the plateau viscosity at high stresses or shear rates *prior to shear thickening*. Also, the volume fraction at which the viscosity diverges is 0.71, which is close to the maximum theoretical packing of an FCC lattice, 0.74. This reflects the fact that the microstructure under shear flow is very different from that at equilibrium (or during shear thickening for that matter). This microstructural rearrangement by the flow will be covered in some detail in subsequent chapters. As a practical application, it is often difficult to determine

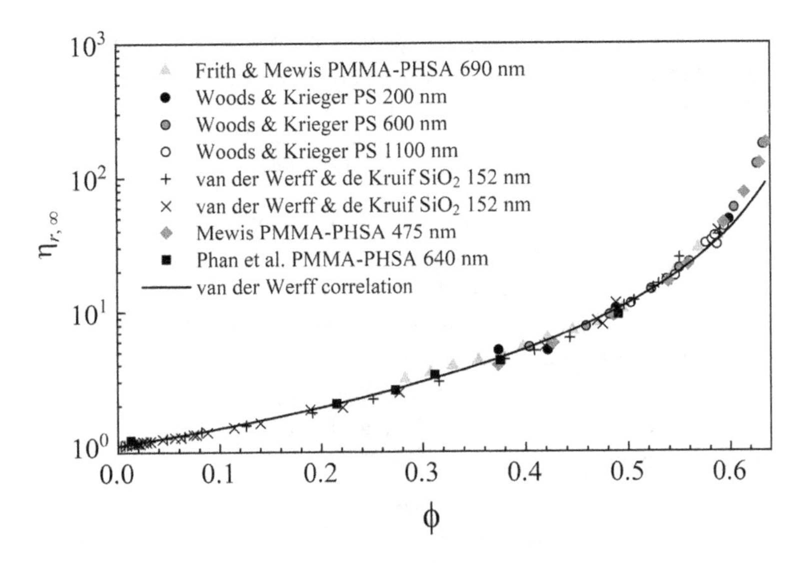

Figure 1.10 High shear relative viscosity as a function of volume fraction, comparing experimental data with the semi-empirical Eq. (1.20). Adapted from Russel et al. [23]

volume fraction as formulation is usually done gravimetrically. The present semi-empirical relationship can be used to determine the volume fraction in concentrated suspensions, in much that same manner that the Einstein relation can be used at low volume fractions [23].

The semi-empirical Eq. (1.20) has the expected quadratic divergence at maximum packing, but does not have the proper limiting form at low volume fractions. For the high shear viscosity, this limiting form for low volume fractions was calculated by Bergenholtz et al. [31] to be:

$$\eta_{r,\infty} = 1 + 2.5\phi + 6.0\phi^2 + \cdots. \tag{1.21}$$

Note that at low volume fractions, while the Einstein term is not affected, the second order term due to interactions is actually slightly higher than the second order term at low shear rates, Eq. (1.14). This points to a nonlinear effect of shearing on the microstructure, whereby at lower volume fractions the viscosity increases slightly under flow due to the increased probability of particles being in close proximity and the resulting increased dissipation under strong shear flow. However, empirically, at low shear rates the viscosity diverges at a much lower volume fraction (0.57 versus 0.71), such that it is possible (and common) to shear melt a colloidal glass, as is also evident in the data presented in Figure 1.2.

1.4.3 Extensional and Bulk Viscosities

The *extensional* viscosity of a Newtonian fluid is three times that of the shear viscosity (see the Appendix for definitions of these different flow types). The extensional viscosity has been calculated by Batchelor and Green [32] for suspensions to second order in particle concentration, which yields an enhancement over that for the shear viscosity as:

$$\eta_{E,r} = 3\left(1 + 2.5\phi + 7.6\phi^2 + \ldots\right), \tag{1.22}$$

Experimental results for noncolloidal suspensions of spheres in a Newtonian matrix suggest that the Trouton ratio ranges from 3.5 to 6 over the volume fraction range up to 50% [33]. However, the extensional viscosity for colloidal suspensions is not known beyond this hydrodynamic contribution and remains an active area of research. Finally, the *bulk* suspension viscosity η_B has also been calculated by Brady and co-workers. The interested reader is referred to their paper on the subject [34]. This enables giving the more general form of Newton's constitutive equation:

$$\boldsymbol{\sigma} = \eta_m\left(\nabla\mathbf{v} + (\nabla\mathbf{v})^{\dagger}\right) - \left(\frac{2}{3}\eta_m - \eta_B\right)(\nabla\bullet\mathbf{v})\mathbf{I}. \tag{1.23}$$

1.4.4 Normal Stress Differences

Newtonian fluids exhibit no normal stress differences and suspensions that have an isotropic microstructure, such as at equilibrium, also exhibit no normal stress differences. Thus, not surprisingly, normal stress differences are typically small for colloidal suspensions in Newtonian media in comparison to those for entangled polymers. Normal stress differences arise because of a shear-induced microstructural rearrangement. Hence, their values are highly sensitive to the deformation of suspension microstructure by the flow. Two new material properties are defined in terms of normal stress difference coefficients for suspensions, as:

$$\begin{aligned} \Upsilon_1 &= \frac{-N_1}{\eta_m|\dot{\gamma}|} \\ \Upsilon_2 &= \frac{-N_2}{\eta_m|\dot{\gamma}|} \end{aligned}. \tag{1.24}$$

Note that these are different from the normal stress coefficients typical of polymeric liquids, which is a consequence of the defining feature of colloidal suspension rheology, namely the ubiquitous effects of hydrodynamic interactions.

The two contributing forces to the suspension stress, hydrodynamic and Brownian forces, yield different signs for the first normal force difference [31]. Here again, Stokesian Dynamics simulations, which will be covered in Chapter 3, provide guidance, as shown in Figure 1.11. There it can be seen that the normal stress difference coefficient for the first or primary normal stress difference goes from negative to slightly positive at higher shear rates or *Pe*. Thus, N_1 starts positive when Brownian stresses dominate, and becomes negative when

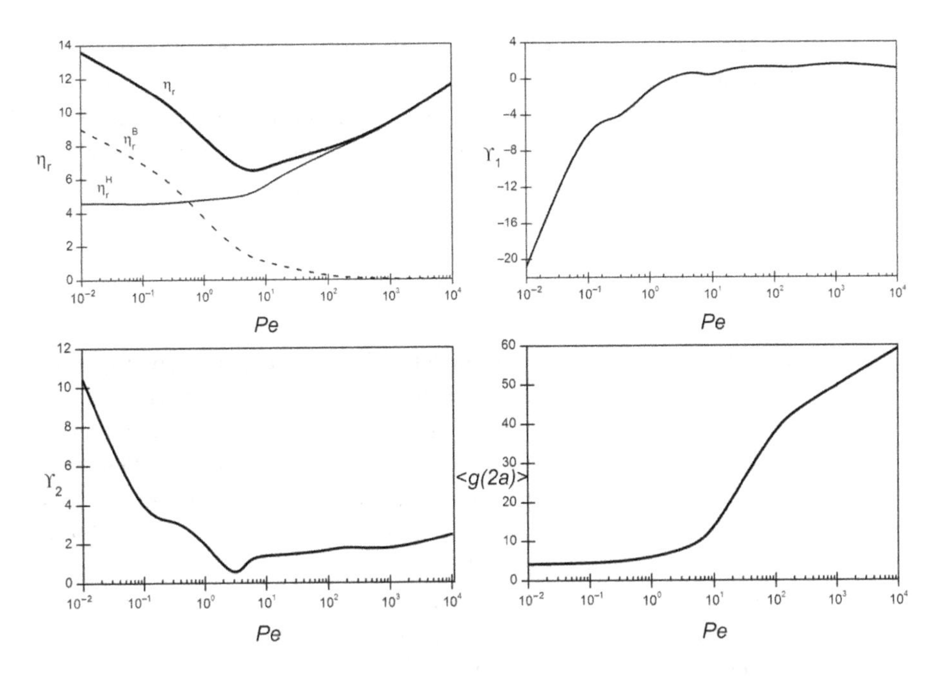

Figure 1.11 Stokesian dynamics simulations of the relative viscosity and normal stress differences as well as the average probability at contact for a 45 vol% suspension of Brownian hard spheres. Data from Foss and Brady [35]

hydrodynamic forces dominate. The second normal stress difference coefficient is always observed to be positive, meaning that the second normal stress difference is negative for both Brownian and hydrodynamic forces. Understanding this requires understanding the shear deformation of the microstructure, which will be discussed in Chapters 2–4.

1.4.5 Shear Thickening and the Shear Thickened State

The shear thickening evident in Figure 1.2 by experiment and Figure 1.11 by simulation is a consequence of shear-induced hydrocluster formation, whereby particles are pushed into close proximity by the shear flow. The strong squeeze flow of the suspending medium between the particles strongly couples the particles together, such that structure in the form of reversible "hydroclusters" appear in the dispersion. These structures have been observed in simulation and experiment. Much of the resultant suspension properties are covered in chapter 8 of *CSR*, and will be discussed in later chapters here. The formation of regions of more concentrated particles in the flow is opposed by the Brownian forces, which act to keep the particles dispersed in suspension. Not surprisingly, when the shear flow forces particles into close proximity, the strong lubrication forces become the dominant force acting between particles and so the suspension adopts a new microstructure evident as nonlinear fluctuations in particle number density, termed "hydroclusters" by Brady, Maranzano, and Wagner [36–38].

A particle-level model for the onset of shear thickening has been derived from this simple concept and a simple, semi-analytical formula has been developed that relates the stress at the onset of shear thickening in colloidal hard sphere suspensions to the particle size and particle volume fraction, as [38]:

$$\frac{\sigma_c^{hs} a^3}{k_B T} = 0.1 e^{\phi/0.153}. \tag{1.25}$$

This critical stress has been predicted for Brownian hard sphere dispersions [37,38,40] by using the equilibrium fluid structure shown in Figure 1.5. The scaling of the critical stress for the onset of shear thickening depends inversely on the particle size cubed for Brownian hard sphere suspensions, as shown experimentally in Figure 1.12, where the additional volume fraction dependence has not been accounted for. This dependence is anticipated from the nature of the Brownian stress, which scales inversely with the volume of the particle (Eq. (1.8)).

As particles are pushed into close proximity during shear thickening, the onset or even presence of shear thickening can be significantly affected by the slightest of nanoscale forces acting between the particles, including actual particle contact or friction. As shown in Figure 1.12, the presence of charge stability, which is nearly ubiquitous in practice, leads to a dramatic increase in the critical stress as well as a change in scale from inverse with the particle volume to inverse with the particle surface area, as might be anticipated. Expressions to predict the onset stress have been derived depending on the form used for the electrostatic interactions [41]. Further predictions and comparisons with experiments [39] and simulations [42] are available for polymer-stabilized particles. For such cases, the lubrication hydrodynamics must be modified to account for the additional resistance to squeeze flow due to the drag on the suspending medium by the polymer brush in the lubrication gap [43]. The critical stresses for polymer coated particles are significantly larger than those expected for hard spherical particles, even if an effective hard sphere size accounting for the brush is used [39]. This is of particular importance when considering the applications such as in cement and large scale processing, where shear thickening has to be suppressed within the operating window [44]. The predictions also show that the critical stress scales with a^{-2} for particles of varying radius, but with constant polymer brush properties, in agreement with experiment. Finally, modeling and experiments also shows that weak interparticle attractions will act to enhance hydrocluster formation and, thus, the onset of shear thickening will occur at lower shear rates [45].

Theory and simulation suggest that below the jamming transition, the suspensions in the shear thickened state (STS) will exhibit a plateau viscosity. This has been observed experimentally and using experimental data and a model for the shear thickened state by Morris and Boulay [46], one finds the following values for the

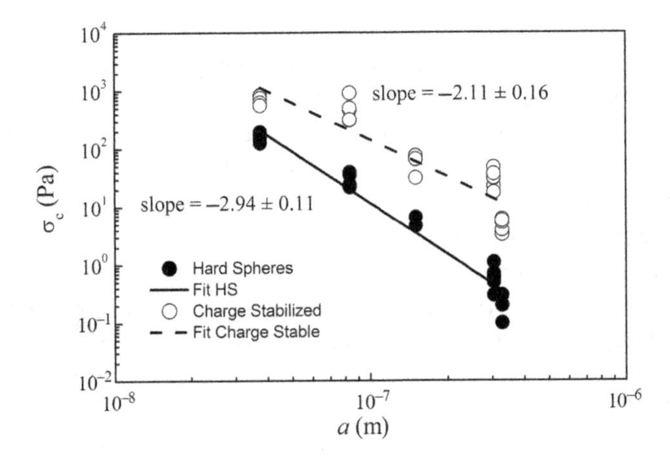

Figure 1.12 Scaling of the critical stress for the onset of shear thickening as a function of particle size (the different experimental points correspond to varying volume fraction). Data from Maranzano and Wagner [37,38]. The scaling for Brownian hard sphere dispersions goes inversely with the cube of the size, as predicted, while the presence of surface forces, such as charge repulsion, changes the scaling to inverse with the square of the size, also as predicted. Also appearing in *CSR*, figure 8.16.

material properties of the shear thickened state, namely the viscosity and first and second normal stress difference coefficients [47]:

$$
\begin{aligned}
\eta_{r,\mathrm{STS}} &= \left(1-\frac{\phi}{\phi_{\max}}\right)^{-2} \\
\Upsilon_{i,\mathrm{STS}} &= K_i\left(\frac{\phi}{\phi_{\max}}\right)^2\left(1-\frac{\phi}{\phi_{\max}}\right)^{-2} \\
\phi_{\max} &= 0.54;\ K_1 = 0.177;\ K_2 = 0.240
\end{aligned}
\quad . \tag{1.26}
$$

The equations are based on a purely hydrodynamic model for shear thickening, which is consistent with the sign of the first normal stress coefficient. Importantly, these material properties accurately describe experimental measurements for non-Brownian suspensions as well as simulations that fully account for hydrodynamic interactions, as shown in Cwalina and Wagner [47]. The measurements show that the flowing, shear thickened state is comprised of fluctuations in particle density that act as non-Brownian particles and dissipate energy through hydrodynamic interactions and short range forces, consistent with experimental measurements [39,41,48–50] and simulations of the microstructure [35,51].

Stokesian Dynamics simulations predict a logarithmic increase in stress in the shear thickened state that is much weaker than what is often observed in experiments [47,50]. The presence of nanoscale forces acting between the particles that goes beyond the simple Brownian hard sphere interaction becomes a critically important consideration in both experiment [52] and simulations [53]. Indeed, at close contact surface roughness becomes an issue as it affects hydrodynamic particle interactions. Also, as the lubrication forces acting microscopically between particles can become very large, the elastic nature of the particles themselves becomes important for fully understanding their rheological properties in the shear thickened state, which is termed *elasto-hydrodynamics* [53]. Once shear thickening has occurred, contact friction can also occur and may be responsible for discontinuous, irreversible shear thickening [42,54]. Furthermore, there has been significant research concerning the role of nanoscale forces on the shear thickened rheology [52] including possible contact forces. Importantly, measurements of the microstructure and complete stress tensor can distinguish between these different, non-HS contributions to the overall mechanics of shear thickening suspensions [50,55].

Discontinuous shear thickening is often associated with *dilatancy*.[5] The contact probability for nearest neighboring particles, $g(2a)$, rises dramatically with increasing volume fraction, as shown in Figure 1.11. This is observed to also rise rapidly at higher shear rates, where shear thickening is evident. The propensity of the shear to drive particles into close proximity and form hydroclusters then also significantly increases the osmotic pressure, as shown by Eq. (1.12). This increase in the particle pressure will lead to dilatancy and limit the degree or amount of shear thickening. As the name implies, dilatancy corresponds to an increase in the volume a sample occupies under stress. A classic example of dilatancy occurs when squeezing a wet suspension, as analyzed by Reynolds in the 1880s [56]. As shown by Metzner and Whitlock [57], a similar effect occurs in colloidal suspensions at high concentrations in strongly shear thickening samples and is commonplace in many commercial suspensions at high particle loadings and stresses. The large increase in particle phase pressure causes the particles to dilate and expand beyond the fluid they are immersed in, as evident by a dramatic roughening of the suspension–air interface, as well as by internal cavitation [58]. Experiments are often limited by the sample rapidly ejecting from the rotational instrument geometry.

The confining pressure holding the particles in the liquid is the capillary force, which scales as the ratio of the surface tension to the particle size, γ_{lv}/a. Brown and Jaeger [59] correlated the maximum extent of shear thickening with this confining capillary pressure for a broad range of colloidal and noncolloidal suspensions, as shown in Figure 1.13, which shows that experimental investigations of strong shear thickening dispersions are limited by the dilatant behavior of the suspension. This effect is generally not accounted for in simulations or theory, which assume constant volume, not constant pressure. Wang and Brady [60] address the effects of confinement in studying concentrated dispersions in recent simulation studies, showing how significant confinement is on jamming. Also, we note that early work by Laun et al. [61] studied the effects of confinement necessary to achieve measurements for discontinuous shear thickening fluids. While it is possible to accurately measure

[5] Dilatancy refers to the expansion of the volume occupied by the particles and has been a source of instability in processing colloidal suspensions.

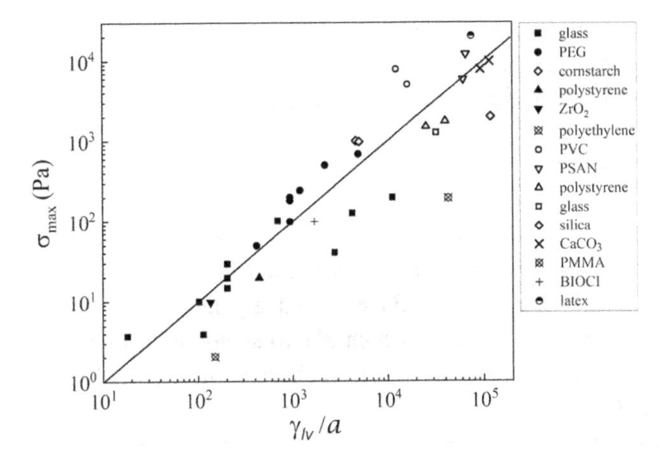

Figure 1.13 Limiting shear stress in the shear thickening regime for a variety of colloidal and noncolloidal suspensions plotted against the ratio of surface tension to particle size. The correlation line is $0.1\gamma_{lv}/a$. Adapted from Brown and Jaeger [59]

strongly shear thickening colloidal dispersions that are reversible and do not dilate [47], most measurements of discontinuous shear thickening are irreproducible and involve extremely large fluctuations in shear viscosity and normal stresses, suggesting that the interpretation of the mechanical signals coming from the rheometer as material properties is suspect at best. Simulations of the true, hard sphere suspension always lead to jamming, which must be relieved through additional effects, such as Brownian motion, short-range "soft" surface forces or particle elasticity, or the suspension must dilate to flow [62]. Therefore, it is important to distinguish dilatancy, which involves a volume expansion, as being distinct from reversible shear thickening [60].

1.5 Colloidal Interaction Potentials

In addition to the thermal and hydrodynamic forces, actual realizable suspensions consist of particles with many additional types of interparticle forces. The ubiquitous dispersion forces nearly always cause a significant attraction force between particles in close proximity. Thus, colloidal stability requires formulation with additional repulsive forces between particles. In aqueous dispersions, electric charges on the particle surface can cause repulsive interactions between particles of like type, or attractions if the particles have different charges. The surface charge is pH and salt dependent. Similarly, the presence of polymers also gives rise to repulsive (steric) or attractive forces (depletion, bridging). A comprehensive discussion of colloidal interactions can be found in numerous textbooks on the subject (see, for example, references [2–5]) and only a brief summary of some model potentials is provided here as a reference for the chapters to come.

The interparticle forces are oriented along the center-to-center line and depend on the separation r between the particle centers. For continuous and differentiable potentials the interparticle forces can be expressed as derivatives of a potential energy Φ:

$$F(r) = -\frac{d\Phi(r)}{dr}. \tag{1.27}$$

Model systems are often described by discontinuous potentials. One limiting case is that of hard spheres, where the particle interaction force is zero until the particles touch and the force becomes immediately infinitely large. This is described by a potential Φ^{hs}, shown in Figure 1.14, which is zero at any finite values of r and jumps to infinity when the particles contact ($r = 2a$) because hard particles cannot overlap, i.e., the *excluded volume* interaction.

Dispersion forces between atoms and molecules of the system usually cause a net attraction force between particles. This is characterized by the Hamaker constant A, which reflects the effect of the nature of the components. The corresponding interparticle force and potential depend also on the geometry of the system. For two spherical particles with radii a_1 and a_2 the potential is given by:

$$\Phi^d(r) = -\frac{A}{6}\left(\frac{2a_1a_2}{r^2-(a_1+a_2)^2}+\frac{2a_1a_2}{r^2-(a_1-a_2)^2}+\ln\frac{r^2-(a_1+a_2)^2}{r^2-(a_1-a_2)^2}\right). \tag{1.28}$$

When the particles are charged and dispersed in an electrolyte solution, electrostatic forces act between the particles. For similarly charged particles the force will be repulsive. Analytical solutions for the electrostatic potential Φ^{el} depend on the geometric specifics and the nature and concentration of the electrolyte. For a symmetric electrolyte with valence z a model interparticle potential can be determined as [3]:

$$\Phi^{el}(h) = 32a\varepsilon\varepsilon_o\left(\frac{kT}{ze}\right)^2\tanh^2\left(\frac{\psi_s ez}{4k_BT}\right)\exp(-\kappa h), \tag{1.29}$$

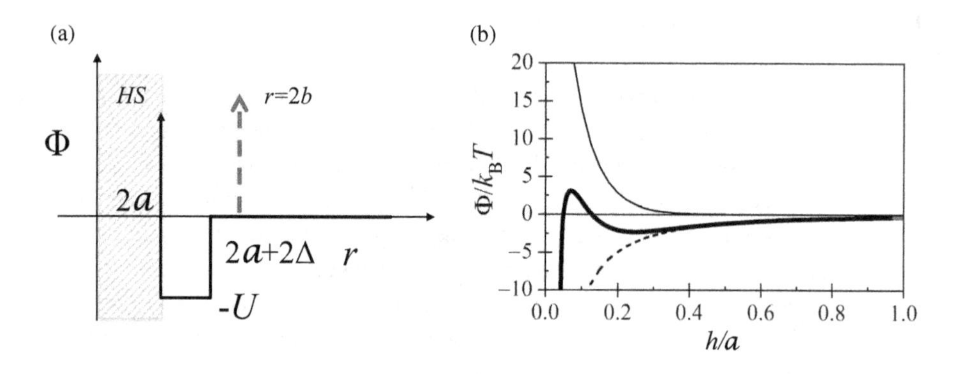

Figure 1.14 Model interparticle potentials: (a) Hard sphere, excluded annulus and square well potentials. (b) DLVO potential long-range attraction due to van der Waals forces and short-range electrostatic repulsion.

where ψ_s is the potential at the particle surface, z the valence of the electrolyte, ε and ε_o the dielectric constant of the liquid and the permittivity of vacuum, respectively, e is the charge of an electron, and $h = r - 2a$, the surface-to-surface distance between the two particles. The Debye length κ^{-1} is a characteristic value for the decay of the potential with distance:

$$\kappa^{-1} = \sqrt{\frac{\varepsilon k_B T}{e^2 \left(\sum_i z_i^2 n_{i,\infty} \right)}}. \tag{1.30}$$

with z_i and $n_{i,\infty}$ the valence and bulk concentration of ions of type i. In the DLVO theory, the effect of dispersion and electrostatic forces are linearly combined to calculate the global potential. The resulting potential goes to $-\infty$ at contact, hence predicting interparticle attraction at close contact. When there is a sufficiently high maximum in $\Phi(h)$ a repulsive force at larger values of h will ensure colloidal stability by preventing particle aggregation. Because of the fluctuations in thermal energy, particles will occasionally have enough energy to overcome the energy barrier and being trapped into close contact. Hence, electrostatic forces provide essentially only *kinetic* and not *thermodynamic* stability.

The sum of these two potentials, electrostatic plus dispersion, is shown in Figure 1.14(b), and is considered the canonical potential for colloidal dispersions: the DLVO potential. The acronym refers to the two groups that proposed it simultaneously, Derjaguin–Landau and Verwey–Overbeek. Depending on the parameters, it can be essentially stable (purely electrostatic), unstable (no energy barrier to prevent contact), or show a very rich behavior as indicated in the figure, with both secondary minimum (flocculation) and primary minimum (aggregation). More will be said about this potential concerning the possible phase behavior and state of the system.

To study attractive systems, often a simplified potential is used consisting of a square well with depth U and width 2Δ:

$$\Phi^{sq}(r) = \begin{matrix} \infty & r < 2a \\ -U & 2a < r < 2(a+\Delta) \\ 0 & r > 2(a+\Delta). \end{matrix} \tag{1.31}$$

This two-parameter model, shown in Figure 1.14(a), can further be simplified when the range of the attractive forces becomes very small while the interaction energy remains finite. In that case, called the *sticky sphere* limit, the only characteristic is the strength of interaction or *Baxter parameter*. It is linked to the parameters of the square-well potential by the equation:

$$\tau_B^{-1} = 4\left(e^{U/k_B T} - 1\right)\left[\left(\frac{a+\Delta}{a}\right)^3 - 1\right]. \tag{1.32}$$

The value of this representation is that it provides a type of *corresponding states*, where many different potentials that are primarily attractive can yield similar phase behavior and rheology when characterized by this single parameter. A similar approach should be familiar to those versed in classical thermodynamics, where the second virial coefficient provides a means of comparing different molecular fluids and colloidal systems. For reference, the second virial coefficients for the HS, SW, and sticky sphere fluids are given here:

$$B_2^{hs} = b_0 = \frac{16\pi a^3}{3}$$
$$B_2^{sq}(T) = b_0\left\{1 + \left(1 - e^{-\epsilon/k_B T}\right)\left[\left(\frac{a+\Delta}{a}\right)^3 - 1\right]\right\} \tag{1.33}$$
$$B_2^{ss}(T) = b_0\left(1 - \frac{1}{4\tau_B}\right).$$

Polymers can be used to produce colloidally stable as well as flocculated or gelled dispersions. Grafting polymer of sufficient molecular mass on the particles with sufficient grafting density ensures thermodynamic stability. A steric repulsion force develops when the polymer layers on neighboring particles start to overlap. According to the Fischer model for stretched, end-grafted polymers the interparticle potential is given by the product of the osmotic pressure Π in the overlap region and the volume of that region: $\Phi^{pol} = \Pi V_0$. The expression for the potential then is:

$$\begin{aligned} &= \infty \quad r < 2a \\ \Phi^{pol}(r) &= \Phi_0\left[-\ln(y) - \frac{9}{5}(1-y) + \frac{1}{3}(1-y^3) - \frac{1}{30}(1-y^6)\right], \\ &\qquad 2a < r < 2(a+L) \\ &= 0 \qquad r > 2(a+L) \end{aligned} \tag{1.34}$$

with

$$y = \frac{r-2a}{2L}$$
$$\Phi_0 = \left(\frac{\pi^3 L\sigma_p}{12N_p l^2}k_B T\right)aL^2.$$

This result is for a grafted layer of polymer molecules with contour length L, degree of polymerization N_p, and segmental length l, σ_p is the graft density of polymer on the particle surface. The assumption of stretched polymers implies that the suspending medium is a good solvent for the polymer. A lower solvent quality reduces the efficiency of the polymer layer and may even result in attractive forces between surface layers. Also, adsorbed polymers can provide steric stability. Plots of examples of this potential are shown in Figure 1.15(a) for various choices of graft density and brush length. This shows the ability to significantly tune this potential by the choice of polymer and how such grafting can protect particles from primary minimum aggregation due to van der Waals forces. The softness of very large brush layers has interesting consequences for suspension rheology, and this softness is the topic of Chapter 6.

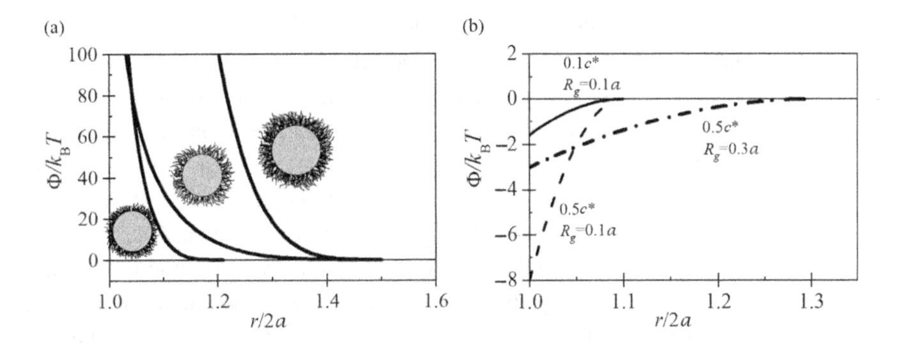

Figure 1.15 Potentials due to added polymer. (a) Repulsive potential due to grafted brush overlap, shown for different values of the brush thickness and graft density. (b) Depletion potential for various polymer concentrations in terms of the overlap concentration c^* and molecular mass, given in terms of the radius of gyration Rg and particle size a. Adapted from *CSR*, figures 1.8 and 1.9

A dissolved polymer can act in a totally different manner. When the separation between two particles decreases below R_g, the radius of gyration of the polymer, the soluble polymer is essentially excluded from the interparticle space based on an entropic effect. This *depletion* effect arises from an imbalance in osmotic pressure around the particle. Polymer in solution generates an osmotic pressure around the colloid, but as the polymer is excluded in the region between the two colloids in close proximity, the pressure is lower there. This imbalance in osmotic pressure surrounding a particle generates an interparticle attraction, which is proportional to the ratio of the actual polymer concentration c_p over that at overlap, c_p*. Asakura and Oosawa [43] calculated the *depletion potential* as being proportional to the product of the osmotic pressure of the polymer in the suspending medium and the volume of the space between the particles:

$$\frac{\Phi^{\text{dep}}(r)}{k_B T} = \begin{cases} \infty & r < 2a \\ -\frac{c_p}{c_p^*}\left(1+\frac{a}{R_g}\right)^3\left[1-\frac{3r}{4(a+R_g)}+\frac{r^3}{16(a+R_g)^3}\right] & 2a < r < a+2R_g \\ 0 & r > 2a+2R_g \end{cases} . \tag{1.35}$$

The polymer concentration determines the osmotic pressure and hence the *magnitude* of the potential. R_g determines the relevant interparticle space and hence also the *range* of the depletion force. Examples of this potential are shown in Figure 1.15(b) and this potential will be used in modeling depletion gels, to be discussed further in Chapter 5.

1.6 Colloidal Phase Behavior beyond Brownian Hard Spheres

As noted previously, the equilibrium phase behavior of colloidal suspensions is not affected by hydrodynamic interactions, however, the latter will affect the kinetics of phase changes. While temperature does not affect the phase behavior of idealized Brownian hard sphere suspensions, the introduction of surface forces introduces an energy scale and so the phase behavior will be sensitive to temperature. At rest, Brownian motion acts to randomize the relative positions of the particles and keep their distribution homogeneous. This corresponds to a fluid state, as shown in Figure 1.5. Increasing the density will induce crystallization. The loss of configurational entropy caused by this change is more than compensated by the gain in local entropy of the particles in the BCC crystal lattice. A crystalline phase starts to appear at $\phi = 0.49$ and the transition is complete at $\phi = 0.54$, where the coexistence region of the two phases end. The particles can still be compressed closer together until $\phi = 0.74$ where the structure transforms in a face-centered crystal (FCC) and where the maximum possible volume fraction of monodisperse spheres is reached.

In real systems deviations from this ideal behavior are common. Interfering factors are sedimentation, interparticle forces, and particle polydispersity. Often a random fluid structure is preserved until $\phi \approx 0.57$–0.58, where a glass transition is observed. Structural changes are still possible but are extremely slow. The structure then still remains fluid-like but Brownian motion is now limited to displacements within the "cage" formed by the surrounding particles. Further increasing the volume fraction in this condition is possible until the *random close packing* is reached at $\phi_{rcp} = 0.638$.

Particle interactions will alter the behavior. Strong attractive forces can drive phase separation and give rise to other structural features such as flocs and gels, resulting in complex *state*[6] diagrams that depend on details of the interaction forces. For example, starting from a colloidally stable dispersion, reducing the electrolyte will increase the range of repulsion and lead to a colloidal crystal (Figure 1.16). Alternatively, increasing the electrolyte concentration will screen this repulsion, revealing a secondary minimum, which will create flocs, and with more electrolyte, aggregation in the primary minimum. While all of these solid states are dynamically arrested, they will have vastly different rheological properties, as will be discussed in later chapters.

To conclude this section on phase behavior, a typical state diagram for colloidal systems with short-range interactions is illustrated in Figure 1.17. The temperature is represented as a dimensionless Baxter stickiness parameter, which plays the role of a reduced temperature, while the concentration is represented in terms of the excluded volume fraction. The hard sphere behavior discussed earlier is evident at higher volume fractions, where a phase transition to a crystal state is often not observed and, rather, a repulsive glass forms. Upon lowering the reduced temperature, i.e., increasing the "stickiness", the crystallization transition and the glass transition shift to higher concentrations. Eventually, the attractions become sufficient for the particles to phase separate to coexisting dilute and concentrated phases, analogous to gas and liquid phases observed for molecular fluids. For many systems, this phase separation is arrested and the suspension aggregates into a nonequilibrium,

[6] We use the term "state" rather than phase because some gels and glasses are considered as kinetically trapped and are not necessarily thermodynamically stable. More about this will be considered in Chapter 5.

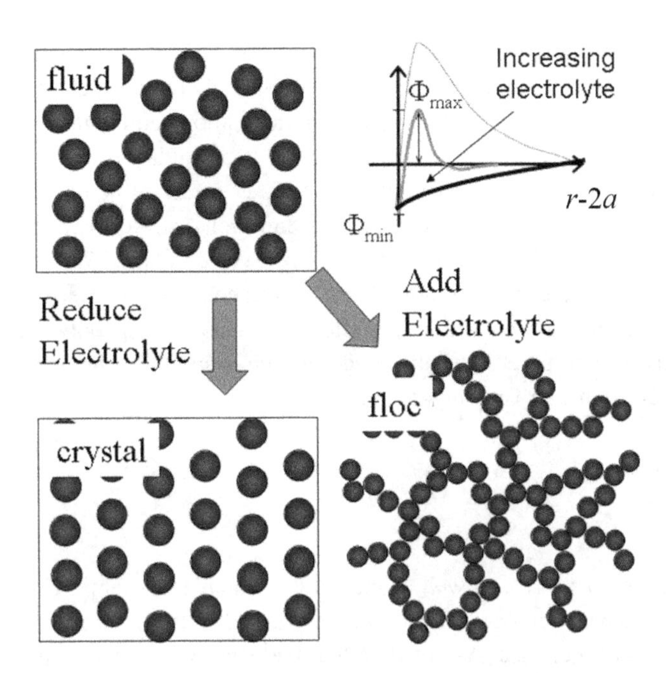

Figure 1.16 Effects of added electrolyte on the DLVO potential. Reducing electrolyte increases the range of the electrostatic repulsion such that stable colloidal crystals can result due to the greater excluded volume. Adding electrolyte can lead to flocculation (secondary minimum) or aggregation (primary minimum). Some of the resultant solid-like states can be gels (homogenous or heterogeneous) or glasses. Also appearing in *CSR*, figure 1.13.

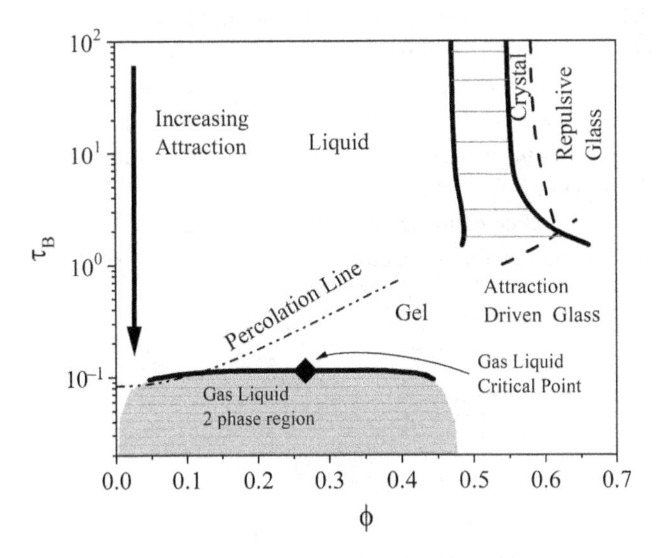

Figure 1.17 Generalized colloidal state diagram for systems with short-range attractions typical of most colloids. Also appearing in *CSR*, figure 1.12.

heterogeneous gel state. Note that this transition can be metastable with respect to the crystallization transition for some systems (not shown here). Percolation can be observed prior to phase separation for some compositions and *rigidity* percolation can drive the formation of a homogeneous gel state. Rigidity percolation requires percolation of rigid bonds and is distinct from other forms of percolation, such as contact or electrical percolation. At high concentrations, sufficiently strong attractions vitrify the suspension and form an attractive driven glass. This state diagram for adhesive hard spheres is not unique because of the very rich variety of potentials discussed in Section 1.5. In addition, particle shape and polydispersity are important considerations. Much more about these various states and their rheological properties will be considered in Chapters 5 and 6.

1.7 Thixotropy

Many colloidal suspensions, as well as many other complex systems, display a time-dependent rheological behavior, called thixotropy, that is distinct from the viscoelasticity identified in Figure 1.2. As noted earlier, there are two distinct contributions to the suspension stress. Brownian motion introduces a statistical interparticle force with an inherent time scale, thus creating a viscoelastic response. As hydrodynamic contributions to the viscosity are directly proportional to the shear rate, upon the sudden stoppage of flow the viscous contributions to the stress must also become zero instantaneously. On the other hand, the elastic stress will not drop instantaneously to zero. Hence, the total stress will rather decrease gradually in time while the microstructure regains its (zero shear) equilibrium. The timescale for such processes is governed by the Brownian diffusion time, given in Eq. (1.4).

Now consider the experiment depicted in Figure 1.18(a), where the shear rate is suddenly decreased to a lower value. Figure 1.18(b) depicts how a viscoelastic fluid, such as a Brownian colloidal suspension, would respond. The stress decays to a lower stress level consistent with this new shear rate. Again, the time constant for this is governed by the Brownian relaxation time.

Thixotropy, on the other hand, arises from the coupling of the viscous, hydrodynamic dissipation to the microstructure. An *ideal* thixotropic system, that is a

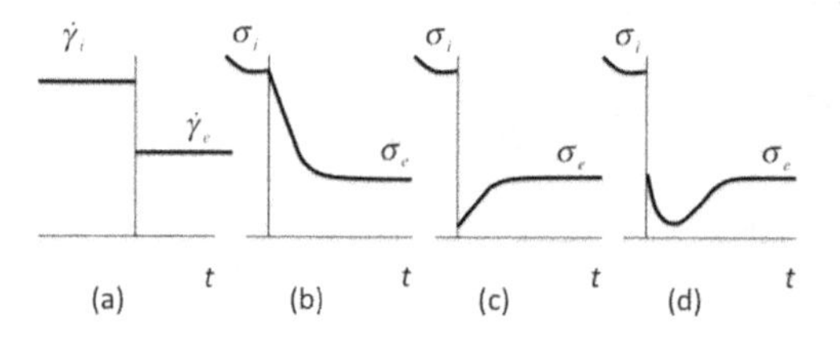

Figure 1.18 Stepwise decrease in shear rate: (a) kinematics; (b) stress response in a viscoelastic fluid; (c) inelastic thixotropic stress response; (d) general stress response for a thixotropic colloidal suspension. Also appearing in *CSR*, figure 7.1.

system with only hydrodynamic contributions to the stress and no viscoelasticity, will have a viscosity that depends on the degree of flocculation or aggregation of the particles that make up this microstructure. Hence, upon performing the experiment depicted in Figure 1.14(a), the stress for an ideal thixotropic suspension would immediately drop in proportion to the shear rate because its microstructure cannot change instantaneously. Rather, after the sudden drop in shear rate, the stress will slowly *increase* in time as the microstructure gradually rebuilds to a value consistent with the new, lower shear rate, see Figure 1.18(c). This is because rebuilding the microstructure leads to an increase in viscosity.

In *thixotropic colloidal suspensions* a more complex time-dependence of the stress is observed due to superposition of thixotropic and viscoelastic effects, as illustrated in Figure 1.18(d). The response of a particular colloidal suspension will depend on the details of the system and the test conditions as to which effect is more prominent. Such behavior is most evident when flocs or aggregation leads to a significant, reversible microstructure, which often also manifests itself as a yield stress. More about these rheological phenomena is discussed in chapters 6, 7 and 9 of *CSR*.

Many consumer and industrial products that contain colloidal particles exhibit some level of time-dependent rheology. Often, this behavior is denoted imprecisely as being "thixotropic" [43,63,64]. As discussed above, most thixotropic colloidal suspensions also exhibit some level of viscoelasticity. They also often display some level of plasticity, the latter being characterized by a yield stress. Interestingly, human blood is thixotropic due to structuring of the red blood cells into rouleaux, and this topic will be covered in detail in Chapter 8. Glasses of dense colloidal suspensions, as nonequilibrium structures, exhibit a limiting case of thixotropy in the sense that the structure can be loosened or "rejuvenated" by shearing whereas at rest it recovers or "ages". The latter, however, will proceed at a rate that keeps on slowing down in time and, hence, without reaching equilibrium within a finite amount of time (see Chapter 5). In other systems at least some of the time dependency is the result of irreversible changes in structure, as in waxy crude oils [65] and cements (see Chapter 9). Such irreversible changes are not actually thixotropic, but formulation and rheological modeling of such complex, irreversible systems requires consideration of thixotropic effects.

A simple and commonly used first test for thixotropy consists of applying a shear rate that linearly increases with time up to a maximum, followed by a gradual reduction at the same rate. The resulting instantaneous shear stress is plotted versus the instantaneous shear rate, an example of which is shown in Figure 1.19. When applying such a test on a sample in which the microstructure has been strongly broken down by preshearing, a complex evolution of the structure ensues. When starting from rest, the increasing shear rate will cause structure formation as the flow brings particles together, as well as break down the developing microstructure when the shear rate continues to increase. This results in a highly nonlinear viscosity and sometimes even a decreasing shear stress. Once the shear rate starts to decrease back to zero, the degree of microstructure and hence the viscosity will gradually rebuild. Structure recovery at high shear rates is slow, and, as a consequence, the decreasing

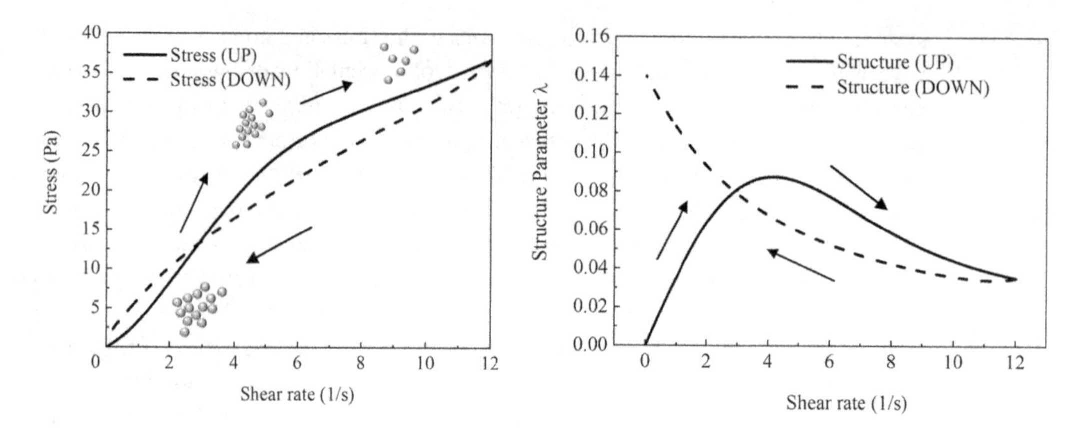

Figure 1.19 Thixotropic loop test for structure parameter $\lambda(t = 0) = 0$ and an acceleration of 0.8 s^{-2} showing how the stress evolves with transient shear rate. Inset shows the shear rate profile with time. Left: Stress evolution. Right: Structure evolution $\lambda(t)$. The system parameters are shown in Table 1.2.

stress curve lies below the increasing one, thus describing a "loop". During this return to the rest state the sample structure rebuilds such that a yield stress is evident at the end of the experiment. This *hysteresis loop* provides a qualitative metric for the degree of thixotropy, but it is highly sensitive to the various parameters of the test. It should be pointed out that a similar experiment, performed sufficiently fast on a viscoelastic fluid, will also display a hysteresis loop, but with some important qualitative and quantitative differences.

Micromechanical modeling for thixotropy requires some model for the shear history dependent microstructure. In principle, the evolving microstructure could be calculated from first principles or simulations based on the details of the particles, the solvent, and the interparticle forces involved. In such an ideal case the detailed rheological behavior would then be automatically generated by suitably summing the forces acting between particles given the instantaneous microstructure. Such modeling is the topic of Chapter 2. This route, however, is still challenging because of the complexity of the microstructural changes that can occur in flocculated colloidal dispersions at different length scales, both at rest and during flow, which are also affected by the various force fields involved. Generally, simulations are required to obtain rheological information, as discussed in Chapter 3. Structural features to be considered include: distributions of floc size, their internal structure, shape, and orientation, as well as connectivity between flocs [66]. Simulations demonstrate that the microstructure can be highly heterogeneous with different kinetics for the structural elements at different length scales [67], which explains some of the complexity of the transient rheology in such systems [63,68].

In the most common approach to modeling thixotropy one selects a nonthixotropic constitutive equation to describe the instantaneous behavior, i.e. corresponding to a given level of microstructure. The parameters of the model are then assumed to depend on the instantaneous microstructure. In the simplest possible case the state of the structure is represented by a single dimensionless structural parameter, often represented by the symbol λ, with values of 0 and 1 for, respectively, the fully broken down and the fully developed structure. In the original formulation by Goodeve [69] this structure parameter was considered a metric of the degree of "bond" formation between the particles in suspension. Structure parameters that go to infinity rather than to one for the maximum structure have been used as well in order to have a diverging viscosity at full structure without using an actual yield stress [70]. The most general form for the basic equation should describe elasto-viscoplasticity, but not all these effects are always included.

As an example, we start with a generalization of the Bingham equation, such that the instantaneous rheological response at time t in one-dimensional form becomes:

$$\sigma(t) = \sigma_{y,0}[\lambda(t)] + \eta_{pl}[\lambda(t)]\dot{\gamma}(t). \tag{1.36}$$

Considering only viscous effects results in models of "nondissipative" [63] or "ideal" [45] thixotropy. This behavior is normally approached in real systems at higher shear rates. Recovery at lower shear rates naturally shows some viscoelastic transients as well, but their timescales are usually much smaller than the thixotropic ones. Elasticity at rest is often associated with the yield stress, as both reflect the presence of a long-range connectivity in the system. The total stress in Eq. (1.36) also has been taken to be the sum of elastic and viscous stress components [71], it has also been substituted by a linear viscoelastic equation such as the Maxwell or Jeffreys models [72]. Tensorial descriptions of the structure can be constructed [73], but most often scalar expressions are used.

Including the memory of the shear history through the microstructure is essential to describe thixotropy. For that purpose an evolutionary equation for the rate of change of the structure is added to the basic rheological model for the instantaneous structure of Eq. (1.36). The rate equation should contain terms for describing the kinetics of breakdown and recovery of structure. When using Eq. (1.36) these are assumed to be functions of the instantaneous values of λ itself and also of the controlling, instantaneous flow parameters, e.g., the shear rate. Breakdown is induced by flow; recovery occurs by means of the Brownian and interparticle forces, but can be accelerated by slow flows (orthokinetic, i.e., flow-induced, flocculation). Hence the kinetic equation can contain three terms that represent these three processes, although often only the first two mechanisms are considered. A general form would be:

$$\frac{d\lambda}{dt} = -k_1\dot{\gamma}^a\lambda^b + k_2(1-\lambda)^c + k_3\dot{\gamma}^d(1-\lambda)^e, \tag{1.37}$$

in which power law functions are assumed for the dependence on structure and shear rate. The first term on the right hand side represents shear breakage of the structure, while the following two terms represent collisional aggregation by shear and by Brownian motion, respectively. The model parameters are determined from available information about the structure kinetics or by fitting experimental data. In Eq. (1.37) the shear rate represents the flow conditions but shear stress or energy have been used

Table 1.2 Thixotropic model parameters

$\sigma_{y,0}$ (Pa)	η_0 (Pa.s)	η_{str} (Pa.s)	k_{-} (-)	k_{+Br} (s^{-1})
8.5	2.0147	29	0.075	0.03

as well [74,75]. It should also be pointed out that, with a linear viscoelastic model as the starting equation, using a structure parameter can generate nonlinear viscoelastic models without real thixotropic features if the thixotropic time scales are not much longer than the viscoelastic relaxation time [45,63].

As an illustration of modeling thixotropic behavior, the following simple model is used (see chapter 7 in *CSR* for more details):

$$\begin{aligned} \sigma(t) &= \lambda(t)\sigma_{y,0} + \lambda(t)\eta_{str}\dot{\gamma}(t) + \eta_0\dot{\gamma} \\ \frac{d\lambda(t)}{dt} &= -k_{-}\dot{\gamma}(t)\lambda(t) + k_{+Br}[1 - \lambda(t)] \end{aligned} \quad (1.38)$$

The first equation assumes Bingham behavior for each instantaneous structure, with a yield stress of $\lambda\sigma_{y,0}$ and a plastic viscosity $(\lambda\eta_{str} + \eta_0)$. The structural viscosity η_{str} multiplies the level of structure, while η_0 represents the background viscosity when $\lambda = 0$. The kinetic equation assumes the most simple dependence on the instantaneous values of λ and $\dot{\gamma}$, using rate constants k_{-} and k_{+} for, respectively, shear-induced structure breakdown and Brownian structure recovery.

Consider the following case study. Starting from a system that has been presheared, so that there is no initial structure, $\lambda = 0$, and using the model parameters of Table 1.2 for a shear rate linearly increasing at a rate of 0.8 s^{-2}, the hysteresis curve shown in Figure 1.19 is obtained. Under this starting condition, the rise in stress is initially enhanced by structure building due to Brownian motion. This is because the preshear has removed all structure. However, the application of shear will break structure and so, as higher shear rates are achieved, this newly formed structure starts breaking down. Only in the limit of a very slow ramp, as compared to the kinetic rates of structure formation and breakage, will the sample reach its equilibrium level of structure and viscosity. More extreme cases can occur, such as a rapid shearing of a sample that has been allowed to rest such that it starts with some structure. Then, a rapid ramp in shear rate can lead to a maximum in the stress, whereupon shear banding [1] may become an issue in the experiments. Once the maximum rate has been achieved and the shear rate is ramped back down, the structure gradually rebuilds as the shear breakage becomes less important. However, the structure is observed to lag behind the equilibrium values corresponding to the instantaneous shear rate and the viscosity lies below that for the ramp-up. As the shear rate is reduced further, the structure continues to build such that the descending stresses become larger than those for the ramp-up at comparable shear rates. As the flow is stopped a yield stress develops. This is observed in Eq. (1.38), where the structure will continue to develop even while the sample is at rest due to Brownian aggregation. In fact, at rest the sample will eventually become fully structured, $\lambda = 1$, and the stress will achieve the value of $\sigma_{y,0}$.

A single scalar measure for the structure parameter cannot completely describe real thixotropic systems. As an example one can compare the recovery of the structure after shearing at a high shear rate under two conditions: arresting the flow and reducing the shear rate to a low value. The short range structures, the flocs, might develop similarly, but at rest a long range connectivity might develop rather early, while this is less the case during shear. Differences in heterogeneity between recovery under different conditions have been demonstrated with rheological and dielectric measurements [77] as well as by simulation [66]. The single λ models also systematically predict transients that are too steep. The actual data can be fitted quite well by stretched exponentials. Incorporating such a term as pre-factor in the kinetic equation enables a better description of transient behavior, at least for one-dimensional models [76]. Another solution is to use multiple structural parameters or a parameter distribution together with suitable nonlinear kinetics [74], but this necessarily introduces more model parameters. Alternatively, the presence of flocs of different sizes can be accounted for by using a population balance that expresses the formation rate for each floc size as the sum of flocculation rates of suitable smaller sizes and the rupture rates of suitable larger sizes [78]. Finally, we note that flow reversals performed on thixotropic materials also indicate additional rheological transients associated with shear-induced orientation of the floc structure, which cannot be described by a scalar structure parameter [71]. To fully address structural anisotropy a tensorial structure parameter is required [45,73,79].

Yield stresses are common in thixotropic suspensions. Their presence in the models entails additional difficulties, in particular when generalizing to three dimensions and when linking yield stress and elasticity [75,80]. In plasticity theory yielding in 3D is characterized by a yield surface in the stress plane. In the case of thixotropy the size of the yield surface then becomes a function of the structure. Simultaneously, a shifting of the yielding surface in the stress plane, i.e., "kinetic hardening" from plasticity theory, can be incorporated [74,81]. It also provides a means to account for shear-induced anisotropy, which is clearly present in most thixotropic suspensions but normally ignored in modeling. Anisotropy clearly shows up in shear reversal experiments and should also affect the LAOS behavior. However, such considerations may not be important for many industrial processes where suspensions are only subject to flow in the same direction during processing.

Finally, we note that the use of an empirical structure parameter does not allow for independent validation of the structure kinetics approach through direct measurement of the structure, as the parameter has no direct physical meaning. In the population balance models the structure parameter has physical meaning in terms of the distribution of particles in flocs, including floc structure information such as fractal dimension [78]. Additional approaches based on the physical principles to be introduced in Chapter 2 promise a more robust approach where the model parameters can be determined from independent measurements of the particle and system properties, leading to a truly *predictive* model for thixotropy.

Story 1.1 Ruth N. Weltmann and Early Studies of Thixotropy

A number of women scientists made important contributions to the emerging field of rheology. Among the first female rheologists in academia was Emma Schalek who, with A. Szegvari as a co-author, described for the first time the phenomenon of thixotropy in 1923 [82]. Ruth Weltmann worked on the same topic and she can apparently be considered the first female industrial rheologist in the literature. In 1937 she started working at Interchemical Corporation with Henry Green, mainly on the rheology of colloidal pigment suspensions. Green had designed and patented a Couette-type viscometer with a variable drive. With this instrument it became possible to subject the samples to arbitrary shear histories. Weltmann developed a recording system for it, which made it possible to readily generate hysteresis loops [83]. Either alone or with Green, she published a series of papers in which this technique method was studied and which contain the first systematic measurements of this type, e.g., [84,85].

Ruth Weltmann had a very eventful and long life, although some details of her career are unclear as the various biographical sources are inconsistent. She was born in 1912 in Berlin where she graduated in physics in 1935. In 1937 she moved to the US and started working at Interchemical. She later moved to what became the NASA Glenn Research Center, but remained for some time associated with Interchemical and continued to publish on rheological topics. She was "discharged of her duties" by NASA, for which she filed a court case claiming sexual discrimination. Afterward she worked for the Ford Motor Company and for Factory Mutual.

Ruth Weltmann was concerned about society as a whole and the Jewish community in particular, violence prevention being a central issue. With her husband, Semi Begun, she established a Society for the Prevention of Violence and later the Semi J. and Ruth W. Begun Foundation, now a supporting foundation of the Jewish Federation of Cleveland. She herself wrote several books on the development of social and violence prevention skills for children. She passed away in 2014, at the age of 102 years, perhaps also another record for a rheologist!

Appendix: Rheological Definitions

An analysis of steady flow in a fluid implies a full description of the kinematics, i.e., the velocity vector $\boldsymbol{v}$ as a function of position $\boldsymbol{r}$. The parameter of interest here is actually the change of velocity with position: the gradient $\nabla \boldsymbol{v}$. In the simplest, one-dimensional case, the liquid is contained in between two parallel planes, the bottom one fixed and the top one sliding at constant velocity V_x in the x-direction. The local velocities in the liquid will change linearly in the vertical or y-direction, starting from zero at the bottom (i.e., no-slip boundary conditions). Hence, there is only a single nonzero velocity gradient: dv_x/dv_y, the shear rate $\dot{\gamma}$. There are in general three velocity components (v_x, v_y, v_z), where each can change in all three directions. Hence, the components of $\nabla \boldsymbol{v}$ are described by a matrix, in Cartesian coordinates it becomes:

$$\nabla \boldsymbol{v} = \begin{pmatrix} \partial v_x/\partial x_x & \partial v_x/\partial x_y & \partial v_x/\partial v_z \\ \partial v_y/\partial x_x & \partial v_y/\partial x_y & \partial v_y/\partial x_z \\ \partial v_z/\partial x_x & \partial v_z/\partial x_y & \partial v_z/\partial x_z \end{pmatrix}. \tag{1.A1}$$

The rheology links the kinematics to the stresses, but $\nabla \boldsymbol{v}$ is not very suitable for that purpose as it can have nonzero components even if there is no flow, e.g., in rigid body rotation. This problem is avoided by using the symmetric part of $\nabla \boldsymbol{v}$: the rate-of-strain tensor **D**:

$$D_{ij} = \left\{ \begin{matrix} \dfrac{\partial v_x}{\partial x} & \dfrac{1}{2}\left(\dfrac{\partial v_x}{\partial y} + \dfrac{\partial v_y}{\partial x}\right) & \dfrac{1}{2}\left(\dfrac{\partial v_x}{\partial z} + \dfrac{\partial v_z}{\partial x}\right) \\ \dfrac{1}{2}\left(\dfrac{\partial v_y}{\partial x} + \dfrac{\partial v_x}{\partial y}\right) & \dfrac{\partial v_y}{\partial y} & \dfrac{1}{2}\left(\dfrac{\partial v_y}{\partial z} + \dfrac{\partial v_z}{\partial y}\right) \\ \dfrac{1}{2}\left(\dfrac{\partial v_z}{\partial x} + \right) & \dfrac{1}{2}\left(\dfrac{\partial v_z}{\partial y} + \dfrac{\partial v_y}{\partial z}\right) & \dfrac{\partial v_z}{\partial z} \end{matrix} \right\}. \tag{1.A2}$$

For incompressible fluids $\nabla \cdot \boldsymbol{v} = \partial v_x/\partial x + \partial v_y/\partial y + \partial v_z/\partial z = 0$. For simple shear flow in the *xy* plane, as shown in Figure 1.A1(a), the only nonzero components are:

$$D_{xy} = D_{yx} = \frac{1}{2}\frac{\partial v_x}{\partial y}. \tag{1.A3}$$

Hence, the shear rate $\dot{\gamma}$ used earlier equals $2D_{xy}$.

In addition to simple shear flow, uniaxial extensional (or "elongational") flow is also a standard reference flow, Figure 1.A1(b). It is the flow that occurs when a cylinder of liquid is stretched in the axial or *x*-direction, as typical in the spinning of fibers. The off-diagonal terms, expressing shear flow, are now equal to zero, while the diagonal terms are interconnected because of the incompressibility requirement. For steady uniaxial extensional flow:

$$\frac{\partial v_y}{\partial y} = \frac{\partial v_z}{\partial z} = \frac{1}{2}\frac{\partial v_x}{\partial x}. \tag{1.A4}$$

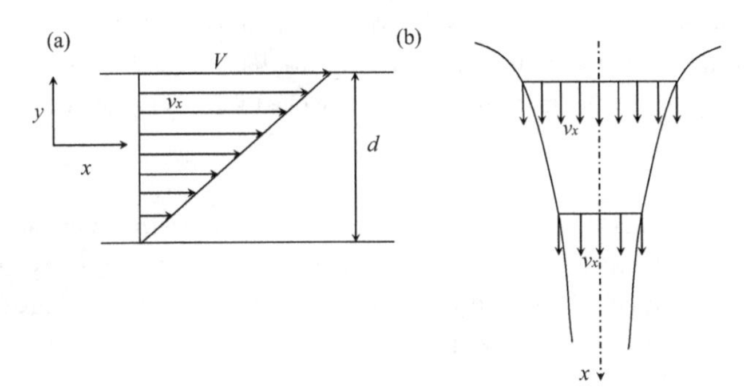

Figure 1.A1 Kinematics of: (a) shear flow and (b) uniaxial flow.

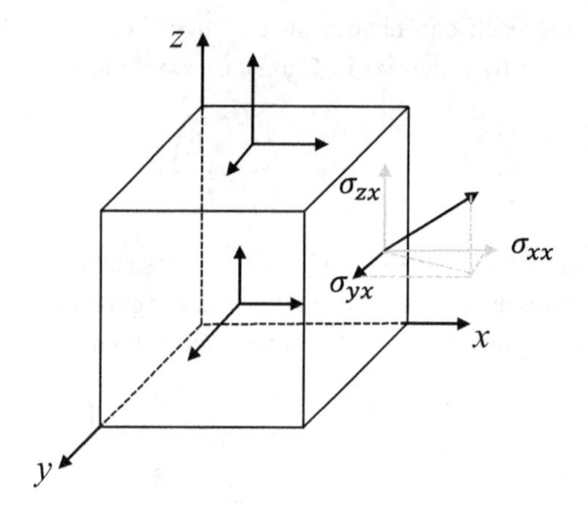

Figure 1.A2 Cartesian coordinate frame showing the components of the stress tensor.

The kinematics of the flow in any point of the fluid are linked to the local stresses. The stress, or force per unit area on a plane with arbitrary orientation at point r, can be computed from the local stresses on the coordinate planes through this point. With each coordinate plane three stress components are associated, e.g., for the plane $dydz$ these are the components σ_{xx}, σ_{yx}, and σ_{zx}. The first index indicates the orientation of the stress and the second one refers to the orientation normal to the plane under consideration, as shown in Figure 1.A2. Hence, the stresses at a point of the liquid are completely described by the following component matrix of the stress tensor $\boldsymbol{\sigma}$:

$$\sigma_{ij} = \begin{pmatrix} \sigma_{xx} & \sigma_{xy} & \sigma_{xz} \\ \sigma_{yx} & \sigma_{yy} & \sigma_{yz} \\ \sigma_{zx} & \sigma_{zy} & \sigma_{zz} \end{pmatrix}. \tag{1.A5}$$

The terms of the first diagonal express stresses normal to the plane on which they act, the off-diagonal terms are stresses within this plane: the shear terms. Even without flow there are nonzero terms in the stress tensor because of the ever present hydrostatic pressure –P (pressure is taken as negative): $\sigma_{xx} = \sigma_{yy} = \sigma_{zz} = P$ or $\sigma = -P\mathbf{I}$, $\mathbf{I}$ being the unit tensor. For incompressible liquids P is not affected by the kinematics. This has to be taken into account when casting the basic definition of a Newtonian fluid, $\sigma_{xy} = \eta \partial v_x / \partial y$, into a three dimensional tensorial form. The latter then becomes:

$$\boldsymbol{\sigma} = -P\mathbf{I} + 2\eta\mathbf{D}, \tag{1.A6}$$

in which the *extra stress* $\sigma + P\mathrm{I}$ is the relevant stress term. As the kinematic tensor $\mathbf{D}$ the stress tensor is symmetric with respect to the first diagonal for ordinary fluids. This equation can now be applied to the kinematics of steady uniaxial extensional flow, Eq. (1.A4). After eliminating P one finds:

$$\sigma_{xx} - \sigma_{yy} = \eta_E \frac{\partial v_x}{\partial x} = 3\eta \frac{\partial v_x}{\partial x}. \tag{1.A7}$$

Table 1.A1 Typical phenomenological constitutive equations used for suspensions

Models without yield stress	
$\sigma = \eta\dot{\gamma}$	Newtonian
$\sigma = k\dot{\gamma}^n$	Power law
$\eta - \eta_\infty = (\eta_0 - \eta_\infty)/(1 + (k'\dot{\gamma})^m)$	Cross
Models with yield stress	
$\sigma = \sigma_y^B + k\dot{\gamma}$	Bingham
$\sigma = \sigma_y^H + k\dot{\gamma}^n$	Herschel–Bulkley
$\sigma^{1/2} = \sigma_y^{1/2} + k\dot{\gamma}^{1/2}$	Casson

This equation defines the extensional viscosity and indicates that the ratio of shear over extensional viscosity or *Trouton ratio* equals 3 for Newtonian fluids.

When only shear flows are discussed, the subindices will be dropped. The most commonly used models for describing generalized Newtonian fluids, which have a nonlinear relation between shear stress and shear rate, are given in Table 1.A1.

In these equations η_0 and η_∞ are the limiting low and high shear viscosities, σ_y represents the yield stress, whereas *k*, *k'*, *n* and *m* are model parameters. Note that the units of *k* depend on the powerlaw exponent for the shear rate specific to the particular model.

The previous model equations describe nonlinear relationships between shear stress and shear rate. They still display the characteristic behavior of viscous fluids in that the energy applied to generate flow is totally dissipated as heat. In colloidal suspensions, as well as in various other fluid-like materials, some of this energy is not dissipated but stored during the motion as is the case in elastic materials. This viscoelastic behavior gives rise to a number of time-dependent effects.

A convenient method to study viscoelasticity is by means of oscillatory flow. This involves applying an oscillatory shear deformation $\gamma(t) = \gamma_0 \sin \omega t$ with peak strain γ_0 and frequency ω. A linear elastic body would cause a stress in phase with and proportional to the instantaneous strain, the proportionality factor being the shear modulus *G*. For a Newtonian fluid the stress would be proportional to the instantaneous shear rate $\dot{\gamma}(t) = \gamma_0\omega \cos \omega t = \gamma_0\omega \sin(\omega t + \pi/2)$, the proportionality factor being η. Hence the viscous stress is shifted by 90° with respect to the strain. As the elastic stress it is also proportional to γ_0, describing linear behavior. When a combination of the two effects occur in a fluid the two stress components are nonzero and have to be added. The global stress can then be written in complex notation as:

$$\sigma = G^* \gamma \tag{1.A8}$$

in which the viscoelastic fluid is described by a complex dynamic modulus $G^* = G' + iG''$, containing a real and imaginary part, expressing, respectively, the elastic and viscous contributions. The stress will be shifted with respect to the strain by a phase angle varying between 0° for purely elastic materials and 90° for the purely viscous ones, thus indicating the relative contribution of the two effects. The material response and hence the dynamic moduli depend on frequency. Linearity, as assumed here, is only satisfied at sufficiently small amplitudes, which is often limited to very small strains (~1%) for colloidal suspensions. Instead of starting from a relation between stress and strain, one could describe linear viscoelasticity starting from a linear relation between stress and strain rate, which then gives rise to a complex viscosity for viscoelastic fluids:

$$\boldsymbol{\sigma} = \boldsymbol{\eta}^* \dot{\boldsymbol{\gamma}}. \tag{1.A9}$$

For frequency dependent experiments, the complex dynamic viscosity also has real and imaginary parts: $\eta^* = \eta' - i\eta''$.

Viscoelasticity in fluids also affects the behavior in steady shear flow. There, the assumption of an isotropic pressure is not satisfied anymore. The three normal stresses σ_{ii} are not necessarily identical anymore and therefore two normal stress differences, i.e., the first (N_1) and the second (N_2) one are defined:

$$\begin{aligned} N_1 &= \sigma_{xx} - \sigma_{yy} \\ N_2 &= \sigma_{yy} - \sigma_{zz}. \end{aligned} \tag{1.A10}$$

Chapter Notation

B_2 second virial coefficient [m^3]
c_p polymer concentration [kg/m^3]
c_p^* overlap concentration [kg/m^3]
$F(r)$ Force between particles, oriented along the center-to-center line [N]
k_i rate terms for structure kinetics in thixotropic models
L contour length of polymer [m]
l segmental length of polymer [m]
N_p degree of polymerization [–]
n_i bulk concentration of ions of type i [1/m^3]
t_D characteristic time for diffusion by Brownian motion [s]
U depth of the square well potential [J]
V_O overlap volume [m^3]

Greek Symbols

γ_lv surface tension [N/m]
Δ half-width of the square well potential [m]
σ_c critical shear stress for shear thinning, Eq. (1.30) [Pa]
σ_c^hs critical shear stress for shear thickening, Eq. (1.24) [Pa]
σ_p grafting density [1/m^2]
Υ_i ith normal stress difference coefficient for suspensions, Eq. (1.23) [–]

Subscripts

str structural, in reference to thixotropic models

Superscripts

pol polymer
sq square well (steric)
ss sticky sphere

References

1. Mewis J, Wagner NJ. *Colloidal Suspension Rheology*. Cambridge: Cambridge University Press; 2012. 393 p.
2. Hiemenz PC, Rajagopalan R. *Principles of Colloid and Surface Chemistry*. 3rd ed. Boca Raton, FL: CRC Press; 1997. 650 p.
3. Russel WB, Saville DA, Schowalter WR. *Colloidal Dispersions*. Cambridge: Cambridge University Press; 1989. 525 p.
4. Berg JC. *An Introduction to Interfaces and Colloids*. Hackensack, NJ: World Scientific Publishing Co. Pte Ltd.; 2010. 785 p.
5. Hunter RJ, White LR. *Foundations of Colloid Science*. Oxford: Clarendon Press; 1987. 673 p.
6. Macosko CW. *Rheology Principles, Measurements, and Applications*. 1st ed. New York: VCH Publications; 1994. 549 p.
7. Laun HM. Rheological properties of aqueous polymer dispersions. *Angewandte Makromolekulare Chemie*. 1984;123(1):335–359.
8. Shikata T, Pearson DS. Viscoelastic behavior of concentrated spherical suspensions. *Journal of Rheology*. 1994;38(3):601.
9. Stokes GG. On the effect of internal friction of fluids on the motion of pendulums. *Transactions of the Cambridge Philosophical Society*. 1851;9(ii):8–106.
10. Sutherland A. A dynamical theory of diffusion for non-electrolytes and the molecular mass of albumin. *Philosophical Magazine*. 1905;9:781–785.
11. Einstein A. Über die von der molekularkinetischen Theorie der Wärme geforderte Bewegung von in ruhenden Flüssigkeiten suspendierten Teilchen. *Annalen der Physik*. 1905;17:549–560.
12. Smoluchowski M. Zur kinetischen Theorie der Brownschen Molekularbewegung und der Suspensionen. *Annalen der Physik*. 1906;21:756–780.
13. Kulkarni PM, Morris JF. Suspension properties at finite Reynolds number from simulated shear flow. *Physics of Fluids*. 2008;20(4):040602.
14. Cheng Z, Chaikin PM, Russel WB, Meyer WV, Zhu J, Rogers RB, et al. Phase diagram of hard spheres. *Materials & Design*. 2001;22(7):529–534.
15. Pusey PN, Vanmegen W. Phase-behavior of concentrated auspensions of nearly hard colloidal spheres *Nature*. 1986;320(6060):340–342.
16. Torquato S. Statistical description of microstructures. *Annual Review of Materials Research*. 2002;32:77–111.

17. Torquato S, Stillinger FH. Jammed hard-particle packings: From Kepler to Bernal and beyond. *Reviews of Modern Physics*. 2010;82(3):2633–2672.
18. Pusey PN, Vanmegen W. Observation of a glass-transition in suspensions of spherical colloidal particles. *Physical Review Letters*. 1987;59(18):2083–2086.
19. Verlet L, Weis J. J. Equilibrium theory of simple liquids. *Physical Review A*. 1972;5 (2):939–951.
20. Silbert LE, Ertaş D, Grest GS, Halsey TC, Levine D. Geometry of frictionless and frictional sphere packings. *Physical Review E*. 2002;65(3):031304.
21. Yurkovetsky Y, Morris JF. Particle pressure in sheared Brownian suspensions. *Journal of Rheology*. 2008;52(1):141–164.
22. Banchio AJ, Nagele G, Bergenholtz J. Viscoelasticity and generalized Stokes-Einstein relations of colloidal dispersions. *Journal of Chemical Physics*. 1999;111(18):8721–8740.
23. Russel WB, Wagner NJ, Mewis J. Divergence in the low shear viscosity for Brownian hard-sphere dispersions: At random close packing or the glass transition? *Journal of Rheology*. 2013;57(6):1555–1567.
24. Bergenholtz J, Wagner NJ. The Huggins coefficient for the square-well colloidal fluid. *I&EC Research*. 1994;33:2391–2397.
25. Einstein A. Eine neue Bestimmung der Molekuldimensionen. *Ann Physik*. 1906;19:289-306.
26. Einstein A. Berichtigung zu meiner Arbeit: "Eine neue Bestimmung der Moleküldimensionen". *Annalen der Physik*. 1911;34:591–592.
27. Banchio AJ, Brady JF. Accelerated Stokesian dynamics: Brownian motion. *Journal of Chemical Physics*. 2003;118(22):10323–10332.
28. Phan SE, Russel WB, Zhu JX, Chaikin PM. Effects of polydispersity on hard sphere crystals. *Journal of Chemical Physics*. 1998;108(23):9789–9795.
29. Lionberger RA, Russel WB. High-frequency modulus of hard-sphere colloids. *Journal of Rheology*. 1994;38(6):1885–1908.
30. de Kruif CG, van Lersel EMF, Vrij A, Russel WB. Hard-sphere colloidal dispersions – viscosity as a function of shear rate and volume fraction *Journal of Chemical Physics*. 1985;83(9):4717–4725.
31. Bergenholtz J, Brady JF, Vicic M. The non-Newtonian rheology of dilute colloidal suspensions. *Journal of Fluid Mechanics*. 2002;456:239–275.
32. Batchelor GK, Green JT. The determination of the bulk stress in a suspension of spherical particles to order c^2. *Journal of Fluid Mechanics*. 1972;56:401–427.
33. Dai SC, Tanner R. Elongational flows of some non-colloidal suspensions. *Rheologica Acta*. 2017;56(1):63–71.
34. Brady J, Khair A, Swaroop M. On the bulk viscosity of suspensions. *Journal of Fluid Mechanics*. 2006;554:109–123.
35. Foss DR, Brady JF. Structure, diffusion and rheology of Brownian suspensions by Stokesian dynamics simulation. *Journal of Fluid Mechanics*. 2000;407:167–200.
36. Wagner NJ, Brady JF. Shear thickening in colloidal dispersions. *Physics Today*. 2009;62 (10):27–32.
37. Maranzano BJ, Wagner NJ. The effects of interparticle interactions and particle size on reversible shear thickening: Hard-sphere colloidal dispersions. *Journal of Rheology*. 2001;45(5):1205–1222.
38. Maranzano BJ, Wagner NJ. The effects of particle-size on reversible shear thickening of concentrated colloidal dispersions. *Journal of Chemical Physics*. 2001;114 (23):10514–10527.

39. Bender JW, Wagner NJ. Optical measurement of the contributions of colloidal forces to the rheology of concentrated suspensions. *Journal of Colloid and Interface Science*. 1995;172 (1):171–184.
40. Krishnamurthy LN, Wagner NJ, Mewis J. Shear thickening in polymer stabilized colloidal dispersions. *Journal of Rheology*. 2005;49(6):1347–1360.
41. Cheng X, McCoy JH, Israelachvili JN, Cohen I. Imaging the microscopic structure of shear thinning and thickening colloidal suspensions. *Science*. 2011;333(6047):1276–1279.
42. Morris JF. Shear thickening of concentrated suspensions: Recent developments and relation to other phenomena. *Annual Review of Fluid Mechanics*. 2020;52(1):121–144.
43. Asakura S, Oosawa F. On interaction between two bodies immersed in a solution of macromolecules. *Journal of Chemical Physics*. 1954;22:1255–1256.
44. Toussaint F, Roy C, Jézéquel P.-H. Reducing shear thickening of cement-based suspensions. *Rheologia Acta*. 2009;48(8):883–895.
45. Krishnamurthy LN, Wagner NJ. Letter to the editor: Comment on: "Effect of attractions on shear thickening in dense suspensions" Journal of Rheology 2004;4, 1321 (2004). *Journal of Rheology*. 2005;49(3):799–803.
46. Morris JF, Boulay F. Curvilinear flows of noncolloidal suspensions: The role of normal stresses. *Journal of Rheology*. 1999;43(5):1213–1237.
47. Cwalina CD, Wagner NJ. Material properties of the shear-thickened state in concentrated near hard-sphere colloidal dispersions. *Journal of Rheology*. 2014;58(4):949–967.
48. Dhaene P, Mewis J, Fuller GG. Scattering dichroism measurements of flow-induced structure of a shear thickening suspension. *Journal of Colloid and Interface Science*. 1993;156(2):350–358.
49. Kalman DP, Wagner NJ. Microstructure of shear-thickening concentrated suspensions determined by flow-USANS. *Rheologica Acta*. 2009;48(8):897–908.
50. Gurnon AK, Wagner NJ. Microstructure and rheology relationships for shear thickening colloidal dispersions. *Journal of Fluid Mechanics*. 2015;769:242–276.
51. Silbert LE, Melrose JR. The rheology and microstructure of concentrated, aggregated colloids. *Journal of Rheology*. 1999;43(3):673–700.
52. Wagner NJ, Bender JW. The role of nanoscale forces in colloid dispersion rheology. *MRS Bulletin*. 2004;29(2):100–106.
53. Jamali S, Boromand A, Wagner N, Maia J. Microstructure and rheology of soft to rigid shear-thickening colloidal suspensions. *Journal of Rheology*. 2015;59(6):1377–1395.
54. Singh A, Mari R, Denn MM, Morris JF. A constitutive model for simple shear of dense frictional suspensions. *Journal of Rheology*. 2018;62(2):457–468.
55. Lee Y-F, Luo Y, Brown SC, Wagner NJ. Experimental test of a frictional contact model for shear thickening in concentrated colloidal suspensions. *Journal of Rheology*. 2020;64(2):267–282.
56. Reynolds O. On the dilatancy of media composed of rigid particles in contact, with experimental illustrations. *Philosophical Magazine [5th Series]*. 1885;20:469–481.
57. Metzner AB, Whitlock M. Flow behavior of concentrated (dilatant) suspensions. *Transactions of the Society of Rheology*. 1958;2:239–253.
58. O'Brien VT, Mackay ME. Stress components and shear thickening of concentrated hard sphere suspensions. *Langmuir*. 2000;16(21):7931–7938.
59. Brown E, Jaeger HM. The role of dilation and confining stresses in shear thickening of dense suspensions. *Journal of Rheology*. 2012;56(4):875–923.
60. Wang M, Brady JF. Constant stress and pressure rheology of suspensions. *Physical Review Letters*. 2015;115(15):158301.

61. Laun HM, Bung R, Schmidt F. Rheology of extremely shear thickening polymer dispersions (passively viscosity switching fluids). *Journal of Rheology*. 1991;35(6):999–1034.
62. Farr RS, Melrose JR, Ball RC. Kinetic theory of jamming in hard-sphere startup flows. *Physical Review E*. 1997;55(6):7203–7211.
63. Mewis J, Wagner NJ. Thixotropy. *Advances in Colloid and Interface Science*. 2009;147–148:214–227.
64. Larson RG, Wei Y. A review of thixotropy and its rheological modeling. *Journal of Rheology*. 2019;63(3):477–501.
65. Geri M, Venkatesan R, Sambath K, McKinley GH. Thermokinematic memory and the thixotropic elasto-viscoplasticity of waxy crude oils. *Journal of Rheology*. 2017;61 (3):427–454.
66. Hipp JB, Richards JJ, Wagner NJ. Structure-property relationships of sheared carbon black suspensions determined by simultaneous rheological and neutron scattering measurements. *Journal of Rheology*. 2019;63(3):423–436.
67. Zia RN, Landrum BJ, Russel WB. A micro-mechanical study of coarsening and rheology of colloidal gels: Cage building, cage hopping, and Smoluchowski's ratchet. *Journal of Rheology*. 2014;58(5):1121–1157.
68. Mewis J, Schoukens G. Mechanical spectroscopy of colloidal dispersions. *Faraday Discussions*. 1978;65:58–64.
69. Goodeve CF. A general theory of thixotropy and viscosity. *Transactions of the Faraday Society*. 1939;35:342–358.
70. Roussel N, Le Roy R, Coussot P. Thixotropy modelling at local and macroscopic scales. *Journal of Non-Newtonian Fluid Mechanics*. 2004;117:85–95.
71. Armstrong MJ, Beris AN, Rogers SA, Wagner NJ. Dynamic shear rheology of a thixotropic suspension: Comparison of an improved structure-based model with large amplitude oscillatory shear experiments. *Journal of Rheology*. 2016;60(3):433–450.
72. Mendes PRD, Thompson RL. A unified approach to model elasto-viscoplastic thixotropic yield-stress materials and apparent yield-stress fluids. *Rheologica Acta*. 2013;52(7):673–694.
73. Stickel JJ, Powell RL. Fluid mechanics and rheology of dense suspensions. *Annual Review of Fluid Mechanics*. 2005;37(1):129–149.
74. Wei YF, Solomon MJ, Larson RG. A multimode structural kinetics constitutive equation for the transient rheology of thixotropic elasto-viscoplastic fluids. *Journal of Rheology*. 2018;62(1):321–342.
75. Mendes PRD, Thompson RL. A critical overview of elasto-viscoplastic thixotropic modeling. *Journal of Non-Newtonian Fluid Mechanics*. 2012;187(4):8–15.
76. Dullaert K, Mewis J. A structural kinetics model for thixotropy. *Journal of Non-Newtonian Fluid Mechanics*. 2006;139(1–2):21–30.
77. Mewis J, Spaull AJB, Helsen J. Structural hysteresis. *Nature*. 1975;253(5493):618–619.
78. Mwasame PM, Beris AN, Diemer RB, Wagner NJ. A constitutive equation for thixotropic suspensions with yield stress by coarse-graining a population balance model. *AIChE Journal*. 2017;63(2):517–531.
79. Stickel JJ, Phillips RJ, Powell RL. A constitutive model for microstructure and total stress in particulate suspensions. *Journal of Rheology*. 2006;50(4):379–413.
80. Denn MM, Bonn D. Issues in the flow of yield-stress liquids. *Rheologica Acta*. 2011;50 (4):307–315.
81. Dimitriou CJ, McKinley GH. A comprehensive constitutive law for waxy crude oil: A thixotropic yield stress fluid. *Soft Matter*. 2014;10(35):6619–6644.

82. Schalek E, Szegvari A. Ueber Eisenoxydgallerten. *Kolloid-Z.* 1923;32:318–319.
83. Weltmann RN, inventor; Interchemical Corporation, assignee. Viscometer Recorder. US patent 2,497,919. 1950.
84. Green H, Weltmann RN. Analysis of the thixotropy of pigment-vehicle suspensions. Basic principles of the hysteresis loop. *Industrial and Engineering Chemistry, Analytical Edition.* 1943;15(3):201–206.
85. Weltmann RN. Breakdown of thixotropic structure as a function of time. *Journal of Applied Physics.* 1943;14(7):343–350.

2 Theory of Colloidal Suspension Structure, Dynamics, and Rheology

Gerhard Nägele, Jan K. G. Dhont, and Thomas Voigtmann

2.1 Introduction

Ever since general expressions have been derived for the flow behavior and rheology of colloidal suspensions, several approaches have been proposed to explicitly evaluate them in terms of concentration- and shear rate dependent viscous and viscoelastic response functions. The explicit evaluation of these general expressions for the suspension stress requires the solution of two different problems.

Recall that the description of colloidal suspensions is usually coarse-grained: one seeks equations of motion of the colloidal configurational degrees of freedom, i.e., their positions and orientations, under the assumption that the momenta relax to equilibrium much faster than it takes the colloid configuration to evolve. Owing to the length-scale separation between a colloidal particle size and the size of solvent molecules, one can treat the solvent as a continuum fluid. Thermal fluctuations in the fluid then cause fluctuating forces on the colloids that are treated as random. In addition, the solvent mediates forces between the colloids, which are essentially instantaneous, and are referred to as hydrodynamic interactions (HIs).

To describe the forces that the flowing solvent exerts onto the surfaces of the colloids, one needs to solve the Navier–Stokes equation for an assembly of colloids with prescribed positions, as well as orientations for the case of anisotropic colloids. This is a complicated hydrodynamics problem, which can be tackled on the basis of the so-called Stokes or "creeping-flow" equations, where the equations are simplified in several ways: (i) inertial forces on fluid elements are neglected, which is accurate due to the much larger, dominant friction forces, (ii) the time resolution is coarse grained to the so-called diffusion (or Brownian) time scale, which is the time required for the thermally averaged velocity of a colloid to relax, (iii) the nonlinear streaming contribution is neglected, which can be done as a result of the small size of the colloids giving rise to a small Reynolds number, and (iv) the solvent and the colloids are approximately incompressible. On the basis of these simplified hydrodynamic equations the forces on the colloid surfaces can be calculated within an approach that is referred to as the method-of-reflections. In practice, this method enables iteratively calculating the flow fields that are hence-and-forth reflected between two colloids, but analytical progress is limited to two-particle interactions. Analytical progress to higher concentrations commonly assumes that higher-order, many-body interactions can be approximated as being pair-wise additive or by invoking mean-field type approximations. The method of reflections leads to expressions for the forces in the form of a series expansion with respect to the inverse distance between the two colloids. Such an expansion is not suited to describe hydrodynamic forces on close approach of the surfaces of the colloids at very high packing fractions. For such so-called lubrication forces an expansion in terms of the distance between the surfaces of two colloids is more appropriate. Full multi-colloid hydrodynamic interactions (HIs) can be taken into account by means of simulations, which will be discussed in Chapter 3.

Once the above hydrodynamics problem is solved to within some approximation, the resulting expression for the stress must be averaged over the positions, and orientations if relevant, of the colloids. The averaging is to be taken with respect to a probability density function (pdf) for the positions. This pdf can be obtained, in principle, as the solution of the so-called generalized Smoluchowski equation (GSmE), which is a Fokker–Planck type equation of motion [1–6]. The input to this equation of motion is the total interaction potential energy and the hydrodynamic interaction functions, which are both commonly assumed to be pair-wise additive. For spherical colloids, the Smoluchowski equation is equivalent to a stochastic equation of motion for the position coordinates, the so-called Langevin equation. This is the overdamped limit of Newton's equation of motion of spheres with a noise contribution that accounts for Brownian motion [1–3,7]. The corresponding equivalent formulation of the Smoluchowski equation is especially useful for simulations. Real-space pdf solutions of the two-particle Smoluchowski equation under steady or oscillatory simple shear allow for calculating viscometric functions of semi-dilute suspensions, and to relate nonlinear rheological effects such as viscosity shear thinning and thickening, as well as normal suspension stress differences to flow-induced distortions of the microstructure of an equilibrium system. The real-space pdf treatment of rheology for nondilute suspensions has been achieved through the invocation of approximate closure relations expressing the three-particles pdf in terms of the two-particles one [8,9]. While in this way HIs are directly accounted for in the form of conditional two-particle averages, real-space rheology methods are commonly not applicable to very concentrated suspensions close to glass transition or gelation points where long-time concentration fluctuation relaxations are substantially slowed down.

Formal solutions of the Smoluchowski equation can be employed in combination with Green–Kubo (GK) relations to derive expressions for linear viscoelastic response functions in terms of thermal averages of the fluctuating, microscopic stress. These expressions are formulated in terms of an exponential time-evolution operator that is found from the generalized many-particles Smoluchowski equation (GSmE). The Green–Kubo relation for the dynamic suspension shear modulus, in particular, is useful in simulation studies of linear viscoelastic properties, and it serves as a starting point for statistical mechanics based methods applicable to concentrated suspensions. These include the ideal mode coupling theory (MCT) developed by Götze [10], summarized in Hansen and McDonald [11] plus a simplified so-called naive version due to Schweizer and coworkers [12–14], and the self-consistent generalized Langevin equation theory (SCGLE) proposed by

Medina-Noyola and coworkers [15–17]. All these methods are approximate and have been developed so far mostly without the inclusion of HIs. Fully analytic results for viscoelastic functions of interacting particles are scarce. A notable exception is the so-called weak coupling limit (WCL) of point-like particles interacting by a screened Coulomb potential for which analytic expressions are available for all viscoelastic functions, as discussed in Section 2.5.4.

Viscoelastic functions can be evaluated numerically using MCT, where approximate expressions are posed for multi-particle correlation functions in terms of two-particle ones. MCT is one of the very few analytical theories for the dynamics and rheology of concentrated suspensions, and in particular for the glass state, where the concentration is sufficiently high to give rise to dynamic arrest on the experimental time scale. The very slow particle dynamics in a glass implies that flow has a significant effect already for very small shear rates. For such systems it is important to account for nonlinear response, that is, for situations where shear-induced microstructural rearrangements are not linear in the applied shear rate anymore. Within the so-called Integration Through Transients (ITT) formalism, where the hydrodynamics related to the reflection of the incident shear field by the colloids is neglected, MCT has been extended by Fuchs, Cates, and coworkers to account for nonlinear rheological effects, as summarized previously [18–20] .

A flow chart of the content of this chapter is shown in Figure 2.1: The suspension rheology is treated on a level where the solvent is described as a structureless continuum. The suspended spherical particles interact by direct forces, e.g., excluded volume, electrostatic, van der Waals, or dispersion forces [21,22], and by the solvent-mediated HIs [1,3,21,23–25]. Direct interaction forces are treated in Chapter 1, while HIs are discussed in this chapter for time and length scales relevant to colloids. In addition, there are stochastic forces acting on the colloidal particles originating from the thermal bombardment by solvent molecules which drive their Brownian motion.

As depicted in the flow chart in Figure 2.1, the suspension rheology can be described theoretically in two alternative methods, which are statistically equivalent. The first one is based on Langevin-type stochastic equations of motion accounting for the various forces noted above. Using a time-stepping numerical integration scheme of the Langevin equations as implemented in widely used Brownian dynamics simulation methods with and without HIs included, particle position trajectories are generated constituting the elements of the many-particles configuration supervector $X(t)$. The macroscopic suspension stress tensor Σ is calculated from time-averaging the associated microscopic stress tensor $\boldsymbol{\sigma}(X)$ depending on $X(t)$, using a representative set of generated trajectories. Brownian dynamics simulation methods are discussed in Chapter 3, in addition to other mesoscale simulation methods.

As depicted in Figure 2.1, the second way to describe suspension rheology is based on the GSmE diffusion-advection equation for the time-dependent configurational pdf, $P(X,t)$, of many-particles positions. This equation is widely used in statistical physics studies of the colloid dynamics where reduced pdfs such as the two-particles pair distribution function $g(\mathbf{r},t)$ play a central role. Here, $g(\mathbf{r},t)$ is proportional to the probability density of finding the centers of two particles at vector distance $\mathbf{r}$ at time t.

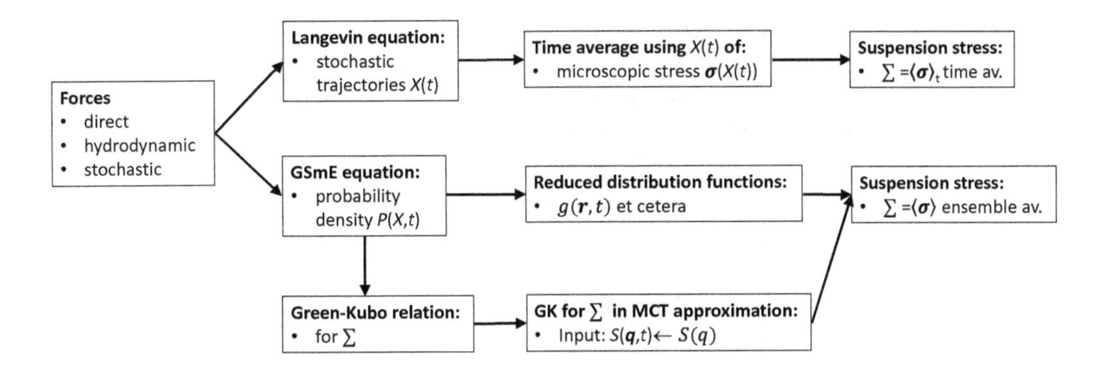

Figure 2.1 Flow chart illustrating the various approaches to rheology discussed in this chapter. The example provided here is for the suspension stress tensor Σ obtained from averaging the microscopic stress tensor $\boldsymbol{\sigma}(X)$.

From the knowledge of reduced pdfs, gained from solving the GSmE using perturbation methods accounting for shear flow, suspension properties such as Σ are calculated as ensemble averages of associated microscopic quantities. For an ergodic system, the ensemble average agrees with the corresponding time average over representative trajectories. This real-space approach to suspension rheology is particularly amenable to moderately concentrated suspensions. As noted before, for very concentrated, homogeneous suspensions, Fourier space methods, such as the approximate MCT, are combined with GK relations to calculate dynamic properties, such as linear viscoelastic functions. In such approaches, instead of $g(\mathbf{r}, t)$ the key input required here is its Fourier-space transform $S(\mathbf{q}, t)$, referred to as the dynamic structure factor. The latter can be related, via additional MCT equations derived from the GSmE, to the static structure factor $S(q) = S(q, t = 0)$ of the unsheared isotropic suspension depending solely on the modulus q of the wavevector $\mathbf{q}$.

2.2 Low Reynolds Number Hydrodynamics

The fundamental quantity in fluid dynamics is the stress tensor Σ, which is defined as follows. Consider an infinitesimally small surface element with surface area dS and unit normal $\hat{\mathbf{n}}$. By definition, $d\mathbf{S} \cdot \Sigma$ with $d\mathbf{S} = \hat{\mathbf{n}}\, dS$ is the force that the fluid just above the surface area element normal to $\hat{\mathbf{n}}$ exerts onto the fluid just below the surface area element. From Gauss's integral theorem it follows that $d\mathbf{r}\nabla \cdot \Sigma$ is the force exerted by the surrounding fluid onto the fluid within an infinitesimally small volume element $d\mathbf{r}$. Newton's equation of motion describing the fluid flow in a Newtonian fluid, i.e., the generalized Navier–Stokes equation, reads thus [24,25]

$$\rho(\mathbf{r}, t)\frac{d\mathbf{u}(\mathbf{r}, t)}{dt} = \rho(\mathbf{r}, t)\left[\frac{\partial \mathbf{u}(\mathbf{r}, t)}{\partial t} + \mathbf{u}(\mathbf{r}, t) \cdot \nabla \mathbf{u}(\mathbf{r}, t)\right] = \nabla \cdot \Sigma(\mathbf{r}, t) + \mathbf{f}^{ex}(\mathbf{r}, t), \quad (2.1)$$

where ρ is the mass density of the fluid. The force density $\mathbf{f}^{ex}$ is the force per unit volume with which a possibly present external field acts onto the fluid. Eq. (2.1) applies on the coarse-grained time resolution $\Delta t \gg \tau_s$, and associated coarse-grained spatial resolution where the fluid can be described as a structureless continuum. Here, $\tau_s \sim 10^{-13}$ s is the mean collision time of fluid molecules. A formal expansion of the stress tensor of an isotropic and isothermal fluid to leading order in gradients of the flow velocity leads to

$$\Sigma = \eta_m \left\{ \nabla\mathbf{u} + (\nabla\mathbf{u})^T \right\} - \hat{\mathbf{I}} \left\{ \mathrm{p} + \left(\frac{2}{3}\eta_m - \eta_B \right) \nabla \cdot \mathbf{u} \right\}, \tag{2.2}$$

where $\hat{\mathbf{I}}$ is the identity tensor with elements $\boldsymbol{\delta}_{\alpha\beta}$ for $\alpha, \beta \in \{x, y, z\}$, η_m is the fluid (medium) shear viscosity and η_B the bulk (volume) viscosity of the fluid, which are both taken as shear rate independent, and p is the total mechanical pressure.

Eqs. (2.1) and (2.2) are referred to as the Navier–Stokes equations for compressible Newtonian fluids. Three unknown field quantities appear in the Navier–Stokes equations, the fluid velocity $\mathbf{u}$, pressure p, and density ρ, so that additional equations are needed. The additional equation of motion is the so-called continuity equation, which expresses conservation of mass,

$$\frac{\partial}{\partial t}\rho(\mathbf{r}, t) + \nabla \cdot [\rho(\mathbf{r}, t)\mathbf{u}(\mathbf{r}, t)] = 0. \tag{2.3}$$

We limit ourselves here to incompressible systems, that is, systems where the density is essentially not affected by flow and pressure. For such incompressible fluids, the continuity Eq. (2.3) reduces to

$$\nabla \cdot \mathbf{u}(\mathbf{r}, t) = 0. \tag{2.4}$$

The bulk viscosity in the stress tensor in Eq. (2.2) therefore plays no role for incompressible fluids.

2.2.1 Time and Length Scales, Creeping-Flow Equations, and Oseen Tensor

The two terms in between the square brackets in the Navier–Stokes Eq. (2.1) can be very different in magnitude, depending on the hydrodynamic problem under consideration. We are interested in fluid flow around small sized objects, such as colloids. A typical value for the fluid flow velocity is the center velocity v of a colloidal particle. The fluid flow velocity decreases from a value v, close to a spherical colloidal particle, to a much smaller value over a distance of the order of a typical linear dimension a of the particle. We thus typically have $| \nabla^2\mathbf{u} | \approx v/a^2$. Similarly, $| \mathbf{u} \cdot \nabla\mathbf{u} | \approx v^2/a$. Let τ_D denote the desired, coarse-grained timescale on which the fluid flow is to be described. Introducing the rescaled variables, $\mathbf{u}' = \mathbf{u}/v$, $\mathbf{r}' = \mathbf{r}/a$, $t' = t/\tau_D$, $\mathrm{p}' = (a/\eta_m v)\mathrm{p}$, and $\mathbf{f}'^{ext} = (a^2/\eta_m v)\mathbf{f}^{ex}$, transforms the Navier–Stokes Eq. (2.1) for incompressible fluids to

$$\rho \frac{a^2}{\eta_m \tau_D} \frac{\partial \mathbf{u}'}{\partial t'} + Re_p \mathbf{u}' \cdot \nabla'\mathbf{u}' = \nabla'^2\mathbf{u}' - \nabla'\mathrm{p}' + \mathbf{f}'^{\mathrm{ext}}. \tag{2.5}$$

The dimensionless number Re_p is the so-called *Reynolds number*, which is equal to

$$Re_{\mathrm{p}} = \frac{\rho a v}{\eta_m}. \tag{2.6}$$

Note that the use of the particle size a as the length scale makes this the *particle Reynolds number* Re_{p}, as defined in Table 1.1. By construction we have $|\mathbf{u}' \cdot \nabla' \mathbf{u}'| \approx |\nabla'^2 \mathbf{u}'| \approx 1$, so that for very small values of the Reynolds number, the term $\sim \mathbf{u}' \cdot \nabla \mathbf{u}'$ in the left hand-side in Eq. (2.5) may be neglected. This renders the Navier–Stokes equation linear, and simplifies its mathematical analysis considerably.

The question now is what the time scale τ_D represents. As will be discussed, our interest is in dynamics on time scales that are much larger than the time required for a colloidal particle to lose momentum due to friction with the surrounding solvent. For sufficiently small objects, e.g., colloids, bacteria, and viruses, embedded in a solvent, the velocity of these objects is lost within a very short time. This momentum relaxation time can be estimated as follows. Consider a colloidal sphere with an initial velocity $\mathbf{v}_0$. The solvent friction force, averaged over thermal collisions with surrounding solvent molecules, is equal to $-\zeta_0 \langle \mathbf{v}(t) \rangle$, where $\langle \mathbf{v}(t) \rangle$ is the thermally averaged velocity at time t, and $\zeta_0 = 6\pi\eta_m a$ is the Stokes friction of a spherical colloid of mass M and radius a to which the fluid sticks at the surface. Newton's equation of motion thus reads, $Md\langle \mathbf{v}(t) \rangle / dt = -\zeta_0 \langle \mathbf{v}(t) \rangle$, the solution of which is [1,26]

$$\langle \mathbf{v}(t) \rangle = \mathbf{v}_0 \exp\left\{ -\frac{\zeta_0}{M} t \right\}. \tag{2.7}$$

The time scale on which the velocity of a colloidal sphere relaxes to zero is thus equal to $\tau_M = M/\zeta_0$. Putting in typical values for the mass and friction coefficient of a colloidal particle, say of radius $a = 0.1$ μm in water, the time scale τ_M is of the order of a nanosecond. The time scale τ_D of interest is the time scale that is at least an order of magnitude larger than τ_M, where momenta of the colloids have been long relaxed. Hence, on the *diffusion* or the *Brownian dynamics time scale* $\tau_D \gg \tau_M$, the time derivative $\partial \mathbf{u}/\partial t$ is essentially zero, since $\mathbf{u}$ relaxes to zero on average as a result of friction forces acting during the time interval τ_D: the prefactor of the time derivative in Eq. (2.5) is very small when $\tau_D \gg \tau_M$. The Navier–Stokes Eq. (2.5) in the original unprimed quantities therefore simplifies to [1,24–26]

$$\nabla \mathrm{p}(\mathbf{r}) - \eta_m \nabla^2 \mathbf{u}(\mathbf{r}) = \mathbf{f}^{\mathrm{ex}}(\mathbf{r}). \tag{2.8}$$

This equation, together with the incompressibility Eq. (2.4), constitute the so-called creeping-flow equations describing the quasi-static flow on the time resolution $\Delta t \gg \tau_M$. Creeping flow refers to the fact that the Reynolds number is small.

An external force density acting only in a single point $\mathbf{r}'$ on the fluid is mathematically described by a delta distribution, $\mathbf{f}^{\mathrm{ex}}(\mathbf{r}) = \mathbf{f}_0 \delta(\mathbf{r} - \mathbf{r}')$. Since the creeping-flow equations are linear, the fluid flow velocity at some point $\mathbf{r}$ in the fluid, due to the point force at $\mathbf{r}'$, is directly proportional to that point force. Hence, $\mathbf{u}(\mathbf{r}) = \mathbf{T}(\mathbf{r} - \mathbf{r}') \cdot \mathbf{f}_0$. The tensor $\mathbf{T}$ is referred to as the *Oseen tensor* , which connects the force concentrated at a point $\mathbf{r}'$ to the resulting fluid flow velocity at a point $\mathbf{r}$. Similarly, the pressure at a point $\mathbf{r}$ is linearly related to the point force: $\mathrm{p}(\mathbf{r}) = \mathbf{T}_\mathrm{p}(\mathbf{r} - \mathbf{r}') \cdot \mathbf{f}_0$. The vector $\mathbf{T}_\mathrm{p}$ is referred to as the *pressure-vector.*

For a general external force distribution acting onto the fluid, due to the linearity of the creeping-flow equations, the fluid flow velocity at some point $\mathbf{r}$ is simply the superposition of the fluid flow velocities and pressures resulting from the (external) force density $\mathbf{f}^{\mathrm{ex}}(\mathbf{r}')$ acting at each point on the fluid,

$$\mathbf{u}(\mathbf{r}) = \int d\mathbf{r}' \, \mathbf{T}(\mathbf{r} - \mathbf{r}') \cdot \mathbf{f}^{\mathrm{ex}}(\mathbf{r}'), \tag{2.9}$$

$$\mathrm{p}(\mathbf{r}) = \int d\mathbf{r}' \, \mathbf{T}_\mathrm{p}(\mathbf{r} - \mathbf{r}') \cdot \mathbf{f}^{\mathrm{ex}}(\mathbf{r}'). \tag{2.10}$$

The Oseen tensor $\mathbf{T}(\mathbf{r})$ and pressure-vector $\mathbf{T}_\mathrm{p}(\mathbf{r})$ can be found from the creeping-flow equations as [1,24,25],

$$\mathbf{T}(\mathbf{r}) = \frac{1}{8\pi\eta_m} \frac{1}{r} \left[\hat{\mathbf{I}} + \frac{\mathbf{r}\mathbf{r}}{r^2} \right], \tag{2.11}$$

$$\mathbf{T}_\mathrm{p}(\mathbf{r}) = \frac{1}{4\pi} \frac{\mathbf{r}}{r^3}. \tag{2.12}$$

The expressions for the flow velocity and pressure in terms of integrals over the applied force field are the integral-equivalent of the creeping-flow differential equations.

The creeping flow equations are at the basis of the description of hydrodynamic interactions between colloidal particles. These interactions are due to the fluid flow perturbations caused by the particle motions. The question then arises as to how much time does it take for the fluid perturbation generated by a colloidal particle to reach another particle. As will be discussed here, the configuration of colloidal particles hardly changes during the time needed for the fluid flow to traverse typical distances between different colloidal particles. Hydrodynamic interactions are thus essentially instantaneous.

There are two types of fluid perturbations which progress on different time scales: shear waves and sound waves. A shear wave is a flow consisting of sliding layers of fluid, while a pressure wave results from spatial variations in local pressure.

First consider shear waves. Imagine a flat plate that is pulled in the direction parallel to itself, as sketched in Figure 2.2(a). The plate spans the xy-plane, and is pulled in the x-direction. At time $t = 0$, the plate is at rest and is then suddenly pulled with constant velocity v. The motion of the plate induces shear waves, that is, sliding layers of fluid, which propagate into the solvent. The Navier-Stokes Eq. (2.1) for the fluid flow velocity $u(z,t)$ of an incompressible fluid in the x-direction at height z and time t, for low Reynolds numbers, is a diffusion-type differential equation,

$$\frac{\partial u(z,t)}{\partial t} = \frac{\eta_m}{\rho} \frac{d^2}{dz^2} u(z,t), \quad z > 0. \tag{2.13}$$

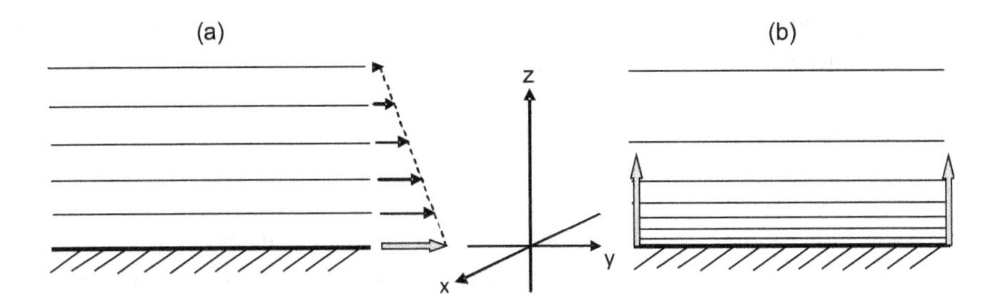

Figure 2.2 (a) Shear waves: The plate at the bottom is suddenly pulled to the right. The arrows indicate the velocities of sliding layers of fluid. (b) Pressure or sound waves: The plate is suddenly moved upward. The lines are used to indicate equal differences in pressure.

The solution of this equation is subject to the no-slip boundary condition $u = v$ at $z = 0$, and the initial condition $u(z > 0, t = 0) = 0$. The corresponding solution reads,

$$u(z,t) = \frac{2v}{\sqrt{\pi}} \int_{z/\sqrt{4\eta_m t/\rho}}^{\infty} dq \exp\left\{-q^2\right\}. \tag{2.14}$$

The typical distance between two correlated colloidal spheres is of the order $10 \times a$ or less, with a the radius of the spheres. According to the above solution, a shear wave diffusively traverses that distance in a time interval of the order,

$$\Delta t \sim 10^2 \tau_h, \quad \tau_h = \frac{a^2 \rho}{\eta_m}, \tag{2.15}$$

where the subscript h stands for hydrodynamic. For water the overdamped shear wave diffusion coefficient, also commonly known as the kinematic viscosity, is $D_h = \eta_m/\rho \approx 10^{-6}$ m^2/s so that for a typical $a = 0.1$ μm sphere the hydrodynamic time scale τ_h is of the order of a nanosecond.

Second, let us consider the propagation of pressure waves, which are more commonly referred to as sound waves. Sound waves are induced by moving the plate in the positive z-direction, as sketched in Figure 2.2(b). The upward velocity of the plate is assumed to be sufficiently small to be able to linearize the Navier–Stokes equation with respect to the change δp of the pressure as well as the solvent flow velocity u, which is now in the z-direction. In addition, the change $\delta\rho$ of the density of the solvent is also assumed to be sufficiently small to allow for linearization of the continuity equation. Viscous effects can be neglected, as these damp the amplitude of sound waves, but do not affect their propagation velocity. At constant temperature, the small change of the pressure is related to the change of the density as, $\delta p(z,t) = (\partial p/\partial \rho)\, \delta\rho(z,t)$. Combining the resulting linearized continuity and Navier–Stokes equations, for an inviscid and incompressible fluid, leads to the following wave equation,

$$\left(\frac{\partial^2}{\partial t^2} - \frac{\partial p}{\partial \rho} \frac{\partial^2}{\partial z^2} \right) \delta\rho(z,t) = 0. \tag{2.16}$$

Any function of the form $f(z,t) \equiv f(z-ct)$ solves this equation, with the sound velocity c being equal to

$$c = \sqrt{\frac{\partial p}{\partial \rho}}. \tag{2.17}$$

For strictly incompressible fluids, for which by definition $\partial p/\partial \rho = \infty$, the sound propagation velocity is therefore infinite. For real fluids the characteristic time that a sound wave requires to propagate over the distance of a particle radius a is $\tau_c = a/c \sim 10^{-10}$s for a 0.1 μm sphere in water. Thus, sound waves progress distinctly faster than shear waves. For $\Delta t \gg \tau_c$ the fluid can be treated as incompressible, justifying thus a posteriori the usage of the Navier–Stokes equations form for an incompressible fluid from which Eq. (2.16) is obtained.

The diffusion displacement of a spherical Brownian particle during a time interval Δt is of the order $\sqrt{\mathcal{D}_0 \Delta t}$, where

$$\mathcal{D}_0 = \frac{k_B T}{6\pi\eta_m a} \tag{2.18}$$

is the Stokes–Einstein–Sutherland diffusion coefficient of an isolated particle inversely proportional to the single-particle translational friction coefficient, $\zeta_0 = 6\pi\eta_m a$, as already introduced in Chapter 1, Eq. (1.5). This allows one to specify $\tau_D = a^2/\mathcal{D}_0$ as the diffusion time scale during which a significant change of the particle position occurs as compaed to its size. For a 0.1 μm particle in water one gets $\tau_D \sim 1$ms, illustrating the time scale separation $\tau_D \gg \tau_M$ by many orders of magnitude. In other words, the diffusional displacement of a colloidal particle in the time span τ_M of particle momentum relaxation is a tiny fraction of its own size.

A crucial observation is that $\tau_h \sim \tau_M$ for neutrally buoyant or weakly sedimenting colloidal particles. This means that the time required for a fluid shear wave to diffuse from one colloid to another correlated one is of the order of the particle momentum relaxation time $\tau_M = M/\zeta_0$, and thus, exceedingly smaller than the particle diffusion time τ_D where a noticeable change in the particle positions takes place. For the diffusion or Brownian dynamics time scale of interest where $\Delta t \gg \tau_h$, the flow mediated so-called hydrodynamic interactions (HIs) between colloidal particles can therefore be regarded as being instantaneous, with the instantaneous flow pattern described by the creeping flow equation that has no dependence on the earlier flow history. Moreover, since $\tau_M \sim \tau_h$, inertial effects of fluid *and* particles motions are not resolved on the Brownian dynamics time scale. Figure 2.3 summarizes the discussed timescales relevant to the dynamics of a colloidal suspension.

2.2.2 Hydrodynamic Interactions of Spheres in Shear Flow

Consider two spheres labeled 1 and 2 moving through an otherwise quiescent solvent, without an externally imposed shear flow, with velocities $\mathbf{v}_1$ and $\mathbf{v}_2$, respectively. Let $\mathbf{F}_1^h$ and $\mathbf{F}_2^h$ denote the forces that the solvent exerts on colloids 1 and 2, respectively,

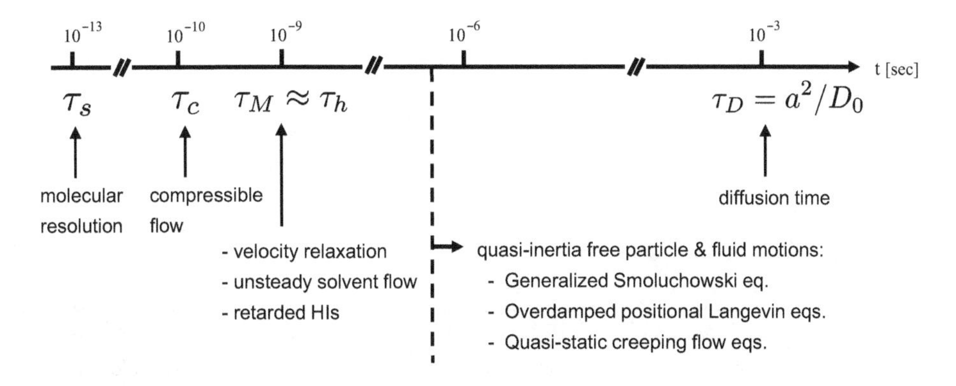

Figure 2.3 Characteristic times for a colloidal sphere of typical radius $a = 0.1$ μm and mass M, suspended in water of mass density ρ at room temperature. τ_s: collision time of solvent molecules; $\tau_c = a/c$: sound propagation time; $\tau_M = M/\zeta_0$: colloid momentum relaxation time; $\tau_h = a^2\rho/\eta_m$: hydrodynamic, shear wave diffusion time; $\tau_D = a^2/\mathcal{D}_0$: colloid diffusion time.

where the superscript h stands for hydrodynamic. For not too large velocities, the linear creeping flow equations imply a linear and instantaneous relation between the velocities and the forces, which can be written as

$$\begin{pmatrix} \mathbf{v}_1 \\ \mathbf{v}_2 \end{pmatrix} = -\beta \begin{pmatrix} \mathbf{D}_{11}^{(2)} & \mathbf{D}_{12}^{(2)} \\ \mathbf{D}_{21}^{(2)} & \mathbf{D}_{22}^{(2)} \end{pmatrix} \cdot \begin{pmatrix} \mathbf{F}_1^h \\ \mathbf{F}_2^h \end{pmatrix}, \tag{2.19}$$

where the microscopic diffusion tensors $\mathbf{D}_{ij}^{(2)}$ are second rank tensors having each 3×3 components. The superscript 2 on the diffusion tensors is used to indicate that only two spheres are considered. A minus sign appears on the right hand-side because the velocity is in the opposite direction of the hydrodynamic force. The nondiagonal tensors $\mathbf{D}_{12}^{(2)}$ and $\mathbf{D}_{21}^{(2)}$ describe how a force on one of the colloids results in a nonzero velocity on the other colloid.

In the absence of hydrodynamic interactions, the force that the otherwise quiescent fluid exerts on a sphere is linearly related to its velocity: $\mathbf{F}_j^h = -\zeta_0 \mathbf{v}_j$, where $\zeta_0 = 6\pi\eta_m a$ is the single-sphere friction coefficient. The diffusion tensors therefore take the following form (with $\hat{\mathbf{I}}$ the identity tensor),

$$\begin{aligned} \mathbf{D}_{ij}^{(2)} &= \mathbf{0} && (i \neq j), \\ \mathbf{D}_{ii}^{(2)} &= \mathcal{D}_0 \hat{\mathbf{I}}, && (\text{no hydrodynamic interactions}). \end{aligned} \tag{2.20}$$

By decomposing the forces $\mathbf{F}^h$ in their components $\hat{\mathbf{r}}_{12}\hat{\mathbf{r}}_{12} \cdot \mathbf{F}^h$ parallel to unit vector $\hat{\mathbf{r}}_{12}$ connecting the centers of the two spheres and their perpendicular components $\left[\hat{\mathbf{I}} - \hat{\mathbf{r}}_{12}\hat{\mathbf{r}}_{12}\right] \cdot \mathbf{F}^h$, it is easily verified that the most general form of the microscopic diffusion tensors reads:

$$\begin{aligned} \mathbf{D}_{11}^{(2)} = \mathbf{D}_{22}^{(2)} &= \mathcal{D}_0 \hat{\mathbf{I}} + \mathcal{D}_0 \left\{ A_s(r_{12}) \hat{\mathbf{r}}_{12}\hat{\mathbf{r}}_{12} + B_s(r_{12}) \left[\hat{\mathbf{I}} - \hat{\mathbf{r}}_{12}\hat{\mathbf{r}}_{12}\right] \right\}, \\ \mathbf{D}_{12}^{(2)} = \mathbf{D}_{21}^{(2)} &= \mathcal{D}_0 \left\{ A_c(r_{12}) \hat{\mathbf{r}}_{12}\hat{\mathbf{r}}_{12} + B_c(r_{12}) \left[\hat{\mathbf{I}} - \hat{\mathbf{r}}_{12}\hat{\mathbf{r}}_{12}\right] \right\}. \end{aligned} \tag{2.21}$$

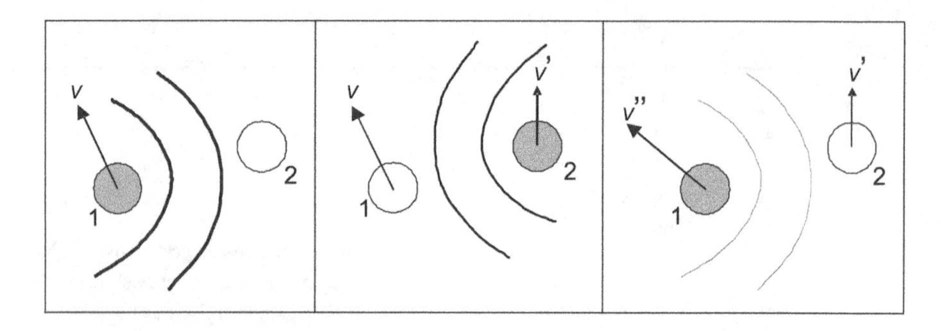

Figure 2.4 Left: Sphere 1 moves with a velocity v and induces a flow field schematically indicated by the two curved lines. Middle: The flow reaches sphere 2, which sets it into motion with a velocity v', and reflects the incident flow field back to sphere 1. Right: The flow field that is reflected by sphere 2 reaches sphere 1, changing its velocity to v''. This flow field is subsequently reflected from sphere 1 back to sphere 2, etc.

The so-called mobility functions $A_{s,c}$ and $B_{s,c}$ (where s and c stand for self and cross, respectively) are functions of the scalar distance r_{12} between the two sphere centers. The term $\mathcal{D}_0 \hat{\mathbf{I}}$ is added in the self-part, the first equation, to assure that all mobility functions tend to zero for infinite separations between the two spheres. For large pair separation, the diffusion tensor must attain its form in Eq. (2.20) where hydrodynamic interactions are absent. Note that the self diffusion tensor $\mathbf{D}_{11}^{(2)}$ is affected by the presence of a second sphere 2. This is due to reflection of the flow field as induced by the moving sphere 1, back from sphere 2 toward sphere 1. A similar physical situation holds for $\mathbf{D}_{22}^{(2)}$, as depicted in Figure 2.4. In fact, flow fields are infinitely reflected back and forth. Each subsequent reflection contribution to the diffusion tensors decays at least a factor $(a/r_{12})^2$ faster with the distance r_{12} between the spheres. By the method of reflections [1,25,27,28], each subsequent term in an expansion with respect to a/r_{12} can be calculated. The first few terms in such an expansion read,

$$
\begin{aligned}
A_s &= -\frac{15}{4}\left(\frac{a}{r_{ij}}\right)^4 + \frac{11}{2}\left(\frac{a}{r_{ij}}\right)^6 + \mathcal{O}\left((a/r_{ij})^8\right), \\
B_s &= -\frac{17}{16}\left(\frac{a}{r_{ij}}\right)^6 + \mathcal{O}\left((a/r_{ij})^8\right), \\
A_c &= \frac{3}{2}\frac{a}{r_{ij}} - \left(\frac{a}{r_{ij}}\right)^3 + \frac{75}{4}\left(\frac{a}{r_{ij}}\right)^7 + \mathcal{O}\left((a/r_{ij})^9\right), \\
B_c &= \frac{3}{4}\frac{a}{r_{ij}} + \frac{1}{2}\left(\frac{a}{r_{ij}}\right)^3 + \mathcal{O}\left((a/r_{ij})^9\right).
\end{aligned}
\tag{2.22}
$$

The numerical values of many hundreds of coefficients for the higher order terms are known [27,28].

Let us now consider a homogeneously sheared suspension, where the shear rate is independent of position. Such a linear flow profile, also referred to as simple linear

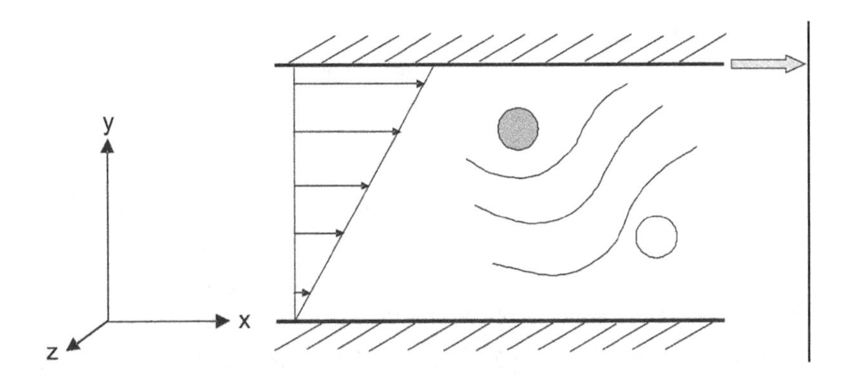

Figure 2.5 The externally imposed shear flow induced here by pulling the upper plate to the right is depicted on the left, where the arrows indicate the local flow velocity. This simple shear flow is reflected by the grey sphere, as schematically indicated by the three curved lines, which affects the motion of another sphere indicated by the open circle.

shear flow, is specified by means of *the velocity gradient tensor* $\mathbf{\Gamma}$, which relates the suspension velocity $\mathbf{u}$ with position $\mathbf{r}$ as,

$$\mathbf{u}(\mathbf{r}) = \mathbf{\Gamma} \cdot \mathbf{r}. \tag{2.23}$$

Considering such flows may seem quite restrictive. However, an arbitrary laminar flow can locally be written in this form, with a position dependent velocity gradient tensor. In case of simple shear flow in the x-direction with increasing velocity in the y-direction, as depicted on the left in Figure 2.5, we have $\mathbf{u} = (\dot{\gamma}, 0, 0)$, where $\dot{\gamma}$ is the shear rate. The x-direction is the *flow direction*, y is along the *gradient direction*, and the left-over z-direction is along the so-called *vorticity direction*. We thus have for simple shear flow,

$$\mathbf{\Gamma} = \dot{\gamma} \begin{pmatrix} 0 & 1 & 0 \\ 0 & 0 & 0 \\ 0 & 0 & 0 \end{pmatrix}. \tag{2.24}$$

Due to the linearity of the Navier–Stokes equation for the solvent at low Reynolds numbers, the microscopic diffusion tensors for a sheared solvent are the same as those for a quiescent solvent. Under shear flow conditions, however, there is an additional contribution to hydrodynamic interactions. The origin of these additional interactions is sketched in Figure 2.5. The externally imposed shear flow as depicted on the left in Figure 2.5 is reflected by the grey colloidal sphere toward the open sphere, which affects its motion. Even for the case when the velocity of the grey sphere is zero, the presence of this sphere causes motion of the open sphere. As before, the flow fields are multiply reflected.

A single sphere, labeled as 1, immersed in a simple shear flow attains the velocity $\mathbf{\Gamma} \cdot \mathbf{r}_1$ of the solvent that exists at its center $\mathbf{r}_1$ before immersing the sphere. In addition, there are contributions to the velocity due to shear-induced HIs with other force-free particles. For small Reynolds numbers, there is a linear relationship between the additional shear-induced velocity, $\Delta \mathbf{v}_1 = \boldsymbol{\Gamma} : \mathbf{C}_1$, attained by sphere 1 (the open sphere in Figure 2.5) due to the applied shear flow with velocity gradient tensor $\boldsymbol{\Gamma}$. By linear superposition, the net velocity is

$$\mathbf{v}_1^c = \boldsymbol{\Gamma} \cdot \mathbf{r}_1 + \boldsymbol{\Gamma} : \mathbf{C}_1, \tag{2.25}$$

where the subscript 1 on $\mathbf{C}$ refers to sphere number 1, and where the double dots denote a contraction, i.e., the *ith* component of $\Delta \mathbf{v}_1$ is equal to $\sum_{m,n=1}^{3} \Gamma_{mn} C_{1,nmi}$. The superscript c is a reminder that $\mathbf{v}_1^c$ is the particle velocity attained in a situation where all particles are freely convected with the imposed shear flow, thus experiencing no hydrodynamic drag forces $\left\{ \mathbf{F}_j^h = 0 \right\}$. Without imposed flow, the convective velocities $\left\{ \mathbf{v}_j^c \right\}$ of all particles are zero. The disturbance tensor $\mathbf{C}_1$ is thus a tensor with three indices, and depends on the distance between sphere 1 and sphere 2, which is the grey colored sphere in Figure 2.5. A similar reflection expansion as for the diffusion tensor can be constructed for $\mathbf{C}_1$ [1,29,30],

$$\begin{aligned} \mathbf{C}_1 = & \left[-\frac{5}{2}\left(\frac{a}{r_{12}}\right)^3 + \frac{20}{3}\left(\frac{a}{r_{12}}\right)^5 \right] \hat{\mathbf{r}}_{12}\hat{\mathbf{r}}_{12}\mathbf{r}_{12} - \frac{4}{3}\left(\frac{a}{r_{12}}\right)^5 \left\{ \hat{\mathbf{I}}\mathbf{r}_{12} + \left(\hat{\mathbf{I}}\mathbf{r}_{12}\right)^{\dagger} \right\} \\ & - \frac{25}{2}\left(\frac{a}{r_{12}}\right)^6 \hat{\mathbf{r}}_{12}\hat{\mathbf{r}}_{12}\mathbf{r}_{12} + \mathcal{O}\left(\left(a/r_{12}\right)^8 \right), \end{aligned} \tag{2.26}$$

where we defined the tensor $\left(\widehat{\mathbf{I}\mathbf{r}}\right)^{\dagger}_{imn} = \delta_{in} r_m$, with δ_{in} the Kronecker delta.

In the above discussion, we considered only two interacting colloidal spheres. The velocity of a colloidal sphere due to hydrodynamic forces, including those resulting from hydrodynamic interactions with *all* other spheres, can be formulated similarly. The velocities are again linearly related to forces,

$$\begin{pmatrix} \mathbf{v}_1 \\ \vdots \\ \mathbf{v}_N \end{pmatrix} = -\beta \begin{pmatrix} \mathbf{D}_{11} & \mathbf{D}_{12} & \cdots & \mathbf{D}_{1N} \\ \mathbf{D}_{21} & \mathbf{D}_{22} & \cdots & \mathbf{D}_{2N} \\ \vdots & \vdots & \cdots & \vdots \\ \mathbf{D}_{N1} & \mathbf{D}_{N2} & \cdots & \mathbf{D}_{NN} \end{pmatrix} \cdot \begin{pmatrix} \mathbf{F}_1^h \\ \vdots \\ \mathbf{F}_N^h \end{pmatrix} + \begin{pmatrix} \boldsymbol{\Gamma} \cdot \mathbf{r}_1 \\ \vdots \\ \boldsymbol{\Gamma} \cdot \mathbf{r}_N \end{pmatrix} + \begin{pmatrix} \boldsymbol{\Gamma} : \mathbf{C}_1 \\ \vdots \\ \boldsymbol{\Gamma} : \mathbf{C}_N \end{pmatrix}, \tag{2.27}$$

where the symmetric tensors $\mathbf{D}_{ij}(X)$ with 3×3 components, and the third rank disturbance tensors $\mathbf{C}_j(X)$ having $3 \times 3 \times 3$ components, each are complicated functions of the $3N$-dimensional column supervector,

$$X = \left(\mathbf{r}_1, \mathbf{r}_2, \cdots, \mathbf{r}_N\right)^T, \tag{2.28}$$

of the position coordinates of all N spheres. Here, the superscript T denotes the matrix transposition operation. In a cluster expansion, the diffusion tensors can be expressed as,

$$\mathbf{D}_{ij}(X) = \mathbf{D}_{ij}^{(2)}\left(\mathbf{r}_{ij}\right) + \sum_{n=1}^{N} \mathbf{D}_{ij}^{(3)}\left(\mathbf{r}_{in}, \mathbf{r}_{nj}\right) + \sum_{m,n=1}^{N} \mathbf{D}_{ij}^{(4)}\left(\mathbf{r}_{in}, \mathbf{r}_{nm}, \mathbf{r}_{mj}\right) + \cdots, \tag{2.29}$$

and similarly for the disturbance tensor. The first term on the right side is the two-particle interaction contribution discussed with respect to Eqs. (2.21) and (2.22). The velocity of sphere j due to the solvent flow from a moving sphere i that is reflected by a third sphere n to sphere j is accounted for by the second term on the right hand-side. The third term includes reflections where two intermediate spheres, other than i and j, are involved. As mentioned earlier, the two-sphere interaction tensors are analytically known as expansions with respect to a/r_{ij}, up to many hundreds of terms [27]. For the three-particle diffusion tensor $\mathbf{D}_{ij}^{(3)}$ only the leading term in an expansion with respect to the inverse distances between the three spheres is analytically known [1]. There are phenomena in colloidal suspensions for which the essence is captured with the total neglect of hydrodynamic interactions. There are, however, also phenomena where hydrodynamic interactions are essential. For the latter, an approximate description of hydrodynamic interactions on the two-particle level can be useful. There are two situations where two-particle, or pair-wise additive, hydrodynamics is sufficient. First of all, when the colloidal spheres interact through long-ranged repulsive interactions the typical distances between the particles are large enough that hydrodynamic interactions through reflections by intermediate spheres are less important. Secondly, at very high packing fractions, where the minimal distances between the surfaces of neighboring spheres are small compared to the sphere radius, hydrodynamic interactions are quite strong provided the solvent sticks to the surfaces, i.e., no-slip. These so-called lubrication forces [25] cannot be described with an expansion with respect to the inverse distance between the spheres, i.e., the expressions in Eqs. (2.22) and (2.26) are not valid for these high packing fractions. The expansion parameter is now the surface-to-surface distance relative to the sphere radius. Such lubrication interactions are pairwise additive, involving interaction contributions between pairs of spheres only. Two-particle hydrodynamics is thus sufficient for dilute suspensions, and for no-slip spheres in case the packing fractions are very high. A full account of HIs in the calculation of suspension properties such as the viscosity can only be achieved through specifically adapted mesoscale simulation methods for Brownian particles, as will be discussed in Chapter 3.

2.3 Smoluchowski Equation for Particles in Shear Flow

Having discussed the hydrodynamic interactions between spheres, we are now in a position to formulate the fundamental evolution equation describing the dynamics of colloidal systems on the noninertial Brownian dynamics level, which is the so-called generalized Smoluchowski equation (GSmE). Elementary derivations of this equation can be found in, for example, Dhont [1] and Pusey [26]. In the case of spherically symmetric colloidal particles, the GSmE is an equation of motion for the probability density function, $P(X, t)$, of the position coordinates X of an assembly of N suspended colloids.

As discussed in Section 2.2.1, the inertia of the colloids can be neglected on the diffusion- or Brownian time scale. The dynamics on such a coarse-grained time scale is referred to as *overdamped dynamics*. Here, coarse-graining refers to averaging over time- and length scales associated with the relaxation of the particles' momenta. After performing this coarse-graining, new properties emerge such as particle friction coefficients.

According to Newton's equation of motion, the neglect of inertial forces implies that the remaining, noninertial forces add up to zero, which is commonly referred to as *force free*. Three noninertial forces must be distinguished: the first one is the force $\mathbf{F}_j = \mathbf{F}_j^I + \mathbf{F}_j^{\mathrm{ex}}$ acting on a colloid as a result of direct interaction forces F_j^I with other colloids due to, e.g., electric surface charges, polymers coated on the colloid surface, and excluded volume, and due to an external force field $\mathbf{F}_j^{\mathrm{ex}}(X)$ such as a buoyancy-corrected gravitational field, or an applied electric field or optical tweezer field. The second one is the hydrodynamic force $\mathbf{F}_j^h$ that the solvent exerts on the colloid. The third one, $\mathbf{F}_j^{\mathrm{Br}}$, is due to time coarse-graining and commonly referred to as *the Brownian force*. Force balance thus implies for each configuration X and instant t that,

$$\mathbf{F}_j + \mathbf{F}_j^h + \mathbf{F}_j^{\mathrm{Br}} = \mathbf{0}. \tag{2.30}$$

The direct interaction force, $\mathbf{F}_j^I(X) = -\nabla_j \Phi(X)$ is related to the total potential interaction energy $\Phi(X)$ of the assembly of N colloids. Commonly the potential energy is taken as pairwise additive,

$$\Phi(X) = \sum_{i<j}^{N} \mathcal{V}(|\mathbf{r}_i - \mathbf{r}_j|), \tag{2.31}$$

where $\mathcal{V}(r)$ is the solvent-averaged isotropic pair potential of colloidal spheres. The hydrodynamic force has been discussed in Section 2.2.2, and the Brownian force will be specified as follows.

Consider an ensemble of identical colloidal suspensions with the same temperature T, the same volume V, and the same number of N colloidal spheres and solvent molecules. Such an NVT-ensemble is depicted in Figure 2.6(a). Although these systems are macroscopically identical, on a microscopic level they are all different in the sense that the position coordinates of all colloidal spheres at a given time are different. The *microscopic state* of each colloidal suspension is defined by the $3N$-dimensional supervector X of all position coordinates of the colloids. The corresponding $3N$-dimensional *phase space* is depicted in Figure 2.6(b), where each point represents the instantaneous microstate of a colloidal suspension. The microscopic state of the ensemble is thus represented by a distribution of points. Each of these points moves through phase space in time, due to Brownian motion and possibly due to an external field. The probability to find a single colloidal suspension in a certain microstate is proportional to the local density of points in phase space. Let $P(X,t)$ denote the probability density function (pdf) at time t of a microstate X. The Smoluchowski equation is the equation of motion for this pdf.

To obtain this equation of motion, consider an arbitrary volume in phase space, as indicated by the grey region in Figure 2.6(b). To find an equation of motion for the pdf of the position coordinates of the colloidal spheres, we first ask for the time rate of change of the number of points inside this phase space volume. Such a general

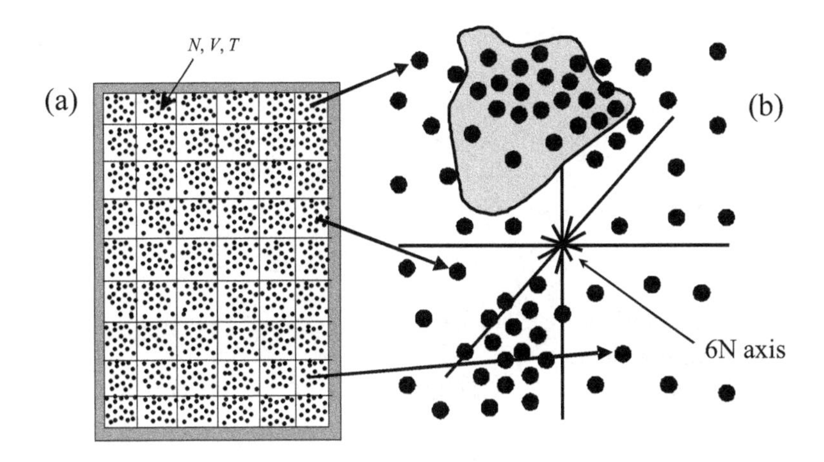

Figure 2.6 (a) An ensemble of macroscopically identical colloidal suspensions, each with a given volume V, number of colloids N, and temperature T. Each dot represents a colloidal sphere. (b) A snapshot of each of the systems in the ensemble is represented by a point in $3N$-dimensional phase space, specifying the instantaneous position coordinates of the N colloidal spheres. The grey area indicates an arbitrary volume in phase space. The continuity equation is obtained from the fact that the in and out fluxes are the only cause for the rate of change of the pdf for the colloid position coordinates.

equation of motion is equivalent to the continuity Eq. (2.3) for fluids, except that each fluid molecule is now a point in phase space, as shown in Figure 2.6(b). The $3N$-dimensional analog of the standard continuity equation reads

$$\frac{\partial}{\partial t} P(X,t) = -\sum_{i=1}^{N} \nabla_i \cdot [\mathbf{v}_i(X,t) P(X,t)]. \tag{2.32}$$

This is an exact equation because it expresses the conservation of the number of points in phase space.

To obtain a closed equation of motion, the velocities $\mathbf{v}_i$ of the spheres must be expressed in terms of P and possibly other known functions of the position coordinates of the spheres. This is the place where approximations are made. Our interest here is in a coarse-grained description on the most relevant Brownian diffusion dynamics scale where $\Delta t \gg \tau_M \sim \tau_h$. As shown in Section 2.2.2, on this level of coarse-graining, HIs are instantaneous, and velocities of colloids are relaxed to equilibrium with the heat bath of solvent molecules. From the noninertial force balance Eq. (2.30), the velocities $\mathbf{v}_i(X,t)$ are obtained as

$$\mathbf{v}_i - \mathbf{v}_i^c = -\beta \sum_{j=1}^{N} \mathbf{D}_{ij} \cdot \left[\mathbf{F}_j^h = -\left(\mathbf{F}_j + \mathbf{F}_j^{\mathrm{Br}}\right)\right], \tag{2.33}$$

where

$$\mathbf{v}_j^c(X) = \mathbf{\Gamma} \cdot \mathbf{r}_j + \mathbf{\Gamma} : \mathbf{C}_j(X) \tag{2.34}$$

are the convective particle velocities that vanish according to Eq. (2.27) without imposed shear flow ($\mathbf{\Gamma} = 0$).

On substituting Eqs. (2.33) and (2.31) into the continuity Eq. (2.32), we obtain a closed equation of motion for the pdf. Without shear flow, the pdf must approach the equilibrium pdf $\sim \exp\{-\beta\Phi\}$. From this requirement imposed on the explicit equation of motion, it follows that [31],

$$\mathbf{F}_j^{\mathrm{Br}} = -k_B T \nabla_j \ln\{P\}. \tag{2.35}$$

Being logarithmically related to the pdf, this force is of an entropic nature: it arises due to time coarse-graining.

We thus finally obtain the GSmE for the pdf $P(x,t)$ of spherical particles in linear shear flow,

$$\frac{\partial}{\partial t} P = \sum_{i,j=1}^{N} \nabla_i \cdot \mathbf{D}_{ij} \cdot \left[\nabla_j P - P\beta\,\mathbf{F}_j\right] - \sum_{i=1}^{N} \nabla_i \cdot \left[P\mathbf{\Gamma}\cdot\mathbf{r}_i + P\mathbf{\Gamma} : \mathbf{C}_i\right]. \tag{2.36}$$

This is the fundamental evolution equation for the pdf, describing phenomena of suspensions of interacting spherical colloidal particles under flow conditions. It is worthwhile to summarize the requirements for its validity: time resolution $\Delta t \gg \tau_M \sim \tau_h$ and associated spatial resolution $\Delta x \gg l_B$, where $l_B = \sqrt{\mathcal{D}_0 \tau_M} \ll a$ is the coasting length along which a sphere loses its momentum by solvent friction. Regarding the shear rate of the imposed shear flow, $\tau_{\dot\gamma} \gg \tau_h$ is required for the creeping flow description of HIs to apply, where $\tau_{\dot\gamma} = 1/\dot\gamma$ is the time scale associated to the flow. This condition is met in sheared suspensions where already for moderate shear rates, e.g., $\dot\gamma \sim \left(10^2\text{–}10^4\right)\ \mathrm{s}^{-1}$, significant changes in the suspension microstructure are induced. Moreover, $l_B \mid \nabla_i \Phi(X) \ll \mid \Phi(X) \mid$ is required, i.e., the interactions and externally applied potential should remain essentially constant across the small coasting length.

An alternative and perhaps more satisfying derivation of the GSmE than the one presented here is to start from the Liouville equation for a highly asymmetric binary mixture of small solvent molecules and in comparison large colloidal particles. The Smoluchowski equation is then found by integrating over the fast phase space variables, i.e., the phase space coordinates of the solvent molecules and the momentum coordinates of the colloidal particles. Such an approach has been taken for spherical colloids without shear flow in other studies [32–34].

Even without hydrodynamic interactions, the distortion of the pdf becomes more severe as the distance between the colloids increases, since the relative velocity of the particles induced by the shear flow increases with increasing distance between the colloids. This is the case, however, only when the distance along the gradient direction increases: for directions $\mathbf{r}$ along the flow- or vorticity direction, we have $\mathbf{\Gamma}\cdot\mathbf{r} = \mathbf{0}$, so that no significant shear-induced distortion is expected even for large relative distances between the colloids in those directions. The distortion of the pdf is thus anisotropic, and in directions along the gradient direction the distortion is singular, i.e., for arbitrary small values of the shear rate the distortion of the pdf becomes significant at sufficiently large distances. A simple Taylor expansion of the pdf in a power series of the shear rate is therefore not possible for sufficiently large distances between the colloids. The pdf is thus a singular function of the shear rate for large distances between the colloids. As a consequence, the corresponding structure factor is singularly perturbed by shear flow at small wave vectors [35,36]. An expansion of the structure factor in terms of spherical harmonics [37] can be employed for the interpretation of experimental data.

As a prerequisite for the derivation in the following section of a positional Langevin equation, which is statistically equivalent to the Smoluchowski equation description, here we recast the GSmE as

$$\frac{\partial}{\partial t} P(X,t) = -\sum_{i=1}^{N} \nabla_i \cdot \left(\mathbf{v}_i^d(X) + \mathbf{v}_i^c(X)\right)P(X,t) + \sum_{i,j=1}^{N} \nabla_i \nabla_j : \mathbf{D}_{ij}(X)P(X,t), \tag{2.37}$$

where $\nabla_i \nabla_j : \mathbf{D}_{ij}\ P = \nabla_i \cdot \left[\nabla_j \cdot \left(\mathbf{D}_{ij}\ P\right)\right]$. It includes the particles convection velocities $\left\{\mathbf{v}_i^c\right\}$ and the drift velocities,

$$\mathbf{v}_i^d(X) = \sum_{j=1}^{N} \left[\beta \mathbf{D}_{ij}(X) \cdot \mathbf{F}_j(X) + \nabla_j \cdot \mathbf{D}_{ij}(X)\right], \tag{2.38}$$

where $\mathbf{F}_j(X)$ is the force on sphere j due to its nonhydrodynamic interactions with other spheres and the action of an external field, and a diffusive part invoking the microscopic diffusion tensors. The drift velocities with HIs include for *Brownian* particles an additional *hydrodynamic drift term*, $\sum_j \nabla_j \cdot \mathbf{D}_{ij}$, which is of importance when particles get close, driving them away from each other to regions of higher hydrodynamic mobility.

It can be shown using the creeping flow equations that the symmetric $3N \times 3N$ diffusion matrix $D(X)$, with $\mathbf{D}_{ij}$ as its elements, is positive definite. This is required for the initially normalized pdf solutions, $P(X,t)$, of the GSmE to remain normalized in the course of time, and to converge to the equlibrium pdf, for zero imposed flow and external force fields [4–6].

2.4 Langevin Dynamics of Brownian Particles

A statistically equivalent description of the motion of colloidal particles that focuses on the random evolution of their degrees of freedom X instead of the deterministic evolution of the associated pdf $P(X,t)$, is provided by the overdamped many-particles Langevin equations discussed in the following. They describe statistically the evolution of the particles trajectories, with the much faster equilibrated particle momenta integrated out of the description. In these stochastic differential or finite difference equations, the influence of the collisions by solvent molecules is represented on a coarse-grained level by stochastic forces or velocities, and by hydrodynamic drag forces. In addition, the particles are subjected to solvent-averaged direct interaction forces.

In Brownian Dynamics (BD) mesoscale simulation methods, such as those discussed in Chapter 3, these Langevin equations are solved numerically by a forward integration scheme with discrete time steps. The stochastic forces are simulated by pseudo-random numbers. In this way, an ensemble of stochastic trajectories is generated. From averaging over these trajectories one can calculate equilibrium and nonequilibrium statistical averages such as self- and collective diffusion properties of a suspension, and rheological properties, including dynamic shear moduli and viscosities. BD simulation methods can be considered lying in between Monte Carlo and Molecular Dynamics methods (see Chapter 3), sharing the element of randomness with the former, and the continuity of trajectories in configuration space with the latter.

As in the equivalent GSmE description, in addition to direct interparticle forces one needs to account for the HIs acting between Brownian particles. These long-range dynamic interactions are transmitted by complicated fluid flow patterns created by the moving particles. In concentrated suspensions, they cause a nonpairwise additive coupling of the particle drift velocities as well as a coupling of the Brownian stochastic forces appearing in the Langevin equations. This coupling causes significant computational difficulties in BD simulations where HIs are included.

2.4.1 Single Microsphere in Shear Flow

Consider first the most simple case of a single force-free spherical particle of mass M immersed in a macroscopically quiescent fluid of small molecules. We explore the motion of this particle first on the more refined time scale τ_M where its inertia is resolved (but hydrodynamic solvent flow is neglected). Subsequently, the sphere motion is reconsidered on the coarser Brownian dynamics scale where its diffusive random excursions are apparent, and where particle and solvent inertial effects are not resolved.

The equation of motion for the translational velocity, $\mathbf{v}(t) = \dot{\mathbf{r}}(t)$, of the sphere with momentary center position $\mathbf{r}(t)$ can be described by the single-sphere Langevin equation [4]

$$M\,\ddot{\mathbf{r}}(t) = -\zeta_0\,\dot{\mathbf{r}}(t) + F^s(t). \tag{2.39}$$

For velocities small enough that inertia plays no role, the colloidal sphere experiences the hydrodynamic friction force, $-\zeta_0\ \dot{\mathbf{r}}$, described by the Stokes drag where $\zeta_0 = 6\pi\eta_m a$, and a rapidly fluctuating, on scale τ_s, stochastic force $\mathbf{F}^s(t)$. The stochastic force describes the accumulative effect of the solvent molecules' collisions on the sphere. It is characterized by its first moment

$$\langle \mathbf{F}^s(t)\rangle = 0, \tag{2.40}$$

and the second moment (3×3 covariance matrix)

$$\langle \mathbf{F}^s(t)\mathbf{F}^s(t')\rangle = \Gamma_s\ \delta(t-t')\ \hat{\mathbf{I}}. \tag{2.41}$$

Here, $\langle\ldots\rangle$ denotes the average over the fast solvent collisions. The first moment relation assures validity of the deterministic Newtonian equation, $M\dfrac{d}{dt}\langle\mathbf{v}\rangle_0 = -\zeta_0\langle\mathbf{v}\rangle_0$, for the mean sphere velocity. In the second moment relation, Γ_s is the strength of the fluctuating force, and $\hat{\mathbf{I}}$ is the unit tensor, expressing spatial isotropy and that the orthogonal Cartesian components $F^s_\alpha(t)$, with $\alpha \in \{x, y, z\}$, are mutually uncorrelated. The delta function indicates that the random forces are uncorrelated on the time scale of the particle motion.

The random force is a time-dependent random variable, i.e., a stochastic process, defined only by its statistical properties. The sphere velocity and position are then likewise random variables for which the pdfs and moments are related by the Langevin equation to those of the stochastic force. As the sum of many independent force contributions, $\mathbf{F}^s$ is a Gaussian distributed process fully determined by its first two moments. Owing to the white noise properties of the random force, this process is also a Markovian process meaning that the future evolution of the random force is determined solely by its present value, i.e., it is independent of history.

This statistical picture given by the Langevin equation relates to an ensemble of independent Brownian spheres, each of which is a realization of the described diffusion process, with $\langle \cdots \rangle$ interpreted as average over this ensemble. Time integration of Eq. (2.39) for a given realization of $\mathbf{F}^s(t)$ leads to

$$\mathbf{v}(t) = \mathbf{v}_0 e^{-t/\tau_M} + \frac{1}{M}\int_0^t du \; e^{-(t-u)/\tau_M}\,\mathbf{F}^s(u), \tag{2.42}$$

where $\mathbf{v}_0$ is the initial velocity. Averaging over the fast random force fluctuations for a subensemble of spheres with equal initial velocity $\mathbf{v}_0$ gives Eq. (2.7). The mean velocity decays to zero due to solvent friction, as quantified by the momentum relaxation time $\tau_M = M/\zeta_0$.

Using the statistical properties of white noise, the mean of the squared velocity follows as:

$$\left\langle \mathbf{v}(t)^2 \right\rangle = \mathbf{v}_0^2 \; e^{-2t/\tau_M} + \frac{3\Gamma_s}{2M\zeta_0}\left(1 - e^{-2t/\tau_M}\right). \tag{2.43}$$

Equipartition of the kinetic energy in equilibrium commands that $\left\langle \mathbf{v}^2(t \to \infty) \right\rangle = 3k_{\mathrm{B}}T/M$ so that Γ_s is determined as :

$$\Gamma_s = 2k_{\mathrm{B}}T\zeta_0. \tag{2.44}$$

This fluctuation-dissipation theorem relates the strength of the fluctuating force in thermal equilibrium to the mean particle friction coefficient, thereby reflecting their common origin in the interaction between mesoscale-particle and solvent molecules.

To describe a suspension of Brownian particles in equilibrium requires additional averaging with respect to a Maxwellian distribution of initial velocities. The double averaging with respect to random force and initial velocities is denoted by $\langle \cdots \rangle_{\mathrm{eq}}$. From the time integrated Eq. (2.42) follows the equilibrium mean-squared displacement

$$\left\langle |\mathbf{r}(t) - \mathbf{r}_0|^2 \right\rangle_{\mathrm{eq}} = 6\mathcal{D}_0 t\left[1 - \frac{\tau_M}{t}\left(1 - e^{-t/\tau_M}\right)\right] \;\to\; \begin{cases} \dfrac{3k_{\mathrm{B}}T}{M}t^2, & \tau_s \ll t \ll \tau_M, \\ 6\mathcal{D}_0\, t, & t \gg \tau_M, \end{cases} \tag{2.45}$$

where the single-particle diffusion coefficient $\mathcal{D}_0$ is given by Eq. (2.18). Eq. (2.45) interpolates between ballistic flight behavior with Maxwell-distributed initial velocities for $t \ll \tau_M$, and linear diffusive behavior for $t \gg \tau_M$.

Importantly, the inertial Langevin Eq. (2.39) is actually not applicable to colloidal particles for $t \sim \tau_M$. It is valid only for times $t \gg \tau_h \sim \tau_M$ where solvent and particle inertia effects are not resolved [38]. As discussed earlier, for $t \sim \tau_h$ the embedding fluid can not instantaneously follow changes in the particle velocity. This leads to an enlarged persistence in the particle velocity auto-correlations not adequately described by δ-correlated random forces. The solvent-enhanced memory in the particle velocities can be accounted for by means of a generalized velocity Langevin equation with so called nonwhite or colored random forces of non-Markovian nature. Calculations based on this equation have revealed, in contrast to Eq. (2.45), that the MSD approaches its limiting form $W(t) \approx \mathcal{D}_0 t$ algebraically slowly rather than exponentially fast, i.e.

$$\left\langle |\mathbf{r}(t) - \mathbf{r}(0)|^2 \right\rangle_{eq} \approx 6\mathcal{D}_0 t \left[1 - \frac{2}{\sqrt{\pi}} \left(\frac{\tau_h}{t} \right)^{1/2} \right], \tag{2.46}$$

for $t \gg \tau_h$. Such an algebraic long-time tail in the mean-squared displacement was observed in dynamic light scattering experiments on colloids [39]. More recently, experiments using colloidal particles in optical tweezers reveal these effects in detail [40].

In most cases of practical relevance, one is interested only in the overdamped or Brownian dynamics regime where $t \gg \tau_M$ and $\Delta x \gg l_B$, and during which significant changes in the particle position take place. In this regime, the particle *and* solvent motion can be regarded as inertia-free. This gives rise to a pure configuration-space description, with the evolution of the positional pdf $P(X,t)$ governed by the generalized Smoluchowski equation discussed earlier, and the stochastic evolution of the particle trajectories described by the overdamped Langevin equations discussed in the following.

On the Brownian dynamics time and length scales, changes in the particle velocity are not resolved. One can therefore neglect the inertia-term $M\dot{\mathbf{v}}(t)$ in Eq. (2.39) as compared to the friction term. Heuristically, this follows from the estimates

$$M \mid \dot{\mathbf{v}}(t) \mid \sim M \frac{(k_B T/M)^{1/2}}{\Delta t} \ll \frac{(M k_B T)^{1/2}}{\tau_M} \sim \zeta_0 \mid \mathbf{v}(t) \mid . \tag{2.47}$$

Consequently, the *positional* Langevin equation for an isolated particle in a quiescent fluid that is valid for $\Delta t \gg \tau_M$ and $\Delta x \gg l_B$ reads

$$\dot{\mathbf{r}}(t) = \frac{1}{\zeta_0} \mathbf{F}^s(t) \equiv \mathbf{v}^s(t). \tag{2.48}$$

This stochastic evolution equation for the particle position expresses a noninertial force-balance between friction and random forces. Here, $\mathbf{v}^s(t)$ is a Gaussian-distributed and δ-correlated, stochastic particle velocity contribution, fully characterized by its first two moments

$$\langle \mathbf{v}^s(t) \rangle = 0, \qquad \langle \mathbf{v}^s(t)\ \mathbf{v}^s(t') \rangle = 2\ \mathcal{D}_0\ \hat{\mathbf{I}}\,\delta(t - t'). \tag{2.49}$$

Integration of Eq. (2.48) over a time interval $\tau \gg \tau_M$ gives the finite-difference positional Langevin equation [7]

$$\Delta\mathbf{r}(\tau) = \mathbf{r}(t_0 + \tau) - \mathbf{r}_0 = \int_{t_0}^{t_0+\tau} du\ \ \mathbf{v}^s(u) = \sqrt{2\mathcal{D}_0\tau}\ \mathbf{n}, \tag{2.50}$$

where $\mathbf{n}$ is a vector of mutually statistically independent Gaussian random numbers n of zero mean and variance one, i.e., with a distribution given by

$$p(n) = (2\pi)^{-1/2} \exp\left\{-n^2/2\right\}, \tag{2.51}$$

so that $\langle \mathbf{n} \rangle = \mathbf{0}$ and $\langle \mathbf{n}\ \mathbf{n} \rangle = \hat{\mathbf{I}}$. All odd moments are zero, and the higher-order even ones are $\langle n^{2m} \rangle = (2m-1)!!$ with $m \in \{2, 3, \ldots\}$.

According to Eq. (2.50), the random displacement during τ is of $O(\tau^{1/2})$ only. From dividing Eq. (2.50) by τ and attempting to take the limit $\tau \to 0$, one notices that the Brownian velocity does not exist in the ordinary sense of calculus. Strictly speaking, the stochastic trajectories generated by the positional Langevin equation are continuous but nowhere differentiable, and they are of fractal nature [5,32]. A mathematically sound interpretation of a Langevin equation with white noise such as Eq. (2.48) can be given using the concept of a Wiener process [5] related to the time integral over $\mathbf{v}^B(t)/\mathcal{D}_0{}^{1/2}$. However, it suffices here to *interpret* the Langevin equation in terms of its finite difference form in Eq. (2.50). From the linearity of the finite difference equation, it follows that $\Delta\mathbf{r}(\tau)$ is a Gaussian random process with zero odd moments, and even moments determined as

$$\left\langle |\Delta\mathbf{r}(\tau)|^2 \right\rangle_0 = 6\mathcal{D}_0\tau, \qquad \left\langle |\Delta\mathbf{r}(\tau)|^{2n} \right\rangle_0 = O(\tau^n). \tag{2.52}$$

The higher-order $2n$ moments of a Gaussian process can be expressed by products of the second moment (mean-squared displacement), explaining their $O(\tau^n)$ dependence. The conditional average, $\langle \cdots \rangle_0$, is performed over a subensemble of particles of equal initial position $\mathbf{r}_0$. The Gaussian *conditional* pdf, $P_c(\Delta\mathbf{r}, \tau)$, for a displacement $\Delta\mathbf{r} = \mathbf{r} - \mathbf{r}_0$ during time step τ is thus

$$P_c(\Delta\mathbf{r}, \tau) = \{4\pi\mathcal{D}_0\tau\}^{-3/2} \exp\left\{-\frac{|\Delta\mathbf{r}|^2}{4\mathcal{D}_0\tau}\right\}. \tag{2.53}$$

This expression is an example of a van Hove time-space correlation function, which describes the Brownian motion of an isolated particle in a solvent. As seen by inspection, this is the solution of the single-particle diffusion equation

$$\frac{\partial}{\partial} P(\Delta\mathbf{r}, \tau) = \mathcal{D}_0 \nabla^2 P(\Delta\mathbf{r}, \tau), \tag{2.54}$$

subject to the initial condition $P(\Delta\mathbf{r}, \tau = 0) = \delta(\Delta\mathbf{r})$. This is the most simple form of a GSmE for the pdf of a continuous Markovian process, realized for zero external field and flow, and no particle interactions. The single-particle GSmE is equivalent to the positional Langevin equation in a statistical sense: The latter is a stochastic differential equation for an ensemble of particle trajectories whereas the former determines the corresponding pdf. Further, multiplication of Eq. (2.54) by $|\Delta\mathbf{r}|^{2n}$, and subsequent spatial integration using the above initial condition shows that the Langevin equation moment relations in Eq. (2.52) are recovered from the GSmE.

The positional Langevin equation for a single particle in simple shear flow, additionally subjected to an external force field $\mathbf{F}^{ex}(\mathbf{r})$, is obtained as a straightforward extension of Eq. (2.48),

$$\dot{\mathbf{r}}(t) = \left[\ \mathbf{v}^d(\mathbf{r}(t)) + \mathbf{v}^c(\mathbf{r}(t))\ \right] + \mathbf{v}^s(t). \tag{2.55}$$

It includes the deterministic velocity contributions $\mathbf{v}^d = \beta\mathcal{D}_0\mathbf{F}^{\text{ex}}(\mathbf{r})$ and $\mathbf{v}^c = \mathbf{\Gamma}\cdot\mathbf{r}$ due to forced drift and flow convection, respectively. The stochastic velocity $\mathbf{v}^s$ is still described as an isotropic Gaussian-Markovian process with moments given in Eq. (2.49), and force strength Γ according to Eq. (2.44), even though the system is driven out of equilibrium. While for $\Delta t \sim \tau_M$, there are shear rate dependent nondiagonal contributions to the random force covariance matrix (see, e.g., Dhont [1]), these contributions are negligible in the Brownian dynamics regime. External force and shear flow fields do not affect $\mathbf{v}^s$ in this regime for which $a/\beta(\mathcal{D}_0|\mathbf{F}^{ex}|) \gg \tau_h$ and $\tau_{\dot{\gamma}} \gg \tau_h$ are necessary requirements guaranteeing inertia-free coupled motion of particles and fluid.

The Langevin Eq. (2.55) is linear in $\mathbf{r}(t)$ for constant or linear external force, such as that for a harmonic potential. The particle position is then a Gaussian–Markovian process whose moments are obtained analytically: For an observer moving with constant drift velocity $\mathbf{v}_0 = \beta\mathcal{D}_0\mathbf{F}^{\text{ex}}$, the force drift term in the Langevin Eq. (2.55) is zero. Time integration for a given realization $\mathbf{v}^s(t)$ gives the result

$$\mathbf{r}(t) = \left(\hat{\mathbf{I}} + t\ \mathbf{\Gamma}\right)\cdot\mathbf{r}_0 + \int_0^t du\ \left(\hat{\mathbf{I}} + (t-u)\ \mathbf{\Gamma}\right)\cdot\mathbf{v}^s(u), \tag{2.56}$$

for an ensemble of independent spheres having same initial position $\mathbf{r}_0$. In its derivation, $\mathbf{\Gamma} = \dot{\gamma}\hat{\mathbf{x}}\hat{\mathbf{y}}$ and hence $\mathbf{\Gamma}^n = \mathbf{0}$ for $n \geq 2$ was used for the n-fold matrix product of $\mathbf{\Gamma}$, with $\hat{\mathbf{x}}$ and $\hat{\mathbf{y}}$ the unit vectors in x and y direction. The first vector moment $\mathbf{M}_1(t)$ and second matrix moment $\mathbf{M}_2(t)$ fully characterizing the process $\Delta\mathbf{r}(t) = \mathbf{r}(t) - \mathbf{r}_0$ are obtained using Eq. (2.56) as:

$$\mathbf{M}_1(t) = \langle\Delta\mathbf{r}(t)\rangle_0 = (\dot{\gamma}t)\ y_0\ \hat{\mathbf{x}} \tag{2.57}$$

$$\mathbf{M}_2(t) = \langle\Delta\mathbf{r}(t)\ \Delta\mathbf{r}(t)\rangle_0 = 2\mathcal{D}_0 t\left[\left(1 + \frac{(\dot{\gamma}t)^2}{3}\right)\hat{\mathbf{x}}\hat{\mathbf{x}} + \hat{\mathbf{y}}\hat{\mathbf{y}} + \hat{\mathbf{z}}\hat{\mathbf{z}} + \frac{\dot{\gamma}t}{2}(\hat{\mathbf{x}}\hat{\mathbf{y}} + \hat{\mathbf{x}}\hat{\mathbf{y}})\right]. \tag{2.58}$$

The mean-squared displacement is the trace of the covariance matrix,

$$\left\langle\ |\Delta\mathbf{r}(t)|^2\ \right\rangle_0 = \text{Tr}\{\mathbf{M}_2(t)\} = 6\mathcal{D}_0 t\left[1 + \frac{(\dot{\gamma}t)^2}{9}\right] = 6\ a^2\ \bar{t}\left[1 + (Pe^2)\frac{\bar{t}^2}{9}\right]. \tag{2.59}$$

Here, $\bar{t} = t/_D$ and

$$Pe = \frac{\tau_D}{\tau_{\dot{\gamma}}} = 6\pi\eta_m\left(\frac{\dot{\gamma}a^3}{k_B T}\right) \tag{2.60}$$

is the Péclet number quantifying the relative importance of shear convection and Brownian diffusion, Eq. (1.2).

The conditional Gaussian pdf, $P_c(\mathbf{r},t|\,\mathbf{r}_0) = P(\Delta\mathbf{r},t)$, of finding a sphere with center at $\mathbf{r}$ at time $t > 0$, given it was at $\mathbf{r}_0$ for $t = 0$ is [1]

$$P_c(\Delta\mathbf{r},t) = \frac{1}{(2\pi)^{3/2}\sqrt{det\ \mathbf{M}_2(t)}} \times \exp\left[-\frac{1}{2}(\Delta\mathbf{r} - \mathbf{M}_1(t))\cdot(\mathbf{M}_2(t))^{-1}\cdot(\Delta\mathbf{r} - \mathbf{M}_1(t))\right], \tag{2.61}$$

and it invokes the inverse of the covariance matrix.

According to Eq. (2.37), the present conditional pdf is the solution of the GSmE for a single sphere in shear flow,

$$\frac{\partial}{\partial t}P(\Delta\mathbf{r},t) = -\dot{\gamma}\ y\frac{\partial}{\partial x}P(\Delta\mathbf{r},t) + \mathcal{D}_0\nabla^2 P(\Delta\mathbf{r},t), \tag{2.62}$$

with initial condition $P(\Delta\mathbf{r},0) = \delta(\Delta\mathbf{r})$. This is most easily verified from the Fourier-transformed GSmE having

$$P_c(\mathbf{q},t) = \int d\mathbf{r} \exp\{i\mathbf{q}\cdot\Delta\mathbf{r}\}\ P_c(\Delta r,t) = \exp\{-i\mathbf{q}\cdot\mathbf{M}_1(t)\}\exp\left\{-\frac{1}{2}\ \mathbf{q}\cdot\mathbf{M}_2(t)\cdot\mathbf{q}\right\} \tag{2.63}$$

as its solution for $P_c(\mathbf{q},0) = 1$.

The smaller the Péclet number, the smaller is the effect of shear flow on the pdf as the shearing force is counterbalanced by the Brownian diffusion acting to restore equilibrium. With increasing *Pe*, the distribution of the particles becomes increasingly anisotropic, as displacements Δx in the x-direction are enhanced by the flow. It is seen from the covariance matrix that, while there is ordinary diffusion along the gradient and vorticity directions, with $\left\langle(\Delta y)^2\right\rangle = \left\langle(\Delta z)^2\right\rangle = 2\mathcal{D}_0 t$, there is a flow-enhanced spreading of the mean-squared displacement in x-direction, $\left\langle(\Delta x)^2\right\rangle = 2\mathcal{D}_0 t\left(1 + (\dot{\gamma}t)^2\right)$, proportional to t^3, referred to as Taylor dispersion. Qualitatively, the diffusional displacement, $|\ \Delta y\ | \sim t^{1/2}$, along the y-direction gives rise to a flow-induced displacement, $|\ \Delta x\ | \sim \dot{\gamma}\ |\ \Delta y\ |\ t \sim t^{3/2}$, in x-direction. This flow-diffusion coupling is quantified by the nondiagonal term $\langle\Delta x\ \ \Delta y\rangle = \mathcal{D}_0\ \ \dot{\gamma}\,t^2$ of the covariance matrix proportional to the Péclet number.

2.4.2 Many-Particles Langevin Equations for Shear Flow

We begin with the free-draining positional Langevin equation for simple shear flow, valid in the Brownian dynamics regime, which is [7]

$$\mathbf{r}_i(t) = \beta\,\mathcal{D}_0\ \ \mathbf{F}_i(X(t)) + \mathbf{\Gamma}\cdot\mathbf{r}_i(t) + \mathbf{v}_i^s(t). \tag{2.64}$$

Free-draining refers to the neglect of HIs, i.e., each particle experiences hydrodynamic drag as if it were alone in the fluid. The position vectors $\mathbf{r}_i(t)$ of spherical particles $i \in 1, 2, \cdots, N$ constituting the elements of the N-particles configuration $X(t)$ are coupled through the direct forces $\mathbf{F}_i$ arising from direct interactions, such as van der Waals attraction and electrostatic repulsion covered in Chapter 1. As flow-induced coupling is neglected, the *additive* Gaussian-Markovian random velocity contributions, $\mathbf{v}_i^s(t)$, are independent for different particles, and determined by their first and second moments

$$\left\langle \mathbf{v}_i^s(t) \right\rangle = 0, \qquad \left\langle \mathbf{v}_i^s(t)\, \mathbf{v}_j^s(t') \right\rangle = 2\, \mathcal{D}_0\, \hat{\mathbf{I}} \delta_{ij} \delta(t - t'). \tag{2.65}$$

A first-order finite-difference form of Eq. (2.64) follows from integrating the Langevin equations over a small time-step $(t_0, t_0 + \tau)$, with $\tau_M \ll \tau \ll \tau_D$, during which the configuration X and the direct forces stay practically constant. This leads to

$$\mathbf{r}_i(t_0 + \tau) = \mathbf{r}_i(t_0) + [\beta \mathcal{D}_0 \mathbf{F}_i(X_0) + \mathbf{\Gamma} \cdot \mathbf{r}_i(t_0)\ ]\tau + \sqrt{2\ \mathcal{D}_0 \tau}\ \mathbf{n} + \mathcal{O}(\tau), \tag{2.66}$$

with an error of order $\mathcal{O}(\tau)$ made in using a finite time-step. A function $Q(\tau)$ is of order $\mathcal{O}(\tau)$ if $\lim_{\tau \to 0} Q(\tau)/\tau = 0$. Eq. (2.66) generalizes Eq. (2.55) to directly interacting particles, referred to as the *Ermak scheme* with shear flow [41]. The forces are calculated in the configuration X_0 at the beginning of the time-step, which is the Itô description of the random process, and the scalar components n_α of the random vector $\mathbf{n}$ are generated independently from a one-dimensional Gaussian distribution of zero mean and variance one. As in the single-sphere case, we can interpret the positional Langevin equations in terms of their finite-difference form, to avoid mathematical difficulties with the δ-correlated random velocity. The Ermak finite difference scheme is a common starting point for free-draining Brownian Dynamics (BD) simulation methods. Eq. (2.66) holds for simple ambient shear flow $\mathbf{u}(\mathbf{r}) = \mathbf{\Gamma} \cdot \mathbf{r}$.

It is seen from the Ermak scheme for linear shear flow that the $3N$-variate stochastic process $X(t)$ is *Markovian*. The continuity of $X(t)$ is reflected in the short-time moments of the N-particles displacement,

$$\Delta X(\tau) = X(t_0 + \tau) - X(t_0), \tag{2.67}$$

obtained from the finite-difference equation by averaging over a subensemble of systems with equal initial configuration X_0:

$$\langle \Delta X(\tau) \rangle_0 = \left[v^d(X_0) + v^c(X_0) \right]\ \tau + o(\tau) \tag{2.68}$$

$$\langle \Delta X(\tau)\ \ \Delta X(\tau) \rangle_0 = 2\ \mathcal{D}_0 \tau\ \hat{\mathrm{I}} + o(\tau) \tag{2.69}$$

$$\langle \Delta X(\tau) \ldots \Delta X(\tau) \rangle_0 = o(\tau). \tag{2.70}$$

All higher-order polyadic moments involving more than two $\Delta X(\tau)$'s are small of order $\mathcal{O}(\tau)$. In Eq. (2.69), $\hat{I}$ denotes the $3N \times 3N$ unit matrix having $\hat{\mathbf{I}}$ as its diagonal entries, and $v^d(X_0)$ and $v^c(X_0)$ are the $3N$-dimensional column vectors of force drift velocities, $\mathbf{v}_i^d = \beta \mathcal{D}_0 \mathbf{F}_i$, and flow convection velocities, $\mathbf{v}_i^c = \mathbf{\Gamma} \cdot \mathbf{r}_i$, respectively.

For the common case of nonlinear forces acting on the particles, $X(t)$ is non-Gaussian distributed, except for its evolution in the short-time step τ where the forces stay constant, and where the short-time Gaussian pdf is determined by the two moments in Eqs. (2.68) and (2.69). Since $\mathbf{D}_{ij} = \mathcal{D}_0 \ \delta_{ij} \ \hat{\mathbf{I}}$ in free-draining approximation, the GSmE determining $P(X,t)$ at all times is given by Eq. (2.37) for zero hydrodynamic drift term, and the second-order diffusion term is simplified to $\mathcal{D}_0 \sum_i \nabla_i^2 P(X,t)$.

Briefly, there are mathematical subtleties involved in formulating many-particles positional Langevin equations with HIs included that need to be clarified. These arise because the delta-correlated random velocities of different particles are now hydrodynamically coupled by nonlinear functions $d_{ij}(X(t))$ appearing as factors multiplying the noise terms. The time integrals of these so-called *multiplicative noise* velocity terms over a short-time interval $[t_0, t_0 + \tau]$ are mathematically ambiguous, unless it is specified which function value should be taken in the interval. The most common specifications are the Itô-convention $d_{ij}(X(t_0))$ where the value at the beginning of the interval is used, and the Stratonovich convention $d_{ij}([X(t_0 + X(t_0 + \tau)]/2)$ where the function value at the arithmetic configuration average is taken [4,5]. Depending on which convention is used, *different* results for the time integral of the multiplicative noise terms and hence a different process $X(t)$ are obtained.

To see which description is applicable in the positional Langevin equation, recall that the GSmE in Eq. (2.37) adequately describes the pdf evolution of directly and hydrodynamically interacting particles in the Brownian dynamics regime. Therefore, we use it as a starting point in deriving an equivalent finite difference Langevin scheme. Consider the short-time pdf, $W_c(X, \tau|X_0)$, for a configurational displacement $X_0 \to X$ during time step τ. Since $\tau \ll \tau_D$, the configuration has changed so little that the microscopic diffusion matrix D and the $3N$-variate total drift supervector $v^D = v^d + v^c$ remain practically constant and equal to their initial values at X_0. The GSmE reduces then to a many-particles diffusion equation with constant coefficients,

$$\frac{\partial}{\partial \tau} W_c(X, \tau|X_0) = -v^D(X_0) \cdot \nabla W_c(X, \tau|X_0) + D(X_0) : \nabla\nabla W_c(X, \tau|X_0), \quad (2.71)$$

where the $3N$-dimensional nabla operator $\nabla = (\nabla_1, \cdots, \nabla_N)$ has been introduced. We seek its solution for the initial condition $W_c(X, 0|X_0) = \delta(X - X_0)$. For this purpose, we change variables from (X, τ) to (Y, τ') according to

$$Y = d^{-1}(X_0) \cdot [X - \ \tau v^D(X_0)], \quad (2.72)$$

and $\tau = \tau'$, implying

$$\begin{aligned} \frac{\partial}{\partial X} &= \left(d^{\ T}\right)^{-1}(X_0) \cdot \frac{\partial}{\partial Y} \\ \frac{\partial}{\partial \tau} &= \frac{\partial}{\partial \tau'} - \left(d^{-1}(X_0) \cdot v^D(X_0)\right) \cdot \frac{\partial}{\partial Y}. \end{aligned} \quad (2.73)$$

This substitution transforms Eq. (2.71) into a standard diffusion equation with unity diffusion matrix, $\hat{I}$, and zero total drift supervector, whose solution is given by

$$W_c(Y,t|Y_0) = \prod_{i=1}^{N} \{4\pi\tau\}^{-3N/2} \exp\left\{ -\frac{|\mathbf{y}_i - \mathbf{y}_{i0}|^2}{4} \right\}, \tag{2.74}$$

where $X_0 = d(X_0) \cdot Y_0$ and $Y_0 = (\mathbf{y}_{10}, \cdots, \mathbf{y}_{N0})$. The first two moments of $\Delta Y = Y - Y_0$ are thus

$$\langle \Delta Y \rangle_0 = 0, \quad \langle \Delta Y(\tau) \Delta Y(\tau) \rangle_0 = 2\tau \hat{\mathrm{I}}. \tag{2.75}$$

We have introduced here the $3N \times 3N$ square-root mobility matrix, $d(X_0)$, of $D(X_0)$ and its transpose $d^T(x_0)$ using the definition

$$D(X_0) = d(X_0) \cdot d^T(X_0). \tag{2.76}$$

The matrix $d(X_0)$ exists and is invertible since $D(X_0)$ is symmetric and positive definite for any physically allowed configuration.

Transforming back from Y to X leads to

$$W_c(X,\tau|X_0) = \{4\pi\tau\}^{-3N/2} \frac{1}{(\det D(X_0))^{1/2}} \cdot \\ \exp\left\{ -\frac{1}{4\tau} \left[X - X_0 - \tau v^D(X_0)\right]^T \cdot D^{-1}(X_0) \cdot \left[X - X_0 - \tau v^D(X_0)\right] \right\}, \tag{2.77}$$

for the short-time transition pdf invoking the positive valued determinant of $D(X_0)$ in its normalization constant. On noting that $P_c(X,\tau|X_0) = W_c(X,\tau|X_0) + o(\tau)$, and that W_c describes a Gaussian–Markovian process, the conditional short-time polyadic moments defined by the $3N$-dimensional integrals

$$\langle \Delta X(\tau) \cdots \Delta X(\tau) \rangle_0 = \int dX \; P_c(X,t|X_0) \; (\Delta X \cdots \Delta X), \tag{2.78}$$

are straightforwardly calculated with the result:

$$\langle \Delta X_i(\tau) \rangle_0 = v^D(X_0)\tau + o(\tau) \tag{2.79}$$

$$\langle \Delta X(\tau) \Delta X(\tau) \rangle_0 = 2D(X_0) + o(\tau). \tag{2.80}$$

All higher-order moments are small of $\mathcal{O}(\tau)$. The hydrodynamic coupling of the short-time displacements of different particles, and the symmetry of the diffusivity matrix, are obviated in the covariance matrix in Eq. (2.80), expressing a fluctuation-dissipation relation.

To arrive at the finite-difference Langevin equation including HIs, we realize first from the comparison of Eq. (2.75) with Eqs. (2.49) and (2.50) describing single-sphere Brownian motion that $\Delta Y(\tau) = \sqrt{2\tau}\tilde{n}$ where $\tilde{n} = \{n, \ldots, n\}$ is the supervector having $3N$ independent Gaussian random variables n of zero mean and variance one as its entries. Together with the above moment relations for ΔX this leads to the *Ermak-McCammon* first-order finite difference scheme [42,43]:

$$X(t_0+\tau) = X(t_0) + v^D(X_0)\tau + \sqrt{2\tau}\ d(X_0)\cdot\tilde{n} + o(\tau) \tag{2.81}$$

$$v^D(X_0) = \beta D(X_0)\cdot F(X_0) + \nabla\cdot D(X_0) + v^c(X_0), \tag{2.82}$$

including a drift term of $\mathcal{O}(\tau)$ and a stochastic diffusion term of $\mathcal{O}(\sqrt{\tau})$. Notice here that the functions on the right side are evaluated at the beginning of the time step. With this Itô convention the Ermak–McCammon scheme is for $\tau \longrightarrow 0$ stochastically equivalent to the GSmE since the same short-time moments are obtained.

A characteristic feature of a continuous Markovian process $X(t)$ is that its pdf $P(X,t)$ can be constructed from an initial distribution $P_{in}(X_0)$, by repeated application of the short-time transition pdf [6]. The knowledge of W_c is sufficient to describe diffusion in multiple steps of τ. Explicitly (see Figure 2.7)

$$P(X,t) = \lim_{n\to\infty}\prod_{i=1}^{n-1}\int dX_{i-1} W_c(X,\tau|X_{n-1}) W_c(X_i,\tau|\, X_{i-1})\ P_{in}(X_0), \tag{2.83}$$

with $\tau = t/n$. We have used here that $W_c(X_i,\tau|\, X_{i-1})$ in Eq. (2.77) depends on the time difference $\tau = t_n - t_{n-1}$ only since $v^D(X)$ is not explicitly time-dependent. The path integral in Eq. (2.83) is the continuous analog of a Markov chain used in so-called biased Monte Carlo simulations to generate representative particle configurations, with the n-fold product corresponding to an n-fold time-step forward integration in Eq. (2.81). It is noticed here that $P(X,t)$ remains positive for an initially positive P_{in}.

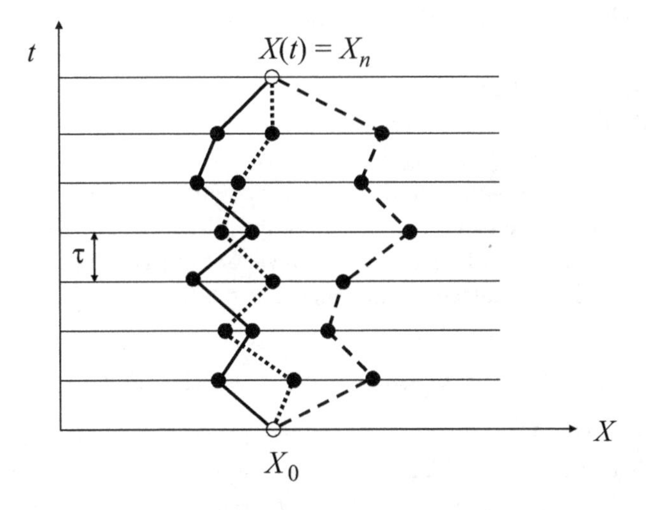

Figure 2.7 Paths in configuration space between initial and final configurations X_0 and X, respectively. The conditional pdf $P_c(X,t|X_0)$ of a continuous Markovian process is the sum over all paths according to Eq. (2.83).

Since W_c is strictly positive for nonzero τ, any physically allowed configuration X can be reached in the course of time.

In common with the free-draining approximation, $X(\tau)$ is Gaussian for short-time displacements. However, with HIs the particle positions are coupled even at small τ, and non-Gaussian corrections to W_c develop earlier than without HIs when τ is increased [44]. For a suspension without imposed flow and external field, these corrections give rise to a sublinear increase of the mean squared displacement at intermediate times.

2.5 Suspension Rheology

2.5.1 Effective Navier–Stokes Equation and Macroscopic Stress

As a first step toward the derivation of the Navier–Stokes equation for colloidal suspensions, we will have to specify what is meant by the suspension stress tensor $\Sigma(\mathbf{r},t)$ and the suspension flow velocity $\mathbf{u}(\mathbf{r},t)$ at position $\mathbf{r}$ and time t. A colloidal suspension is a two-phase system, consisting of an immiscible mixture of two materials: the core-material of the colloidal particles and the solvent. There is no generally valid microscopic expression for the stress tensor for a two-phase system that could be ensemble averaged to obtain the macroscopic stress, while the evaluation of the acceleration of a volume element in such a system is also not straightforward. It is therefore a highly nontrivial problem to derive the Navier–Stokes equation that describes the flow of a two-phase system, where equations of motion depend on how averages are defined (see, for example, Lhuillier and Nozières [45] and Wang and Prosperetti [46]). Due to the relatively small size of the colloidal particles, approximations can be made that do not apply to two-phase systems in general.

First, we have to ask about the precise definition of the suspension velocity $\mathbf{u}$. The definition of the rate-of-change of the velocity of the volume element is a nontrivial matter, since the two phases may have different accelerations and the local proportions of the two phases may change during time. However, the specific mass densities of the colloids and solvent are usually similar, while the colloidal particles attain velocities that are close to that of the solvent as a result of the relatively large friction forces of the solvent onto the colloids. In that case the Navier–Stokes equation retains its form as in Eq. (2.1), where the velocity is interpreted as the local average velocity of the suspension,

$$\mathbf{u}(\mathbf{r},t) \quad = \quad \varphi(\mathbf{r},t)\mathbf{u}_c(\mathbf{r},t) + [1-\varphi(\mathbf{r},t)]\,\mathbf{u}_s(\mathbf{r},t), \tag{2.84}$$

where φ is the local volume fraction of colloidal material, and where $\mathbf{u}_c$ and $\mathbf{u}_s$ are the average velocities of colloidal particles and solvent within an infinitesimally small volume element at position $\mathbf{r}$ and time t, respectively. It should be kept in mind, however, that for general two-phase systems, other definitions are possible [45,46]. The specific mass density can be similarly formulated as the velocity in Eq. (2.84).

Second, consider the divergence of the suspension stress tensor for force- and torque-free particles, i.e., when there is no external forcing of the particles, e.g., by electric, magnetic, or gravitational fields. By definition, the divergence of the suspension stress tensor is the force per unit volume with which the surrounding material acts on a volume element. For a suspension consisting of solid colloidal particles embedded in a solvent, there are three types of forces to be distinguished: (i) forces with which colloidal particles outside the volume element act onto those inside the volume element, (ii) forces of solvent molecules outside the volume element that are exerted onto the colloidal particle inside, and (iii) forces of solvent molecules outside acting on solvent inside. The corresponding contributions to the total stress tensor of the suspension are denoted by (i) Σ^{pp} (where the index p stands for colloidal particle), (ii) Σ^{ps} (where the index s stands for solvent), and (iii) Σ^{ss}. Assuming incompressibility (see Eq. (2.4)), it is found that [47]:

$$\begin{aligned}
\nabla \cdot \Sigma^{pp} &= -\sum_{j=1}^{N} < \delta\left(\mathbf{r}-\mathbf{r}_j\right)\mathbf{F}_j^h > , \\
\nabla \cdot \Sigma^{ps} &= < \sum_{j=1}^{N} \delta\left(\mathbf{r}-\mathbf{r}_j\right)\mathbf{F}_j^h > - < \sum_{j=1}^{N} \oint_{\partial A_j} dS'\,\delta(\mathbf{r}-\mathbf{r}')\,\mathbf{f}_j^h(\mathbf{r}') >, \\
\nabla \cdot \Sigma^{ss} &= \eta_m \nabla^2 \mathbf{u}(\mathbf{r},t) - \nabla p^{ss}(\mathbf{r},t),
\end{aligned} \tag{2.85}$$

where N is the number of colloidal particles under consideration, $\delta(\mathbf{r}-\mathbf{r}_j)$ is the delta distribution with $\mathbf{r}_j$ the center position of the colloidal particle j, ∂A_j is the surface area of colloidal particle j, η_m is the shear viscosity of the solvent, while p^{ss} is the pressure resulting from solvent–solvent interactions. Since the microstructural order of solvent molecules is not affected by the flow (because they are much smaller and exhibit a much faster dynamics as compared to the colloidal particles), this is the mechanical pressure that is given by the equation-of-state of the solvent. Furthermore, $\mathbf{f}_j^h$ is the force per unit area that the solvent exerts onto the surface of colloid j, and,

$$\mathbf{F}_j^h = \oint_{\partial A_j} dS'\,\mathbf{f}_j^h(\mathbf{r}'), \tag{2.86}$$

is the total force that the solvent exerts on the jth colloidal particle. The divergence of the total stress tensor Σ is obtained by addition of these three contributions, leading to:

$$\nabla \cdot \Sigma = \eta_m \nabla^2 \mathbf{u}(\mathbf{r},t) - \nabla p^{ss}(\mathbf{r},t) - \sum_{j=1}^{N} < \oint_{\partial A_j} dS'\,\delta\left(\mathbf{r}-\mathbf{r}'\right)\mathbf{f}_j^h(\mathbf{r}') >. \tag{2.87}$$

This expression is valid for solid colloidal particles, and for both homogeneous and inhomogeneous systems, where the shear rate and/or the concentration of colloids varies with position. It also applies to both steady and oscillatory shear rates. In an evaluation of this expression for the stress tensor, the explicit shear rate and frequency dependence originates from both the hydrodynamic forces as well as the probability density function with respect to which the ensemble averages are evaluated.

An expansion with respect to spatial gradients in density and shear rate is obtained by expanding the delta-distribution in Eq. (2.87) according to:

$$\delta(\mathbf{r}-\mathbf{r}') = \delta\left(\mathbf{r}-\mathbf{r}_j\right) + \sum_{n=1}^{\infty} \frac{(-1)^n}{n!} \left(\mathbf{r}'-\mathbf{r}_j\right)^n \odot \nabla^n \delta\left(\mathbf{r}-\mathbf{r}_j\right), \tag{2.88}$$

where the $\odot$ denotes contraction with respect to the polyadic products $\nabla^n = \nabla\nabla\cdots\nabla$ and $\left(\mathbf{r}'-\mathbf{r}_p\right)^n = \left(\mathbf{r}'-\mathbf{r}_p\right)\left(\mathbf{r}'-\mathbf{r}_p\right)\cdots\left(\mathbf{r}'-\mathbf{r}_p\right)$. Substitution into Eq. (2.87) leads to a spatial-gradient expansion of the divergence of the stress tensor. For a homogeneous system, gradient contributions of order ∇^2 and higher can be neglected, which leads to

$$\begin{aligned}\nabla\cdot\Sigma &= \eta_m \nabla^2 \mathbf{u}(\mathbf{r},t) - \nabla p^{ss}(\mathbf{r},t) \\ &+ \sum_{j=1}^{N} \left\{ - < \delta\left(\mathbf{r}-\mathbf{r}_j\right) \mathbf{F}_j^h > + \nabla\cdot < \delta\left(\mathbf{r}-\mathbf{r}_j\right) \oint_{\partial A_j} dS' \; \left(\mathbf{r}'-\mathbf{r}_j\right) \mathbf{f}_j^h(\mathbf{r}') > \right\}.\end{aligned} \tag{2.89}$$

It can be shown that the above result leads to the following expression for the suspension stress tensor [47],

$$\Sigma = \eta_m \left\{ \nabla\mathbf{u} + (\nabla\mathbf{u})^T \right\} - p^{ss}(\mathbf{r},t)\hat{\mathbf{I}} + \frac{1}{V} \sum_{j=1}^{N} \left\{ < \oint_{\partial A_j} dS' \left(\mathbf{r}'-\mathbf{r}_j\right) \mathbf{f}_j^h(\mathbf{r}') > + < \mathbf{r}_j \mathbf{F}_j^h > \right\}, \tag{2.90}$$

where V is the volume that contains the N colloids. This result for the stress tensor of a homogeneously sheared suspension has been derived before by Batchelor [29,30] in a different way.

From this general stress tensor expression, hydrodynamic, Brownian, and interaction stress contributions are identified and expressed in terms of hydrodynamic mobility/ diffusion matrices. To this end, we introduce the symmetric hydrodynamic force dipole tensor (stresslet) exerted on sphere j for no-slip surface boundary condition,

$$\mathbf{S}_j^h(X) = \frac{1}{2} \oint_{\partial A_j} dS' \left[\left(\mathbf{r}'-\mathbf{r}_j\right) \mathbf{f}_j^h(\mathbf{r}') + \mathbf{f}_j^h(\mathbf{r}') \left(\mathbf{r}'-\mathbf{r}_j\right) \right]. \tag{2.91}$$

The stresslet has a mathematical form that is analogous to an electric dipole where the hydrodynamic force densities play the role of the electric charge density. It has a nonzero trace, and is linearly related to the hydrodynamic forces and the symmetric and traceless, i.e., incompressible, rate-of-strain tensor of the applied simple shear flow,

$$\mathbf{E} = \frac{1}{2}\left(\mathbf{\Gamma} + \mathbf{\Gamma}^\dagger\right) = \frac{\dot{\gamma}}{2}(\hat{\mathbf{x}}\hat{\mathbf{y}} + \hat{\mathbf{y}}\hat{\mathbf{x}}), \tag{2.92}$$

by

$$\mathbf{S}_j^h(X) = -\sum_{n=1}^{N} \boldsymbol{\mu}_{jn}^{dt}(X)\cdot\mathbf{F}_n^h + \sum_{n=1}^{N} \boldsymbol{\mu}_{jn}^{dd}(X) : \mathbf{E}. \tag{2.93}$$

Note that in Chapter 1 and some other places in this book the tensor **E** is denoted by **D**, but we use the notation here to be consistent with the theoretical literature cited in this chapter.

The sum $\sum_n \boldsymbol{\mu}^{dt}_{nj} = \mathbf{C}_j$ of hydrodynamic translation-dipole tensors $\boldsymbol{\mu}^{dt}_{jn}$ is equal to the disturbance tensor $\mathbf{C}_j(X)$. It holds that $\left(\boldsymbol{\mu}^{td}_{jn}\right)_{\alpha\beta\gamma} = \left(\boldsymbol{\mu}^{dt}_{nj}\right)_{\beta\gamma\alpha}$ for the associated translation-dipole tensors $\boldsymbol{\mu}^{td}_{jn}$ relating stresslets (force dipoles) to translational velocities, with the greek indices labeling the Cartesian components. The dipole–dipole tensor, $\boldsymbol{\mu}^{dd}_{jn}(X)$, has four Cartesian indices and relates the rate-of-strain tensor, taken at the center of a particle n, to the stresslet that particle j attains if all N particles are force- and torque-free. The number of its different Cartesian components is reduced by its symmetry in the first two and last two indices, owing to the symmetry of $\mathbf{S}^h_j$ and $\mathbf{E}$.

Substitution of Eqs. (2.91) and (2.93) into Eq. (2.90) leads, with Eq. (2.30) and $\nabla\mathbf{u} = \boldsymbol{\Gamma}^T$, to

$$\Sigma = 2\eta_m \mathbf{E} - p^{ss}\hat{\mathbf{I}} + \left\{\Sigma^h + \Sigma^I + \Sigma^{Br}\right\}, \tag{2.94}$$

where $\Sigma^p = \Sigma^{pp} + \Sigma^{ps} = \{\cdots\}$ in Eq. (2.94) is the particle stress part with contributions:

$$\Sigma^h = \frac{1}{V}\left\langle \sum_{j,n=1}^{N} \boldsymbol{\mu}^{dd}_{jn} \right\rangle : \mathbf{E}$$

$$\Sigma^I = -\frac{1}{V}\left\langle\left(\sum_{j=1}^{N} \mathbf{r}_j\mathbf{F}^I_j - \sum_{j,n=1}^{N} \boldsymbol{\mu}^{dt}_{jn}\cdot\mathbf{F}^I_n \right)\right\rangle$$

$$\Sigma^{\rm Br} = -\frac{1}{V}\left\langle\left(\sum_{j=1}^{N} \mathbf{r}_j\mathbf{F}^{\rm Br}_j - \sum_{j,n=1}^{N} \boldsymbol{\mu}^{dt}_{jn}\cdot\mathbf{F}^{\rm Br}_n \right)\right\rangle = -nk_{\rm B}T\ \hat{\mathbf{I}} - \frac{k_BT}{V}\left\langle \sum_{n=1}^{N} \nabla_n\cdot\mathbf{C}_n \right\rangle, \tag{2.95}$$

due to hydrodynamic, direct, and Brownian interactions, respectively. These contributions are additive for Stokes flow. The second equality for the Brownian stress Σ^{Br} follows from partial integration, with $n = N/V$ denoting the mean particle concentration. For time-independent shear flow, the average over the particle positions,

$$\langle(\cdots)\rangle = \int dX\, P(X;\dot{\gamma})(\cdots), \tag{2.96}$$

is taken using as pdf the stationary solution, $P(X;\dot{\gamma})$, of the GSmE in Eq. (2.36) describing the shear-distorted microstructure. The shear flow related hydrodynamic mobility tensors $\boldsymbol{\mu}^{dd}_{jn}$, $\mathbf{C}_j$ and $\boldsymbol{\mu}^{dd}_{jn}$ are zero for vanishing particle radius $a \to 0$. Consequently, the hydrodynamic stress contribution Σ^h vanishes for interacting point particles, and the particle stress reduces to the simpler form known for simple liquids:

$$\Sigma^p = -nk_{\rm B}T\ \hat{\mathbf{I}} - \frac{1}{V}\left\langle \sum_{j=1}^{N} \mathbf{r}_j\mathbf{F}^I_j \right\rangle. \tag{2.97}$$

For colloidal hard spheres, the nonoverlap condition is taken care of by hydrodynamic lubrication interactions so that $\Sigma^I = \mathbf{0}$, provided no additional soft potential interactions are present such as van der Waals attraction or electric repulsion. For vanishing macroscopic flow ($\mathbf{\Gamma} = \mathbf{0}$), the suspension is isotropic without off-diagonal stress components. The suspension stress then reduces to [48,49]:

$$\Sigma(\dot{\gamma} = 0) = -(p^{ss} + \Pi_{\text{eq}})\hat{\mathbf{I}}, \tag{2.98}$$

with equal normal stress components $\Sigma_{\alpha\alpha}$ for $\alpha \in \{x, y, z\}$. Here,

$$\Pi_{\text{eq}} = nk_\text{B}T + \frac{1}{3V}\left\langle \sum_{j=1}^{N} \mathrm{r}_j \cdot \mathbf{F}_j^I \right\rangle_{\text{eq}} = nk_\text{B}T\left[1 - \frac{2\pi}{3}n\beta \int_0^\infty dr\, r^3 \frac{d\mathcal{V}(r)}{dr}\, g_{\text{eq}}(r)\right] \tag{2.99}$$

is the equilibrium osmotic particle pressure, and $\langle \cdots \rangle_{\text{eq}}$ is the average invoking the equilibrium pdf $P_{\text{eq}}(X) \propto \exp\{-\beta\Phi(X)\}$ of the nonsheared (i.e., quiescent) suspension. The second equality for Π_{eq} is valid for a pairwise additive N-particles interaction energy $\Phi(X)$ according to Eq. (2.31). Here, $g_{\text{eq}}(r)$ is the isotropic pair distribution function of the quiescent system, being equal to the conditional probability of finding a particle at center-to-center distance r from another one, as depicted in Figure 1.5. For monodisperse hard spheres, the direct interactions are exactly pairwise additive, and the osmotic pressure,

$$\Pi_{\text{eq}} = nk_\text{B}T\left[1 + 4\,\phi\, g_{\text{eq}}(r = 2a^+)\right] \approx nk_\text{B}T\left[\frac{1 + \phi + \phi^2 - \phi^3}{(1-\phi)^3}\right], \tag{2.100}$$

is solely determined by the two-spheres contact value $g_{\text{eq}}(2a^+)$, where $\phi = (4\pi/3)na^3$ is the particle volume fraction. The hard sphere osmotic pressure is plotted in Figure 1.5. The second equality is the Carnahan–Starling hard sphere pressure result [11], valid to good accuracy in the fluid phase up to the freezing concentration $\phi_f = 0.494$, and additionally inside the metastable isotropic phase up to $\phi \approx 0.57$. In the presence of flow where Σ^P attains nonzero off-diagonal components, a *nonequilibrium* generalized osmotic pressure Π can be defined by the trace,

$$\Pi(\dot{\gamma}) = -\frac{\text{Tr}(\Sigma^P)}{3} = -\frac{1}{3}\left(\Sigma^P_{xx} + \Sigma^P_{yy} + \Sigma^P_{zz}\right), \tag{2.101}$$

with $\Pi(\dot{\gamma} = 0) = \Pi_{\text{eq}}$. The leading order, flow-induced normal particle stress contributions to $\Pi - \Pi_{\text{eq}}$ are quadratic in $\dot{\gamma}$, and signal the onset of non-Newtonian flow behavior.

The HIs are often disregarded in theory and simulation, either for simplicity or to elucidate their effects in comparison to simulation results where HIs are included [50]. Without HIs, the suspension stress in Eq. (2.94) reduces to:

$$\Sigma = -(p^{ss} + nk_\text{B}T)\hat{\mathbf{I}} + 2\eta_m\left(1 + \frac{5}{2}\phi\right)\mathbf{E} + \Sigma^I, \tag{2.102}$$

with

$$\Sigma^I = -\frac{1}{V}\left\langle \sum_{j=1}^{N} \mathbf{r}_j \mathbf{F}_j^I \right\rangle = -\frac{n^2}{2}\int d\mathbf{r}(\hat{\mathbf{r}}\,\hat{\mathbf{r}})r\frac{d\mathcal{V}}{dr}g(\mathbf{r};\dot{\gamma}), \tag{2.103}$$

where $g(\mathbf{r};\dot{\gamma})$ is the nonisotropic (nonequilibrium) pair distribution function established under steady shear, and $\mathbf{r}$ the vector distance between a pair of particles. There is no Brownian stress contribution without HIs, except for the isotropic kinetic stress. The hydrodynamic stress Σ^h is here equal to the Einstein term $5\eta_m\phi\mathbf{E}$ also for nonsmall ϕ. Only the direct interaction stress remains nontrivial without HIs, and for larger shear rates it gives rise to non-Newtonian flow behavior including shear thinning and nonzero normal stress differences related to the shear distorted pair distribution function.

2.5.2 Rheological Properties and Flow Microstructure

For imposed simple shear flow according to Eq. (2.24), with the flow, flow gradient, and flow vorticity directions along the x,y, and z axes, respectively, the (zero frequency) shear viscosity $\eta(\dot{\gamma})$ is determined by the off-diagonal suspension stress component $\Sigma_{xy} = \Sigma_{yx}$, i.e.,

$$\eta(\dot{\gamma}) = \frac{\Sigma_{xy}(\dot{\gamma})}{\dot{\gamma}} = \frac{\mathbf{E}:\Sigma(\dot{\gamma})}{\dot{\gamma}^2}. \tag{2.104}$$

Using Eqs. (2.90) and (2.104), the viscosity expression,

$$\eta(\dot{\gamma}) \quad = \eta_m + \Delta\eta_h(\dot{\gamma}) + \Delta\eta(\dot{\gamma}), \tag{2.105}$$

is obtained with the hydrodynamic viscosity contribution,

$$\Delta\eta_h(\dot{\gamma}) = \frac{1}{\dot{\gamma}^2}\left\langle \frac{1}{V}\sum_{i,n=1}^{N}\left[\mathbf{E}:\left(\boldsymbol{\mu}_{in}^{dd}\right):\mathbf{E}\right]\right\rangle, \tag{2.106}$$

and the contribution due to the direct and Brownian forces,

$$\Delta\eta(\dot{\gamma}) = \Delta\eta_I(\dot{\gamma}) + \Delta\eta_{\mathrm{Br}}(\dot{\gamma}) = -\frac{1}{\dot{\gamma}^2}\left\langle \frac{1}{V}\sum_{j=1}^{N}\left[\mathbf{v}_j^c\cdot\left(\mathbf{F}_j^I + \mathbf{F}_j^{\mathrm{Br}}\right)\right]\right\rangle. \tag{2.107}$$

Here, $\mathbf{v}_j^c$ is the velocity of a freely flow-convected particle j introduced in Eq. (2.34). According to Eq. (2.107), $\dot{\gamma}^2\Delta\eta(\dot{\gamma})$ is the rate of dissipated energy per unit volume supplied by Brownian and direct forces in their attempt to restore isotropic equilibrium perturbed by the applied shear flow. Since $\mathbf{v}_j^c$ and $\left(\mathbf{F}_j^I + \mathbf{F}_j^{\mathrm{Br}}\right)$ are of $\mathcal{O}(\dot{\gamma})$, the low shear limiting and equivalently, zero frequency, viscosity,

$$\eta_0 = \eta(\dot{\gamma}\to 0), \tag{2.108}$$

is well defined.

Two additional viscometric functions characterizing non-Newtonian flow behavior under nonweak, steady shear flow are the first (N_1) and second (N_2) normal stress differences:

$$\begin{aligned} N_1(\dot{\gamma}) &= \Sigma^p_{xx} - \Sigma^p_{yy} \\ N_2(\dot{\gamma}) &= \Sigma^p_{yy} - \Sigma^p_{zz}, \end{aligned} \tag{2.109}$$

which are of $\mathcal{O}(\dot{\gamma}^2)$ at small shear rates so that the associated coefficients $\Psi_i(\dot{\gamma}) = N_i/\dot{\gamma}^2$ tend to constant values at low shear.

The calculation of $\eta(\dot{\gamma})$ requires first the determination of the associated pdf from the GSmE in Eq. (2.36). Qualitative rheological features can be illustrated on the two-particle level, for a dilute suspension where HIs are pairwise additive. Integration of the GSmE equation over the positions of $(N-2)$ particles while keeping the relative vector distance $\mathbf{r} = \mathbf{r}_2 - \mathbf{r}_1$ of two particles fixed and ignoring triplet and higher-order correlations, leads for a homogeneous system to the two-particle Smoluchowski equation,

$$\frac{\partial g(\mathbf{r},t)}{\partial t} + \nabla \cdot \mathbf{J}(\mathbf{r},t) = 0, \tag{2.110}$$

for the flow-distorted and time-dependent pair distribution function $g(\mathbf{r},t) = V^2 P_2(\mathbf{r}_1,\mathbf{r}_2,t)$, where P_2 is the reduced two-particle pdf and $\nabla = \nabla_{\mathbf{r}}$. The relative two-particle flux is

$$\mathbf{J}(\mathbf{r},t) = -\mathbf{D}^{(2)}(\mathbf{r}) \cdot \nabla g(\mathbf{r},t) + \beta \mathbf{D}^{(2)}(\mathbf{r}) \cdot \mathbf{F}^I(\mathbf{r}) g(\mathbf{r},t) + \mathbf{v}^c(\mathbf{r})\ g(\mathbf{r},t), \tag{2.111}$$

with two-particle relative convective velocity $\mathbf{v}^c = \mathbf{\Gamma} \cdot \mathbf{r} + 2\ \mathbf{E} : \mathbf{C}^{(2)}(\mathbf{r})$ featuring the disturbance tensor $\mathbf{C}^{(2)}(\mathbf{r})$ in Eq. (2.26), and relative diffusion tensor $\mathbf{D}^{(2)} = 2\left(\mathbf{D}^{(2)}_{11} - \mathbf{D}^{(2)}_{12}\right)$ where $\mathbf{D}^{(2)}_{ij}(\mathbf{r})$ according to Eq. (2.21). Furthermore, $\mathbf{F}^I = -\nabla V(r)$ is the pair interaction force. As noticed from Eq. (2.111), the flux is the sum of a diffusion, direct force drift, and shear flow convection contribution.

The two accompanying boundary conditions are:

$$(\hat{\mathbf{r}} \cdot \mathbf{J})_{r=2a^+} = 0 \quad \text{and} \quad g \to 1 \text{ as } r \to \infty. \tag{2.112}$$

The zero radial flux condition at two-sphere contact distance ensures impenetrability of particles, while the second condition imposes fluid-like structure without long-range order. For nonstationary flow conditions such as for start-up shear flow, for which $g(\mathbf{r}, t = 0) = g_{eq}(r)$, and shear flow cessation, an initial condition for g at $t = 0$ is additionally required. For zero flow and in equilibrium, $\mathbf{J} = \mathbf{0}$ and $g(\mathbf{r}) = g_{eq}(r) = \exp\{-\beta \mathcal{V}(r)\}$ according to Eq. (2.111). The zero concentration form of $g_{eq}(r)$ shows up here since three-body and higher-order correlation and interaction contributions are neglected in Eqs. (2.110) and (2.111). The deviatoric, i.e., traceless, Brownian and interaction stress contributions, whose $x{-}y$ components give the viscosity contributions $\Delta\eta_{\mathrm{Br}}$ and $\Delta\eta_I$ after division by $\dot{\gamma}$, are [51]:

$$\begin{aligned}
\left[\Sigma^{\rm Br}(\dot\gamma)\right]_{\rm dev} &= -\frac{k_{\rm B}T}{2}\, n^2 \int d\mathbf{r} \left(\hat{\mathbf{r}}\hat{\mathbf{r}} - \frac{1}{3}\,\hat{\mathbf{I}}\right) W(r)\, \left[g(\mathbf{r};\dot\gamma) - g_{\rm eq}(r)\right] \\
\left[\Sigma^{I}(\dot\gamma)\right]_{\rm dev} &= \frac{1}{2}\, n^2 \int d\mathbf{r} \left(\hat{\mathbf{r}}\hat{\mathbf{r}} - \frac{1}{3}\,\hat{\mathbf{I}}\right) [1 - A(r)]\, r\frac{d\mathcal{V}}{dr}\, \left[g(\mathbf{r};\dot\gamma) - g_{\rm eq}(r)\right],
\end{aligned} \tag{2.113}$$

with $W(r) = \nabla \cdot \left[\mathbf{C}^{(2)}\right]_{\rm dev}$. Here, $\left[\mathbf{C}^{(2)}\right]_{\rm dev}$ is the traceless part of $\mathbf{C}^{(2)}(\mathbf{r})$ regarding its first two Cartesian indices, and the divergence is taken with respect to the third index. Moreover, $A(r)$ is a hydrodynamic function related to the two-sphere disturbance tensor which as for $W(r)$ is a tabulated function of pair separation r. At large separation, W and A decay proportional to $(a/r)^6$ and $(a/r)^3$, respectively. Owing to lubrication, $A(r = 2a) = 1$ at pair contact, implying according to Eq. (2.113) zero direct interaction stress for hard spheres for which $g(\mathbf{r};\dot\gamma)d\mathcal{V}/dr \propto \delta(r - a)$. In suspensions of particles with long-range, soft repulsion such as in charge-stabilized colloids where near-contact pairs are very unlikely, Σ^I gives instead the major viscosity contribution, whereas Brownian stress is small now in comparison, owing to the rapid decay of $W(r)$. Notice here the dipole-related traceless tensor $\left(\hat{\mathbf{r}}\hat{\mathbf{r}} - \hat{\mathbf{I}}/3\right)$ whose orientational average is zero. Thus, only the nonisotropic shear-distorted part, $g(\mathbf{r}) - g_{eq}(r)$ contributes to the integrals in Eq. (2.113). As concerns Σ^h in Eq. (2.95), $g_{eq}(r)$ itself contributes to this stress in linear order in $\dot\gamma$. Note here that taking $g(\mathbf{r};\gamma)$ to zeroth order in ϕ yields the stress and viscosity to quadratic order in ϕ.

For weak shear flow, i.e., to linear order in the Péclet number defined in Eq. (2.60), a regular linear response perturbation of $g(\mathbf{r};\dot\gamma)$ around equilibrium can be used,

$$g(\mathbf{r};\dot\gamma) = g_{eq}(r)\left[1 - Pe(\hat{\mathbf{r}}\cdot\hat{\mathbf{E}}\cdot\hat{\mathbf{r}})f(r)\right] + \mathcal{O}(Pe^2), \tag{2.114}$$

where $\hat{\mathbf{E}} = \sqrt{2}\mathbf{E}/\dot\gamma$ is shear rate independent. Eq. (2.114) reflects the affine character of $g(\mathbf{r};\dot\gamma)$ to linear order in Pe, triggered by the extensional, affine contribution $\mathbf{E}\cdot\mathbf{r}$ of the applied simple shear flow. The spherically symmetric pair distribution is slightly deformed, with particle pairs brought together along the compressional axis $y = -x$ of the $(x - y)$ shear velocity-gradient plane for $\mathbf{r}\cdot\mathbf{E}\cdot\mathbf{r} < 0$, i.e., in shear plane quadrants $xy < 0$, and with pairs convected away along the extensional axis $y = x$ for $\mathbf{r}\cdot\mathbf{E}\cdot\mathbf{r} > 0$, i.e., inside shear plane quadrants $xy > 0$. The compression and extensional axes are illustrated by the dashed lines in the right part of Figure 2.8. Substitution into Eq. (2.110) results in an ordinary differential equation for $f(r)$ which must be solved numerically in the presence of HIs. For hard spheres with fully neglected HIs, the analytic solution

$$f(r) = \frac{32}{3}\left(\frac{a}{r}\right)^3 \tag{2.115}$$

is obtained. Substitution of this solution into Eq. (2.103) gives

$$\Sigma^p = (nk_{\rm B}T)4\phi\ \hat{\mathbf{I}} + 2\eta_m\left(1 + \frac{5}{2}\phi + \frac{12}{5}\phi^2\right)E + \mathcal{O}(Pe^2), \tag{2.116}$$

with the isotropic stress part due to $g_{\mathrm{eq}}(r) = \Theta(r - 2a)$ in Eq. (2.114) where $\Theta(r)$ is the unit step function, and the stress contribution proportional to $\phi^2 \mathbf{E}$ due to the perturbation function $f(r)$. Consequently,

$$\eta_0 = \eta_m \left(1 + \frac{5}{2}\phi + \frac{12}{5}\phi^2 \right) + \mathcal{O}(\phi^3) \tag{2.117}$$

is obtained for the zero shear limiting viscosity of hard spheres without HIs, up to cubic order in concentration. The effect of HIs is to enlarge viscous dissipation with the consequence that at larger ϕ the actual viscosity is severely underestimated without HIs. The numerically precise quadratic in concentration coefficient with HIs is 5.9 instead of $12/5$, as presented in Eq. (1.14), with contribution 5.0 from Σ^h and 0.9 from Σ^{Br} in Eq. (2.113).

Bergenholtz et al. [51] solved the two-particle GSmE Eq. (2.110) for simple shear flow numerically over a wide range of *Pe* numbers, both for hard spheres and excluded annulus spheres, as also discussed in section 3.4 of *CSR*. Non-Newtonian rheology is observed already in quadratic order in ϕ, with shear thinning (viscosity decrease) at intermediate *Pe* and continuous shear thickening (viscosity increase) at large *Pe*, plus nonzero normal stress differences and an osmotic pressure increase above Π_{eq}. These nonlinear features are due to the flow-induced distortion of $g(\mathbf{r};\dot{\gamma})$ beyond the linear response level in Eq. (2.114), where in competition with the equilibrium-restoring Brownian motion the fore-aft symmetry of the pair distribution (dictated by pure hydrodynamics at low Reynolds number) is broken.

The right panel of Figure 2.8 shows simulation results with many-body HIs included for the pair distribution function of Brownian hard spheres at $\phi = 0.35$ and $Pe = 25$ [52–54]. Similar to the affine structure predicted at low *Pe*, there is an accumulation of particle pairs at contact in the compressional quadrants ($xy < 0$), and a depletion in the extensional quadrant ($xy > 0$). However, at large *Pe* one can see from the figure that the maximum of $g(\mathbf{r};\dot{\gamma})$ is shifted from the compression to the flow direction while the long-range structure of the suspension remains fluid-like [53]. A narrow boundary layer of thickness $\sim 1/Pe$ forms. Inside the boundary layer, the hydrodynamic squeezing force pushing the particles together along the compressional flow axis is balanced by Brownian motion, i.e., by the drift term $\sum_j \nabla_j \cdot D_{ij}$ in Eq. (2.38) attempting to re-disperse the particles. This nearest neighbor distribution under flow has been verified experimentally [55] using techniques discussed in Chapter 4.

Simulation results [52] for the hydrodynamic contribution, $\eta_m + \Delta\eta_h(\dot{\gamma})$, and the Brownian contribution, $\Delta\eta_{Br}(\dot{\gamma})$, to the zero frequency shear viscosity $\eta(\dot{\gamma})$ of hard spheres are shown in the left part of Figure 2.8, as functions of the Péclet number $Pe \propto \dot{\gamma}a^3$. Note that $\Delta\eta_I = 0$ for hard spheres, provided near-contact lubrication interaction is included. As seen in the figure, the total viscosity η goes through a shear thinning region at lower *Pe*, and a shear thickening region at high *Pe*. Shear thinning is due to the decrease of the Brownian viscosity contribution $\Delta\eta_{Br}(\dot{\gamma})$ with increasing *Pe*, because diffusional re-dispersion becomes progressively less effective. In contrast, the hydrodynamic viscosity contribution remains roughly constant and

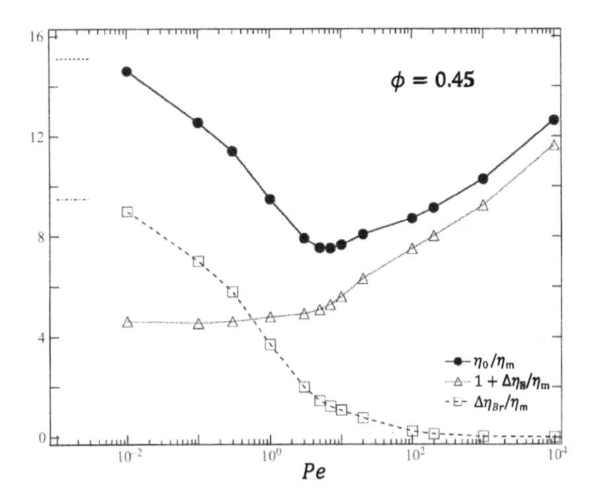

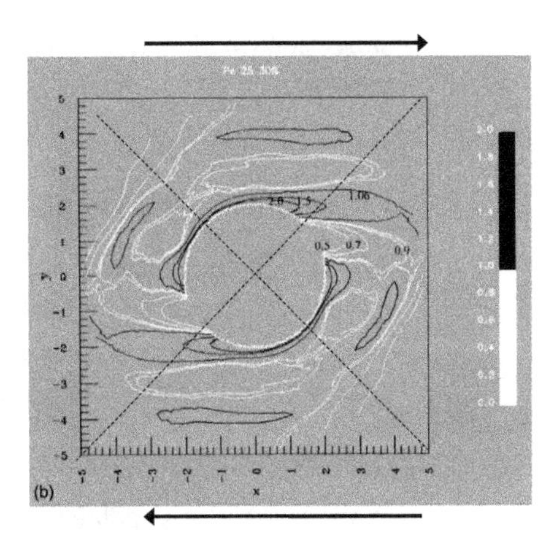

Figure 2.8 Left: Stokesian dynamics simulation results for the hydrodynamic and Brownian viscosity contributions Δ_h and Δ_{Br}, respectively, to the zero frequency viscosity η for a hard sphere suspension at $\phi = 0.45$. The horizontal line segments to the far left are zero shear limits determined using the equilibrium Green-Kubo Eq. (2.137). Reproduced with permission from Foss and Brady, Structure, diffusion and rheology of Brownian suspensions by Stokesian dynamics simulation. Journal of Fluid Mechanics 2000;407:167 [52]. Right: Hard sphere pair distribution function $g(x, y; \dot{\gamma})$ in the x–y velocity-gradient plane for simple shear flow at $Pe = 25$ and $\phi = 0.3$. Length scale is the hard sphere radius a, and the countours indicate the magnitude of $g(x, y; \dot{\gamma})$. The dashed diagonal lines indicate the compression ($y = -x$) and extension ($y = x$) axis, respectively. The arrows indicate the direction of simple shear flow. Reproduced with permission from Morris and Katyal, Microstructure from simulated Brownian suspension flows at large shear rate. Physics of Fluids 2002;14:1920 [53]

equal to its zero shear rate limit η'_∞ in the shear thinning regime of η, while for $Pe > 10$ it increases, causing η to shear thicken. This continuous hydrodynamic shear thickening of concentrated hard sphere suspensions at large Pe is attributed to the formation of lubrication-induced, anisotropic, and fractal-like hydroclusters of particles [54,56,57]. While shear thinning is qualitatively recovered in simulations of Brownian hard spheres without HIs, hydrodynamic shear thickening is not captured [58]. The second normal stress difference is always compressive for hard spheres ($N_2 < 0$). The first normal stress difference N_1 crosses over from positive values at lower Pe, through a maximum at $Pe \sim 1$, to negative values at larger Pe. Normal stress differences cease at low $Pe \ll 1$, since, due to fore–aft reflection symmetry of relative pair trajectories with respect to $x = 0$, the normal stress differences in Eq. (2.109) balance to zero.

For colloidal particles having additional longer-range soft repulsion, or solvent-permeable hard spheres where the hydrodynamic radius a_h is smaller than the excluded volume one, there is no strong lubrication stress contribution any more to Σ^h. Moreover, owing to the faster decay of the function $W(r)$ in the Brownian stress expression in Eq. (2.113) than that of the function $A(r)$ in the direct interaction stress, and considering the low likelihood of near-contact configurations, also the Brownian stress contribution is small. The (non-Newtonian) rheology and microstructure in these systems is thus mainly due to Σ^I [59].

2.5.3 Linear Rheology and Equilibrium Green–Kubo Relation

Another means of calculating suspension rheology in the linear response regime is through the Green–Kubo relationship, as presented in this section. Consider a homogeneous suspension subjected to a weak and oscillatory simple shear flow of frequency ω and small strain rate amplitude $\dot{\gamma}$,

$$\mathbf{u}(\mathbf{r},t) = \mathbf{\Gamma}\cdot\mathbf{r}\ \ \cos(\omega t). \tag{2.118}$$

While being time-dependent, the rate-of-strain tensor $\mathbf{\Gamma}(t) = \mathbf{\Gamma}\cos(\omega t)$ and its symmetric part $\mathbf{E}(t) = \mathbf{E}\cos(\omega t)$ are still taken as spatially constant. The suspension flow and microstructure are characterized now, in addition to the dimensionless shear rate Pe and the particles volume fraction ϕ, by the dimensionless frequency (bare Deborah number) $De = \omega\tau_D \propto \omega a^3/k_{\rm B}T$.

To first order in $Pe \propto \dot{\gamma}$, the macroscopic shear stress is linearly and causally related to the driving shear rate, $\dot{\gamma}(t) = \dot{\gamma}\cos(\omega t)$, by:

$$\begin{aligned}\Sigma_{xy}(\dot{\gamma},t) = \int_{-\infty}^{t} d\tau G(t-\tau)\dot{\gamma}(\tau) &= G'(\omega)\gamma(t) + \frac{G''(\omega)}{\omega}\dot{\gamma}(t)\\ &= \omega\ \eta''(\omega)\gamma(t) + \eta'(\omega)\dot{\gamma}(t),\end{aligned} \tag{2.119}$$

with imposed strain $\gamma(t) = (\dot{\gamma}/\omega)\sin(\omega t)$. In the considered linear response limit superposition applies, and $\dot{\gamma}(\tau)$ in the time integral could be in principle any function having a time variation compatible with the assumed Smoluchowski dynamics. It is further assumed that the imposed flow was switched on in the "infinite" past so that the associated transient stress response in the fluid phase is fully relaxed to zero at time t. The time-dependent linear shear modulus, $G(t)$, quantifies the shear stress relaxation induced by the slightly shear-distorted microstructure. According to Eq. (2.119), the linear stress response to $\dot{\gamma}(t)$ is likewise sinusoidal and of the same frequency ω, but with a frequency-dependent phase shift $\delta = \tan[G''(\omega)/G'(\omega)]$ relative to the strain $\gamma(t)$. Here,

$$\begin{aligned}G'(\omega) &= \omega\eta''(\omega) = \omega\int_0^{\infty} dt\ \ \sin(\omega t)\ \ G(t)\\ G''(\omega) &= \omega\eta'(\omega) = \omega\int_0^{\infty} dt\ \ \cos(\omega t)\ \ G(t),\end{aligned} \tag{2.120}$$

are the elastic storage modulus G' and the viscous loss modulus G'', respectively, which as $G(t)$ are shear rate independent. The viscosities (η',η'') are equivalent measures of the moduli (G'',G'). Eq. (2.119) describes linear viscoelastic suspension behavior: At high frequencies $\omega\to\infty$ (meaning physically $1/\tau_h \gg \omega \gg 1/\tau_D$) and thus short relaxation times $\tau_h \ll t \ll \tau_D$ on the level of the GSmE, Brownian, and direct interaction stress contributions to Σ^p_{xy} have no time to relax during a cycle, different from the hydrodynamic stress which, viewed on time scale $t \gg \tau_h$, relaxes instantaneously (cf. Figure 2.3). The short-time stress response is thus

$$\Sigma_{xy}(\dot{\gamma}, t) \sim G'_\infty \gamma(t) + \eta'_\infty \dot{\gamma}(t) \tag{2.121}$$

where $\eta'_\infty = \eta'(\omega \to \infty) = \eta_m + \Delta\eta_h(\dot{\gamma} = 0)$ is the high frequency viscosity, and $G'_\infty = G'(\omega \to \infty)$ the high frequency elastic modulus. Thus, except for the hydrodynamic dissipated energy quantified by η'_∞, the energy of Brownian and direct interaction stresses remain elastically stored for $De \gg 1$, akin to a Hookean spring where stress and strain are in phase ($\delta = 0$). The described time-scale separation of hydrodynamic and nonhydrodynamic stresses implies

$$G(t) = 2\eta'_\infty \ \delta(t) + \Delta G(t), \tag{2.122}$$

i.e., there is a noninstantaneous shear modulus part, $\Delta G(t)$, related to $\Sigma^{\mathrm{Br}}_{xy}$ and Σ^{I}_{xy}, and a delta function contribution expressing that the hydrodynamic stress response $\Sigma^h_{xy} = \left(\eta'_\infty - \eta_m\right)\dot{\gamma}(t)$ follows $\dot{\gamma}(t)$ instantaneously. For regular $\Delta G(t)$, the high frequency elastic modulus is given by the initial value

$$G'_\infty = \Delta G(t = 0). \tag{2.123}$$

With decreasing frequency, $G'(\omega)$ decreases and the suspension behaves increasingly viscous. For $\omega \to 0$, i.e., at large response times, the suspension behaves as a viscous Newtonian fluid characterized by the (zero frequency) low shear viscosity $\eta_0 = \eta'(\omega = 0)$, with:

$$\eta_0 - \eta'_\infty = \Delta\eta = \int_0^\infty dt \ \Delta G(t). \tag{2.124}$$

The viscosity part $\Delta\eta = \Delta\eta_{Br}(\dot{\gamma} = 0) + \Delta\eta_I(\dot{\gamma} = 0)$ is the time integral of $\Delta G(t)$.

The frequency-dependent viscosity functions $\eta'(\omega) - \eta'_\infty$ and $\eta''(\omega)$ are one-sided Fourier cosine and sine transforms of $\Delta G(t)$, respectively, implying that each of them includes the same viscoelastic information as $\Delta G(t)$. They can be mutually expressed in terms of each other in form of the Kramers-Kronig relations:

$$\eta'(\omega) - \eta'_\infty = \frac{2}{\pi}\int_0^\infty d\alpha \ \frac{\alpha \ \eta''(\alpha)}{\omega^2 - \alpha^2} \tag{2.125}$$

$$\eta''(\omega) = -\frac{2}{\pi}\int_0^\infty d\alpha \ \frac{\omega \ \eta'(\alpha)}{\omega^2 - \alpha^2}, \tag{2.126}$$

where the integrals are interpreted as Cauchy principal values.

In the fluid phase, $\eta'(\omega)$ decays monotonically with increasing ω from its low frequency, Newtonian, plateau value η toward the high frequency limiting value η'_∞, while $\eta''(\omega)$ related to the elastically stored energy increases from $\eta''(\omega = 0) = 0$ to its peak value at about $\omega\tau_{MW}(\phi) \approx 1$, subsequently decaying toward zero with further increasing ω. Here,

$$\tau_{MW}(\phi) = \int_0^\infty dt \, t \ \frac{\Delta G(t)}{\Delta\eta} \tag{2.127}$$

is the mean, Maxwell, shear stress relaxation time not to be confused with the particle momentum relaxation time τ_M. On considering the reduced dynamic viscosities $R(\omega) = (\eta' - \eta'_\infty)/\Delta\eta$ and $I(\omega) = \eta''/\Delta\eta$, the scale differences between different ϕ are removed so that these two functions collapse approximately on two single master curves for all ϕ when plotted as a function of the *dressed* Deborah number $\omega\tau_{MW}(\phi)$ [60]. As discussed in Banchio et al. [60] using mode coupling theory (MCT), and in Foss and Brady [58] using BD simulations, $\tau_{MW}(\phi)$ decreases first as ϕ grows past the semi-dilute regime, reaching for hard spheres a shallow minimum at about $\phi \approx 0.3$ of value $\sim 0.1 \times \tau_D$, owing to the increasing number of collisions. With further increasing concentration beyond $\phi \approx 0.3$, particle caging becomes important, giving rise to a slower shear stress relaxation with τ_{MW} strongly increasing, diverging eventually at the ideal glass transition concentration ϕ_g (≈ 0.58 for hard spheres) due to dynamic arrest as predicted by MCT, or at random close packing $\phi_{rcp} \approx 0.64$ owing to jamming.

The high frequency viscosity η'_∞ can be written as an equilibrium average $\langle \cdots \rangle_{\text{eq}}$. This renders this routinely measured quantity particularly amenable to analytic calculations. On treating HIs as pairwise additive also for concentrations beyond the semi-dilute regime, a useful analytic expression for η'_∞ in terms of $g_{eq}(r)$ is obtained from Σ^h_{xy} in Eq. (2.95) as [61]

$$\frac{\eta'_\infty}{\eta_m} = 1 + \frac{5}{2}\phi\left[1 + \phi + 3\phi \int_2^\infty ds\, s^2\, g_{\text{eq}}(s)\, J(s)\right], \tag{2.128}$$

which includes a rapidly and monotonically decaying hydrodynamic shear function $J(s)$ with $J(s) \sim (15/2)s^{-6}$ for $s = r/a \gg 2$, linearly related to elements of the dipole–dipole tensors $\boldsymbol{\mu}^{dd}_{jn}$ and known algebraically [27]. The contribution $(5/2)\phi^2$ to η'_∞ independent of $g_{eq}(r)$ derives from a regularization of Σ^h_{xy} in Eq. (2.95) originally due to Batchelor and Green [61], since divergent and conditionally convergent integrals are otherwise encountered owing to the long-range HIs. Various hydrodynamic regularization methods are discussed in the literature. A versatile hydrodynamic regularization scheme for suspension properties including viscosities, sedimentation, and diffusion coefficients is given in Szymczak and Cichocki [62].

Eq. (2.128) is the so-called hydrodynamic pairwise additivity approximation (PA) of the high frequency viscosity. For hard spheres, the Verlet-Weiss corrected Percus-Yevick (VW–PY) solution can be used for $g_{eq}(r)$, accurate up to the freezing concentration ϕ_f [11]. While the numerically exact second-order concentration expansion, $\eta'_\infty/\eta_m = 1 + 2.5\phi + 5.0021\phi^2 + \mathcal{O}(\phi^3)$ is recovered in PA, hard sphere simulation data of η'_∞ are underestimated for $\phi > 0.2$, as noticed in Figure 2.9. This can be ascribed to the neglected three-body and higher-order HIs contributions coming into play at larger ϕ.

For a given $g_{\text{eq}}(r)$, the PA is extended to the so-called excluded annulus suspension model consisting of colloidal hard spheres whose hydrodynamic radius a_h is smaller than the excluded volume radius a, as introduced in Chapter 1. The hydrodynamic volume fraction $\phi_h = \hat{b}^3\phi$, with the reduced hydrodynamic radius $\hat{b} = a_h/a$, is thus smaller than ϕ. The excluded annulus model is often used as a simplistic description of

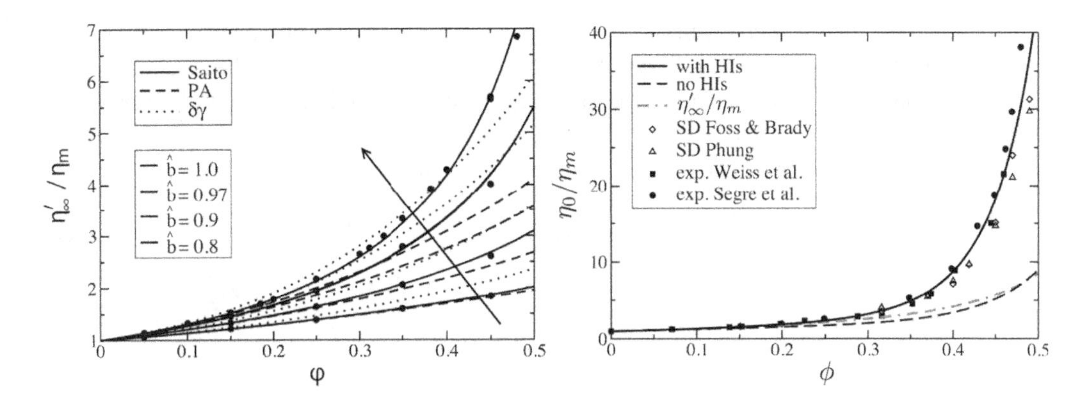

Figure 2.9 Left: Reduced high frequency viscosity, $\eta_{\infty'}/\eta_m$, of excluded annulus particles for indicated values of $\hat{b} = a_h/a$. The arrow points into the direction of increasing $\hat{b}$. Symbols: simulation data; Solid lines: generalized Saito formula in Eq. (2.129); Dashed lines: PA scheme results; Dotted lines: $\delta\gamma$ scheme results. Right: Reduced low shear limiting, zero frequency viscosity, η_0/η_m, of hard spheres (b = 1), with and without HIs. Solid black, dashed black and solid red lines are according to Eqs. (2.157), (2.158), and (2.129), respectively. Open symbols: Stokesian Dynamics (SD) simulation results for Brownian hard spheres by Foss and Brady [52], and Phung [72]. Closed symbols: Experimental data by Segrè et al. [74] and Weiss et al. [75]. Adapted from Riest et al. [65]

particles with long-range, soft repulsion such as charged colloidal spheres where a characterizes the length scale of soft repulsion. More recently, it has been used as a realistic model for hydrodynamically structured spherical particles such as homogeneously solvent-permeable spheres [63], and core-shell spheres with a dry spherical core and a solvent-permeable shell [64], as realized, e.g., in nonionic spherical microgel systems [65]. It is shown in theory, simulation, and experiment [65] that such particles can be described to good accuracy as excluded annulus particles, where $\hat{b}$ is analytically related to the (mean) Darcy permeability K characterizing the solvent permeability. Typical K values correspond to $\hat{b} > 0.9$.

Again assuming pair-wise additive HIs, results for the excluded annulus model viscosity $\eta'_{\infty}(\phi, \hat{b})$ for PA are shown in Figure 2.9 (left), in comparison with high precision simulation results, and viscosity predictions by the so-called $\delta\gamma$ method due to Beenakker [66] adapted to excluded annulus particles [65]. In the $\delta\gamma$ method, near-contact lubrication is disregarded different from the PA, but many-body HIs are approximately included in a mean-field way. As for the PA scheme, the $\delta\gamma$ method requires $g_{eq}(r)$ as its only input. The virtue of the *PA* and $\delta\gamma$ methods is their analytic simplicity, and the straightforward applicability to different pair potentials $\mathcal{V}(r)$ embodied in $g_{\text{eq}}(r)$. Both methods have been profitably used, e.g., for the high frequency viscosity calculations of charge-stabilized suspensions [67–69] and protein solutions [70,71].

According to the left part of Figure 2.9, the $\delta\gamma$ predictions for η'_{∞} at larger ϕ are closer to the simulation data [52,72] than the PA ones. At smaller $\hat{b}$, the fluid can flow through the annulus particles which relieves the hydrodynamic stress built-up in the suspension so that η'_∞ is reduced. Many-body HIs become weaker since ϕ_h is smaller, and the applicability of the PA extends to larger ϕ. As regards the excluded annulus model, a very useful generalized Saito formula for $\eta'_\infty(\phi, \hat{b})$ is available, which for $\phi < 0.5$ agrees well with simulation data, including those for hard spheres for which $\hat{b} = 1$. Explicitly,

$$\frac{\eta'_\infty(\phi, \hat{b})}{\eta_m} = 1 + [\eta](\hat{b})\phi \frac{1 + \hat{S}(\phi, \hat{b})}{1 - \frac{2}{5}\,[\eta](\hat{b})\phi\left[1 + \hat{S}(\phi, \hat{b})\right]}, \tag{2.129}$$

with the Saito function

$$\hat{S}(\phi, \hat{b}) = \phi\left(\frac{\lambda_V(\hat{b})}{[\eta](\hat{b})} - \frac{2}{5}\,[\eta](\hat{b})\right), \tag{2.130}$$

intrinsic viscosity $[\eta](\hat{b}) = (5/2)\hat{b}^3$. The 2nd-order virial coefficient for $\hat{b} \geq 2/3$ is approximated by:

$$\lambda_V(\hat{b}) = 5.0021 - 39.279\,(1 - \hat{b}) + 143.179\,(1 - \hat{b})^2 - 288.202\,(1 - \hat{b})^3 + 254.581\,(1 - \hat{b})^4. \tag{2.131}$$

The solid lines in the figure are the generalized Saito formula results for η'_∞. While the formula is exact at low volume fractions and works well at intermediate volume fractions, it diverges at very small volume fractions, and formulas such as the Lionberger–Russel expression in Chapter 1 should be used [73].

It is interesting to compare η'_∞ with the zero frequency viscosity $\eta_0 = \eta'_\infty + \Delta\eta$ of hard spheres, with and without HIs considered. This comparison is made for the fluid phase region in the right part of Figure 2.9, where also experimental data [74,75] are shown. Results for η_0 with and without HIs are shown as obtained from an accurate analytic expression, given in Section 2.5.5 in the context of generalized Stokes–Einstein relations. With HIs, $\Delta\eta$ for hard spheres is entirely due to Brownian stress, and for $\phi > 0.4$ it makes the major contribution to η_0. Without HIs, $\Delta\eta = \Delta\eta_I$, and there is no hydrodynamic stress contribution any more except for the single-particle stresslets, so that $\eta'_\infty/\eta_m = 1 + (5/2)\phi$.

Eq. (2.128) has been generalized to concentrated suspensions using concentration-dependent *effective* hydrodynamic pair mobility functions, derived for isotropic hard sphere suspensions [76] with an efficient method denoted Accelerated Stokesian Dynamics, which is discussed in Chapter 3. This work demonstrates additionally that there is no reduction in the range of the HIs, i.e., no hydrodynamic screening, in concentrated suspensions of freely diffusing particles.

The frequency-dependent linear viscometric functions can be calculated on the pair level by generalizing the time-independent pair perturbation function $f(r)$ in Eq. (2.114) to [8,9,73]

$$f(r) \;\rightarrow\; f_{in}(r, \omega)\cos(\omega t) + f_{out}(r, \omega)\sin(\omega t), \tag{2.132}$$

and solving for the perturbation contributions f_{in} and f_{out} to $g(\mathbf{r}, t)$ oscillating in phase and out of phase with $\dot{\gamma}(t)$, respectively. We follow instead a different route based on a very useful exact Green-Kubo relation for $\Delta G(t)$ due to Nägele and Bergenholtz [77]. A Green–Kubo (GK) relation expresses a transport property in terms of the time integral of an associated time autocorrelation function. GK relations are well-suited starting points for introducing statistical physics based calculation schemes for transport properties [2].

This relation is obtained from writing the GSmE (Eq. 2.37) formally as:

$$(\partial/\partial t)P(X,t) = \Omega(X,t)P(X,t) = [\Omega_0(X) + \delta\Omega(X,t)]P(X,t), \tag{2.133}$$

where the linear differential operator $\Omega(X,\tau)$ is known as the Smoluchowski operator. It consists of an unperturbed time-independent part, $\Omega_0(X)$, associated with a quiescent suspension and a time-dependent perturbation operator, $\delta\Omega(X,t)$, due to oscillatory shear flow, which is linear in Pe. Explicitly:

$$\begin{aligned} \Omega_0(X) &= \sum_{i,j=1}^{N} \nabla_i \cdot \mathbf{D}_{ij}(X) \cdot \left[\nabla_j - \beta \mathbf{F}_j^I(X)\right] \\ \delta\Omega(X,t) &= \cos(\omega t) \sum_{i=1}^{N} \nabla_i \cdot \mathbf{v}_i^c(X), \end{aligned} \tag{2.134}$$

where $\mathbf{v}_i^c(X)$ is the convective velocity. Recall that $\mathbf{D}_{ij}(X)$ are positive definite matrices; from this, one easily shows that Ω_0 is an operator with only nonpositive real eigenvalues, guaranteeing an approach to equilibrium in a quiescent system as mentioned above. By flow symmetry, the disturbance tensor $\mathbf{C}_i(X)$ is symmetric in its first two indices so that $\mathbf{\Gamma} : \mathbf{C}_i = \mathbf{E} : \mathbf{C}_i$. On taking as an initial condition that the suspension was in flow-unperturbed equilibrium in the infinite past, the *formal* solution of the GSmE is given by [77]:

$$P(X,t) = P_{eq}(X)\left\{1 - \int_{-\infty}^{t} d\tau \, \cos(\omega\tau) \, e^{(t-\tau)\,\Omega_0^\dagger(X)} \sum_{i=1}^{N} \left[\beta \mathbf{F}_i^I(X) - \nabla_i\right] \cdot \mathbf{v}_i^c(X)\right\} + o(Pe^2), \tag{2.135}$$

where

$$\Omega_0^\dagger(X) = \sum_{i,j=1}^{N} \left[\nabla_i + \beta \mathbf{F}_i^I\right] \cdot \mathbf{D}_{ij}(X) \cdot \nabla_j \tag{2.136}$$

is the unperturbed adjoint Smoluchowski operator [38]. One can formulate a general pdf solution for arbitrary Pe, in which the full adjoint operator $\Omega^\dagger(X,)$ appears now in a time-ordered exponent, since $\delta\Omega(X,\tau)$ at different times is not self-commuting in general. As discussed in Section 2.6.3, the pdf for arbitrary Pe has been used as a starting point for an Integration-Through-Transients (ITT) approach to nonlinear rheology, developed so far without HIs.

Using Eq. (2.135) for the linearly perturbed pdf in conjunction with Eq. (2.95) for Brownian and interaction stresses, the following GK relation for the dynamic shear modulus is obtained [77]:

$$\Delta G(t) = \frac{1}{k_B T \ V} \left\langle \sigma_{xy} e^{\Omega_0^\dagger \ t} \sigma_{xy} \right\rangle_{\text{eq}}, \tag{2.137}$$

where the transversal Cartesian component $\sigma_{xy}(X)$ of the *microscopic* shear stress tensor $\boldsymbol{\sigma}(X)$ is the sum of potential interaction and hydrodynamic parts, i.e., $\sigma_{xy} = \sigma_{xy}^{\text{pot}} + \sigma_{xy}^{\text{hyd}}$, with:

$$\begin{aligned} \sigma_{xy}^{\text{pot}}(X) &= -\left[\sum_{i=1}^{N} \mathbf{r}_i \ \mathbf{F}_i^I(X) \right]_{xy} \\ \sigma_{xy}^{\text{hyd}}(X) &= -\left[\sum_{i=1}^{N} \mathbf{C}_i(X) \cdot \mathbf{F}_i^I + k_B T \ \nabla_i \cdot \mathbf{C}_i(X) \right]_{xy} . \end{aligned} \tag{2.138}$$

The GK relation expresses $\Delta G(t)$ as a time autocorrelation function of the microscopic shear stress $\sigma_{xy}(t) = \exp\left\{\Omega_0^\dagger \ t\right\}\sigma_{xy}$ in unperturbed equilibrium. Just as for all equilibrium time autocorrelation functions in Smoluchowski dynamics, owing to the nonpositivity of the spectrum of the operator Ω_0, $\Delta G(t)$ is strictly monotonically decaying in the fluid phase and hence expressable as a positive superposition of purely relaxing exponentials [78]. The maximum of $\Delta G(t)$ at $t = 0$ determines the high frequency elastic modulus as:

$$G'_\infty = \frac{1}{k_B T \ V} \left\langle \left|\sigma_{xy}\right|^2 \right\rangle_{\text{eq}}. \tag{2.139}$$

The moduli $G'(\omega)$ and $G''(\omega)$ follow from $G(t)$ by Fourier transformation, and the low shear viscosity part $\Delta\eta$ is equal to the time integral of $\Delta G(t)$. The particles in the above GK relation can belong to different components, allowing for its straightforward application to colloidal mixtures [79] and concentrated electrolyte solutions [80,81].

The GK relation Eq. (2.137) is routinely used in Stokesian dynamics simulations for calculating the low shear viscosity of Brownian suspensions with many-particles HIs included. Because the equilibrium time evolution operator is involved, $\Delta G(t)$ and hence $\Delta\eta$ are directly obtained in BD simulations from averaging over representative particle trajectories in equilibrium.

2.5.4 Applications of the Green–Kubo Relation

From a theoretical viewpoint, the GK relation is a convenient starting point for various applications, and for implementing statistical mechanics based approximations such as the mode coupling theory (MCT) scheme, discussed in Section 2.6. As a first application, G'_∞ is obtained for low concentration by approximating $\sigma_{xy}(X)$ by the sum, $\sum_{i<j}\sigma_{xy}^{(2)}(\mathbf{r}_{ij})$, of pair stresses, valid at all concentrations for pairwise additive direct interactions and without HIs. Substitution of this stress sum into Eq. (2.139) leads to [77]:

$$\frac{a^3 \ G'_\infty}{k_B T} = \frac{3\phi^2}{40\pi} \int_0^\infty ds \ s^2 \ g_{\text{eq}}(s) \left\{ W(s) - \beta[1 - A(s)]s \ \frac{dV(s)}{ds} \right\}^2 + O(\phi^3), \tag{2.140}$$

where $s = r/a$. Without HIs, the Mountain–Zwanzig formula is obtained:

$$\frac{a^3\ G'_\infty}{k_B T} = \frac{3\phi^2}{40\pi} \int_0^\infty ds\ g_{eq}(s) \frac{d}{ds}\left(s^4 \frac{d\mathcal{V}(s)}{ds}\right), \tag{2.141}$$

up to an additional kinetic contribution $(3\phi/4)$, which holds for all ϕ in the fluid phase [82]. With regard to G'_∞, the influence of HIs is mainly in a thin lubrication layer near the particle surfaces. When lubrication is accounted for, G'_∞ is finite owing to the hydrodynamic factor $(1 - A(s))$ in Eq. (2.140) which is zero at $s = 2$, whereas a diverging elastic modulus is predicted without HIs for genuine hard spheres, and for hard particles having additional soft repulsion not strong enough to prevent in-contact pairs. This peculiar behavior without HIs is due to the singular nature of the hard sphere potential, causing $G(t)$ to diverge as $1/\sqrt{t}$ for $t \to 0$, and consequently $G'(\omega)$ to diverge as $\sqrt{\omega}$ at large frequencies [73,83]. For soft particles (e.g., microgel and high functional star polymers), and rigid spheres which are solvent permeable to a certain degree or show surface roughness (so that $a_h < a$), the high frequency modulus G'_∞ is finite.

Exact expressions for rheological and diffusion properties can be obtained from the GK relation in Eq. 2.137, for the simplifying model of weakly interacting point-like particles with Fourier-integrable pair potential, referred to as the weak coupling limit (WCL) model. For neglected HIs, the shear relaxation function becomes:

$$\Delta G(t) = \frac{k_B T\ n^2}{60\pi^2} \int_0^\infty dq\ q^4 \left[\beta \frac{d\mathcal{V}(q)}{dq}\right]^2 \exp\left\{-2\ \mathcal{D}_0\ q^2 t\right\}, \tag{2.142}$$

where $\mathcal{V}(q)$ is the spatially Fourier-transformed pair potential, and $\mathcal{D}_0$ the single-particle Stokes–Einstein–Sutherland diffusion coefficient with hydrodynamic radius $a_h > 0$. Consider for example the screened Coulomb potential,

$$\beta V(r) = \lambda_B Z^2\ \frac{\exp\{-\kappa r\}}{r}, \quad r > 0, \tag{2.143}$$

describing the effective electrostatic interaction of charge-stabilized colloids in the limit of the point-particle approximation. The particle charge Z in this potential is in units of the positive elementary charge e, the electric screening length $1/\kappa$ sets the range of electric repulsion, and the Bjerrum length $\lambda_B = e^2/(\epsilon k_B T)$ characterizes the solvent of dielectric permeability ϵ. Weak particle coupling amounts here to a small electric coupling strength $\Gamma_{el} = (\lambda_B/\bar{r})Z^2 \ll 1$, where $\bar{r} = n^{-1/3}$ is the mean inter-particle distance.

For this potential, analytic expressions for $G(t)$ and its associated frequency-dependent moduli are obtained from Eq. (2.142), as functions of t/τ_{MW} and $\omega\tau_{MW}$, respectively. The mean Maxwell time $\tau_{MW} = 1/(2\kappa^2 \mathcal{D}_0)$ obtained from Eq. (2.127) is approximately equal to the diffusion time required to traverse the screening length distance. In the WCL, $\Delta G(t)$ has a broad, continuous distribution of stress relaxation times according to:

$$\frac{\Delta G(t)}{G'_\infty} = \frac{16}{5\pi} \int_0^\infty d\lambda \frac{\lambda^{5/2}}{(1+\lambda)^4}\ e^{-\lambda t/\tau_{MW}}, \tag{2.144}$$

with high frequency modulus and zero frequency viscosity:

$$\frac{\tau_{MW} G'_{\infty}}{\eta_m} = \frac{(\pi \Gamma_{el})^2}{2\bar{\kappa}^3} \left(\frac{a_h}{\bar{r}}\right) \tag{2.145}$$

$$\eta_0 - \eta'_{\infty} = \tau_{MW} \ G'_{\infty}/5, \tag{2.146}$$

respectively, where $\bar{\kappa} = \kappa \ \bar{r}$. Different from hard spheres treated without HIs, $G'_{\infty} = \Delta G(0)$ is finite in WCL. A broad distribution of relaxation times is observed also in semi-dilute hard sphere suspensions [83]. We do not give here the analytic expression for $G(t)$ following from Eq. (2.144) but only note that:

$$\frac{\Delta G(t \gg \tau_{MW})}{G'_{\infty}} \sim \frac{6}{\sqrt{\pi}} \left(\frac{\tau_{MW}}{t}\right)^{7/2}. \tag{2.147}$$

The same algebraic long-time tail exponent $7/2$ is observed in the hard sphere $\Delta G(t)$ [83]. Fourier-cosine and sine transformation of Eq. (2.144) lead to the (reduced) dynamic viscosities:

$$\begin{aligned} R(\omega) &= \frac{\eta'(\omega) - \eta'_{\infty}}{\Delta\eta} = \frac{1 + 4\ \sqrt{2}\ \omega^{1/2} + 16\ \omega + 20\ \sqrt{2}\ \omega^{3/2} + 29\ \omega^2 + 8\ \sqrt{2}\ \omega^{5/2}}{\left(1 + \sqrt{2}\ \omega^{1/2} + \omega\right)^4} \\ I(\omega) &= \frac{\eta''(\omega)}{\Delta\eta} = \frac{\omega\left(1 + 4\ \sqrt{2}\ \omega^{1/2} + 16\ \omega + 12\ \sqrt{2}\ \omega^{3/2} + 5\ \omega^2\right)}{\left(1 + \sqrt{2}\ \omega^{1/2} + \omega\right)^4}, \end{aligned} \tag{2.148}$$

which are related to dissipated and elastically stored mechanical energy, respectively, and expressed as rational functions of the square-root of the frequency ω taken in units of $1/\tau_{MW}$. Figure 2.10 displays the WCL dynamic viscosities with the algebraic high frequency asymptotic decay indicated by the dashed lines. There is a narrow Newtonian plateau region for $\eta'(\omega)$ at $\omega\tau_{MW} < 0.01$ where the suspension behaves purely viscous and $\eta''(\omega)$ is practically zero. Linear viscoelastic behavior with decreasing, frequency-thinning $\eta'(\omega)$ comes into play at higher frequencies with $\eta''(\omega)$ having its maximum at $\omega \ \tau_{MW} \approx 1.8$.

2.5.5 Generalized Stokes–Einstein Relations

Microrheological techniques often depend on interpreting diffusive particle motion in terms of local rheological properties [84,85]. Numerous attempts have been made over the past decades to relate viscoelastic properties of a concentrated suspension to associated diffusion properties, in form of so-called generalized Stokes–Einstein (GSE) relations. GSE relations are fundamental to the field of microrheology [84]. Theoretical and experimental studies revealed that the approximate validity of a specific GSE relation is strongly dependent on the form and range of the particles interaction potential, and the proximity of the considered suspension to a vitrification or gelation point.

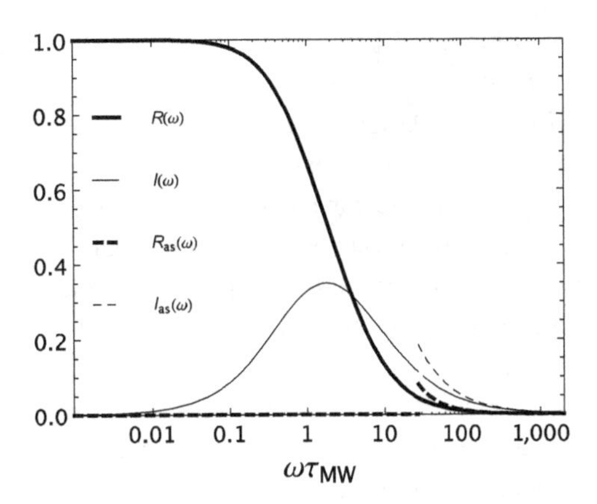

Figure 2.10 Frequency-dependent reduced viscosities $R(\omega)$ and $I(\omega)$ in the WCL of point-Yukawa particles according to Eq. (2.148) as functions of frequency ω in units of $1/\tau_{MW}$ (solid lines). Note the high frequency asymptotic behavior $R(\omega) \sim R_{as}(\omega) = 8\sqrt{2}/(\omega\tau_{MW})^{3/2}$ (lower dashed curve segment) and $I(\omega) \sim I_{as}(\omega) = 5/(\omega\tau_{MW})$ for $\omega \ \tau_{MW} \gg 1$ (upper dashed curve segment).

Consider here a generic form of a GSE relation in a one-component colloidal suspension,

$$\frac{\hat{\eta}(\phi)}{\eta_m} \approx \frac{\mathcal{D}_0}{\hat{D}(\phi)}, \tag{2.149}$$

relating a linear viscoelastic property, $\hat{\eta}(\phi)$, of a concentrated suspension to a diffusion property, $\hat{D}(\phi)$. In short-time GSE relations, $\hat{\eta}$ is identified here with η'_∞, and for $\hat{D}$ various short-time diffusion coefficients have been considered, including in particular the short-time self-diffusion coefficient $D_s(\phi) = \lim_{t\to 0} W(t)/t$, where $6\ W(t) = \left\langle [\mathbf{r}(t) - \mathbf{r}(0)]^2 \right\rangle_{\text{eq}}$ is the particle mean-squared displacement (MSD) in three dimensions. We denote $W(t)$ as the MSD for short. In long-time GSE relations, $\hat{\eta}$ is identified with the zero shear viscosity η_0, and for $\hat{D}$ the long-time self-diffusion coefficient $D_l(\phi) = \lim_{t\to\infty} W(t)/t$ is used among other long-time diffusion coefficients. Notice here that a valid GSE relation generalizes the single-particle Stokes–Einstein–Sutherland relation to finite concentrations.

A frequency-dependent GSE was proposed by Mason and Weitz [86] relating the Laplace transform, $\tilde{G}(s)$, of $G(t)$ to the Laplace transform, $\tilde{W}(s)$, of the MSD according to

$$\frac{\tilde{G}(s)}{\eta_m} \approx \frac{\mathcal{D}_0}{s^2\tilde{W}(s)}, \quad \tilde{G}(s) = \int_0^\infty dt\ e^{-st}\ G(t). \tag{2.150}$$

The frequency-dependent viscosities (moduli) follow from $\tilde{G}(s)$ by analytic continuation using $\tilde{G}(-i\omega) = \eta'(\omega) + i\eta''(\omega)$. The Mason–Weitz GSE is based on the assumption that $G(t)$ can be linearly expressed in terms of the microscopic friction kernel function associated with $W(t)$. Eq. (2.150) includes the short-time GSE between D_s and η'_∞, and the long-time GSE between D_l and η_0 as its high frequency and zero frequency limit, respectively.

The approximate validity of a GSE relation is very useful because linear viscoelasticity can be determined from a dynamic scattering measurement that probes diffusion, often over a higher frequency range that cannot be accessed by conventional mechanical rheometers. From a theoretical viewpoint, GSE relations are valid only approximately, if at all, since the physical mechanisms of viscosity and self diffusion are clearly distinguishable in the fluid phase, at least away from a glass or jamming transition point.

Short-time GSE relations have been thoroughly scrutinized in elaborate simulation and theory studies, for suspensions of permeable spheres and charge-stabilized particles, and particles with competing attractive and repulsive interactions describing low salinity protein solutions [60,65,67,68,87–89]. These studies show that the validity of a GSE relation is strongly dependent on the form and range of the interaction potential.

We consider here the following short-time GSE relations expressed in terms of the GSE functions Λ_s, Λ_{cge} and Λ_{KD} given by:

$$\Lambda_s(\phi) \equiv \frac{D_s(\phi)}{\mathcal{D}_0} \times \frac{\eta'_\infty(\phi)}{\eta_m}, \tag{2.151}$$

$$\Lambda_{cge}(\phi) \equiv \frac{D_{cge}(\phi)}{\mathcal{D}_0} \times \frac{\eta'_\infty(\phi)}{\eta_m}, \tag{2.152}$$

$$\Lambda_{\mathrm{KD}}(\phi) \equiv \frac{D_c(\phi)\sqrt{S(0)}}{\mathcal{D}_0} \times \frac{\eta'_\infty(\phi)}{\eta_m}. \tag{2.153}$$

If $\Lambda_s \approx 1$ is valid independently of ϕ, η'_∞ scales with the inverse of D_s. Likewise, if $\Lambda_{cge} \approx 1$ holds then η'_∞ scales with the inverse of the short-time cage diffusion coefficient $D_{cge} = \mathcal{D}_0 H(q_m)/S(q_m)$. The cage diffusion coefficient is equal to the ratio of hydrodynamic function $H(q)$ and static structure factor $S(q)$, both taken at the wavenumber q_m associated with the extension ($\sim 2\pi/q_m$) of the dynamic cage of nearest neighbors formed around each particle. The coefficient D_{cge} characterizes the slowest decay of sinusoidal concentration fluctuations occurring at the wavenumber q_m. The GSE relation involving the collective diffusion coefficient, $D_c = \mathcal{D}_0 H(q=0)/S(q=0)$ multiplied by the square root of the suspension osmotic compressibility factor $S(q=0)$, was proposed by Kholodenko and Douglas [90] and has been applied in particular to protein solutions. The positive valued hydrodynamic function $H(q)$ quantifies the influence of the hydrodynamic interactions on the colloidal short-time diffusion in a quiescent fluid. In the unrealistic case of vanishing hydrodynamic interactions in a concentrated suspension it would attain its constant infinite dilution value of one. Undulations in $H(q)$ are thus a hallmark of the influence of hydrodynamic interactions. The zero wavenumber value $H(0)$ is smaller than one for interacting colloidal particles and has the physical meaning of a sedimentation coefficient [69].

For colloidal hard spheres and excluded annulus particles, analytic expressions are available for $H(0)$ and $H(q_m)$, additionally to the generalized Saito formula for η'_∞ in Eq. (2.129). These expressions are in excellent agreement with simulation data for $\phi \le \phi_f = 0.494$. We quote them for hard spheres only [65],

$$\begin{aligned} H(q_m) &= 1 - \phi/\phi_{cp}, \\ H(0) &= 1 - 6.5464\phi\ \left(1 - 3.348\phi\ + 7.426\phi^2 - 10.034\ \phi^3 + 5.882\ \phi^4\right), \end{aligned} \tag{2.154}$$

where $\phi_{cp} = \pi/(3\sqrt{2}) \approx 0.74$ is the closest packing volume fraction for a FCC or HCP lattice. In combination with the hard sphere structure factor expressions [67]:

$$\begin{aligned} S(q_m) &= 1 + \phi_{\text{rcp}}\left[\phi g_{\text{eq}}(2a^+;\phi)\right], \quad \text{with} \ \ g_{\text{eq}}(2a^+;\phi) = \frac{1 - \phi/2}{(1-\phi)^3}, \\ S(0) &= \frac{(1-\phi)^4}{(1+2\phi)^2 + \phi^3(\phi - 4)}. \end{aligned} \tag{2.155}$$

By invoking the Carnahan–Starling expression for the pair distribution function at contact, $g_{\text{eq}}(2a^+;\phi)$, and the osmotic compressibility factor, $S(0)$, of colloidal hard spheres, $D_{cge}(\phi)$ and $D_c(\phi)$ are readily obtained. Here, $\phi_{rcp} = 0.644$ is the random close packing volume fraction, and $S(q_m, \phi_f) = 2.85$ holds in accord with the Hansen-Verlet freezing criterion [11]. The above expressions apply to the equilibrium fluid phase only, but not to the metastable fluid regime for which $\phi_f < \phi < \phi_{rcp}$.

The short-time GSE functions obtained from the above expressions are pictured in the left part of Figure 2.11. For $\phi > 0.4$, the self-diffusion GSE is violated by more than 30%. The KD-GSE for D_c is more accurate at smaller ϕ, but it deteriorates at larger concentrations. In contrast, the GSE for the short-time cage diffusion coefficient, D_{cge}, is quite accurate for all ϕ, with less than 7% deviation throughout. While the cage-diffusion GSE is the best choice for hard spheres, it was found in simulations that it is violated for charge-stabilized particles dispersions at lower-salinity conditions [67]. This reflects that different from η'_∞, D_{cge} is distinctly sensitive to the direct interaction range.

Analytic expressions for $\phi \le \phi_f$ are also available for the long-time self-diffusion coefficient, D_l, and the zero frequency viscosity, η_0, of Brownian hard spheres. As shown in Riest et al. [65] by the comparison with simulation and experimental data, D_l with HIs is well described by the short-time factorization expression:

$$\begin{aligned} \frac{D_l(\phi)}{\mathcal{D}_0} &= \frac{D_l^{wo}(\phi)}{\mathcal{D}_0} \times \frac{D_s(\phi)}{\mathcal{D}_0}, \\ \frac{D_l^{wo}(\phi)}{\mathcal{D}_0} &= 1 - 2\phi\ + 1.272\phi^2 - 1.951\ \phi^3, \\ \frac{D_s(\phi)}{\mathcal{D}_0} &= 1 - 1.8315\phi\ \left(1 + 0.12\phi - 0.70\phi^2\right), \end{aligned} \tag{2.156}$$

where D_s is the short-time self-diffusion coefficient with HIs, and D_l^{wo} is the long-time self-diffusion coefficient without HIs. Eq. (2.156) conforms to an argument that

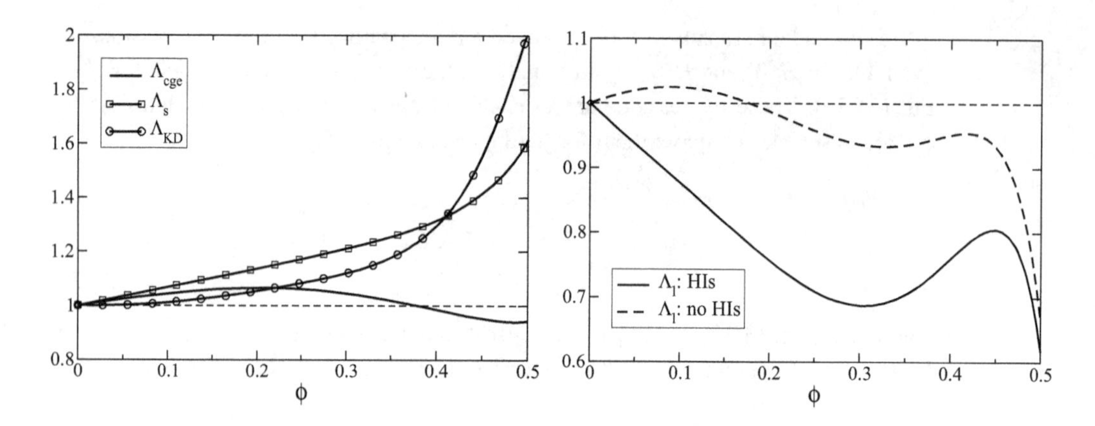

Figure 2.11 Test of GSE relations for a hard sphere suspension using the analytic expressions in Eqs. (2.154)–(2.158). Left: Short-time GSE functions as indicated. Right: Long-time self-diffusion GSE function $\Lambda_l(\phi)$ with and without HIs. A useful GSE relation is characterized by a GSE function close to one (dashed horizontal lines).

pre-averages the HIs as first put forward by Medina-Noyola [91], and subsequently elaborated and extended by Brady [92,93] for the zero frequency viscosity. The approximation is that, for hard spheres, a short-time (high frequency) transport coefficient can be factored out of the associated long-time coefficient with the remaining long-time factor being approximately independent of the HIs. Notice that $D_l(\phi_f)/D_s(\phi_f) = D_l^{wo}(\phi_f)/\mathcal{D}_0 \approx 0.1$, in accord with the Löwen–Palberg dynamic freezing criterion [94] related to the static Hansen–Verlet criterion [95], and $D_l(\phi) < D_l^{wo}(\phi)$. The hydrodynamically induced slowing of long-time self-diffusion described by Eq. (2.156) applies only to particle suspensions for which the near-distance part of the HIs is dominant. On the contrary, long-time self-diffusion is enhanced for particles with long-ranged soft repulsion such as in charge-stabilized colloids [69]. The self-diffusion enhancement triggered by the far-field part of the HIs was first theoretically predicted using the HIs-extended MCT.

The analytic factorization expression for the zero frequency low shear viscosity of hard spheres reads [65]:

$$\eta_0(\phi) = \eta'_\infty(\phi)\left[1 + \frac{1}{\Lambda_s(\phi)}\frac{(\Delta\eta)^{wo}(\phi)}{\eta_m}\right], \tag{2.157}$$

with the stress relaxation part without HIs, $(\Delta\eta)^{wo}$, given by:

$$\frac{(\Delta\eta)^{wo}(\phi)}{\eta_m} = \frac{12\phi^2}{5}\left(\frac{1 - 7.085\phi + 20.182\phi^2}{1 - \dfrac{\phi}{\phi_{rcp}}}\right). \tag{2.158}$$

The above expression for $(\Delta\eta)^{wo}$ combines the exact leading-order quadratic concentration dependence, i.e., Eq. (2.117), with a linear-order divergence at random close packing, and it is in good agreement with no-HIs simulation data for $\phi < \phi_f$. The term inside the bracket of Eq. (2.157) is only moderately dependent on HIs through the short-time self-diffusion GSE function $\Lambda_s(\phi)$. Results for η obtained from Eq. (2.157) are shown in the right part of Figure 2.9. While in good overall agreement with available hard sphere simulation data, the above factorization expressions for d_l and η are not exact because the exact low concentration limiting forms of both transport properties are not recovered.

Eqs. (2.156)–(2.158) allow us to scrutinize the long-time self-diffusion SE function for hard spheres,

$$\Lambda_l(\phi) \equiv \frac{D_l(\phi)}{\mathcal{D}_0} \times \frac{\eta_0(\phi)}{\eta_m}, \tag{2.159}$$

with the result shown in the right part of Figure 2.11. While the self-diffusion GSE without HIs is fulfilled within 7% accuracy up to $\phi \approx 0.45$, larger deviations in $\Lambda_l(\phi)$ appear for $\phi > 0.45$. The effect of the HIs is to enlarge the deviations of $\Lambda_l(\phi)$ to about 30% already at intermediate ϕ values. Note that $\Lambda_S(\phi) > 0$, i.e., η'_∞/η_m, is larger than $\mathcal{D}_0/D_s(\phi)$, while the opposite ordering $\Lambda_l(\phi) < 1$ is observed for the long-time GSE with HIs. Figure 2.11 does not include the long-time GSE functions related to D_{cge} and D_c since the associated GSE relations are more strongly violated for $\phi > 0.3$ than the self-diffusion GSE and are thus of no practical use. A conclusion from this theoretical and experimental work is that the validity of a specific GSE relation is dependent on the type and range of the interaction potential, and the proximity to a vitrification or gelation transition.

2.6 Mode Coupling Theory of Dense Suspension Flow

The Green–Kubo relation, Eq. (2.137), relates the viscosity of a suspension to the equilibrium dynamic shear modulus $\Delta G(t)$, i.e., the auto-correlation function of the fluctuating microscopic stresses. At high colloid concentrations and provided that the suspension does not crystallize, these fluctuations typically decorrelate very slowly because their relaxation requires structural rearrangements of the colloids that are strongly hindered due to steric repulsion between the colloids. Caging provides an intuitive physical picture to rationalize slow structural relaxation: each particle is transiently caged by its neighbors. Motion over a length scale of a particle diameter or more is only possible if these neighbors move, but they themselves are caged by their neighbors, as shown in Figure 1.4. A collective feedback mechanism emerges that slows down the motion. Ultimately, if the concentration is high enough, cages persist essentially indefinitely, and structural relaxation becomes ineffective. Consequently, the viscosity of the suspension diverges, at least in a mathematical idealization, and the system turns from a fluid into an amorphous solid, the glass, at a density given by $\phi = \phi_g$.

Even if particles are merely caged transiently, during this time the stress correlations can only relax partially, so that $\Delta G(t)$, for times large compared to τ_D, develops a pronounced plateau. This plateau is called the low frequency shear modulus or plateau modulus ΔG_∞, which is not to be confused with the high frequency elastic modulus G'_∞. Only once the local configuration of particles relaxes sufficiently, i.e., when the transient cages break, $\Delta G(t)$ decays to zero. The time scale where this occurs is referred to as the time scale of structural relaxation, $\tau_{\rm struc}$. This two-step decay of $\Delta G(t)$ explains linear viscoelasticity: Over time scales where particles are caged, the suspension responds as an elastic solid with a shear modulus ΔG_∞. The scale $\tau_{\rm struc}$ marks the time beyond which the suspension reacts as a viscous fluid. In essence, this picture linking structural relaxation to the viscosity was already suggested by Maxwell [96] in his famous treatise on the kinetic theory of gases, identifying $\tau_{\rm struc} \sim \tau_{\rm MW}$.

A very influential theory of the dynamics of dense suspensions and in particular their rheology is the Mode Coupling Theory of the glass transition (MCT) [10], developed originally in the 1980s [97,98]. It focuses on the cage effect as the main physical mechanism of structural relaxation. The theory was originally developed to describe the glass transition of low molecular weight fluids, but has been extended to colloidal suspensions neglecting HIs, based on the generalized Smoluchowski equation (GSmE) [99]. More recently, MCT has been extended to deal with the nonlinear rheology of colloidal suspensions, as we will outline next. MCT is an approximate theory, and a large body of work is devoted to clarifying those aspects of the glass transition that are not captured by MCT, and to putting into context the predictions of the theory in the wider framework of the statistical physics of disordered systems [100]. For typical experiments on colloidal suspensions, the accessible time scales are comparable to those where MCT is most accurate, and thus we will focus on that theory in the following. A mathematically closely related approach to the high density dynamics of colloidal suspensions is the self-consistent generalized Langevin equation (SCGLE) theory by Medina-Noyola and coworkers [15–17]. Note that both SCGLE and MCT result in a very similar mathematical structure describing the same salient features of the colloidal glass transition.

2.6.1 MCT Description of Linear Rheology

The starting point for a description of the dynamics is an equation of motion for the dynamic density-correlation function $S(\mathbf{q},t) = \left\langle \delta n_{\mathbf{q}}^* \exp\left[\Omega^\dagger t\right] \delta n_{\mathbf{q}} \right\rangle$, where $\delta n_{\mathbf{q}} = \sum_{k=1}^{N} \exp\left[i\mathbf{q}\cdot\mathbf{r}_k\right]/\sqrt{N}$ is the microscopic density fluctuation for nonzero wave vector $\mathbf{q}$ in an N-particle suspension of identical colloidal particles. Note that this is directly related to the Fourier transform of the real-space, time-dependent normalized pair distribution function $g(\mathbf{r},t)$ discussed in relation to Eq. (2.110). Its zero-time value $S(\mathbf{q},0) = S(q)$ is the static structure factor of the colloidal suspension [11]. Angular brackets denote a canonical average, and here and in the following we use the convention that operators inside such averages act to the right but not including the canonical distribution function. The operator $\Omega^\dagger$ that drives the time evolution of the microscopic fluctuations is the adjoint, with respect to the standard scalar product of the Smoluchowski differential operator Ω that appears in the GSmE, Eq. (2.133). To describe the linear-response rheology, one assumes that the correlation functions are unaffected by the weak shear flow used to drive the suspension and to measure its flow response. The operator Ω can then be replaced by the quiescent fluid equilibrium one, Ω_0, as given in Eq. (2.134). Assuming further that the suspension remains statistically homogeneous and isotropic, $S(q,t)$ depends on the wave vector $\mathbf{q}$ of the density fluctuations only through its magnitude $q = |\mathbf{q}|$.

A formal way to derive an exact equation of motion for such correlation functions is the projection-operator formalism by Zwanzig and Mori, as presented in Hansen and McDonald [11]. In essence, one picks a set of "relevant" variables in the Hilbert space of the fluctuations, say $\delta n_{\mathbf{q}}$, and splits the time evolution into parts that remain in the relevant subspace and those that lie in the orthogonal subspace. Some formal algebra in rewriting the time-evolution operator without HIs leads to [10,99]:

$$\tau_0(q)\partial_t S(q,t) + S(q)^{-1}S(q,t) + \int_0^t dt' M(q,t-t')\partial_{t'}S(q,t')\,dt' = 0, \tag{2.160}$$

where $\partial_t = \partial/\partial_t$ denotes the time derivative. Here, $\tau_0(q) = 1/q^2\mathcal{D}_0 = \tau_D/(aq)^2$ sets the time scale of the initial relaxation for the density fluctuations due to single-particle diffusion without HIs.

Eq. (2.160) results from the structure of a generalized Langevin equation for the density fluctuation dynamics quantified by $\exp\left[\Omega^{\dagger}t\right]\delta n_{\mathbf{q}}$, where the fluctuating force drops out in performing the canonical average to obtain an equation for the correlation function. The slow structural relaxation is encoded in the "friction memory kernel" $M(q,t)$ for which a formally exact expression can be given. It is an auto-correlation function of the fluctuating forces, evolving with a reduced dynamics driven by the one-particle irreducible Smoluchowski operator [101] projected onto the subspace of the irrelevant fluctuations. Starting from the exact Eq. (2.160), theories such as MCT and SCGLE proceed by approximations for the time-delayed friction. The central approximation of MCT consists of two steps: (i) the fluctuating stresses are assumed to be dominated by density-fluctuation pairs, which formalizes the cage-effect picture, and (ii) the reduced dynamics of the corresponding four-point correlation function of density-fluctuation pairs is assumed to factorize into a product of density-correlation functions involving the full dynamics. This approximation scheme is somewhat ad-hoc and can only be justified a posteriori. For colloidal suspensions, this can be understood formally as the result of a resummation of a class of scattering diagrams for density fluctuations [102]; however, a systematic discussion of the errors involved in the MCT approximation has still not been possible.

As a result, within MCT one approximates the memory kernel to be given by a bilinear form of the correlation functions, which provides a self-consistent closure to Eq. (2.160). Specifically,

$$M(q,t) \approx \frac{n}{2}\int \frac{d^3k}{(2\pi)^3} V_{\mathbf{qkp}}V_{\mathbf{qkp}}S(k,t)S(p,t), \tag{2.161}$$

restricted to $\mathbf{q} = \mathbf{k} + \mathbf{p}$ to ensure momentum conservation in the statistical average. Here, n is the average colloid number concentration. The vertices $V_{\mathbf{qkp}}$ are entirely given in terms of the static equilibrium structure functions of the fluid; this information is assumed to be known from liquid-state theory [11] and can for most applications be reduced to the static structure factor $S(q)$: recalling the definition of the direct correlation function, $c(q) = 1/n - 1/nS(q)$, one obtains, after a technical approximation concerning static triplet correlations:

$$V_{\mathbf{qkp}} = \frac{(\mathbf{q}\cdot\mathbf{k})}{q^2}c(k) + \frac{(\mathbf{q}\cdot\mathbf{p})}{q^2}c(p). \tag{2.162}$$

Through the direct correlation function, the relevant thermodynamic control parameters and the direct particle interactions enter the theory.

The MCT equations Eqs. (2.160) and (2.161) form a nonlinear set of integro-differential equations for $S(q,t)$. They can be solved numerically, but a number of features can be understood analytically. Due to their nonlinearity, the equations contain bifurcation points in the long-time behavior: the asymptotic value $F_q = \lim_{t\to\infty} S(q,t)$ is found to obey a set of nonlinear algebraic equations,

$$\left(S(q) - F_q\right)^{-1} = S(q)^{-1} + M_q[F], \tag{2.163}$$

where $M_q[F] = \lim_{t\to\infty} M(q,t)$ is, within MCT, a quadratic functional of the F_q. A detailed analysis of this equation [103,104] reveals that, among the possible solutions, $F_q = 0$ holds for sufficiently weak coupling coefficients $V_{\mathbf{qkp}}$, e.g., at low colloid concentration, while for strong coupling, e.g., at high concentration, a solution $F_q > 0$ emerges. The latter implies that density correlations do not relax fully, and these solutions characterize a nonergodic state of the system. Thus, $F(q)$ is called the nonergodicity parameter. The bifurcation points of Eq. (2.163), where nonzero real solutions first appear, hence indicate the (idealized) transition from a fluid ($F_q = 0$) to a glass ($F_q > 0$). The ordinary case described by MCT is a discontinuous transition driven by increasing colloid packing fraction ϕ, where right at the bifurcation point ϕ_c, the long-time limit jumps from zero to a finite value $F_q^c > 0$. These points are called ideal glass-transition points or dynamic transition points, and are identified with the colloidal glass transition point ϕ_g. Approaching such a point from the liquid side, the structural relaxation time τ_{struc} diverges. In principle, Eq. (2.163) may contain additional bifurcation points, where a non-zero $F_q^{(1)} > 0$ changes discontinuously to another solution $F_q^{(2)} > F_q^{(1)}$. Such points are termed glass–glass transitions, and they describe for example the transition between repulsion-dominated and attraction-dominated glasses in colloid-polymer mixtures, as will be discussed in Chapter 5.

To describe the linear rheology of a suspension, MCT is used in combination with an approximation to the Green–Kubo relation, Eq. (2.137), to express the dynamic shear modulus in terms of density correlation functions. One obtains $\Delta\eta = \int_0^\infty dt\,\Delta G(t)$ with:

$$\Delta G(t) \approx \frac{n^2 k_B T}{60\pi^2}\int_0^\infty dk\,k^4 (c'(k))^2 S(k,t)^2 \equiv (nk_B T)\int \frac{d^3k}{(2\pi)^3}\tilde{V}_{\mathbf{0k}} S(k,t)^2, \tag{2.164}$$

or, for the case of an m-component mixture of Brownian particles [77]:

$$\Delta G(t) \approx \frac{n^2 k_B T}{60\pi^2} \int_0^\infty dk\, k^4 \,\mathrm{Tr}\left\{ \left(\frac{d\tilde{\mathbf{c}}(k)}{dk} \cdot \mathbf{S}(k,t) \right)^2 \right\}, \tag{2.165}$$

together with a suitable extension of Eqs. (2.160) and (2.161) for the symmetric $m \times m$ matrix $S(k,t)$ of partial dynamic structure factors $S_{ij}(k,t)$ [105]. Here, the matrix elements are labeled by the colloidal species. Moreover, $\tilde{\mathbf{c}}(k)$ is the symmetric $m \times m$ matrix of partial direct correlation functions augmented by concentration factors, i.e., $\tilde{c}_{ij}(k) = \sqrt{x_i x_j} c_{ij}(k)$, where i and j label the species, $x_i = n_i/n$ is the molar fraction of species i of concentration n_i, and n is the total concentration of particles. It is worth noting that the MCT approximation preserves exactly the fact that $\Delta G(t)$ is a completely monotonic function in quiescent Brownian dynamics [77,103,104]; given the approximations made to the memory kernel, this is a nontrivial statement.

In the above expressions for $\Delta G(t)$, HIs are assumed to be negligible. Before turning to a description of the features of $\Delta G(t)$ close to the glass transition, where this assumption can be justified, we briefly discuss the inclusion of HIs, that can in principle be achieved through the disturbance tensor $\mathbf{C}_i(X)$. Combined with HIs-extended MCT equations for the dynamic matrix $\mathbf{S}(k,t)$ derived in Nägele et al. [106] and a suitable (long-distance) approximation of $\mathbf{C}_i(X)$ applicable in particular to charged particles, a closed set of integro-differential equations for $\Delta G(t)$ and $\mathbf{S}(k,t)$ is obtained, with the matrix of partial static structure factors, $\mathbf{S}(k) = \mathbf{S}(k, t = 0)$, constituting the external input. Owing to its complexity, a fully self-consistent solution of this HIs-extended MCT set has not been attempted so far. For moderately concentrated dispersions of charged colloidal particles, and solutions of globular proteins or electrolytes, the HIs-extended MCT expression for $\Delta G(t)$ is more simply evaluated by using as an additional approximation step the known short-time forms of $\mathbf{S}(q,t)$ as input, and by treating the HIs on the long-distance level. With these simplifications, the viscosity of strong binary electrolytes, e.g., was calculated to good accuracy even for elevated concentrations up to 2 Molar [80,81]. Owing to the smallness of salt ions and their accordingly strong Brownian motion (since $Pe \propto a^3$), non-Newtonian effects are physically irrelevant in electrolyte solutions.

For fluid-phase, monodisperse suspensions of hard spheres as well as for suspensions of charge-stabilized spherical particles, Nägele and collaborators made extensive comparisons of the fully self-consistent MCT predictions of equilibrium diffusion and linear viscoelastic properties with exact theory, simulation, and experiment [60,87,106]. While these MCT studies neglect HIs, ways of including HIs approximately have been explored, through an appropriate scaling of short-time properties akin to the one for η_0 in Eq. (2.157). Such a scaling approach works reasonably well for hard spheres. As an example, Figures 2.12(a) and (b) include a MCT analysis of the frequency-dependent Mason–Weitz GSE relation in Eq. (2.150) for colloidal hard spheres, and for salt-free aqueous suspensions of charge-stabilized colloidal spheres. In the latter case, the charged particles interact by an effective screened Coulomb potential [60]. For hard spheres, the Laplace transform $\tilde{G}(s)$ of $G(t)$, and the Laplace

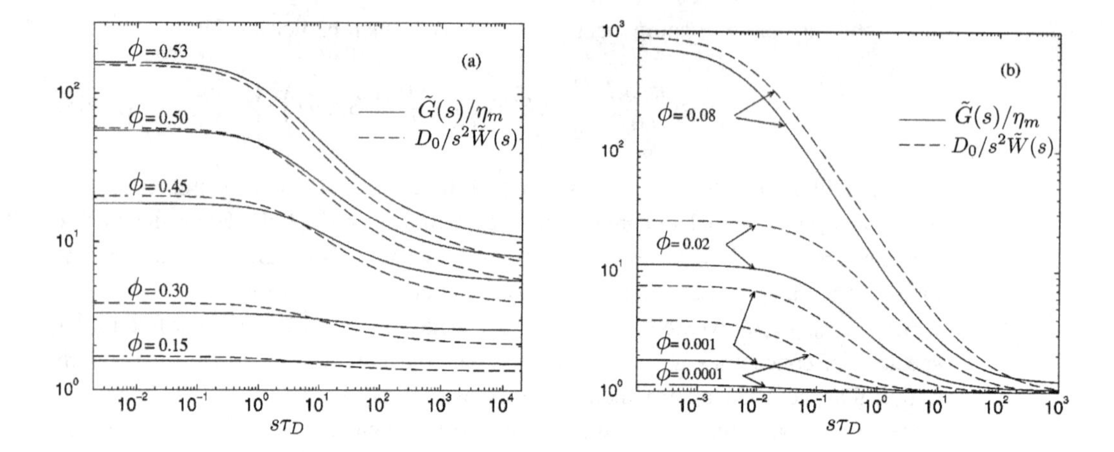

Figure 2.12 MCT based validity check of the frequency-dependent Mason-Weitz relation in Eq. (2.150), for (a) colloidal hard spheres at indicated volume fractions, and (b) deionized aqueous suspensions of charge-stabilized colloidal spheres of radius a = 50 nm and effective charge number Z = 500 in units of the elementary charge. The HIs factorization rescaling is applied to case (a) only. Frequency parameter s is scaled using τ_D. Adapted from Banchio et al. [60]

transform $\tilde{W}(s)$ of the MSD $W(t)$ with HIs included are obtained from the MCT results $\Delta G_{MCT}(t)$ and $W_{MCT}(t)$ without HIs using the factorization approximations $\Delta G(t) \approx \Delta G_{MCT}(t)(\eta'_\infty/\eta_m)$ and $W(t) \approx W_{MCT}(t)(D_s/\mathcal{D}_0)$, respectively. These approximations predict, in particular, a hydrodynamically induced slowing of long-time self-diffusion which is actually observed for hard spheres. However, they can not be used for charge-stabilized suspensions where long-time self-diffusion is enhanced by the HIs [69].

As has been shown [60,87], the HIs-rescaled MCT predicts values for the the zero frequency viscosity $\eta_0(\phi)$ of hard spheres in good agreement with experimental results. According to the HIs-rescaled MCT results shown in Figure 2.12(a), the Mason–Weitz GSE is approximately valid for hard spheres, with better accuracy at larger volume fractions and small reduced frequencies $s\tau_D \ll 1$ corresponding to the long-time regime. The deviations of the long-time GSE function $\Lambda_l(\phi)$ displayed in Figure 2.11 from the reference value one are comparable in magnitude, but overall somewhat larger than those by the HIs-rescaled MCT results for hard spheres. It should be borne in mind that MCT is not an exact method. In particular, it tends to overrate the slowing effect of dynamic particle caging [69].

According to Figure 2.12(b), MCT without HIs predicts a significant violation of the Mason–Weitz GSE for charge-stabilized suspensions at low salinity. This violation is expected to persist when HIs are accounted for, for the reason that HIs are weaker for charged particles which are well separated due to long-range electrostatic repulsion [69]. For large frequencies $s\tau_D \gg 1$ in the short-time range, this has been confirmed in experiments [107] and simulations [67] where D_s and η'_∞ have been determined independently.

The studies referred to above have established MCT as a predictive tool not only for exploring dynamic long-time arrest in concentrated systems as it is mainly used for, but also for equilibrium fluid-phase diffusion and linear viscoelastic transport properties. The Fourier-space based MCT does not reproduce the exact low density viscoelastic behavior without HIs, owing to its violation of the real-space zero relative flux condition in Eq. (2.112). Rather, its realm of application is for concentrated dispersions where dynamic caging is strong. In particular, MCT predicts the correct algebraic long-time tail exponents of $G(t)$ and $S(q,t)$ except for not reproducing the exact numerical prefactors [60,77]. Moreover, MCT recovers the Mountain–Zwanzig expression for G'_∞ in Eq. (2.141), however with $-\beta\mathcal{V}(r)$ approximated by the direct correlation function $c(r)$ [77].

2.6.2 Linear Rheology at the Glass Transition

The shear modulus, $\Delta G(t)$, in the MCT approximation Eq. (2.164), inherits the qualitative time dependence of the density correlation function $S(k,t)$. Close to ϕ_c, this time dependence can be specified analytically: as $\phi \to \phi_c$, there develops an increasingly large time window where the correlation functions remain close to the plateau value if right at the transition point, $S(q,t) \approx F_q^c$. A two-step decay emerges: relaxation to the plateau on a time scale comparable to τ_D is followed by slow structural relaxation from the plateau set by the nonergodicity parameter on an increasingly large time scale τ_{struc}. Thus, Eq. (2.164) describes linear viscoelasticity for $t \gg \tau_D$, but $t \ll \tau_{\text{struc}}$. The response of the system is linear elastic in the strain $\gamma = \int \dot{\gamma}\, dt$, i.e., $\Sigma^P_{xy} = \Delta G_\infty \gamma$ with a modulus given by

$$\Delta G_\infty \approx (nk_{\text{B}}T) \int \frac{d^3k}{(2\pi)^3} \tilde{V}_{\mathbf{0},\mathbf{k}} \left(F_k^c\right)^2. \tag{2.166}$$

This follows from Eq. (2.164) because in the relevant time window, the t-integral does not decay and is dominated by the constant $S(k,t) \approx F_k^c$. On the liquid side of the glass transition, the time window $t \gg \tau_{\text{struc}}$ is characterized by linear viscous response because the t-integral in Eq. (2.164) remains finite, since $\Delta G(t)$ ultimately decays to zero for $t \to \infty$. Note that the natural units of $\Delta G(t)$ and hence ΔG_∞ are $nk_{\text{B}}T$, or, for the case of hard spheres $k_{\text{B}}T/a^3$.

In the plateau regime, the appearance of $S(q,t) - F_q^c$ as a small parameter allows one to derive asymptotic laws for the density relaxation [10]. In essence, there appear two power laws, $S(q,t) - F_q^c \sim (t/t_0)^{-a}$ for $\tau_D \ll t \ll t_\sigma$, and, for $\phi < \phi_c$, $S(q,t) - F_q^c \sim -(t/t_\sigma)^b$ for $t_\sigma \ll t \ll t'_\sigma$, with two exponents, $0 < a < 1/2$ and $0 < b < 1$, that are nonuniversal and determined by the static structure functions. Here, t_0 is a microscopic time scale that is of $\mathcal{O}(\tau_D)$ and separates the short-time window of the relaxation from the generic power-law regime at long times. The time scales $t_\sigma \sim |\varepsilon|^{-1/(2a)}$ and $t'_\sigma \sim |\varepsilon|^{-1/(2a)-1/(2b)}$ both diverge as the distance to the glass transition point, defined as $\varepsilon = (\phi - \phi_c)/\phi_c$, approaches zero. In fact their ratio t'_σ/t_σ

also diverges, and this explains the increasingly large time window of intermediate elastic response in the fluid. The structural relaxation time scale τ_{struc} is controlled by t'_σ, and, as a result, MCT predicts that the shear viscosity of the suspension diverges according to

$$\Delta\eta \sim \left|\phi - \phi_g\right|^{-\gamma} \tag{2.167}$$

with $\gamma = 1/(2a) + 1/(2b) > 3/2$. Typical values for hard sphere-like suspensions are $a \approx 0.3$, $b \approx 0.58$, and thus $\gamma \approx 2.47$. This behavior has been verified experimentally, as shown in Chapter 1.

Experiments on colloidal hard sphere suspensions have identified the glass-transition point $\phi_g \approx 0.58$ [26,108]. Close to this point, the MCT predicts power law scaling with the expected relation between the exponents a and b. This finding supports the identification of ϕ_c with ϕ_g. In a first-principles calculation based on the Percus–Yevick approximation to the static structure factor, the value $\phi_c \approx 0.516$ is obtained from the theory [109], somewhat below the experimentally determined ϕ_g. It is customary to regard this as a numerical error of the theory, and to compare the predictions of MCT with data from experiment and simulation at the same reduced distance ε to the transition point. With this procedure, MCT is quantitatively accurate in a typical time window of colloidal experiments [110–112] and Brownian dynamics simulations [113,114].

From Eq. (2.167), MCT predicts the low shear viscosity to diverge at $\phi_g < \phi_{rcp}$, in contrast to some empirical formulas that suggest the viscosity to diverge only at the random-close packing density ϕ_{rcp}. Existing viscosity data support a divergence below ϕ_{rcp} that is in good agreement with the MCT prediction [115]. MCT neglects some relaxation processes that cause the viscosity to remain large but finite at least in some regime inside the ideal glass; these are well known from low molecular glass-forming fluids and probably are also present in colloidal suspensions [116,117]. Still the theory provides a large number of reliable detailed predictions for the dynamics of dense colloidal suspensions over relevant experimental time scales.

Figure 2.13 shows a typical result for the modulus $\Delta G(t)$ and the corresponding Fourier representation, i.e., the frequency-dependent storage and loss moduli $\Delta G'(\omega)$ and $\Delta G''(\omega)$. One identifies in $\Delta G(t)$ the two-step decay pattern with an intermediate plateau value ΔG_∞, and structural relaxation whose time scale τ_{struc} increases strongly as the density increases toward the glass transition. The two-step decay pattern causes a distinct low frequency maximum in $\Delta G''(\omega)$. The MCT power laws cause this structural-relaxation peak in $\Delta G''(\omega)$ to be significantly broader than a Debye peak (which would correspond to single-exponential relaxation), and the minimum between the peaks to be strongly enhanced in comparison to a mere superposition of two resonances. Data from typical hard sphere-like colloidal suspensions are in good agreement with the theory [118], up to a numerical prefactor regarding the absolute value of the elastic modulus. The MCT equations have been solved numerically for a number of reference models, such as the hard sphere model or the square-well model (to mimic colloid-polymer mixtures); for the latter, the theory predicts a rich glass-transition scenario that involves

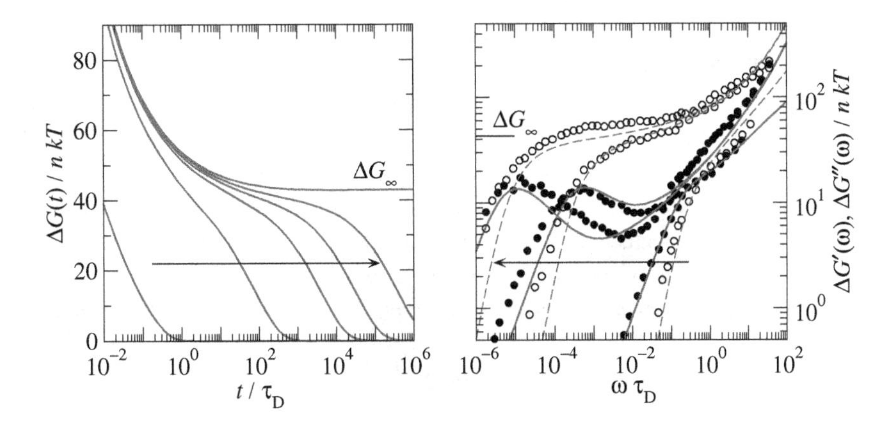

Figure 2.13 Left: Regular part of the dynamic shear modulus $\Delta G(t)$ in linear-response from MCT for a HS suspension with packing fractions approaching the glass transition (scaled by 2.2 to match experimental data). One notes the emergence of a viscoelastic plateau ΔG_∞ as the packing fraction increases. Right: Corresponding storage (dashed lines, open circles) and loss (solid lines, filled circles) moduli $\Delta G'(\omega)$ and $\Delta G''(\omega)$. Data points are for suspensions of thermosensitive core-shell microgel particles [118]. Arrows indicate the evolution with increasing density.

distinct repulsive and attractive glasses separated by glass–glass transitions that are examples of additional bifurcation points mentioned in connection with Eq. (2.163) [119,120]. This topic is covered in Chapter 5.

The self-consistent generalized Langevin equation theory SCGLE provides a similar description of the colloidal glass transition. It proceeds from Eq. (2.160) by a related, but somewhat simpler approximation for the friction memory kernel. The main difference compared to MCT is that SCGLE treats both the collective and the tagged-particle density fluctuations on equal footing (by what is referred to as a generalized Vineyard approximation). For typical hard sphere-like suspensions and below hydrodynamic length scales, this is not an essential difference. In a similar self-consistent closure of Eq. (2.160), SCGLE again arrives at a bifurcation equation akin to Eq. (2.163), conventionally formulated as a self-consistent equation for the particle-localization length instead of the nonergodicity parameters [15]. Hence, the predictions of SCGLE and MCT regarding the glass-transition diagrams and the dynamics close to it usually are in qualitative agreement. By extending the generic framework of Onsager by connecting it with SCGLE theory [121,122], it has been possible to extend that theory to include nonequilibrium "aging" effects that arise when a thermodynamic control parameter suddenly changes [123]. This addresses in a quantitative way a relevant feature of experiments close to the glass transition: since the time τ_{struc} for density fluctuations to relax becomes increasingly large, equilibration of an initially prepared sample becomes exceedingly slow. The slow evolution of the system toward thermal equilibrium causes quantities such as the mechanical moduli to depend on the waiting time t_w since the sample was prepared, and in principle to depend on the whole history of preparation as long as $t_w \lesssim \tau_{\text{struc}}$. This slow evolution is referred to as aging.

Simplified versions of MCT have also been proposed, notably the so-called naive MCT by Schweizer and coworkers [12–14], allowing to attach the mathematical description of the glass transition that is suggested by MCT to more standard notions of condensed-matter physics. In naive MCT, one constructs a free-energy function of the cage localization length, such that the stable minimum is given by a bifurcation equation similar to Eq. (2.163) but based on the tagged-particle dynamics following approximations that are similar in spirit to those made in the SCGLE approach. For the shear modulus of the glass, one gets an expression similar to Eq. (2.166), where the glass form factor F_k in Eq. (2.166) is replaced by a modulated Gaussian density profile,

$$\Delta G_\infty \approx (nk_\mathrm{B}T)\int \frac{d^3k}{(2\pi)^3}\,\tilde{V}_{\mathbf{0},\mathbf{k}}e^{-k^2/(2\alpha S(k))}, \tag{2.168}$$

where α is a "local order parameter" that is related to the cage-localization length, $r_\mathrm{loc}^2 = 3/2\alpha$ [12–14]. Under these simplifications, a scaling law for $\Delta G'$ has been derived [124],

$$\frac{\Delta G'_\infty a^3}{k_\mathrm{B}T} = 0.29\frac{\phi a^2}{r_\mathrm{loc}^2}. \tag{2.169}$$

This has been discussed in chapter 6 of *CSR* (see Eq. (6.30)) and will be used in Chapter 5.

In Section 2.5.5 we discussed the validity of GSE relations for equilibrium fluid-phase systems. The validity of the Mason–Weitz GSE relation near the ideal glass transition was explored by Fuchs and Mayr [125] using MCT. Since the theory predicts that the dynamics close to the glass transition is dominated by slow structural relaxation that couples all particle-transport mechanisms strongly, both the dynamic shear modulus and the single-particle MSD are governed by the same asymptotic behavior close to ϕ_g. This is the basis for the Mason–Weitz GSE in glass-forming suspensions, because it relates the friction kernel function associated with $W(t)$ to the frequency-dependent viscosities, as in Eq. (2.150). Coincidentally, even the pre-factor was found in MCT to be close to the one expected from the GSE, for a hard sphere model. This complies with the fluid-phase MCT study in Banchio et al. [60] according to which the Mason–Weitz GSE holds decently well for hard spheres, but is violated for charge-stabilized particles. While MCT predicts the approximate validity of the Mason–Weitz GSE relation and its zero frequency form in Eq. (2.159) close to a glass transition point, there are conflicting Molecular Dynamics simulation results showing a strong violation of the GSE relation [126]. Experimental hard sphere data for $D_l(\phi)$ and $\eta(\phi)$ for volume fractions $0.52 < \phi < 0.58$ in the metastable isotropic phase region are also indicative of the breakdown of the Mason–Weitz GSE at low frequencies [127]. The violations of the GSE relation close to the glass transition are commonly attributed to "dynamic heterogeneities", such that there is no longer homogeneity in structure and dynamics throughout the suspension [100]. Extensions of MCT including fluctuations of the dynamics around the averaged correlation functions would be required to describe such effects in the framework of the theory [128].

2.6.3 Integration through Transients Approach to Nonlinear Rheology

If one considers states close to the glass transition, the structural relaxation time τ_{struc} becomes increasingly large. It therefore interferes with the flow-induced time scale $\tau_{\dot{\gamma}}$ already for relatively slow flow: even if the shear flow is slow compared to the Brownian relaxation time, i.e. $Pe = \tau_D/\tau_{\dot{\gamma}} \ll 1$, it may be fast compared to structural relaxation. This is quantified by the "dressed Péclet number" or Weissenberg number $Wi = \tau_{\text{struc}}/\tau_{\dot{\gamma}}$. Close to the glass transition, $\tau_{\text{struc}} \gg \tau_D$, a window of nonlinear glassy rheology opens [20] that is characterized by $Wi \gg 1$ but $Pe \ll 1$. Here, the static structure of the system is still only weakly distorted by the flow but the breaking of cages due to flow advection becomes a significant factor and nonlinear rheology effects such as shear thinning emerge, even in the absence of shear-induced ordering. This is shown schematically in Figure 2.14. To describe such effects, the MCT formalism presented in this section needs to be extended.

Formally, density correlation functions decorrelate not only due to caged Brownian motion as encoded in the Mori–Zwanzig equations for linear viscoelasticity, but also due to flow advection: a density fluctuation $\delta n_{\mathbf{q}}$ at time t originates from an earlier one $\delta n_{\mathbf{q}_0(t)}$ at time 0, where $\mathbf{q}_{t'}(t) = \mathbf{q} \cdot \mathbf{F}_{tt'}$ is the affine deformation acting due to flow. Here, the deformation gradient tensor $\mathbf{F}_{tt'}$ describes the accumulated affine distortion due to flow between times t' and $t \geq t'$. It is known from the finite-strain theory of continuum mechanics and is, in homogeneous flow, related to the velocity-gradient tensor by $\partial_t \mathbf{F}_{tt'} = \mathbf{\Gamma}(t) \cdot \mathbf{F}_{tt'}$.

If $Wi \gg 1$, the linear-response assumption made in deriving the Green–Kubo relation for the viscosity, Eq. (2.137), is no longer valid. One route to proceed is based on the so-called Integration Through Transients (ITT) formalism that was developed for colloidal rheology by Fuchs and Cates [129]. One splits the Smoluchowski operator into its equilibrium part, and the perturbation due to shear flow $\delta\Omega(X,t) = -\sum_{i=1}^{N} \nabla_i \cdot [\mathbf{\Gamma}(t) \cdot \mathbf{r}_i + \mathbf{C}_i(t) : \mathbf{\Gamma}(t)]$. A tractable theory is obtained if one drops the second term involving the disturbance tensor $\mathbf{C}_i$, with the caution that it

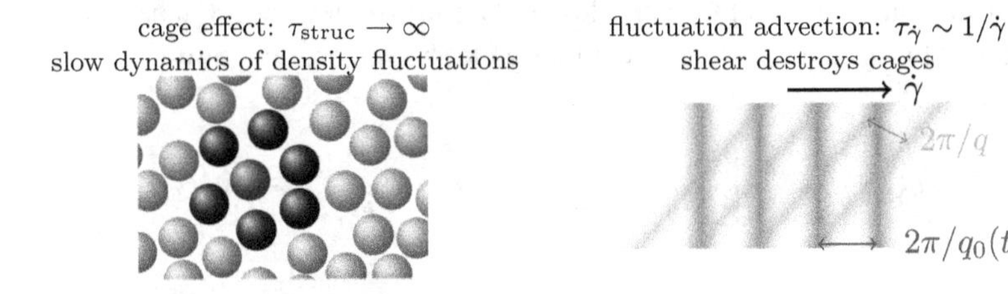

Figure 2.14 Schematic representation of the main effects described by MCT for the nonlinear rheology close to the colloidal glass transition: The cage effect (left) causes particle motion to collectively slow down due to excluded volume constraints imposed by the neighboring particles; it leads to dynamic arrest of structural relaxation ($\tau_{\text{struc}} \to \infty$). Shear advection (right) smears out density fluctuations by increasing their wave number from $q_0(t)$ to the advected wavenumber q (see text), and it imposes a relaxation time $\tau_{\dot{\gamma}}$ proportional to the inverse shear rate.

may be significant at high concentrations. We will also assume homogeneous incompressible flow throughout, i.e., $\mathrm{Tr}\,\boldsymbol{\Gamma} = 0$.

The solution of the generalized Smoluchowski equation, $\partial_t P(X,t) = \Omega(X,t)P(X,t)$, can be written formally as:

$$P(X,t) = P(X,t_0) + \int_{t_0}^{t} dt' \exp_+\left[\int_{t'}^{t} \Omega(X,s)\,ds\right]\Omega(X,t')P(X,t_0), \tag{2.170}$$

which is easily checked by differentiation. Here, the (positively) time-ordered exponential $U_{tt'} = \exp_+\left[\int_{t'}^{t} \Omega(s)\,ds\right]$ is the operator that satisfies $\partial_t U_{tt'} = \Omega(t)U_{tt'}$ with $U_{tt} = 1$. It propagates forward in time and lets operators Ω corresponding to smaller times act first; it appears because in nonstationary flow, $\Omega(X,t)$ might not commute with itself at different times. Equation (2.170) is the central starting point of the ITT formalism. One now assumes that for $t_0 \to -\infty$, the system starts in quiescent (stress-free) equilibrium. The nonequilibrium stress tensor then can be written as an ITT integral, recalling that $\Sigma^p = \langle\boldsymbol{\sigma}\rangle(t) = \int dX\,P(X,t)\boldsymbol{\sigma}(X)$, where $\boldsymbol{\sigma} = -(1/V)\sum_{k=1}^{N} \mathbf{r}_k \mathbf{F}_k^I$ is the microscopic stress tensor without hydrodynamic interactions, Eq. (2.138), that enters the particle stress in Eq. (2.94). From Eq. (2.170) one obtains a generalized Green–Kubo relation,

$$\Sigma^p(t) = \int_{-\infty}^{t} dt'\,\beta V\left\langle \boldsymbol{\Gamma}(t') : \boldsymbol{\sigma}\,\exp_-\left[\int_{t'}^{t} \Omega^\dagger(X,s)\,ds\right]\boldsymbol{\sigma}\right\rangle_{\mathrm{eq}}, \tag{2.171}$$

as under our assumptions one evaluates $\Omega(X,t)P_{\mathrm{eq}}(X) = \delta\Omega(X,t)P_{\mathrm{eq}}(X) = \beta\mathcal{V}P_{\mathrm{eq}}(X)\boldsymbol{\Gamma}(t) : \boldsymbol{\sigma}(X)$. The correlation function that appears in Eq. (2.171) is formally identical to the one that features in the linear Green–Kubo relation, except that the time evolution is driven by the full nonequilibrium dynamics through the reverse-time-ordered exponential operator. In particular, the statistical average is still performed over the quiescent equilibrium ensemble, i.e., the canonical distribution. Correlation functions of this type are called transient correlation functions.

In the same spirit, MCT can be extended to account for nonlinear effects arising from shear advection. This has been performed by Fuchs and Cates and coworkers, first for stationary simple shear flow [129,130] (reviewed in Fuchs [19]) and later for arbitrarily time-dependent incompressible homogeneous flow [18,131–134]. One obtains:

$$\Sigma^p/(nk_BT) = -n\int_{-\infty}^{t} dt' \int \frac{d^3k}{32\pi^3}\left[\mathbf{k}\cdot\partial_{t'}\mathbf{B}_{tt'}\cdot\mathbf{k}\right]\frac{\mathbf{kk}}{kk_{t'}(t)}c'(k)c'(k_{t'}(t))\Phi_{\mathbf{k}_{t'}(t)}(t,t')^2. \tag{2.172}$$

Here, $\Phi_{\mathbf{k}_{t'}(t)}(t,t') = \left\langle \delta n^*_{\mathbf{k}_{t'}(t)} \exp_-\left[\int_{t'}^{t} \Omega^\dagger(s)\,ds\right]\delta n_{\mathbf{k}}\right\rangle_{\mathrm{eq}}$ is the transient density correlator, which reduces to the equilibrium correlation function $S(k,t)$ in the absence of flow. In Eq. (2.172), the flow-induced distortion appears in terms of the Finger tensor, or left Cauchy–Green tensor, known from continuum mechanics, $\mathbf{B}_{tt'} = \mathbf{F}_{tt'}\cdot\mathbf{F}_{tt'}^T$; it is the rotation-invariant measure of deformation in Euler coordinates. In simple shear flow, the expression $-\partial_{t'}\mathbf{B}_{tt'}$ reduces to the scalar shear rate $\dot\gamma(t')$ for the corresponding off-diagonal components of the stress tensor. Similarly, the x-y component of the expression in Eq. (2.172) reduces to the time-integral in Eq. (2.164) in linear response.

For a description of the nonlinear rheology of dense colloidal suspensions, neglecting HIs, the MCT equation Eq. (2.160) is extended and can be written using the reverse-advected wave vector $\bar{\mathbf{q}}_{t'}(t) = \mathbf{q} \cdot \mathbf{F}_{tt'}^{-1}$ as:

$$\tau_{\bar{\mathbf{q}}_{t'}}(t)^0 \partial_t \Phi_{\mathbf{q}}(t,t') + S(\bar{\mathbf{q}}_{t'}(t))^{-1} \Phi_{\mathbf{q}}(t,t') + \int_{t'}^{t} dt'' M_{\mathbf{q}}(t,t'',t') \partial_{t''} \Phi_{\mathbf{q}}(t'',t') = 0; \tag{2.173}$$

additional time dependencies appear because no assumptions have been made about stationarity of the dynamics. In stationary flow, $\Phi_{\mathbf{q}}(t,t')$ reduces to the time-translational invariant form $\Phi_{\mathbf{q}}(t-t')$. The memory kernel is:

$$M_{\mathbf{q}}(t,t'',t') = \frac{\varrho}{2} \int \frac{d^3k}{(2\pi)^3} V_{\bar{\mathbf{q}}_{t'}(t)\bar{\mathbf{k}}_{t'}(t)\bar{\mathbf{p}}_{t'}(t)} V_{\bar{\mathbf{q}}_{t'}(t'')\bar{\mathbf{k}}_{t'}(t'')\bar{\mathbf{p}}_{t'}(t'')} \Phi_{\bar{\mathbf{k}}_{t'}(t'')}(t,t'') \Phi_{\bar{\mathbf{p}}_{t'}(t'')}(t,t''), \tag{2.174}$$

which contains a suppression of the coupling vertices due to flow advection. Specifically, the term $V^2_{\mathbf{q},\mathbf{k},\mathbf{p}}$ that was present in the linear version of the theory decorrelates due to advection to $V_{\bar{\mathbf{q}}_{t''}(t),\bar{\mathbf{k}}_{t''}(t),\bar{\mathbf{p}}_{t''}(t)} V_{\mathbf{q},\mathbf{k},\mathbf{p}}$. In effect, advection causes the dynamics to lose memory, and thus to speed up structural relaxation. Confocal-imaging experiments of sheared colloidal suspensions directly demonstrate this speed-up of the dynamics [135]. As a consequence, the viscoelastic contribution to the stress tensor in the generalized Green–Kubo relation is strongly reduced by the flow: the system shows shear thinning. In a glass under steady shear, density correlations always decay because flow-induced cage breaking remains possible. Due to the coupling of density fluctuations in all spatial directions that is encoded in Eq. (2.174), shear fluidizes the system in all directions, not just trivially in the flow direction. This effect is called shear melting of the glass. The fluidization also in directions perpendicular to the applied flow has been directly verified in experiment [136].

A qualitative analysis of the solutions of the ITT-MCT equations is possible using further ad-hoc simplifications of the theory. The time-dependent nonstationary stresses have been investigated using a so-called schematic model. In this model one sets: [134]

$$\Sigma^p = v_\sigma \int_{-\infty}^{t} dt' [-\partial_{t'} B_{tt'}] \phi^{(\dot{\gamma})}(t,t')^2, \tag{2.175}$$

with some parameter v_σ to provide the scale for the stresses, and introduces a scalar correlation function $\phi^{(\dot{\gamma})}(t,t')$ obeying:

$$\tau_0 \partial_t \phi^{(\dot{\gamma})}(t,t') + \phi^{(\dot{\gamma})}(t,t') + \int_{t'}^{t} m^{(\dot{\gamma})}(t,t'',t') \partial_{t''} \phi^{(\dot{\gamma})}(t'',t') dt' = 0, \tag{2.176}$$

which is a simplified ad-hoc version of the extended MCT Eq. (2.173).

The decorrelation effect of shear advection in the memory kernel is empirically modeled by setting $m^{(\dot{\gamma})}(t,t'',t') = h(t,t'')h(t,t')m(t,t'')$, where $m(t,t'') = v_1\phi^{(\dot{\gamma})}(t,t'') + v_2\phi^{(\dot{\gamma})}(t,t'')^2$ is an empirical form that allows to reproduce the quiescent glass transition qualitatively. In this latter case, time-translational invariance causes the solutions of Eq. (2.176) to only depend on a single time, allowing to set $t' = 0$. The model then reduces to the so-called F_{12} model of MCT [10]. Its solution $\phi(t)$ reproduces for sufficiently large (v_1, v_2) a finite nonergodicity parameter, $\lim_{t\to\infty}\phi(t) = f > 0$, with a critical value f^c at the bifurcation points (v_1^c, v_2^c) that are the glass transition points of the model. To mimic flow-induced decorrelation of density fluctuations, the function $h(t,t')$ is chosen ad hoc as a decaying function of the accumulated strain, in analogy to the damping function introduced in some empirical constitutive equations of nonlinear rheology. Eq. (2.175) is superficially similar in form to a class of empirical rheological models, called K-BKZ constitutive equations. However, in nonstationary flow the correlation functions depend on two separate times, and this double time dependence causes Eq. (2.175) to be much richer in behavior than is known from standard empirical models [137]. A main advantage of the schematic model of ITT-MCT is that it can be numerically solved for a number of nonstationary flow conditions, such as large amplitude oscillatory shear (LAOS) [131], where it was found to be in excellent agreement with simulations and experiments. Similarly, the tensorial structure of the ITT-MCT equations allows to address nonlinear superposition rheology [138], where a stationary base flow is combined with orthogonal oscillatory flow in order to probe the frequency-dependent response around the nonequilibrium stationary sheared state.

Shear advection introduces $\tau_{\dot{\gamma}}$ as a relevant time scale for stationary flow. For cases where $Wi \gg 1$, the transient density correlation functions then decay as $\phi^{(\dot{\gamma})}(t) \approx f^c \exp\left[-t/t_{\dot{\gamma}}\right]$ where $t_{\dot{\gamma}} = |\dot{\gamma}|/\gamma_c \propto \tau_{\dot{\gamma}}$ and γ_c marks the strain scale that is required to break cages [130]. One finds $\gamma_c \approx 0.1$ for the yield strain; this is intuitively expected on the basis that the average localization length of a particle in a cage is about 10% of its diameter, which is generally referred to as the Lindemann criterion [139], so that a 10% distortion of cages due to flow is sufficient to break them.

For $Wi \gg 1 \gg Pe$, the only relevant time scale for the decay of correlation functions is proportional to the inverse of the shear rate. Thus, the correlation function entering the generalized Green–Kubo relation obeys a scaling function, $f(|\dot{\gamma}|t)$, in stationary flow. Therefore, Eq. (2.175) entails that even as $\dot{\gamma} \to 0$, the stress remains a nonzero constant, $\Sigma^p \sim \int_0^\infty \dot{\gamma}dtf(|\dot{\gamma}|t) = \text{const.}$ in the glass (where $Wi \gg 1$ is always fulfilled due to letting $\tau_{\text{struc}} \to \infty$ first); a yield stress appears. The limit $\dot{\gamma} \to 0$ is thus, in the ITT-MCT idealization, singular: the quiescent glass approached by $\lim_{\tau_{\text{struc}}\to\infty} \lim_{\dot{\gamma}\to 0}$ does not contain stresses, but the arrest of flow inside the glass by $\lim_{\dot{\gamma}\to 0} \lim_{\tau_{\text{struc}}\to\infty}$ produces remnant stresses. This singularity is also the basis of the ITT-MCT prediction of residual stresses in the glass, i.e., stresses that remain indefinitely after the cessation of the shear molten flow of a glass [140]. For the case of simple shear, the dynamic yield stress in the schematic model is

$\Sigma_y = \Sigma^p_{xy}(\dot\gamma \to 0) \approx \Delta G_\infty \gamma_c$. This is in good agreement with experimental observation: the dynamic yield stress is about $1/10$ of the plateau modulus. The appearence of a dynamic yield stress and a regime where density correlation functions decay on a time scale $t_{\dot\gamma} \sim \tau_{\dot\gamma}$ explains a mechanism of shear thinning in which laning of particles and the formation of inhomogeneous density and flow profiles are still absent. Asymptotically, $\eta \sim 1/\dot\gamma$, although pre-asymptotic effects often cause $\eta(\dot\gamma)$ to exhibit an effective power law with a shear thinning exponent less than 1 [135].

A defining feature of the MCT scenario for nonlinear rheology is the discontinuous emergence of a yield stress: while just below the quiescent glass transition, no stress remains for $\dot\gamma \to 0$, right at the transition, a nonzero value Σ_y is obtained. This is in contrast to other types of nonlinear rheology close to dynamically arrested states, where a yield stress may emerge continuously across the quiescent fluid–solid transition. Materials showing the latter behavior include disordered foams and emulsions and are termed soft glassy materials, after an influential empirical model, the soft-glassy rheology (SGR) model [141]. Within SGR, the effect of a deformation is expressed through a free-energy-like functional whose local minimum defines the state of the material, and whose shape gets tilted by the applied stress. As a result, stress facilitates relaxation by lowering the thermal-energy barrier for transitions between local minima of a rough free-energy landscape. It is not straightforward to compare this approach to MCT, because in the formulation of ITT-MCT, the concept of a rough free-energy landscape does not enter in an obvious way. Within naive MCT, the notion that applied stress changes the free-energy barriers could be incorporated, but the interpretation of the results is quite different from that in standard MCT [14].

Figure 2.15 shows typical flow curves obtained for hard sphere-like colloidal suspensions [118] and the corresponding MCT analysis. The yield stress that emerges

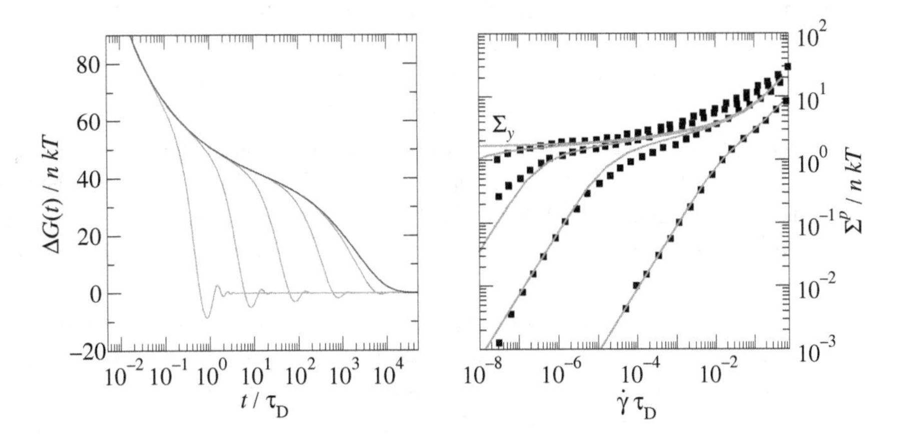

Figure 2.15 Left: Nonlinear transient dynamic shear modulus obtained from MCT at a fixed packing fraction ϕ close to and below the ideal glass transition (corresponding to Figure 2.13), with increasing shear rate $\dot\gamma$ (right to left). Right: Flow curves resulting from MCT for various different ϕ below and at the glass transition (top to bottom); symbols are experimental data for the suspensions as in Figure 2.13 [118].

for $\dot{\gamma} \to 0$ in the glass is clearly identified; to the regime where $\Sigma^p(\dot{\gamma}) \approx$ const., there corresponds the regime where $\Delta G(t)$ decays as a function of $|\dot{\gamma}|t$. The data thus support the scenario of a discontinuous emergence of the yield stress. $\Delta G(t)$ displays some typical undershoots in the final relaxation to zero: these are a signature of the nonequilibrium nature of cage breaking. Further, this can be linked to a phenomenon observed under start-up shear, the so-called stress overshoot , because the undershoot in $\Delta G(t)$ causes $\sigma(t)$ to increase nonmonotonically when starting shear at $t = 0$ [142].

Further applications of the ITT formalism allow us to provide MCT-based expressions for the flow-distorted static structure functions $S(\mathbf{q};\dot{\gamma})$ and $g(\mathbf{r};\dot{\gamma})$ [143,144]. Also, the theory can be extended to describe tagged-particle correlation functions under shear, and hence the effect of flow on the mean-squared displacement of a single particle in a dense sheared suspension. This description extends the discussion of Taylor dispersion to dense suspensions [145], showing that the shear-enhanced superlinear asympotic growth of the mean-squared displacement at long times, as discussed in Section 2.4.1, still holds at high colloid densities.

ITT-MCT has the advantage that it still only requires the quiescent equilibrium static structure functions of the suspension as input. It is not the only way to extend MCT to nonlinear rheology. Miyazaki and Reichman proposed an extension of MCT based on ideas of fluctuating hydrodynamics [146,147] and starting from the driven stationary state. They thus require in principle the stationary-state static structure factor $S(\mathbf{q};\dot{\gamma})$ as input, which is generally unknown, but for the purpose of MCT is close enough to the quiescent one to warrant plausible approximations. The main qualitative result of faster relaxation of density fluctuations as the result of flow-induced cage breaking remains the same. An issue arises because in the driven system, the fluctuation-dissipation theorem is no longer satisfied, and hence one needs to distinguish response and correlation functions. The link from the nonequilibrium density correlation functions to the transport coefficients of rheological interest is therefore not straightforward. The ITT-based approach circumvents this issue; there the violation of FDT emerges as a separate result [148,149]. Further differences arise in the treatment of nonstationary shear flow, most notably large amplitude oscillatory shear (LAOS). The ITT-MCT approach arrives at a rather intricate dependence of the memory kernel on three time points (cf. Eq. (2.174)), but this is needed in order to capture the physics of higher-harmonic generation and temporary yielding in LAOS [131] or other time-dependent rheological protocols. Notably in strain-driven oscillatory shear, ITT-MCT predicts that the regime of linear response vanishes at the glass transition [150].

The tensorial structure of Eq. (2.171) incorporates typical non-Newtonian flow effects such as normal-stress differences, $N_1 \neq 0$ and $N_2 \neq 0$, see Eq. (2.109), and how these depend on the interactions of the colloidal particles. A well-known effect from polymer melts is that of rod-climbing or Weissenberg effect, whereupon the fluid will "climb up" a rotating rod quite contrary to what is expected from centrifugal forces in Newtonian fluids. It can be shown that the fluid surface becomes unstable to rod climbing when a specific combination of normal-stress coefficients becomes positive, $\Psi_1 + 4\Psi_2 > 0$; this condition is for example fulfilled for typical dense colloidal suspensions according to ITT-MCT [151].

2.7 Summary and Outlook

We have discussed two theoretical methods for describing the rheology of colloidal suspensions. The method based on the GSmE focuses on the many-particles pdf, and microscopic hydrodynamic diffusion tensors specifying the HIs. The Langevin equation method considers instead stochastic trajectories of individual particles subjected to direct, hydrodynamic, and solvent-induced fluctuating forces. While formally exact, these equations can only be solved approximately for concentrated suspensions, owing to the complicated many-particles interactions which as far as the HIs part is concerned are, in general, long-range and nonpairwise additive. The Langevin equation is fundamental to simulation schemes, as discussed in Chapter 3. On the other hand, the GSmE is very useful for theoretical explorations, and as a starting point for developing approximate methods. In the framework of the GSmE, rheological properties can be described either in real or Fourier space. In the real space description, reduced two- and three-particles distribution functions of a sheared suspension derived from the GSmE are of key importance. A useful starting point in Fourier space for introducing approximative rheological schemes are linear and nonlinear Green–Kubo relations for the shear stress relaxation function. MCT is such a scheme that is particularly useful to examine concentrated systems near to a glass or gel transition. It enables systematically exploring the effect of the microstructure on the macroscopic rheology of suspensions and has been successfully generalized to nonequilibrium states for predicting shear melting and flow of glasses.

The discussion of rheology in this chapter was centered around suspensions of rigid, spherical particles. Generalizations to more complex systems are still a challenge owing to the more complicated HIs, orientation-dependent direct interactions, and a consequentially strong coupling of translational and rotational motions affecting the rheology. There are extensions of MCT to nonspherical particles [152–156], and to active particles [157–160], so far without the inclusion of HIs. Future exploration of such challenges are warranted, where focusing on reduced distribution function solutions derived from the GSmE with HIs, and on MCT and SCGLE approaches with HIs are promising. It will be particularly interesting to explore the influence of HIs on the glass transition and gelation.

References

1. Dhont JKG. *An Introduction to Dynamics of Colloids*, 2nd ed. Amsterdam: Elsevier; 2003.
2. McQuarrie DA. *Statistical Mechanics*. New York: Harper & Row; 1976.
3. Graham MD. *Microhydrodynamics, Brownian Motion, and Complex Fluids*. Cambridge: Cambridge University Press; 2018.
4. van Kampen NG. *Stochastic Processes in Physics and Chemistry*, 3rd ed. Amsterdam: North Holland; 2007.
5. Gardiner CW. *Handbook of Stochastic Methods*, 3rd ed. Berlin: Springer; 2004.

6. Risken H. *The Fokker-Planck Equation*, 2nd ed. Berlin: Springer; 1996.
7. Nägele G. Brownian Dynamics simulations. In Blügel S, Gompper G, Koch E, Müller-Krumbhaar H, Spatschek R, Winkler RG (eds.) *Computational Condensed Matter Physics*, vol. 32 of *Schriften des Forschungszentrums Jülich*. Jülich, Germany: Forschungszentrum Jülich; 2006; pp. B4.1–B4.34.
8. Lionberger RA, Russel WB. A Smoluchowski theory with simple approximations for hydrodynamic interactions in concentrated dispersions. *Journal of Rheology*. 1997;41 (2):399–425.
9. Lionberger RA, Russel WB. Microscopic theories of the rheology of stable colloidal dispersions. *Advances in Chemical Physics*. 2000;111:399–474.
10. Götze W. *Complex Dynamics of Glass-Forming Liquids*. Oxford: Oxford University Press; 2009.
11. Hansen JP, McDonald IR. *Theory of Simple Liquids*, 3rd ed. London: Academic Press; 2006.
12. Schweizer KS, Saltzman EJ. Entropic barriers, activated hopping, and the glass transition in colloidal suspensions. *Journal of Chemical Physics*. 2003;119(2):1181–1196.
13. Saltzman EJ, Schweizer KS. Transport coefficients in glassy colloidal fluids. *Journal of Chemical Physics*. 2003;119(2):1197–1203.
14. Kobelev V, Schweizer KS. Strain softening, yielding, and shear thinning in glassy colloidal suspensions. *Physical Review E*. 2005;71(2):021401.
15. Yeomans-Reyna L, Campa HA, de Jesús Guevara-Rodríguez F, Medina-Noyola M. Self-consistent theory of collective Brownian Dynamics: Theory versus simulation. *Physical Review E*. 2003;67(2):021108.
16. Chávez-Rojo MA, Medina-Noyola M. Self-consistent generalized Langevin equation for colloidal mixtures. *Physical Review E*. 2005;72(3):031107.
17. Juárez-Maldonado R, Medina-Noyola M. Theory of dynamic arrest in colloidal mixtures. *Physical Review E*. 2008;77(5):051503.
18. Brader JM. Nonlinear rheology of colloidal dispersions. *Journal of Physics: Condensed Matter*. 2010;22(36):363101.
19. Fuchs M. Nonlinear rheological properties of dense colloidal dispersions close to a glass transition under steady shear. *Advances in Polymer Science*. 2010;236(6):55–115.
20. Voigtmann Th. Nonlinear glassy rheology. *Current Opinion in Colloid and Interface Science*. 2014;19(6):549–560.
21. Russel WB, Saville DA, Schowalter WR. *Colloidal Dispersions*. Cambridge: Cambridge University Press; 1989.
22. Israelachvili JN. *Intermolecular and Surface Forces*, 3rd ed. Amsterdam: Academic Press; 2011.
23. Guazzelli E, Morris JF. *A Physical Introduction to Suspension Dynamics*. Cambridge: Cambridge University Press; 2012.
24. Happel J, Brenner H. *Low Reynolds Number Hydrodynamics: With Special Applications to Particulate Media*. Dordrecht: Martinus Nijhoff Publishers; 1973.
25. Kim S, Karrila SJ. *Microhydrodynamics: Principles and Selected Applications. Series in Chemical Engineering*. Boston: Butterworth-Heinemann; 1991.
26. Pusey PN. Colloidal suspensions. In Hansen JP, Levesque D, Zinn-Justin J (eds.) *Liquids, Freezing and Glass Transition*. Amsterdam: North Holland; 1991; pp. 765–942.
27. Jeffrey DJ, Onishi Y. Calculation of the resistance and mobility functions for two unequal rigid spheres in low-Reynolds-number flow. *Journal of Fluid Mechanics*. 1984;139:261–290.

28. Cichocki B, Felderhof BU. Short-time diffusion coefficients and high frequency viscosity of dilute suspensions of spherical Brownian particles. *Journal of Chemical Physics*. 1988;89(2):1049–1054.
29. Batchelor GK. The stress system in a suspension of force-free particles. *Journal of Fluid Mechanics*. 1970;41(3):545–570.
30. Batchelor GK. The effect of Brownian motion on the bulk stress in a suspension of spherical particles. *Journal of Fluid Mechanics*. 1977;83(1):97–117.
31. Batchelor GK. Brownian diffusion of particles with hydrodynamic interaction. *Journal of Fluid Mechanics*. 1976;74(1):1–29.
32. Mazo RM. On the theory of Brownian motion. III. Two-body distribution function. *Journal of Statistical Physics*. 1969;1:559–562.
33. Deutch JM, Oppenheim IJ. Molecular theory of Brownian motion for several particles. *Journal of Chemical Physics*. 1971;54(8):3547–3555.
34. Murphy TJ, Aguirre JL. Brownian motion of n interacting particles. I. Extension of the Einstein diffusion relation to the n-particle case. *Journal of Chemical Physics*. 1972;57 (5):2098–2104.
35. Dhont JKG. On the distortion of the static structure factor of colloidal fluids in shear flow. *Journal of Fluid Mechanics*. 1989;204:421–431.
36. Blawzdziewicz J, Szamel G. Structure and rheology of semidilute suspensions under shear. *Physical Review E*. 1993;48(6):4632–4636.
37. Wagner NJ, Ackerson BJ. Analysis of nonequilibrium structures of shearing colloidal suspensions. *Journal of Chemical Physics*. 1992;97(2):1473–1483.
38. Nägele G. On the dynamics and structure of charge-stabilized suspensions. *Physics Reports*. 1996;272(5–6):215–372.
39. Paul GN, Pusey PN. Observation of a long-time tail in brownian motion. *Journal of Physics A*. 1981;14(12):3301–3327.
40. Franosch T, Grimm M, Belushkin M, Mor F, Foffi G, Forró L, et al. Resonances arising from hydrodynamic memory in Brownian motion. *Nature (London)*. 2011;478:85–88.
41. Ermak DL. A computer simulation of charged particles in solution. I. Technique and equilibrium properties. *Journal of Chemical Physics*. 1975;62(10):4189–4196.
42. Ermak DL, McCammon JA. Brownian Dynamics with hydrodynamic interactions. *Journal of Chemical Physics*. 1978;69(4):1352–1360.
43. Allen MP, Tildesley DJ. *Computer Simulation of Liquids*, 2nd ed. Oxford: Oxford University Press; 2017.
44. Tough RJA, Pusey PN, Lekkerkerker HN, van den Broeck C. Stochastic descriptions of the dynamics of interacting Bownian particles. *Molecular Physics*. 1986;59 (3):595–619.
45. Lhuillier D, Nozières P. Volume averaging of slightly non-homogeneous suspensions. *Physica A Statistical Mechanics and Its Applications*. 1992;181(3):427–440.
46. Wang W, Prosperetti A. Flow of spatially non-uniform suspensions. part III: Closure relations for porous media and spinning particles. *International Journal of Multiphase Flow*. 2001;27(9):1627–1653.
47. Dhont JKG, Briels WJ. Gradient and vorticity banding. *Rheologica Acta*. 2008;47(3):257.
48. Brady JF. Brownian motion, hydrodynamics, and the osmotic pressure. *Journal of Chemical Physics*. 1993;98(4):3335–3341.
49. Jeffrey DJ, Morris JF, Brady JF. The pressure moments for two rigid spheres in low Reynolds number flow. *Physics of Fluids A: Fluid Dynamics*. 1993;5(10):2317–2325.

50. Marenne S, Morris JF. Nonlinear rheology of colloidal suspensions probed by oscillatory shear. *Journal of Rheology*. 2017;61(4):797–815.
51. Bergenholtz J, Brady JF, Vicic M. The non-Newtonian rheology of dilute colloidal suspensions. *Journal of Fluid Mechanics*. 2002;456:239–275.
52. Foss DR, Brady JF. Structure, diffusion and rheology of Brownian suspensions by Stokesian dynamics simulation. *Journal of Fluid Mechanics*. 2000;407:167–200.
53. Morris JF, Katyal B. Microstructure from simulated Brownian suspension flows at large shear rate. *Physics of Fluids*. 2002;14(6):1920–1937.
54. Morris JF. A review of microstructure in concentrated suspensions and its implications for rheology and bulk flow. *Rheologica Acta*. 2009;48(8):909–923.
55. Gurnon A, Wagner NJ. Microstructure and rheology relationships for shear thickening colloidal dispersions. *Journal of Fluid Mechanics*. 2015;769(23):242–276.
56. Maranzano BJ, Wagner NJ. The effects of particle size on reversible shear thickening of concentrated colloidal dispersions. *The Journal of Chemical Physics*. 2001;114 (23):10514–10527.
57. Wagner NJ, Brady JF. Shear thickening in colloidal dispersions. *Physics Today*. 2009;62 (October):27–32.
58. Foss DR, Brady JF. Brownian dynamics simulation of hard-sphere colloidal dispersions. *Journal of Rheology*. 2000;44(3):629–651.
59. Nazockdast E, Morris JF. Effect of repulsive interactions on structure and rheology of sheared colloidal dispersions. *Soft Matter*. 2012;8(8):4223–4234.
60. Banchio AJ, Nägele G, Bergenholtz J. Viscoelasticity and generalized Stokes-Einstein relations of colloidal dispersions. *Journal of Chemical Physics*. 1999;111(18):8721–8740.
61. Batchelor GK, Green JT. The determination of the bulk stress in a suspension of spherical particles to order c2. *Journal of Fluid Mechanics*. 1972;56(3):401–427.
62. Szymczak P, Cichocki B. A diagrammatic approach to response problems in composite systems. *Journal of Statistical Mechanics: Theory and Experiment*. 2008;2008(01): P01025.
63. Abade GC, Cichocki B, Ekiel-Jezewska ML, Nägele G, Wajnryb E. High-frequency viscosity of concentrated porous particles suspensions. *The Journal of Chemical Physics*. 2010;133(8):084906.
64. Abade GC, Cichocki B, Ekiel-Ježewska ML, Nägele G, Wajnryb E. Diffusion, sedimentation, and rheology of concentrated suspensions of core-shell particles. *Journal of Chemical Physics*. 2012;136(10):104902.
65. Riest J, Eckert T, Richtering W, Nägele G. Dynamics of suspensions of hydrodynamically structured particles: Analytic theory and applications to experiments. *Soft Matter*. 2015;11 (14):2821–2843.
66. Beenakker C. The effective viscosity of a concentrated suspension of spheres (and its relation to diffusion). *Physica A: Statistical Mechanics and its Applications*. 1984;128(1):48–81.
67. Banchio AJ, Nägele G. Short-time transport properties in dense suspensions: From neutral to charge-stabilized colloidal spheres. *The Journal of Chemical Physics*. 2008;128 (10):104903.
68. Heinen M, Banchio AJ, Nägele G. Short-time rheology and diffusion in suspensions of Yukawa-type colloidal particles. *The Journal of Chemical Physics*. 2011;135(15):154504.
69. Banchio AJ, Heinen M, Holmqvist P, Nägele G. Short- and long-time diffusion and dynamic scaling in suspensions of charged colloidal particles. *The Journal of Chemical Physics*. 2018;148(13):134902.

70. Das S, Riest J, Winkler RG, Gompper G, Dhont JKG, Nägele G. Clustering and dynamics of particles in dispersions with competing interactions: Theory and simulation. *Soft Matter*. 2018;14(1):92–103.
71. Riest J, Nägele G, Liu Y, Wagner NJ, Godfrin PD. Short-time dynamics of lysozyme solutions with competing short-range attraction and long-range repulsion: Experiment and theory. *The Journal of Chemical Physics*. 2018;148(6):065101.
72. Phung T. Behaviour of Concentrated Colloidal Suspensions by Stokesian Dynamics Simulation [PhD thesis]. Pasadena: California Institute of Technology; 1995.
73. Lionberger RA, Russel WB. High frequency modulus of hard sphere colloids. *Journal of Rheology*. 1994;38(6):1885–1908.
74. Segrè PN, Meeker SP, Pusey PN, Poon WCK. Viscosity and structural relaxation in suspensions of hard-sphere colloids. *Physical Review Letters*. 1995;75(5):958–961.
75. Weiss A, Dingenouts N, Ballauff M, Senff H, Richtering W. Comparison of the effective radius of sterically stabilized latex particles determined by small-angle x-ray scattering and by zero shear viscosity. *Langmuir*. 1998;14(18):5083–5087.
76. Su Y, Swan JW, Zia RN. Pair mobility functions for rigid spheres in concentrated colloidal dispersions: Stresslet and straining motion couplings. *The Journal of Chemical Physics*. 2017;146(12):124903.
77. Nägele G, Bergenholtz J. Linear viscoelasticity of colloidal mixtures. *The Journal of Chemical Physics*. 1998;108(23):9893–9904.
78. Widder DV. *The Laplace Transform*, 2nd ed. Princeton: Princeton University Press; 1946.
79. Viehman DC, Schweizer KS. Theory of gelation, vitrification, and activated barrier hopping in mixtures of hard and sticky spheres. *The Journal of Chemical Physics*. 2008;128(8):084509.
80. Contreras-Aburto C, Nägele G. Viscosity of electrolyte solutions: A mode-coupling theory. *Journal of Physics: Condensed Matter*. 2012;24(46):464108.
81. Aburto CC, Nägele G. A unifying mode-coupling theory for transport properties of electrolyte solutions. II. Results for equal-sized ions electrolytes. *The Journal of Chemical Physics*. 2013;139(13):134110.
82. Wagner NJ. The high-frequency shear modulus of colloidal suspensions and the effects of hydrodynamic interactions. *Journal of Colloid and Interface Science*. 1993;161(1):169–181.
83. Cichocki B, Felderhof BU. Linear viscoelasticity of semidilute hard-sphere suspensions. *Physical Review A*. 1991;43(10):5405–5411.
84. Furst EM, Squires TM. *Microrheology*. Oxford: Oxford University Press; 2017.
85. Puertas AM, Voigtmann Th. Microrheology of colloidal systems. *Journal of Physics: Condensed Matter*. 2014;26(24):243101.
86. Mason TG, Weitz DA. Optical measurements of frequency-dependent linear viscoelastic moduli of complex fluids. *Physical Review Letters*. 1995;74(7):1250–1253.
87. Banchio AJ, Bergenholtz J, Nägele G. Rheology and dynamics of colloidal suspensions. *Physical Review Letters*. 1999;82(8):1792–1795.
88. Abade GC, Cichocki B, Ekiel-Jezewska ML, Nägele G, Wajnryb E. High-frequency viscosity and generalized Stokes–Einstein relations in dense suspensions of porous particles. *Journal of Physics: Condensed Matter*. 2010;22(32):322101.
89. Riest J, Nägele G. Short-time dynamics in dispersions with competing short-range attraction and long-range repulsion. *Soft Matter*. 2015;11(48):9273–9280.
90. Kholodenko AL, Douglas JF. Generalized Stokes-Einstein equation for spherical particle suspensions. *Physical Review E*. 1995;51(2):1081–1090.

91. Medina-Noyola M. Long-time self-diffusion in concentrated colloidal dispersions. *Physical Review Letters*. 1988;60(26):2705–2708.
92. Brady JF. The rheological behavior of concentrated colloidal dispersions. *The Journal of Chemical Physics*. 1993;99(1):567–581.
93. Brady JF. The long-time self-diffusivity in concentrated colloidal dispersions. *Journal of Fluid Mechanics*. 1994;272:109–134.
94. Löwen H, Palberg T, Simon R. Dynamical criterion for freezing of colloidal liquids. *Physical Review Letters*. 1993;70(10):1557–1560.
95. Nägele G, Banchio AJ, Kollmann M, Pesche R. Dynamic properties, scaling and related freezing criteria of two- and three-dimensional colloidal dispersions. *Molecular Physics*. 2002;100(18):2921–2933.
96. Maxwell JC. On the dynamical theory of gases. *Philosophical Transactions of the Royal Society of London*. 1867;157:49–88.
97. Bengtzelius U, Götze W, Sjölander A. Dynamics of supercooled liquids and the glass transition. *Journal of Physics C*. 1984;17(33):5915–5934.
98. Leutheusser E. Dynamical model of the liquid-glass transition. *Physical Review A*. 1984;29(5):2765–2773.
99. Szamel G, Löwen H. Mode-coupling theory of the glass transition in colloidal systems. *Physical Review A*. 1991;44(12):8215–8219.
100. Berthier L, Biroli G. Theoretical perspective on the glass transition and amorphous materials. *Reviews of Modern Physics*. 2011;83(2):587–645.
101. Kawasaki K. Irreducible memory function for dissipative stochastic systems with detailed balance. *Physica A*. 1995;215(1–2):61–74.
102. Szamel G. Dynamics of interacting Brownian particles: A diagrammatic formulation. *Journal of Chemical Physics*. 2007;127(8):084515.
103. Götze W, Sjögren L. General properties of certain non-linear integro-differential equations. *Journal of Mathematical Analysis and Applications*. 1995;195(1):230–250.
104. Franosch T, Voigtmann Th. Completely monotone solutions of the mode-coupling theory for mixtures. *Journal of Statistical Physics*. 2002;109(1):237–259.
105. Götze W, Voigtmann Th. Effect of composition changes on the structural relaxation of a binary mixture. *Physical Review E*. 2003;67(2):021502.
106. Nägele G, Bergenholtz J, Dhont JKG. Cooperative diffusion in colloidal mixtures. *The Journal of Chemical Physics*. 1999;110(14):7037–7052.
107. Bergenholtz J, Horn FM, Richtering W, Willenbacher N, Wagner NJ. Relationship between short-time self-diffusion and high-frequency viscosity in charge-stabilized dispersions. *Physical Review E*. 1998;58(4):R4088–R4091.
108. Pusey PN, van Megen W. Phase behaviour of concentrated suspensions of nearly hard colloidal spheres. *Nature (London)*. 1986;320:340–342.
109. Franosch T, Fuchs M, Götze W, Mayr MR, Singh AP. Asymptotic laws and preasymptotic correction formulas for the relaxation near glass-transition singularities. *Physical Review E*. 1997;55(6):7153–7176.
110. van Megen W, Underwood SM. Glass transition in colloidal hard spheres: Mode-coupling theory analysis. *Physical Review Letters*. 1993;70(18):2766–2769.
111. Voigtmann Th. Dynamics of colloidal glass-forming mixtures. *Physical Review E*. 2003;68(5):051401.
112. Sperl M. Nearly logarithmic decay in the colloidal hard-sphere system. *Physical Review E*. 2005;71(6):060401(R).

113. Voigtmann Th, Puertas AM, Fuchs M. Tagged-particle dynamics in a hard-sphere system: Mode-coupling theory analysis. *Physical Review E*. 2004;70(6):061506.
114. Weysser F, Puertas AM, Fuchs M, Voigtmann Th. Structural relaxation of polydisperse hard spheres: Comparison of the mode-coupling theory to a Langevin dynamics simulation. *Physical Review E*. 2010;82(1):011504.
115. Russel WB, Wagner NJ, Mewis J. Divergence in the low shear viscosity for Brownian hard-sphere dispersions: At random close packing or the glass transition? *Journal of Rheology*. 2013;57(6):1555–1567.
116. Szamel G, Flenner E. Independence of the relaxation of a supercooled fluid from its micrsocopic dynamics: Need for yet another extension of the mode-coupling theory. *Europhysics Letters*. 2004;67(5):779–785.
117. Brambilla G, Masri DE, Pierno M, Berthier L, Cipelletti L, Petekidis G, et al. Probing the equilibrium dynamics of colloidal hard spheres above the mode-coupling glass transition. *Physical Review Letters*. 2009;102(8):085703. Comments and replies in *Physical Review Letters*. 2010;104(16):169601; 2010;104(16):169602; 2010;105(19):199604; 2010;105(19):199605.
118. Siebenbürger M, Fuchs M, Winter H, Ballauff M. Viscoelasticity and shear flow of concentrated, noncrystallizing colloidal suspensions: Comparison with mode-coupling theory. *Journal of Rheology*. 2009;53(3):707–726.
119. Dawson K, Foffi G, Fuchs M, Götze W, Sciortino F, Sperl M, et al. Higher order glass-transition singularities in colloidal systems with attractive interactions. *Physical Review E*. 2000;63(1):011401.
120. Pham KN, Puertas AM, Bergenholtz J, Egelhaaf SU, Moussaïd A, Pusey PN, et al. Multiple glassy states in a simple model system. *Science*. 2002;296(5565):104–106.
121. Ramírez-González P, Medina-Noyola M. General nonequilibrium theory of colloid dynamics. *Physical Review E*. 2010;82(6):061503.
122. Ramírez-González P, Medina-Noyola M. Aging of a homogeneously quenched colloidal glass-forming liquid. *Physical Review E*. 2010;82(6):061504.
123. Pérez-Ángel G, Sánchez-Díaz LE, Ramírez-González PE, Juárez-Maldonado R, Vizcarra-Rendón A, Medina-Noyola M. Equilibration of concentrated hard-sphere fluids. *Physical Review E*. 2011;83(6):060501(R).
124. Chen YL, Schweizer KS. Microscopic theory of gelation and elasticity in polymer–particle suspensions. *Journal of Chemical Physics*. 2004;120(15):7212–7222.
125. Fuchs M, Mayr MR. Aspects of the dynamics of colloidal suspensions: Further results of the mode-coupling theory of structural relaxation. *Physical Review E*. 1999;60(5):5742–5752.
126. Puertas AM, De Michele C, Sciortino F, Tartaglia P, Zaccarelli E. Viscoelasticity and Stokes-Einstein relations in repulsive and attractive colloidal glasses. *Journal of Chemical Physics*. 2007;127(14):144906.
127. Bonn D, Kegel WK. Stokes-Einstein relations and the fluctuation-dissipation theorem in a supercooled colloidal fluid. *Journal of Chemical Physics*. 2003;118(4):2005–2009.
128. Rizzo T, Voigtmann T. Qualitative features at the glass crossover. *Europhysics Letters*. 2015;111(5):56008.
129. Fuchs M, Cates ME. Theory of nonlinear rheology and yielding of dense colloidal suspensions. *Physical Review Letters*. 2002;89(24):248304.
130. Fuchs M, Cates ME. A mode coupling theory for Brownian particles in homogeneous steady shear flow. *Journal of Rheology*. 2009;53(4):957–1000.

131. Brader JM, Siebenbürger M, Ballauff M, Reinheimer K, Wilhelm M, Frey SJ, et al. Nonlinear response of dense colloidal suspensions under oscillatory shear: Mode-coupling theory and fourier transform rheology experiments. *Physical Review E*. 2010;82(6):061401.
132. Brader JM, Cates ME, Fuchs M. First-principles constitutive equation for suspension rheology. *Physical Review Letters*. 2008;101(13):138301.
133. Brader JM, Cates ME, Fuchs M. First-principles constitutive equation for suspension rheology. *Physical Review E*. 2012;86(2):021403.
134. Brader JM, Voigtmann T, Fuchs M, Larson RG, Cates ME. Glass rheology: From mode-coupling theory to a dynamical yield criterion. *Proceedings of the National Academy of Science, USA*. 2009;106(36):15186–15191.
135. Besseling R, Weeks ER, Schofield AB, Poon WCK. Three-dimensional imaging of colloidal glasses under steady shear. *Physical Review Letters*. 2007;99(2):028301.
136. Ovarlez G, Barral Q, Coussot P. Three-dimensional jamming and flows of soft glassy materials. *Nature Materials*. 2010;9(2):115–119.
137. Voigtmann Th, Brader JM, Fuchs M, Cates ME. Schematic mode coupling theory of glass rheology: Single and double step strains. *Soft Matter*. 2012;8(15):4244–4253.
138. Farage TFF, Brader JM. Three-dimensional flow of colloidal glasses. *Journal of Rheology*. 2012;56(2):259–278.
139. Lindemann FA. Ueber die Berechnung molekularer Eigenfrequenzen. *Physikalische Zeitschrift*. 1910;11(14):609–612.
140. Ballauff M, Brader JM, Egelhaaf SU, Fuchs M, Horbach J, Koumakis N, et al. Residual stresses in glasses. *Physical Review Letters*. 2013;110(21):215701.
141. Cates ME, Sollich P. Tensorial constitutive models for disordered foams, dense emulsions, and other soft nonergodic materials. *Journal of Rheology*. 2004;48(1):193–207.
142. Zausch J, Horbach J, Laurati M, Egelhaaf SU, Brader JM, Voigtmann Th, et al. From equilibrium to steady state: The transient dynamics of colloidal liquids under shear. *Journal of Physics: Condensed Matter*. 2008;20(40):404210.
143. Henrich O, Weysser F, Cates ME, Fuchs M. Hard discs under steady shear: Comparison of Brownian dynamics simulations and mode coupling theory. *Philosophical Transactions of the Royal Society of London A*. 2009;367(1909):5033–5050.
144. Amann CP, Denisov D, Dang MT, Struth B, Schall P, Fuchs M. Shear-induced breaking of cages in colloidal glasses: Scattering experiments and mode coupling theory. *Journal of Chemical Physics*. 2015;143(3):034505.
145. Krüger M, Weysser F, Fuchs M. Tagged-particle motion in glassy systems under shear: Comparison of mode coupling theory and Brownian dynamics simulations. *European Physical Journal E*. 2011;34(9):88.
146. Miyazaki K, Reichman DR. Molecular hydrodynamic theory of supercooled liquids and colloidal suspensions under shear. *Physical Review E*. 2002;66(6):050501(R).
147. Miyazaki K, Wyss HM, Weitz DA, Reichman DR. Nonlinear viscoelasticity of metastable complex fluids. *Europhysics Letters*. 2006;75(6):915–921.
148. Krüger M, Fuchs M. Fluctuation dissipation relations in stationary states of interacting Brownian particles under shear. *Physical Review Letters*. 2009;102(13):135701.
149. Krüger M, Fuchs M. Nonequilibrium fluctuation-dissipation relations of interacting Brownian particles driven by shear. *Physical Review E*. 2010;81(1):011408.
150. Seyboldt R, Merger D, Coupette F, Siebenbürger M, Ballauff M, Wilhelm M, et al. Divergence of the third harmonic stress response to oscillatory strain approaching the glass transition. *Soft Matter*. 2016;12(43):8825–8832.

151. Farage TFF, Reinhardt J, Brader JM. Normal-stress coefficients and rod climbing in colloidal dispersions. *Physical Review E.* 2013;88(4):042303.
152. Schilling R, Scheidsteger T. Mode coupling approach to the ideal glass transition of molecular liquids: Linear molecules. *Physical Review E.* 1997;56(3):2932–2949.
153. Letz M, Schilling R, Latz A. Ideal glass transitions for hard ellipsoids. *Physical Review E.* 2000;62(4):5173–5178.
154. Schilling R. Reference-point-independent dynamics of molecular liquids and glasses in the tensorial formalism. *Physical Review E.* 2002;65(5):051206.
155. Chong SH, Götze W, Singh AP. Mode-coupling theory for the glassy dynamics of a diatomic probe molecule immersed in a simple liquid. *Physical Review E.* 2000;63 (1):011206.
156. Chong SH, Götze W. Idealized glass transitions for a system of dumbbell molecules. *Physical Review E.* 2002;65(4):041503.
157. Szamel G, Flenner E, Berthier L. Glassy dynamics of athermal self-propelled particles: Computer simulations and a nonequilibrium microscopic theory. *Physical Review E.* 2015;91(6):062304.
158. Szamel G. Theory for the dynamics of dense systems of athermal self-propelled particles. *Physical Review E.* 2016;93(1):012603.
159. Flenner E, Szamel G, Berthier L. The nonequilibrium glassy dynamics of self-propelled particles. *Soft Matter.* 2016;12(34):7136–7149.
160. Liluashvili A, Ónody J, Voigtmann Th. Mode-coupling theory for active Brownian particles. *Physical Review E.* 2017;96(6):062608.

3 Methods of Colloidal Simulation

Ronald G. Larson

3.1 Introduction

Over the past three decades, methods of simulating the dynamics of suspensions have expanded enormously, to the point that there are now dozens of methods and variants thereof available, whose diversity rivals that of the problems of interest. While the proliferation of methods affords flexibility and choice, it also leads to bewilderment and uncertainty regarding the "best" method to use for a problem of interest. What counts as an "optimal" method depends on the problem of interest as well as on the desired accuracy and available computational resources. This chapter is intended to provide the reader some guidance in the choice of method, based on their limitations, and strengths and weaknesses, and to provide references that would allow the reader to dig deeper into a method of interest. The reader should be warned that improvements and expansions of existing methods, as well as entirely new methods, are constantly being developed, and so any assessment of their relative merits is open to continual update. The present review is at best a snapshot.

The major features of the rheology of suspensions of hard spherical particles suspended in Newtonian media, commonly referred to as "solvents", at low Reynolds number that one might wish to model are described in Stickel and Powell [1] and throughout this book. Suspensions of particles of other shapes, flexibilities, and in the presence of inertia or viscoelastic effects, are also of interest, as are nonrheological transport and structural properties. While we will mention throughout the versatility of various methods to consider a wide range of suspension transport phenomena, to keep the content within bounds and provide a focus to this chapter, we will maintain a focus on hard sphere suspension rheology at low Reynolds number. For this problem, the rheology is controlled by two independent dimensionless groups, namely the volume fraction of spheres ϕ, and the dimensionless shear rate, or Péclet number, Eq. (1.7). A schematic dependence of the dimensionless shear viscosity or relative viscosity $\eta_r = \eta/\eta_m$ on Pe and ϕ is shown in Figure 3.1. This figure shows the key features that one wishes to predict, quantitatively, if possible: (1) the steep, and eventually singular, increase in zero shear viscosity as volume fraction rises toward maximum random packing at around $\phi = 0.6$; (2) the steep shear thinning for high volume fractions $\phi > 0.4$ when $Pe \geq 0.01$ or so; and (3) the shear thickening that becomes pronounced for $\phi > 0.4$ when $Pe \geq 1$.

Beyond predicting this basic behavior accurately, one would like to relate variations in this behavior from one system to another to the presence of particle charge,

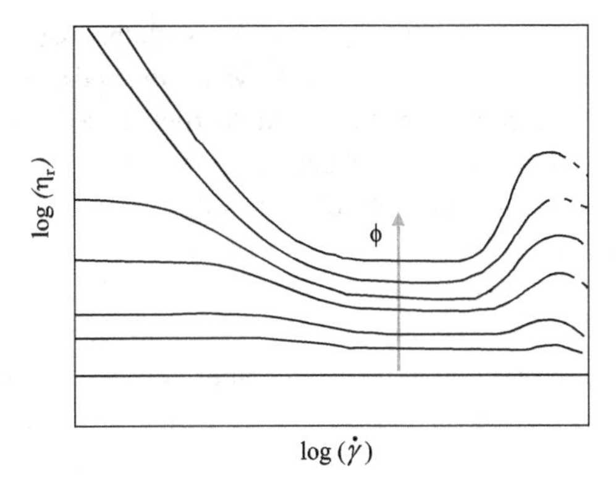

Figure 3.1 Typical flow curves for the relative viscosity versus shear rate for a suspension of hard spheres.

grafted stabilizing layers, particle attractive or repulsive interactions, modulus and deformation of the particles, particle–particle direct contacts (i.e., lubrication breakdown), particle roughness, and other factors present in experimental systems. One would also like to be able to expand beyond hard spherical particles to soft particles, other particle shapes such as disks, rods, and ellipsoids, droplets and bubbles, biological cells, swimming micro-organisms, as well as nonrheological phenomena such as sedimentation and electrokinetics, and influences of inertia and fluid viscoelasticity.

To even begin an assessment of methods of colloidal simulation, we need initially to inventory in more detail the features that a simulation method may contain, or lack. A method that excels in one application might perform dismally in another, depending on which aspects of the problem are dominant. A major reason that suspension dynamics have inspired so many disparate methods is that the problem has multiple length scales, ranging from small gaps h between particles, to the larger radius R of the particles, to the still-larger length scale L of a suspension fluid. For dense suspensions, $L \gg R \gg h$, making the ratio L/h very large indeed. Thus, a large range of distance scales must be covered and methods are strained to model all such scales accurately, especially outside of the limits of spherical rigid particles, no inertia, etc. Hence, depending on which aspect of the problem is deemed most crucial, different methods are called for. Thus, future efforts should include development of multi-scale modeling schemes, in which methods for resolving small-scale interactions are applied to small numbers of particles (or even just two), and the results used to tune parameters and functions deployed in lower-resolution models. Such multi-scale coarse-graining schemes are now a prominent part of simulations of molecular systems. At the present time, multi-scale modeling is rare in suspension simulations, and most researchers settle for choosing a method appropriate to their application.

Well-established continuum methods for solving Navier–Stokes equations, such as Finite Difference (FD) and the Finite Element (FE) methods, when applied to suspensions, are challenged by the many fluid–solid surfaces present in suspension flow, and especially by their movement during flow or transport. Over the past 30 years, many specialized simulation methods have been developed to cope with this issue. These can be classed into the broad categories of "continuum solvent" methods, in which the suspending medium is treated as a continuum fluid, and "particle based" in which the solvent momentum is carried by discrete particles. Among the continuum-based methods, one can take advantage of the linearity of Stokes flow and various theorems derived therefrom to determine the forces, torques, stresslets, and velocities of particles and their surfaces without ever solving explicitly for the flow of the solvent. Such methods have limitations, including greater difficulty (or even impossibility) of including inertia, and the effects of non-Newtonian media. Another option is to solve for the flow of the solvent explicitly by discretizing its velocity on a mesh, while dealing with boundaries between particles and fluid in some efficient way. The mesh enclosing the fluid can either move with the particle (a Lagrangian meshing scheme), or allow the particles to move through fixed mesh points (an Eulerian meshing scheme). An organization of many of the major methods along the lines sketched out above is given here:

I. Continuum Solvent
    A. Unmeshed Solvent: Brownian Dynamics (BD), Stokesian Dynamics (SD), Accelerated Stokesian Dynamics (ASD), Fast Lubrication Dynamics (FLD), and the Boundary Element Method (BEM)
    B. Meshed Solvent (aka : "Meshed Continuum Solvent (MCS) Methods"):
        1. Moving Mesh: Arbitrary Lagrangian–Eulerian (ALE) method
        2. Fixed Mesh: Distributed Lagrange Multiplier/Fictitious Domain (DLM/FD) Method, Force Coupling Method (FCM), Immersed Boundary Method (IBM), Fluid Particle Dynamics (FPD), Smoothed Profile Method (SPM)
II. Particle Solvent
    A. Unmeshed: Smoothed Particle Hydrodynamics (SPH), Dissipative Particle Dynamics (DPD)
    B. Meshed: Multi-Particle Collision (MPC) Dynamics, Lattice Boltzmann (LB) method

Note the large number of methods above that use a continuum, meshed, solvent. For simplicity, and because the scope of this review does not permit describing each of these methods in detail, we here coin the term "Meshed Continuum Solvent" (MCS) to include all such methods. These include the Arbitrary Lagrangian–Eulerian (ALE) method, in which the mesh moves with the particle, and the fixed-mesh methods, including the Distributed Lagrange Multiplier/Fictitious Domain (DLM/FD) method, the Force Coupling Method (FCM), the Immersed Boundary Method (IBM), and the Smoothed Profile Method (SPM). Also, since Stokesian Dynamics (SD), Accelerated Stokesian Dynamics (ASD), and Fast Lubrication Dynamics (FLD) all involve setting up grand mobility or resistance tensors, using additivity of hydrodynamic interactions between particle pairs, we will collectively refer to these methods as "Stokesian particle" (SP) methods. Before describing these and other methods in more detail, we first summarize some of the key factors that can be either essential, or unimportant, for a particular application, and, for each factor, list simulation methods that (in our view) are particularly well suited to capture it.

Details and references are given later in the discussion of each method.

1. *Large number of particles*. Some methods, such as simple Brownian dynamics, can handle thousands or even nearly a million, particles, while others, such as some meshed-continuum-solvent (MCS) methods like the Smoothed Profile Method, have generally been restricted to a limited number of particles, or to two-dimensional flows. Methods that incorporate hydrodynamic interactions (HI) but can readily handle many hundreds or even thousands of particles include Fast Lubrication Dynamics (FLD), Accelerated Stokesian Dynamics (ASD), the Distributed Lagrange Multiplier/Fictitious Domain (DLM/FD) method, and the Force Coupling Method (FCM). The Particle Solvent methods, namely the Lattice Boltzmann (LB) method, Multi-Particle Collision (MPC) dynamics, Dissipative Particle Dynamics (DPD), and Smoothed Particle Hydrodynamics (SPH), are more expensive, especially for refined rather than crude calculations, but have favorable linear scaling properties with increasing numbers of particles, unlike regular Stokesian Dynamics, which can be pushed to around 100 particles, but rapidly becomes unaffordable beyond that scale.
2. *Far-field hydrodynamics*. Given the great expense of accounting for them, inclusion of far-field HI is avoided if possible. If one is only interested in equilibrium structures, then far-field (or even near-field) hydrodynamics is unimportant; when dynamics and kinetics is of interest, the inclusion of hydrodynamics becomes important, but must be weighed against its high cost and its importance. Methods that handle far-field hydrodynamics well include the meshed continuum solvent (MCS) methods, Stokesian Dynamics (SD), Accelerated Stokesian Dynamics (ASD), the Boundary Element Method (BEM), and the particle-solvent methods (SPH, DPD, MPC, and LB).
3. *Near-field hydrodynamics and particle contact*. Near-field hydrodynamics, especially lubrication hydrodynamics, can be accounted for by relatively simple pair-wise interactions between near-neighbor particles only, and hence can be much cheaper to implement for some methods than is far-field hydrodynamics. However, near-field hydrodynamics are sensitive to gap and therefore often require very small time steps to calculate their effects accurately, increasing the expense and difficulty of introducing them, especially in dense suspensions, where the inclusion of near-field hydrodynamics is most important. Methods that describe near-field hydrodynamics well include some of the Meshed Continuum Solvent (MCS) methods especially the Smoothed Profile Method (SPM), as well as the Stokesian Particle (SP) methods. The Boundary Element Method (BEM) can also be adapted to allow closely spaced surfaces, and a number of other methods, including Dissipative Particle Dynamics (DPD), the Force Coupling Method (FCM), the Distributed Lagrange Multiplier/Fictitious Domain (DLM/FD) method, and the Lattice Boltzmann (LB) method have versions in which lubrication forces are included. An important special case occurs when lubrication layers are squeezed out altogether and particles come into contact and experience contact friction. These contact forces, which are

similar to those dominating granular flows, can be added to a number of the methods above, such as the Stokesian Particle (SP) methods.

4. *Brownian motion.* Depending on particle size and solvent viscosity, Brownian effects may be dominant or completely unimportant. In most methods, Brownian forces can be introduced in some way, perhaps at significant expense (in the case of Stokesian dynamics or Lattice Boltzmann dynamics), while in others, such as Dissipative Particle Dynamics (DPD), Brownian forces are intrinsic to the method and the non-Brownian regime can only be accessed approximately for fast flows. Methods especially well suited to inclusion of Brownian forces include Brownian Dynamics (BD), particle-based-solvent methods, such as Dissipative Particle Dynamics (DPD) and Multi-Particle Collisions (MPC) dynamics, as well as the Stokesian Particle (SP) methods. Brownian forces can be added to the Lattice Boltzmann (LB) and Smoothed Particle Hydrodynamics (SPH) methods. There are also methods for adding Brownian motion to Meshed Continuum-Solvent (MCS) methods, such as to the Immersed Boundary Method (IBM).
5. *Nonspherical Particle Shape.* Many of the fastest suspension simulation methods are primarily intended for spherical particles. Methods that can be applied to other shapes, such as ellipsoids, disks, or more arbitrary shapes, are of increasing interest. Methods suited for this include the meshed continuum-solvent (MCS) methods, as well as the Boundary Element Method (BEM), and the Lattice Boltzmann (LB) method. Other methods typically require complex shapes to be made from composites of spheres, which then give irregular and imprecise approximations to the actual particles. Methods for which this can be done include Dissipative Particle Dynamics (DPD), Multi-Particle Collision (MPC) dynamics, and the Stokesian Particle (SP) methods.
6. *Flexibility of particles.* Hard, nondeformable particles are normally considered in most simulations, but applications to deformable spheres, red blood cells, vesicles, and other deformable objects are of increasing interest. Methods that handle flexible particles well include the Immersed Boundary Method (IBM) and the Boundary Element Method (BEM). Particle flexibility can also be modeled using composite particles in the Dissipative Particle Dynamics (DPD) method. The Lattice Boltzmann (LB) method can also be used, when it is combined with a method that tracks deformable interfaces, such as the Immersed Boundary Method (IBM). These methods are used for studying systems discussed in Chapters 6–9.
7. *Inertial effects.* Methods designed to gain speed by taking advantage of Stokes-flow theorems, such as the Stokesian Particle (SP) methods and the Boundary Element Method (BEM) become much less usable if particle or fluid inertia becomes important. Methods that use solvent particles to transmit momentum can only approximate Stokes flow in the limit of slow flows and always contain at least some inertia, and so their parameters, including velocity, can be set to achieve a desired Reynolds number. If the particles are soft, however, as in Dissipative Particle Dynamics (DPD), or are ghost particles, as in Multi-Particle Collision (MPC) dynamics, artifacts due to compressibility may arise if the Reynolds number becomes too high. Methods that allow inertial effects while maintaining incompressibility include the meshed continuum-solvent (MCS) methods.

8. *Non-Newtonian solvent.* There is considerable interest in the behavior and rheology of suspensions in non-Newtonian fluids, including viscoelastic ones. Usually, inclusion of non-Newtonian solvent requires use of a method that solves for the flow in the solvent phase explicitly, and therefore includes the Meshed Continuum Solvent (MCS) methods. The Lattice Boltzmann (LB) method can also be adapted to allow inclusion of special cases of non-Newtonian solvents.
9. *Flow domain.* The flow domain significantly affects the choice of method. Some methods, such as the Stokesian Particle (SP) methods, work well mostly in a quasi-infinite system with periodic boundary conditions, or in simple geometries, such as slits or tubes. Other methods, including the Lattice Boltzmann (LB) method, are more easily adapted to complex domains with complex surfaces. Generally, grid-based methods, such as the Lattice Boltzmann (LB) method and Multi-Particle Collision (MPC) Dynamics, are fine for straight boundaries or ones with right angles, but have more difficulty with curved geometries, which have to be approximated using stepped profiles. Off-lattice solvent particle methods, such as Brownian Dynamics (BD), Dissipative Particle Dynamics (DPD), and Stokesian Particle (SP) methods can handle complex curved geometries better, by use of wall particles. For Stokesian Dynamics, where computer time scales with total number of particles as N^3, the wall particles add great expense. Other methods that handle complex boundaries well are the Boundary Element Method (BEM) and the Meshed Continuum Solvent (MCS) methods. We now describe the various methods in more detail, providing references to the source material.

3.2 Continuum Solvent Methods – Unmeshed Solvent

3.2.1 Brownian or Langevin Dynamics

The simplest and fastest method of simulating colloidal dynamics and rheology is simple Brownian dynamics, originating from the work of Ermak and McCammon [2]. In the absence of hydrodynamic interactions among the particles, the motion of a particle in a structureless, viscous Newtonian solvent, modeled as a continuum fluid with constant viscosity, is described by a simple stochastic differential equation:

$$\dot{\mathbf{r}}_i(t) = \mathbf{v}(\mathbf{r}_i) + \frac{\mathcal{D}_0}{k_B T}\mathbf{F}_i(t) + \xi_i(t), \tag{3.1}$$

where $\mathbf{r}_i(t)$ is the position vector of bead i, $\mathbf{F}_i(t)$ is the total force on the particle (exerted by other particles and external fields), $\mathbf{v}(\mathbf{r}_i)$ is the solvent velocity at the position of the bead, and $\xi_i(t)$ a random force satisfying $<\mathbf{X}_i(t)\mathbf{X}_i(t) \geq 2\mathcal{D}_0\delta(t)\mathbf{I}$, where $\mathcal{D}_0$ is the particle's diffusivity, $\delta(t)$ is the Kronecker delta function, and $\mathbf{I}$ is the unit tensor. Eq. (3.1) omits hydrodynamic interaction (HI). Note that the inertial term proportional to $\ddot{\mathbf{r}}_i(t)$ is missing from Eq. (3.1), implying a noninertial integration. Even when inertia is unimportant, it is sometimes numerically helpful to include the inertial term, in which case the simulation method is sometimes called *Langevin dynamics*.

To prevent particle overlap during the simulations the force $\mathbf{F}_i(t)$ can contain a steep repulsive term, but for hard particles this necessitates using tiny time steps to allow the steeply increasing repulsion to push colliding particles apart before they overlap significantly. Strating [3] avoided this by using the *elastic collision method*, in which the time and position of particle collisions are updated based on constant velocities between elastic collisions. A closely related method was proposed earlier by Heyes and Melrose [4]. This method allows for correct determination of particle–particle pair distribution functions at equilibrium. Under shearing flow, because of its neglect of HI, the prediction of the shear viscosity with this method fails at intermediate and high shear rates, and also fails in the low shear limit at small and moderate volume fractions, as illustrated in Figures 3.2 and 3.3. Note in Figure 3.3 that the BD simulations without HI under-predict the data at lower volume fractions but over-predict at higher particle concentrations. The same can be seen in Figure 3.2, which

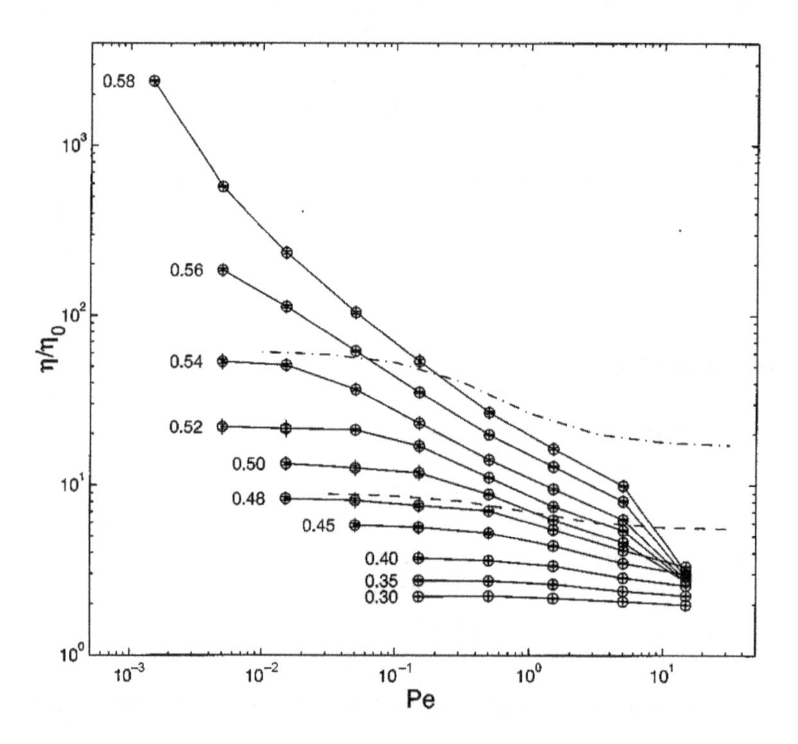

Figure 3.2 Steady state shear viscosity as a function of the Péclet number, here defined as $Pe = \dot{\gamma}a^2/\mathcal{D}_0$ where a is the particle radius and $\mathcal{D}_0$ the particle diffusion coefficient of an isolated sphere, for various volume fractions given, from elastic-collision BD simulations of Strating for more than 1,000 particles. Data of Van der Werff and de Kruif [5] for silica particles in cyclohexane are plotted for volume fractions of 0.419 (dashed line) and 0.538 (dash-dotted line). The viscosity drop at high volume fraction and high Pe results from a transition to hexagonally ordered strings in the direction of the flow. Reproduced with permission from Strating, Brownian dynamics simulation of a hard-sphere suspension. *Physical Review E* 1999;59(2):2175–2187 [3] by the American Physical Society, DOI 10.1103/PhysRevE.59.2175

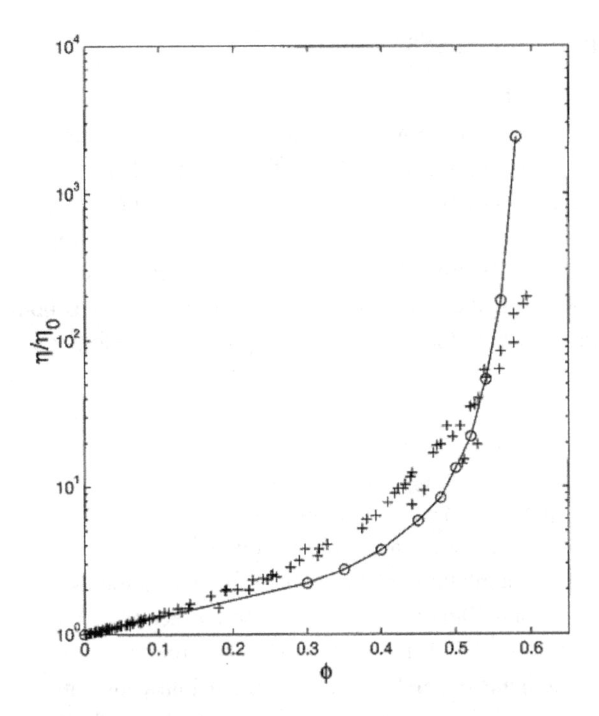

Figure 3.3 Relative shear viscosity (normalized by solvent viscosity η_0) as a function of volume fraction from elastic-collision Brownian dynamics computer simulations at the lowest shear rates from Figure 3.2 (open circles), compared to the measurements of Van der Werff and de Kruif [5]. Reproduced with permission from Strating, Brownian dynamics simulation of a hard-sphere suspension. *Physical Review E* 1999;59(2):2175–2187 [3] by the American Physical Society, DOI 10.1103/PhysRevE.59.2175

also shows that the simulations exaggerate the shear thinning and show an additional drop in viscosity at high concentrations and high shear rates. This is due to particles lining up into "strings" in the flow direction, a result that is likely sensitive to the finite box size as well as to HI. The Strating model does not show shear thickening (as depicted in Figure 3.1) at high rates, due to lack of HI.

While the Brownian Dynamics method without HI does not adequately describe transport properties of suspensions, especially at high volume fractions, the model is perfectly adequate for equilibrium properties, for which transport rates are irrelevant and only the energetic interparticle interactions matter. In these cases, the speed of the method and ability to equilibrate quickly and for large numbers of particles is attractive, although of course other methods, such as Metropolis sampling, can also deliver thermodynamic properties. Nevertheless, BD methods have been used to determine phase diagrams of colloids with attractive interactions, and to give at least a qualitative picture of nonequilibrium dynamics, such as colloidal gelation [6]. The speed of the BD method has allowed it recently to be used to examine in detail the coarsening dynamics of some 750,000 attractive spheres, albeit without HI [7].

3.2.2 Stokesian Particle (SP) Methods

Stokesian Dynamics (SD)

The key feature missing from the above simulations is hydrodynamic interaction (HI). One of the first, and still most popular, methods for dealing realistically with HI (both near- and far-field), is the "Stokesian Dynamics" method presented first by Durlofsky et al. [8] and used frequently since for problems of rheology at low Reynolds number [9–11], sedimentation [12,13], colloidal aggregation [14], electrophoresis [15], swimming of microparticles [12], and many others. The method combines both near- and far-field hydrodynamics, where the far-field hydrodynamics accounts for many body effects out to high order, and so provides an accurate, though not completely rigorous, method for describing dynamics, rheology, and transport properties of dense suspensions of dense spherical particles at a vanishingly small Reynolds number. The method exploits the multipole expansion in inverse separation (i.e., in powers of $1/r$) for force and torque (or linear and rotational velocity) of two spheres in the Stokes-flow limit. The various terms in this expansion had been developed over many years, culminating in the tabulations of the various matrix elements by Jeffrey and Onishi [16] and by Kim and Karrila [17]), with some corrections summarized in the appendix of Mohammadi et al. [18]. The multipole expansion fails for particles in near contact (since it is an expansion in $1/r$), but near-field lubrication terms were also developed, and Jeffrey and Onishi provided a means of combining both near- and far-field terms to produce an accurate method of describing hydrodynamic interactions between two spheres at any separation.

In the form presented by Jeffrey and Onishi, however, the method only permitted the description of two-sphere problems. The extension to many spheres is the triumph of the Stokesian dynamics methods, which accomplishes this by adding pairwise the effects on the *mobility* of a given particle induced by forces on each particle, and then inverting the resulting grand mobility matrix for interactions of all particles to produce a grand resistance matrix to which pairwise lubrication forces are added. The summation of the pairwise mobility matrices itself does not contain multi-body effects; that is, the effects on the hydrodynamic interactions between a pair of spheres by other surrounding spheres is not included in the pairwise summation that yields the grand mobility matrix. Yet, amazingly, once the matrix is inverted into the grand resistance matrix, these many-body effects are captured, at least for the lower third-to-fifth order multipoles in the expansion [8,19]. The Stokesian Dynamics method is summarized briefly by Foss and Brady [20]; in a nutshell, both the diffusivity and random force term in Eq. (3.1) must be replaced by matrix-multiplied terms to account for the hydrodynamic interactions. An early assessment of multipole moment expansion methods and their accuracy and cost is provided by Ladd [21].

While Stokesian Dynamics is much more accurate than simpler methods, the required matrix manipulations are expensive; the computational cost scales as the third power (N^3) of the number of particles [20], limiting the original method to simulation of only a few hundred particles or so. However, centrosymmetric interparticle forces are readily included, as well as torques that might be generated when noncentrosymmetric forces (such as axisymmetric forces) are present. Brownian motion can be included as well, with added expense. As SD simulations are based on analytical hydrodynamic interaction tensors valid for Stokes flow, flow domain boundaries require either the use of modified interaction tensors that account for the boundaries, or the use of boundary particles to account for momentum transfer to or from walls. The need to add wall particles increases the expense of the method. Periodic geometries can be handled by using a particle mesh Ewald scheme to sum hydrodynamic influences over periodic images. The results are satisfying: Figure 2.8 shows the predicted curve of shear viscosity versus Péclet number, which not only gives a result in agreement with experiments (as exemplified by the schematic in Figure 3.1), but shows the origins of the shear thinning and thickening phenomena in the Brownian and hydrodynamic contributions, respectively, to the stress. Note in Figure 3.2 the failure of simple Brownian Dynamics (without HI) to capture the shear thickening is obviously a result of the absence of the hydrodynamic interactions.

Accelerated Stokesian Dynamics (ASD)

The great expense of Stokesian dynamics and its rapid decline in efficiency with increasing numbers of particles (notice the small number of particles used to obtain the data in Figure 2.8) motivated the development of accelerated Stokesian dynamics (ASD) [9,22]. ASD removes the most costly step for large numbers of particles N, which is the computation and inversion of the far-field grand mobility or resistance matrices, which include interactions between every pair of particles in the simulation. ASD avoids this expense by taking advantage of the fact that the far-field hydrodynamic force (and not the elements of the matrix itself) suffices for the *iterative* calculation of the particle velocities, rather than using the expensive direct inversion. Hence, using a particle-mesh approach [23], an Ewald-summed wave-space contribution is calculated as a Fourier transform sum, and the iterative inversion can then be calculated at order $O(N \ln N)$ rather than $O(N^3)$ as in regular SD. The usual near-field lubrication forces are added again, as in the regular SD technique. For the sedimentation velocity in a model problem of a simple cubic array of spheres, which has an exact solution due to Zick and Homsy [24], the ASD method appears to be just as accurate as SD, although both begin to deviate from the exact solution as particle density becomes high. A related acceleration technique, the Spectral Ewald Accelerated Stokesian Dynamics (SEASD) method was recently proposed by Wang and Brady [25]; it becomes more efficient than regular Stokesian dynamics when the number of particles exceeds around 300, and can be sped up another factor of 10 by using GPU processors. Various versions of it include speeded-up methods for handling Brownian motion using a mean-field approximation.

An extension of Stokesian dynamics to nonspherical particles is possible through treating nonspherical particles as aggregates of spherical particles [26], at the cost of much greater expense, especially when the nonaccelerated SD is used with its N^3 scaling of simulation time with the number of spherical subparticles.

Fast Lubrication Dynamics (FLD)

Ball and Melrose [27] noted that, for dense suspensions, near-field lubrication hydrodynamics becomes more important than far-field hydrodynamics at least in shear flows. Moreover, lubrication hydrodynamics is limited to pairs of neighboring particles, while far-field HI requires summing over all pair interactions between each sphere and every other sphere. Thus, a great savings in cost is achieved if the far-field interactions are neglected, while retaining the near-field lubrication hydrodynamics. This "pair-drag" approximation is able to predict qualitatively the shear-dependent viscosity of spheres in dense suspensions. Kumar and Higdon [28,29] suggested an improvement, called Fast Lubrication Dynamics, in which, rather than ignoring completely the far-field hydrodynamics as the "pair-drag" model does, instead replaces them by an isotropic far-field resistance tensor, resembling an "effective viscosity" calculation by Beenakker [30]. The resistances in this isotropic tensor to translation, rotation, and to velocity gradients are obtained from short-time diffusion and viscosity coefficients from full Stokesian dynamics simulations [31]. The far-field tensor in this method is thus independent of time-dependent or shear-dependent suspension structure. The short-range pair-wise lubrication resistance matrix is retained and carries a strong dependence on suspension structure. An open-source FLD code exists within the LAMMPS software suite. The zero shear viscosity for FLD matches that of full Stokesian Dynamics for particle concentrations less than 40%, and falls below SD results by about 20% when the concentration reaches 50% [32].

Bolintineanu et al. [31] carefully compared FLD with both Dissipative Particle Dynamics (DPD) and Multi-Particle Collision (MPC) dynamics simulations for dense suspensions, using LAMMPS codes for each of these methods. They concluded that, when the equations are integrated "inertially", that is with the acceleration term included for the colloids (the solvent is Stokesian of course, and responds instantly to changes in colloidal forces), the FLD method yields late-time diffusion coefficients in excellent agreement with those from the more rigorous Stokesian dynamics method, while significant deficiencies emerge for both DPD and MPC dynamics. It is also much easier to obtain simulation input parameters from physical parameters for the FLD method than for DPD or MPC.

However, given that the far-field isotropic resistance in FLD was tuned to capture equilibrium suspension properties, the largest deviations in FLD might well occur under strong flows that distort both near- and far-field suspension structure away from equilibrium. A comparison of the predictions of FLD with Dissipative Particle Dynamics (DPD) and Multi-Particle Collision (MPC) dynamics (also known as Stochastic Rotation Dynamics, or SRD), using LAMMPS algorithms, can be found in Schunk et al. [33].

Following prescient work by Melrose and Ball [34], it has been increasingly recognized in recent years that in dense suspensions lubrication layers will at moderate stress be squeezed down to thicknesses comparable to particle roughness or even molecular dimensions, leading to "lubrication breakdown", including direct contacts between particle surfaces with frictional contact forces. A "minimal" model that includes such forces is that of Seto and colleagues [35,36], who modified the "pair-drag" model of Melrose and Ball [34] to include static and sliding frictional forces that

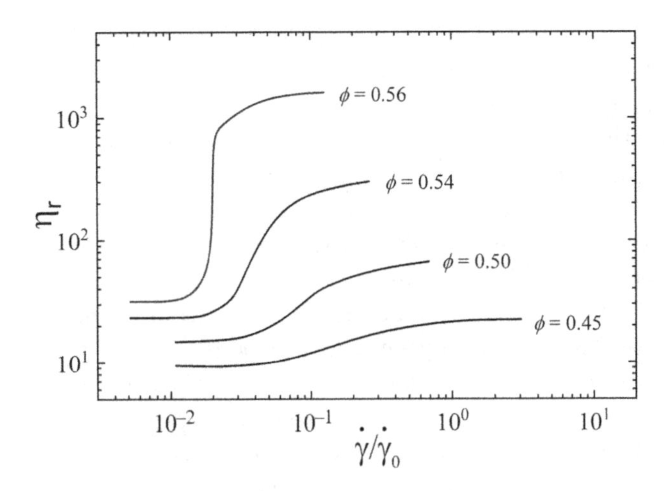

Figure 3.4 Dependence of the relative viscosity on reduced shear rate for lubrication dynamics simulations with frictional contacts and 1,000 particles. Redrawn using data from Mari et al. [36]

occur upon attainment of a critical normal force pressing the two particles together. This model, while phenomenological, is able to predict the discontinuous shear thickening that is observed in many dense colloids, but is not predicted by models containing only traditional lubrication forces in the near field; see Figure 3.4. The area is currently of great interest and controversy, and since these contact forces arise as solvent is squeezed to the molecular level, the phenomenon illustrates well the challenge in suspension dynamics of handling length scales spanning literally from atomic to continuum. Simulation methods, or combinations of them, that can realistically encompass such a range are needed, and should inspire much imaginative and exciting work in the coming years, work that is likely to have wide application in many areas outside of suspension rheology, such as tribology, printing, and others.

3.2.3 Boundary Element Analysis

The Boundary Element Method (BEM) (or Boundary Integral Method) uses the linearity of the Stokes equations (at zero Reynolds number) to determine a Green's function that allows derivation of integral equations from which forces and velocities on surfaces can be obtained, without the need to determine velocities in the bulk fluid (although these can be computed afterward by post-processing boundary data) [37]. Linearity is similarly used in Stokesian Dynamics (SD) to avoid solving explicitly for the flow. However SD also takes advantage of explicit formulas available for hydrodynamic interactions (multipole expansions for the far field and lubrication formulas for the near field) between pairs of spheres. BEM uses only the point-force limit of these formulas, but integrates this point force over particle surfaces that can be of complex and/or flexible shape. Such shapes include capsules [38,39], biological cells [40], and swimming micro-organisms [41]. While the BEM can be more efficient than solving the flow field in the bulk fluid, especially for three dimensional flows, it is still computationally demanding, and often limited to just one or two such particles, although Zhao et al. [42] simulated around 50 red blood cells in a capillary, but with equal viscosity for fluid inside and outside of the membrane to reduce costs. BEM can be used to obtain pairwise interaction matrices for particles of complex but rigid shape, which are then used to solve many-particle problems in periodic geometries, albeit in semi-dilute conditions where two-body interactions dominate. Such an approach has been used for example for swimming micro-organisms by Ishikawa and Pedley [43]. Further, Brownian forces were included in a "fluctuating boundary element" method [44].

Finally, Swan and Wang [45] developed a fast method to compute transport particles of complex shape. It resembles Stokesian Dynamics in that it uses resistance matrices to compute forces and velocities of spherical particles, but resembles the Boundary Element Method in that the resistance matrices are based on the low order Rotne–Prager tensor and the spherical particles are used as points to outline the shape of nonspherical particles. It gains speed and favorable scaling with system size through the use of GPU implementation of Particle Mesh Ewald (PME) summation of long-range interactions. Its intended use includes the hydrodynamics of complex particles such as proteins.

3.3 Continuum Solvent Methods – Meshed Solvent

Our review of these methods will be very brief, but a more detailed review has been published recently by Maxey and Patel [46] of these and other methods to which interested readers are referred.

3.3.1 Arbitrary Lagrangian–Eulerian Method (ALE)

This method uses a Lagrangian moving mesh, and is a versatile technique for carrying out finite element analysis in problems with moving surfaces. It has been applied to suspension-flow problems, especially by Hu [47] and Hu et al. [48], and others since then. As an indication of its versatility, it has been applied to the 2D simulation of the migration of a particle in Poiseuille flow of a viscoelastic fluid at finite Reynolds number [49] and in a viscoelastic Couette flow [50].

Methods that allow the particles to move through a fixed mesh include the Force Coupling Method (FCM), the Distributed Lagrange Multiplier/Fictitious Mesh (DLM/FM) method, the Immersed Boundary Method (ICM), the Fluid Particle Dynamics (FPD) method, and the Smoothed Profile Method (SPM). A brief summary of most of these methods can be found in Luo et al. [51]; an even briefer summary follows here.

Force Coupling Method (FCM)

The FCM [46,52–54] avoids applying no-slip boundary conditions directly, but instead represents each particle in the flow by a low-order expansion of force

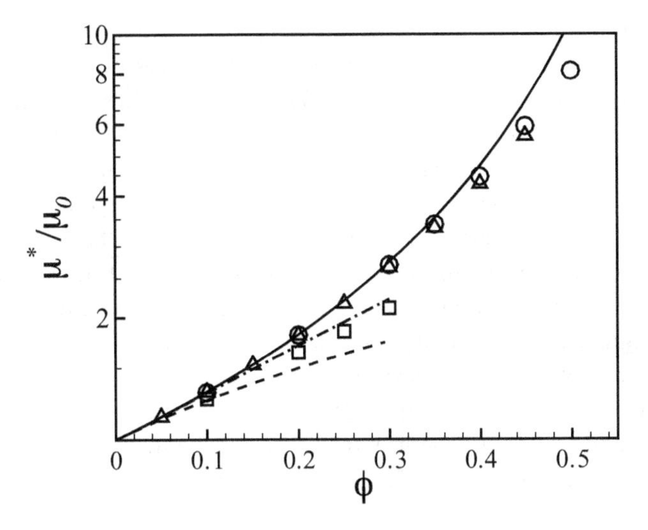

Figure 3.5 Zero shear viscosity, relative to the viscosity of the suspending medium η_m, as a function of volume fraction, where the squares are from the Force Coupling Method (FCM) with monopole and dipole terms, while the circles also include the lubrication terms. The solid line gives the experimental results of Krieger and Dougherty [57]; the dashed line is the Stokes–Einstein estimate, and the dash-dot line that of Batchelor and Green [58]. The triangles are from a multi-pole simulation of Ladd and Frenkel [59]. Reprinted with permission from Yeo and Maxey, Simulation of concentrated suspensions using the force-coupling method. *Journal of Computational Physics* 2010;229(6):2401–2421 [55] by Elsevier, DOI 10.1016/j.jcp.2009.11.041

multipoles. This is then applied as a distributed body force on the flow field, gridded on a lattice. It captures the far-field solution accurately and can typically resolve body forces well but does not fully resolve the flow field close to the particle surfaces. The lack of resolution in this method is compensated by its relative speed, with the cost scaling as $O(N\ln(N)$ [55]; simulations of sedimentation of up to 10,000 particles have been carried out with the FCM [56]. In addition, lubrication forces have been added to the FCM method [55], while retaining its favorable scaling, and allowing 1,000 particles to be simulated. The viscosity versus particle concentration thereby obtained is similar to that of the most accurate methods; see Figure 3.5.

Distributed Lagrange Multiplier/Fictitious Domain (DLM/FD)

In this method, particles are constrained to rigid-body motion by the use of a well-chosen field of Lagrange multipliers [60–62]. Lubrication forces have recently been added to the method to permit accurate near-field hydrodynamics, and the method was recently applied to the simulation of 10,000 sedimenting particles [63]. The DLM/FD method has also been adapted to simulating suspensions in viscoelastic (Oldroyd-B) fluids [64], and to particles that collide [65]. By combining DLM/FD with the Lattice Boltzmann (LB) method, fluid interactions with elastic solids can be simulated without remeshing [66].

Immersed Boundary Method (IMB)

For this method [67,68], the fluid–solid boundary moves with the local fluid velocity, and in return exerts a force on the fluid. In the original version, this force is obtained from the elastic properties of the immersed boundary, but the method has been extended to flows with rigid bodies [69,70]. Brownian motion has been included in the IMB method using the method of fluctuating hydrodynamics [71], where the thermal fluctuations are directly incorporated in the governing fluid equations [72]. A similar approach was used to incorporate Brownian dynamics into the Lattice Boltzmann (LB) method [73]. The IBM is very widely used in the mechanical engineering community, especially for particle-containing flows with strong turbulence [74], including gas/solid suspensions such as fluidized beds, and including heat transfer effects [75]. It is also often combined with the Lattice Boltzmann method for solving momentum transfer while tracking particle surfaces [76], including soft particles [77]. The IBM and its applications were reviewed by Deen et al. [78].

Fluid Particle Dynamics (FPD)

In this method, Tanaka and Araki [79] treat the particle as a fluid with very high viscosity, thereby removing the fluid–solid boundaries, rendering the entire domain (including the solids) a "fluid" with inhomogeneous viscosity, making it essentially a hydrodynamic phase-field model [80]. The method was used to simulate colloidal aggregation in a liquid crystal [81], and recently for a collection of dumbbell swimmers [82].

Smoothed Profile Method (SPM)

This method [83,84] uses a fixed computational mesh that is not shaped to the geometry of the particles. Instead, the particle surfaces are replaced smoothed-out density profiles and used to construct a body force term that is added into the Navier–Stokes equations. The accuracy of the method has been validated for flow around a single particle, a single rigid chain, and two colliding particles [85]. The method has also been applied to two-dimensional problems with many particles (2D disks) [83].

As a general comment, fluctuating forces; i.e., Brownian forces, can be introduced into continuum solvent in Eulerian finite volume schemes [86], similar to what is done in Lattice Boltzmann simulations by Ladd and coworkers [87,88].

3.4 Particle Solvent Methods – Unmeshed Solvent

3.4.1 Smoothed Particle Hydrodynamics (SPH)

The Smoothed Particle Hydrodynamics (SPH) method introduced by Monaghan [89,90], and reviewed by Liu et al. [54], Hoover [91], and Li and Liu [92], is a mesh-free method that discretizes the Navier–Stokes equation by treating both colloids and fluid as "particles" with smoothed density distribution whose interactions

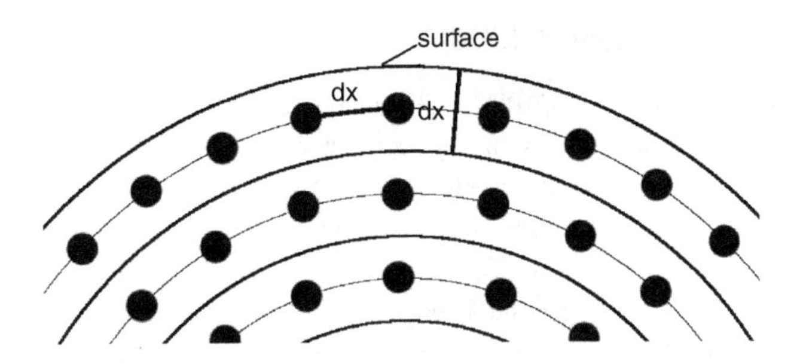

Figure 3.6 Representation of a spherical solid particle in Smoothed Particle Hydrodynamics. Boundary particles are placed parallel to the surface with the same neighboring distance *dx*, and constrained to move together to impose rigid motion of the solid object. Reproduced with permission from Bian et al., Hydrodynamic shear thickening of particulate suspension under confinement. *Journal of Non-Newtonian Fluid Mechanics* 2014;213:39–49 [97]

generate accelerations that produce both pressure gradients and viscous damping functions. The key step in the method is to assign a velocity to boundary particles that ensures that the no-slip velocity is imposed on the surface of the prescribed boundary; see Figure 3.6. This establishes better control over the boundary condition of a suspended object than is possible with some other particle-based methods. Inclusion of near-field lubrication generally requires explicit inclusion of analytical forms for lubrication forces [93], and this has been applied with success to the prediction of the zero shear viscosity of dense hard sphere suspensions [94]. A review of the method can be found in Harting et al. [95] along with a comparison to Lattice Boltzmann dynamics and some example problems involving sedimentation. While the original method does not include Brownian motion, in an extension of it called "Smoothed Dissipative Particle Dynamics" (SDPD) the SPH method is modified to allow fluctuating hydrodynamics [96], in a manner consistent with the Landau–Lifshitz theory for fluctuating hydrodynamics [72]. The method then bears similarity to Dissipative Particle Dynamics (DPD), but with a more careful control of the solid-liquid boundary, extendible to nonspherical particles without needing to compose them of aggregated spheres. A number of problems involving one or two rotating spheres, and a sphere interacting with a wall or interacting with another sphere hydrodynamically, were solved accurately by SDPD [93]. The method requires an interaction cut-off radius r_c that is 2–4 times longer than required by DPD, with a consequent increase in cost, but with more accurate solution. Recent simulations with this method were able to solve up to 1,000 or so non-Brownian 2D disks at high concentration with the inclusion of short-range repulsive forces between particles, allowing simulation of shear thickening [97].

In addition to Smoothed Particle Hydrodynamics, other methods of using non-Brownian particles to replace continuum solvent include the element-free Galerkin method (EFGM) [98], the reproducing kernel particle method (RKPM) [99], and the finite point method (FPM) [100]. Since these methods have not as of this writing been applied to the study of colloidal suspensions [31], we consider them no further here.

3.4.2 Dissipative Particle Dynamics (DPD)

Dissipative particle dynamics (DPD) [101–103] was developed as a mesoscopic particle simulation method that uses *soft* conservative repulsive forces to regulate relative interactions among the components and dissipative and random thermal forces to control the friction. The method allows larger time steps than is possible with atomistically realistic interactions. The restriction of conservative forces to purely repulsive forms is workable for condensed liquid phases, where volume changes on mixing are small, and fluid structure is controlled by *relative* interactions among the species. Thus a *relative attraction* between two species can be created by making their interaction *less repulsive* than that of other interactions. The basic equations of the DPD method are:

$$\frac{d\mathbf{r}_i}{dt} = \mathbf{v}_i, \quad m_i \frac{d\mathbf{v}_i}{dt} = \mathbf{F}_i, \tag{3.2}$$

where $\mathbf{r}_i$, m_i, and $\mathbf{v}_i$ are the position, mass, and velocity of the ith particle; and $\mathbf{F}_i$ is the total force experienced by the ith particle. The total force acting on the ith particle by all other particles (j) is given by

$$\mathbf{F}_i = \sum\nolimits_{j \neq i} \left(\mathbf{F}_{ij}^C + \mathbf{F}_{ij}^D + \mathbf{F}_{ij}^R \right), \tag{3.3}$$

where $\mathbf{F}_{ij}^C$, $\mathbf{F}_{ij}^D$, and $\mathbf{F}_{ij}^R$ represent the conservative, dissipative, and random forces, respectively. These pairwise additive forces are given as:

$$\mathbf{F}_{ij}^C = a_{ij} w(r_{ij}) \hat{\mathbf{r}}_{ij} \tag{3.4}$$

$$\mathbf{F}_{ij}^D = -\gamma w^D(r_{ij}) (\hat{\mathbf{r}}_{ij} \cdot \mathbf{v}_{ij}) \hat{\mathbf{r}}_{ij} \tag{3.5}$$

$$\mathbf{F}_{ij}^R = \sigma w^R(r_{ij}) \theta_{ij} (\Delta t)^{-1/2} \hat{\mathbf{r}}_{ij} \tag{3.6}$$

where $\mathbf{r}_{ij} = \mathbf{r}_i - \mathbf{r}_j, \hat{\mathbf{r}}_{ij} = \mathbf{r}_{ij} / |\mathbf{r}_{ij}|$, $\mathbf{v}_{ij} = \mathbf{v}_i - \mathbf{v}_j, \Delta t$ is the timestep, a_{ij} is the repulsion parameter between beads i and j, γ is the friction coefficient, σ is noise amplitude, $w^D(r)$ and $w^R(r)$ are dimensionless weight functions for the dissipative and random processes, and $\theta_{ij} = \theta_{ji}$ is a white-noise Gaussian random variable: $< \theta_{ij}(t) >= 0$ and $< \theta_{ij}(t)\theta_{kl}(t') >= (\delta_{ik}\delta_{jl} + \delta_{il}\delta_{jk})\delta(t - t')$. The fluctuation-dissipation theorem requires that $w^D(r) = [w^R(r)]^2$ and $\sigma^2 = 2\gamma k_B T$. A common choice for weight function of the conservative force is the soft interaction given by:

$$w(r) = \begin{cases} 1\text{-}r/r_c & (r \leq r_c) \\ 0 & (r > r_c) \end{cases}. \tag{3.7}$$

A similar weight function is often used for the random force $w^R(r)$. Since the solvent is composed of particles that exchange momentum through both the conservative and dissipative forces, DPD allows hydrodynamics to be captured for arbitrary spatial distributions of particles and arbitrary boundaries. The soft potentials and dissipative forces make no attempt at molecular realism, but allow mass and momentum transfer to occur with time steps that are much larger than those in molecular dynamics, with a consequent speed-up in simulation. A key aspect of the dissipative force is that it conserves linear momentum, and so preserves hydrodynamic interactions. The dissipative force does not conserve mechanical energy, however, and the energy lost is replaced by the random force, for which momentum is conserved pair-by-pair. Especially given that publically available DPD codes are available, such as in the LAMMPS suite, DPD has become a versatile way of solving many problems in soft-matter thermodynamics and transport, involving polymers, surfactants, colloids, biological materials, and more [104]. The computational cost of DPD scales simply linearly with the number of DPD particles, $O(N)$. Colloidal particles can be simulated in DPD by lumping hundreds of small DPD particles into each larger, rigid, colloid [105].

Because the method relies on collisions of DPD particles to transfer momentum, the speed of momentum transfer within the fluid is finite, and hence the Stokes-flow limit cannot be attained, although it can be approached by decreasing the dimensionless velocity, with a consequent increase in cost. The softness of the particle interactions also makes the fluid somewhat compressible. Thus both the Mach number $Ma = U/U_s$, where U_s is the speed of sound, and the Reynolds number $\mathrm{Re} = \rho UL/\eta_m$ are nonzero, and if their values are not made small enough, results of DPD simulations can show spurious effects. The softness of the potential also leads to rather slow momentum transfer relative to mass transfer, leading to values of the Schmidt number $Sc = \eta_m/\rho D$, i.e. the ratio of these two rates, to be artificially low, unless countermeasures are taken [106]. The use of particles with a thermostat regulating their velocities means that DPD particles are Brownian, so that the fluid bead Péclet number (Pe_f) is finite, and fluctuations in stress must be averaged out, making the method less suitable for problems where Brownian forces are negligible. The colloid particle Péclet number Pe in DPD simulations is the DPD fluid particle Péclet number Pe_f times the ratio of the diffusivity of the colloidal particle to the diffusivity of the DPD fluid particle [107]. These sources of possible error can be assessed from the following formulas [102]:

$$Ma = \frac{U}{\sqrt{(k_\mathrm{B}T/m) + (\lambda a_{ii} n r_c^4/m)}}; \quad \mathrm{Re} = \frac{mnUL}{\eta}; \quad Sc = \frac{\eta}{mnD}; Pe_f = \frac{UL}{D} = Re{\cdot}Sc, \tag{3.8}$$

where m is the mass of the DPD particle, n is the number of particles per unit volume, λ is a numerical constant typically equal to $\pi/15$, r_c is the range of the repulsive potential, a_{ii} is a typical bead–bead repulsion parameter, η the fluid viscosity, D the DPD fluid particle diffusivity, and U and L are the characteristic velocity and length scale of the flow. Formulas for η and D can be obtained in Marsh et al. [108].

A key challenge in DPD simulations is that one cannot control directly the dimensionless groups above (Ma, Re, Sc, Pe_f), but only indirectly through choice of DPD parameters (m, $k_\mathrm{B}T$, n, a_{ii}). In particular, the final expression in Eq. (3.7) implies that non-Brownian ($Pe_f \gg 1$), noninertial ($\mathrm{Re} \ll 1$) flows require very large Schmidt numbers *Sc*. As mentioned, there are means of achieving high *Sc* through creating very strong dissipative interactions between DPD beads, but these lead to poor resolution of flows in thin boundary layers [107]. On the other hand, if high *Re* is desired, the Mach number *Ma* rises with it, eventually destroying the accuracy of the flow through artificial compressibility of the fluid. For example, this generally puts turbulent flow with DPD out of reach. The DPD method is thus "mesoscopic" not only in using fluid particles or mesoscopic size, but also in the sense that all dimensionless groups are of intermediate value, often neither vanishingly small nor asymptotically high. This makes it essential to control DPD parameters carefully so that the dimensionless groups one wishes to be asymptotically small or large, be "small or large enough" to give acceptably accurate results. Additionally, integration of the DPD equations must be done carefully to avoid either inaccuracies on the one hand, or excessively small time steps on the other. A brief review of methods can be found in Mai-Duy et al. [106].

An alternative version of DPD is the Fluid Particle Model (FPM) (not to be confused with Fluid Particle Dynamics (FPD), in which the solvent is a continuum) introduced by Español and Revenga [96], and its modification by Pan et al. [109]. This method uses a large DPD particle for each colloidal sphere, rather than an aggregate of DPD solvent particles, and includes a noncentrosymmetric dissipative force between colloidal particles or between solvent and colloidal particle as well as including particle torques. The conservative force is made relatively hard for the sphere–sphere interactions. The result is a rather successful prediction of zero shear viscosity versus particle concentration, as shown in Figure 3.7 [109]. Since the solvent must carry hydrodynamic information from each colloidal particle, there must be many (around 200 solvent particles per colloid particle), and thus DPD simulations with hundreds of colloid particles require tens of thousands of DPD solvent particles to simulate a dense suspension. Nevertheless, Pan et al. [109] estimate that DPD simulations of colloidal suspension are around three orders of magnitude faster than accelerated Stokesian Dynamics (ASD). While regular DPD seems to give viscosities and normal stress differences roughly consistent with SD simulations, precise comparisons of these methods and clear benchmarks to establish the relative accuracy and speed of these methods do not seem to have been developed yet. Also, standard DPD suspension simulations do not seem able to yield the shear thickening regime as they do not resolve thin lubricating layers.

Whether the colloid particle is simulated as an aggregate of solvent beads or as a separate larger particle, the size of the solvent particles is an appreciable fraction of the size of the colloidal particle. This creates two problems when the colloidal particles approach each other closely. First, the few, or no, solvent particles remaining in the gap between the colloidal particles are unable to properly represent the lubricating hydrodynamic forces between the particles. Second, when there is no room between the colloids for the solvent particles, a depletion force arises that drives the colloids together to free up more room for the excluded solvent particles outside of the gap. These problems are addressed by the "core-modified" DPD method of Whittle and Travis [110], which, like the method of Pan et al. [109],

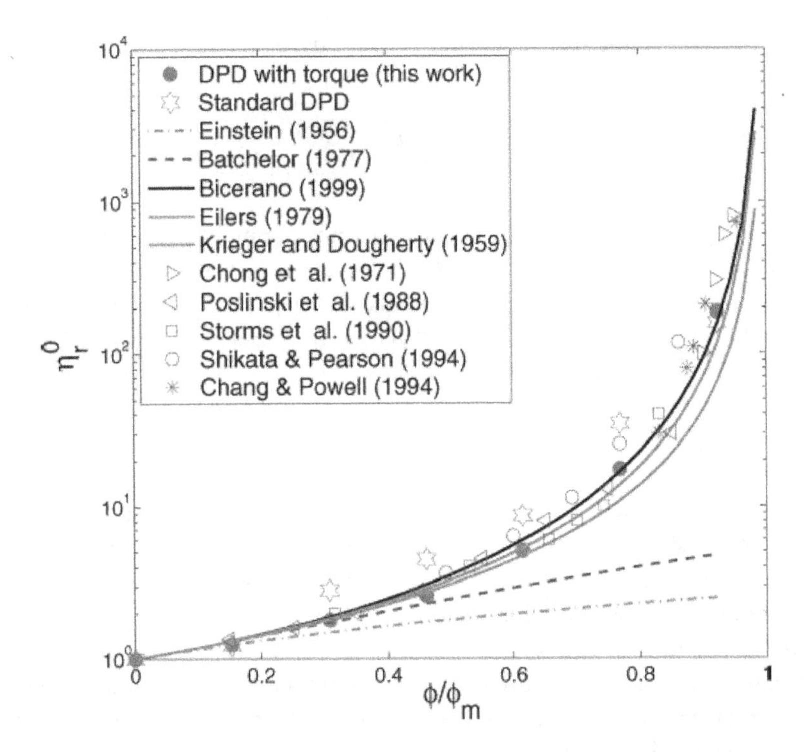

Figure 3.7 Zero shear relative viscosity determined by DPD simulations with torque balance (red filled symbols) as a function of volume fraction normalized by the maximum packing volume fraction ϕ_m and fitted by empirical formulas (lines) and compared with experimental data (open symbols and asterisks). Reprinted with permission from Pan et al., Rheology, microstructure and migration in Brownian colloidal suspensions. *Langmuir* 2010;26(1):133–142 [109] by the American Chemical Society

uses noncentrosymmetric, dissipative interactions, which occur over a "shell" layer surrounding a particle core that is prevented from overlapping other particle cores through a strong core–core repulsion. Another dissipative force between DPD colloids is used to represent the lubrication force, using analytical theory to provide the form of this force. Issues with including depletion can be offset by including an additional counteracting force that is based on an analytical expression for dilute suspensions [110].

Jamali et al. [111] and Boromand et al. [112] have very recently extended the method of Whittle and Travis [110] by including elastohydrodynamic deformation resulting from compressive forces acting on the particles as they are driven together by flow. Thus, they solved the following equation:

$$m_i \frac{\mathrm{d}\mathbf{v}_i}{\mathrm{d}t} = \sum \mathbf{F}_{ij}^{C} + \mathbf{F}_{ij}^{D} + \mathbf{F}_{ij}^{R} + \mathbf{F}_{ij}^{el} + \mathbf{F}_{ij}^{H} + \mathbf{F}_{ij}^{\mathrm{Core}-(n,t)}, \tag{3.9}$$

where the conservative, dissipative, and random force terms (the first three on the right side) are similar to those of conventional DPD, discussed earlier. The fourth term is an electrostatic interaction (if needed). The lubrication (pair-drag) term is given by:

$$\mathbf{F}_{ij}^{H} = -\mu_{ij}^{H}\left(\mathbf{v}_{ij}\cdot\mathbf{e}_{ij}\right)\mathbf{e}_{ij}; \quad \mu_{ij}^{H} = \begin{cases} 3\pi\eta_0 a^3/2\delta, & 0 < h_{ij} \le \delta \\ 3\pi\eta_0 a^3/2h_{ij}, & h_{ij} > \delta \end{cases}, \tag{3.10}$$

which regularizes the normal lubrication term by limiting the force to a maximum one at a lubrication layer thickness of δ. Here a is the colloidal particle radius. η_m is the viscosity of the suspending medium and the gap between particles i and j is h_{ij}. Finally, the "core" force includes both a normal (n) and a tangential (t) force, the latter of which can support both static and sliding stress, is influenced by particle roughness Δ, and is described by Jamali et al. [111] and in most recently modified form by Boromand et al. [112]. In the simulations of Jamali et al. [111], δ was set to $10^{-6}a$, while Δ was set to $10^{-3}a$, where a is the particle radius. The resulting DPD method gives similar results to those obtained using frictional Stokesian dynamics [112]. Figure 3.8 shows that the predictions of these "core-modified" DPD simulations are in qualitative agreement with experimental data for ϕ around 0.50. The model shows a transition from negative first normal stress difference N_1 in the lubricated shear thickening regime to positive values in the frictional regime where a percolated pathway of frictional contacts emerges, a normal-stress transition seen in recent experiments [113]. Such features of shear thickening in hard sphere colloidal suspensions are discussed in Chapter 1.

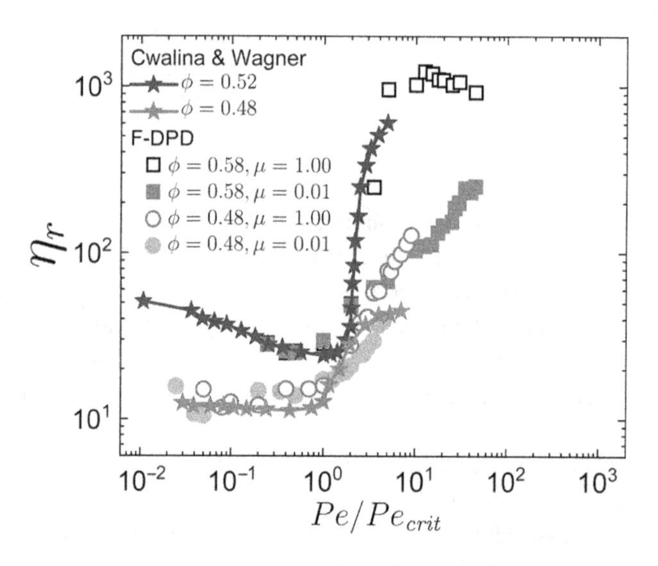

Figure 3.8 Comparison between prediction of the core-modified "frictional" DPD (F-DPD) model of Boromand et al. [112] (symbols) with experimental studies of silica nanoparticles by Cwalina and Wagner [114] (star-line). The x-axis has been re-scaled to account for interparticle interactions. Reprinted with permission from Boromand et al., A generalized frictional and hydrodynamic model of the dynamics and structure of dense colloidal suspensions. *Journal of Rheology* 2018;62(4):905–918 [112], by The Society of Rheology

These various corrections to DPD, while yielding results similar to those from rigorous Stokesian dynamics, do not yet provide quantitative agreement to date to experimental data for hard sphere dispersions over the complete range of particle concentration and shear rate. In addition, Whittle and Travis [110] find, for their core-modified method, that the short-time diffusivity increases with increasing particle volume fraction above $\phi = 0.4$, rather than decreasing, as it does in experiments and in SD simulations. (They believe this result to be general to DPD simulations, although it was not reported on in the work of Jamali et al. [111] or Boromand et al. [112]). However, it appears that with recent modifications the DPD method is approaching the accuracy of Stokesian dynamics for both dilute and dense suspensions [106]. Spheroidal and rod- and disk-like particles have also been simulated by DPD simulations [115–117].

The DPD method is very versatile and can be applied readily to flow of dispersed droplets, as well as nonspherical particles such as spheroids, and can readily accommodate the effects of inertia. It has recently been used to compute inertial focusing of droplets and particles in channel flows, for example [107,117,118].

3.5 Particle Solvent Methods – Meshed Solvent

3.5.1 Multi-Particle Collision (MPC) Dynamics, or Stochastic Rotation Dynamics (SRD)

A method formerly known as Stochastic Rotation Dynamics (SRD) [119,120], but now frequently called Multi-Particle Collision (MPC) dynamics [121], is both lattice and particle based, sharing the advantages and disadvantages of both. The method has two basic steps: (1) a simple constant-velocity streaming flow and (2) an en masse "collision" of solvent particles, in which momentum is exchanged among all particles within each of the cubic cells into which the domain is gridded. In the simplest version of MPC dynamics, this momentum exchange is carried out as follows: the center-of-mass velocity of the particles in each lattice cell is subtracted from the velocities of all particles in that cell. The resulting relative velocities are then rotated through a random angle, which preserves linear momentum within the cell since these rotated velocities retain zero mean velocity. The streaming and collisions steps of the method are given by:

$$\mathbf{r}_i(t+\Delta t) = \mathbf{r}_i(t) + \mathbf{v}_i(t)\Delta t \tag{3.11}$$

$$\mathbf{v}_i(t+\Delta t) = \mathbf{u}(t) + \Omega[\mathbf{v}_i(t) - \mathbf{u}(t)], \tag{3.12}$$

where Ω is the rotation matrix and $\mathbf{u}(t)$ is mean velocity in the MPC cell. The position and velocity of the solvent particle i are $\mathbf{r}_i$ and $\mathbf{v}_i$. The rotation process exchanges momentum of particles in the cell, permitting momentum diffusion. Because the momentum is exchanged among many particles (5–10) within a single cell in one step, pair-wise addition of momentum exchange is avoided, as are neighbor lists, making the method much faster, for a given number of solvent particles, than DPD in particular [122]. For the solvent flow, MPC simulations have nonzero inertia and fluid compressibility as potential artifacts, if one is attempting to simulate incompressible Stokes flow. The compressibility of an MPC fluid is greater than for typical DPD fluids, since in MPC the solvent particles have no repulsive interactions and therefore have the compressibility of an ideal gas rather than of a liquid. The Mach number can therefore become high enough to induce significant density changes [122]. The compressibility of the fluid could of course be countered by applying repulsive interactions between the solvent particles, but this would introduce pairwise momentum exchange during "collisions" and the advantageous speed of MPC dynamics would be lost. So the velocity must be kept low, relative to the thermal velocity of the solvent beads, to prevent unacceptable compression of the fluid. The low velocities mean that runs must be longer to average out Brownian noise [122]. It also means that the Schmidt number of an MPC fluid, measuring the ratio of momentum to mass diffusivities, is near unity, typical of a gas rather than of a liquid. For objects which are large compared to the typical grid spacing, the small Schmidt number need not be a serious problem.

MPC dynamics has been applied to colloids by introducing repulsive interactions between large colloidal particles and between solvent and particles [123]. The range of the colloid–solvent repulsion sets the range of depletion interactions that induce short-range attraction between the colloids – a limitation of the method for hard sphere fluids. A similar limitation exists for DPD fluids, which also use solvent particles to carry momentum. This depletion artifact can be limited by shortening the range of the colloid–solvent interaction [123]. Another, related, problem with MPC fluids, similar to that in DPD fluids, is the limit on its ability to capture short-range lubrication interactions, again set by the range of the colloid–solvent repulsive interaction, which sets the gap between colloidal particles below which solvent particles are squeezed out and lubrication fails. Padding and Louis [123] very thoroughly discuss the dimensionless groups controlling the behavior of colloids in MPC simulations, including relationships among Reynolds number, Mach number, Knudsen number, and how they relate to colloid size, grid size, and solvent mean-free path. The time scales of the method are also thoroughly discussed in the same publication. Schunk et al. [33] have compared predictions of colloidal rheology predicted by LAMMPS versions of MPC with that of Dissipative Particle Dynamics and found much better performance with MPC simulations. MPC simulations have been used to study the transition to a nematic phase in a suspension of rod-like particles [124], as well as sedimentation dynamics [125].

3.5.2 Lattice Boltzmann Method

The Lattice Boltzmann (LB) method was born out of earlier work with lattice gas simulations [126–129] but with a simple empirical collision operator, such as the Bhatnagar–Gross–Krook operator, that greatly increases the flexibility of the LB method [128,129]. The novelty in LB simulations relative to other methods is the discretization in momentum space as well as in real space and time, wherein fictitious "particles" carry momentum only in a discrete set of directions with discrete values of the momentum. While resolution of ordinary finite differencing or finite elements is limited by time and space discretization, LB goes "whole-hog" by also losing local resolution in momentum space. Since accuracy is seriously limited anyway at the small length and time scales, the additional loss due to discretization in momentum space is not particularly serious, at least not in comparison to the gain in speed, relative to particle-based methods that retain a continuous distribution of particle momentum, such as in the DPD method or MD simulations. In this sense, as explained by Li et al. [130] in a chart of simulation methods with resolution of velocity on one axis and computational intensity on the other, LB resides between Direct Numerical Simulation (DNS) of the Navier–Stokes equation and DPD on the other, with MD simulations on the extreme of high resolution/high computational cost. If one wishes to use the cheapest particle-based method, and does not need or want Brownian fluctuations, the LB method becomes a prime candidate to consider.

Of course, there are challenges and limitations to the method. Care is required to perform the discretization properly, and there is a substantial literature on this. For high Reynolds number flows, the BGK (single-relaxation-time) operator leads to artificial instabilities [131], which can be ameliorated by use of a multiple-relaxation-time (MRT) collision operator [130]. The LB method seems to be technically more demanding than some of the other particle-based methods such as DPD and MPC simulations. However, commercial codes for LB simulations exist, and the LB method is now routinely used in simulations of fluid mechanics problems as a substitute for other standard methods [129], and is especially good at allowing tortuous boundary conditions, such as that of porous media. Brief reviews of the method are available elsewhere [95,128,130].

To apply the method to suspensions, Brownian motion can be added, which is not native to the original method. Ladd [87,132] accomplished this by adding a stochastic contribution to the collision operator, to which improvements were added by Dunweg et al. [127] and Adhikari et al. [133] to ensure equipartition of energy on all length scales. In addition, for suspensions, the transport of momentum across a fluid–solid boundary must be managed. This was initially done by modifying the "bounce-back" rule for stationary surfaces [59,87] to moving surfaces. Later, it was performed by coupling the motion of the colloidal particle to the fluid through a frictional drag [134,135], similar to the method used in the Immersed Boundary (IB) method [136,137]. Discretization of a spherically curved surface is carried out by finding the links between lattice nodes that the curved surface cuts; see Figure 3.9. LB fluid "particles" (which are actually momentum packets) moving along these links interact with the solid at the halfway point along the links, thus producing a discretization that, while not smooth locally, represents the particle more accurately as the grid is refined. Even when the particle diameter is only five times the grid spacing, the hydrodynamic interactions are within 1% of the exact result [128]; except when particles come into close proximity; see Figure 3.10 open symbols.

Typically sized particles used in LB simulations have radii less than five lattice spacings and, for these sizes, hydrodynamics is not properly modeled when particle

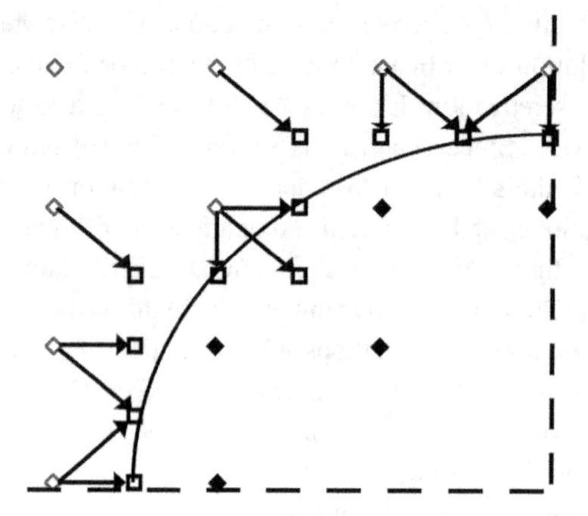

Figure 3.9 Location of boundary nodes for a curved surface. The velocities along links cutting the boundary surface are indicated by arrows. The locations of the boundary nodes are shown by filled squares, and the fluid nodes by filled circles. The open circles indicate nodes in the solid adjacent to fluid nodes. Adapted from Dünweg and Ladd [128]

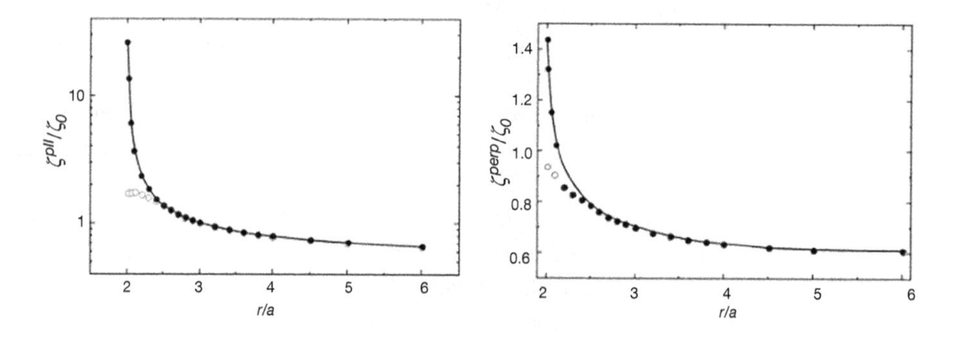

Figure 3.10 Effect of lubrication hydrodynamic interactions on friction coefficients for motion of two spheres parallel to the line of centers (left) and perpendicular to the line of centers (right) from LB simulations (symbols) with particles of radius $a = 2.5b$, where b is the lattice spacing, both with (solid) and without (open) lubrication hydrodynamics and compared to nearly exact results (lines). Data from Dünweg and Ladd [128]

separations decrease to 10% of the particle radius or less, which occurs commonly in dense suspensions. Small-separation lubrication corrections have, however, been introduced into LB simulations, resulting in the improvement shown in Figure 3.10 (closed symbols) [128]. Various schemes also exist for introducing non-Newtonian solvent effects into the LB method [138,139]. Ladd [140] has provided a recent review of the methods and limitations of LB methods for suspension flows.

Beyond suspension flows, the Lattice Boltzmann method has been applied to a huge range of applications; a recent sampling ranges from sedimentation of blood cells [141], microchannel suspension flow [142], granular flows [143], the flow of non-Newtonian nanofluids, nanoparticle-laden interfaces [95], and many others. The LB method is frequently used as a momentum transport solver in combination with other methods, such as the Immersed Boundary Method (IBM), for tracking compositional fields or interfaces.

3.6 Summary

Many methods are now available for the simulation of suspensions and these are often publicly available. The most numerous are available for spherical hard particles in low Reynolds number flows in Newtonian solvents. For these problems, the linearity of Stokes flow is exploited to great advantage and thousands of particles can be simulated accurately with full hydrodynamic interactions in a variety of geometries using "Stokesian particle" methods, of which there are now multiple codes, under continuous improvement. These methods avoid the need to either grid the solvent or replace it by discrete fluid particles, which conveys the great advantage of being able to handle very thin lubrication layers without excessive grid refinement, which is incurred when using methods requiring meshing of a continuum solvent, or the use of huge numbers of small solvent particles when using particle-solvent methods.

"Particle-solvent" methods, which capture hydrodynamics through tracking of discrete solvent particles rather than continua, can however be exploited for spherical particles in Stokes flow, either at lower colloid density, or by using counter-measures to allow lubrication films, as, for example, in the "Core-Modified" Dissipative Particle Dynamics method, which simulates both lubricated and frictional contacts. In addition, particle-solvent methods can become favorable over Stokesian particle methods when inertia is present, when the suspended phase is a liquid or a deformable solid, the solvent is viscoelastic, the particles are nonspherical, or multiple of these features are present simultaneously. If the fluid is complex, and high accuracy is not required, solvent particle methods, especially Brownian dynamics (BD), Dissipative Particle Dynamics (DPD), Multi-Particle Collision (MPC) dynamics, and Lattice Boltzmann (LB) dynamics, are versatile, and can quickly be adapted to problems with great complexity, such as mixtures containing polymers, colloids, emulsions, swimmers, etc. Establishing the accuracy of such methods and which method to use for a given problem remain outstanding problems.

Methods in which the fluid flow between particles is resolved using a meshed continuum have been covered here in less detail. These include the Arbitrary Lagrangian–Eulerian (ALE) method, the Distributed Lagrange Multiplier/Fictitious Domain (DLM/FD) method, the Force Coupling Method (FCM), the Immersed Boundary Method (IBM), Fluid Particle Dynamics (FPD), and the Smoothed Profile Method (SPM). It seems unclear at present how to judge the relative strengths and weaknesses of these methods relative to the others covered here in more detail. It would be valuable in the future to establish benchmark problems by which the relative merits and demerits of the various methods can be assessed.

Chapter Notation

a_{ij}	repulsion parameter between DPD particles i and j, Eq. (3.4) [N]
b	Lattice-Boltzmann lattice spacing [m]
$\mathbf{F}_i$	total force on particle i by other particles and external fields [N]
$\mathbf{F}_{ij}^C$	conservative force acting on particle i by particle j in DPD simulations [N]
$\mathbf{F}_{ij}^{\text{Core}-(n,t)}$	core force acting on particle i by particle j influenced by particle roughness in DPD simulations [N]
$\mathbf{F}_{ij}^D$	dissipative force acting on particle i by particle j in DPD simulations [N]
$\mathbf{F}_{ij}^{el}$	electrostatic force acting on particle i by particle j in DPD simulations [N]
$\mathbf{F}_{ij}^H$	lubrication force acting on particle i by particle j [N]
$\mathbf{F}_{ij}^R$	random force acting on particle i by particle j in DPD simulations [N]
h_{ij}	gap between particles i and j [m]
L	length scale large enough to represent the whole suspension [m]
m	mass of the DPD particle, Eq. (3.8) [kg]
n	number of DPD particles per unit volume [m^{-3}]
$\mathbf{r}_i$	position vector of particle i [m]
r_c	range or cut-off radius of interparticle attraction potential [m]
U	characteristic velocity of the flow, Eq. (3.8) [m/s]
$\mathbf{u}(\mathrm{t})$	mean velocity in a cell in MPC (or SRD) method [m/s]

Greek Symbols

γ	friction coefficient (Eq. 3.5) in DPD method [N/m s]
δ	lubrication layer thickness in DPD method [m]
$\boldsymbol{\theta}_{ij}$	white-noise Gaussian random variable [–]
λ	numerical constant in Eq. (3.8) [–]
ξ	random force in Eq. (3.1) [N]
σ	noise amplitude in Eq. (3.6) [N $s^{1/2}$]
Ω	rotation matrix [–]

Abbreviations

ALE	Arbitrary Lagrangian–Eulerian method
ASD	Accelerated Stokesian Dynamics method
BD	Brownian Dynamics method
BEM	Boundary Element method
DLM/FD	Distributed Lagrange Multiplier/Fictitious Domain method
DNS	Direct Numerical Simulation method
DPD	Dissipative Particle Dynamics method

F-DPD	Frictional Dissipative Particle Dynamics method
FCM	Force Coupling method
FD	Finite Difference method
FE	Finite Element method
FLD	Fast Lubrication Dynamics method
FPD	Fluid Particle Dynamics method
HI	hydrodynamic interactions
IBM	Immersed Boundary method
LB	Lattice Boltzmann method
MCS	Meshed Continuum Solvent method
MD	Molecular Dynamics
MPC	Multi-Particle Collision method
SD	Stokesian Dynamics method
SP	Stokesian Particle method
SPH	Smoothed Particle Hydrodynamics method
SPM	Smoothed Profile method
SRD	Stochastic Rotation Dynamics method

References

1. Stickel JJ, Powell RL. Fluid mechanics and rheology of dense suspensions. *Annual Review of Fluid Mechanics*. 2005;37:129–149.
2. Ermak DL, McCammon JA. Brownian dynamics with hydrodynamic interaction. *Journal of Chemical Physics*. 1978;69(4):1352–1360.
3. Strating P. Brownian dynamics simulation of a hard-sphere suspension. *Physical Review E*. 1999;59(2):2175–2187.
4. Heyes DM, Melrose JR. Brownian dynamics simulations of model hard-sphere suspensions. *Journal of Non-Newtonian Fluid Mechanics*. 1993;46(1):1–28.
5. Van der Werff JC, de Kruif CG. Hard-sphere colloidal disperisons – the scaling of rheolgical properties with particle-size, volume fraction, and shear rate. *Journal of Rheology*. 1989;33(3):421–454.
6. Foffi G, De Michele C, Sciortino F, Tartaglia P. Arrested phase separation in a short-ranged attractive colloidal system: A numerical study. *Journal of Chemical Physics*. 2005;122(22):13.
7. Zia RN, Landrum BJ, Russel WB. A micro-mechanical study of coarsening and rheology of colloidal gels: Cage building, cage hopping, and Smoluchowski's ratchet. *Journal of Rheology*. 2014;58(5):1121–1157.
8. Durlofsky L, Brady JF, Bossis G. Dynamic simulation of hydrodynamically interacting particles *Journal of Fluid Mechanics*. 1987;180:21–49.
9. Sierou A, Brady JF. Accelerated Stokesian dynamics simulations. *Journal of Fluid Mechanics*. 2001;448:115–146.
10. Phung TN, Brady JF, Bossis G. Stokesian Dynamics simulation of Brownian suspensions. *Journal of Fluid Mechanics*. 1996;313:181–207.
11. Wagner NJ, Brady JF. Shear thickening in colloidal dispersions. *Physics Today*. 2009;62 (10):27–32.

12. Swan JW, Brady JF, Moore RS, ChE. Modeling hydrodynamic self-propulsion with Stokesian Dynamics. Or teaching Stokesian dynamics to swim. *Physics of Fluids*. 2011;23(7):19.
13. Kopp M, Hofling F. GPU-accelerated simulation of colloidal suspensions with direct hydrodynamic interactions. *European Physical Journal-Special Topics*. 2012;210 (1):101–117.
14. Cao XJ, Cummins HZ, Morris JF. Hydrodynamic and interparticle potential effects on aggregation of colloidal particles. *Journal of Colloid and Interface Science*. 2012;368 (1):86–96.
15. Drews AM, Cartier CA, Bishop KJM. Contact charge electrophoresis: Experiment and theory. *Langmuir*. 2015;31(13):3808–3814.
16. Jeffrey DJ, Onishi Y. Calculation of the resistance and mobility functions for two unequal rigid spheres in low-Reynolds-number flow. *Journal of Fluid Mechanics*. 1984;139:261–290.
17. Kim S, Karrila SJ. *Microhydrodynamics: Principles and Selected Applications*. Boston: Butterworth-Heinemann; 1991. xxiii, 507 p.
18. Mohammadi M, Larson ED, Liu J, Larson RG. Brownian dynamics simulations of coagulation of dilute uniform and anisotropic particles under shear flow spanning low to high Peclet numbers. *Journal of Chemical Physics*. 2015;142(2):16.
19. Weinbaum S, Ganatos P, Yan ZY. Numerical multipole and boundary integral-equation techniques in stokes-flow. *Annual Review of Fluid Mechanics*. 1990;22:275–316.
20. Foss DR, Brady JF. Structure, diffusion and rheology of Brownian suspensions by Stokesian Dynamics simulation. *Journal of Fluid Mechanics*. 2000;407:167–200.
21. Ladd AJC. Hydrodynamic transport-coefficients of random dispersions of hard-spheres *Journal of Chemical Physics*. 1990;93(5):3484–3494.
22. Banchio AJ, Brady JF. Accelerated Stokesian dynamics: Brownian motion. *Journal of Chemical Physics*. 2003;118(22):10323–10332.
23. Hockney RW, Eastwood JW. *Computer Simulation using Particles*. Special student ed. Bristol, UK: A. Hilger; 1988. xxi, 540 p.
24. Zick AA, Homsy GM. Stokes-flow through periodic arrays of spheres. *Journal of Fluid Mechanics*. 1982;115:13–26.
25. Wang M, Brady JF. Spectral Ewald acceleration of Stokesian Dynamics for polydisperse suspensions. *Journal of Computational Physics*. 2016;306:443–477.
26. Kutteh R. Rigid body dynamics approach to Stokesian dynamics simulations of nonspherical particles. *Journal of Chemical Physics*. 2010;132(17):18.
27. Ball RC, Melrose JR. A simulation technique for many spheres in quasi-static motion under frame-invariant pair drag and Brownian forces. *Physica A*. 1997;247(1–4):444–472.
28. Kumar A, Higdon JJL. Origins of the anomalous stress behavior in charged colloidal suspensions under shear. *Physical Review E*. 2010;82(5):7.
29. Kumar A, Higdon JJL. Particle mesh Ewald Stokesian dynamics simulations for suspensions of non-spherical particles. *Journal of Fluid Mechanics*. 2011;675:297–335.
30. Beenakker CWJ. The effective viscosity of a concentrated suspension of spheres (and its relation to diffusion). *Physica A*. 1984;128(1–2):48–81.
31. Bolintineanu DS, Grest GS, Lechman JB, Pierce F, Plimpton SJ, Schunk PR. Particle dynamics modeling methods for colloid suspensions. *Computational Particle Mechanics*. 2014;1(3):321–356.
32. Bybee MD. Hydrodynamic Simulation of Colloidal Gels: Microstructure, Dynamics and Rheology [PhD]. Urbana, IL: University of Illinois at Urbana-Champagne; 2009.

33. Schunk PR, Pierce F, Lechman JB, Grillet AM, Veld PJI, Weiss H, et al. Performance of mesoscale modeling methods for predicting rheological properties of charged polystyrene/water suspensions. *Journal of Rheology*. 2012;56(2):353–384.
34. Melrose JR, Ball RC. The pathological behavior of sheared hard-spheres with hydrodynamic interactions. *Europhysics Letters*. 1995;32(6):535–540.
35. Seto R, Mari R, Morris JF, Denn MM. Discontinuous shear thickening of frictional hard-sphere suspensions. *Physical Review Letters*. 2013;111(21):5.
36. Mari R, Seto R, Morris JF, Denn MM. Shear thickening, frictionless and frictional rheologies in non-Brownian suspensions. *Journal of Rheology*. 2014;58(6):1693–1724.
37. Youngren GK, Acrivos A. Stokes flow past a particle of arbitrary shape – numerical-method of solution. *Journal of Fluid Mechanics*. 1975;69(MAY27):377–403.
38. Ramanujan S, Pozrikidis C. Deformation of liquid capsules enclosed by elastic membranes in simple shear flow: Large deformations and the effect of fluid viscosities. *Journal of Fluid Mechanics*. 1998;361:117–143.
39. Lac E, Morel A, Barthes-Biesel D. Hydrodynamic interaction between two identical capsules in simple shear flow. *Journal of Fluid Mechanics*. 2007;573:149–169.
40. Pozrikidis C. Numerical simulation of cell motion in tube flow. *Annals of Biomedical Engineering*. 2005;33(2):165–178.
41. Ishikawa T, Simmonds MP, Pedley TJ. Hydrodynamic interaction of two swimming model micro-organisms. *Journal of Fluid Mechanics*. 2006;568:119–160.
42. Zhao H, Shaqfeh ESG, Narsimhan V. Shear-induced particle migration and margination in a cellular suspension. *Physics of Fluids*. 2012;24(1):21.
43. Ishikawa T, Pedley TJ. The rheology of a semi-dilute suspension of swimming model micro-organisms. *Journal of Fluid Mechanics*. 2007;588:399–435.
44. Bao YX, Rachh M, Keaveny EE, Greengard L, Donev A. A fluctuating boundary integral method for Brownian suspensions. *Journal of Computational Physics*. 2018;374:1094–1119.
45. Swan JW, Wang G. Rapid calculation of hydrodynamic and transport properties in concentrated solutions of colloidal particles and macromolecules. *Physics of Fluids*. 2016;28(1):20.
46. Maxey MR, Patel BK. Localized force representations for particles sedimenting in Stokes flow. *International Journal of Multiphase Flow*. 2001;27(9):1603–1626.
47. Hu HH. Direct simulation of flows of solid-liquid mixtures. *International Journal of Multiphase Flow*. 1996;22(2):335–352.
48. Hu HH, Patankar NA, Zhu MY. Direct numerical simulations of fluid-solid systems using the arbitrary Lagrangian–Eulerian technique. *Journal of Computational Physics*. 2001;169(2):427–462.
49. Trofa M, Vocciante M, D'Avino G, Hulsen MA, Greco F, Maffettone PL. Numerical simulations of the competition between the effects of inertia and viscoelasticity on particle migration in Poiseuille flow. *Computers & Fluids*. 2015;107:214–223.
50. Choi YJ, Hulsen MA, Meijer HEH. An extended finite element method for the simulation of particulate viscoelastic flows. *Journal of Non-Newtonian Fluid Mechanics*. 2010;165(11–12):607–624.
51. Luo X, Maxey MR, Karniadakis GE. Smoothed profile method for particulate flows: Error analysis and simulations. *Journal of Computational Physics*. 2009;228(5):1750–1769.
52. Lomholt S, Maxey MR. Force-coupling method for particulate two-phase flow: Stokes flow. *Journal of Computational Physics*. 2003;184(2):381–405.

53. Liu D, Maxey M, Karniadakis GE. A fast method for particulate microflows. *Journal of Microelectromechanical Systems*. 2002;11(6):691–702.
54. Liu D, Maxey M, Karniadakis GE. FCM-spectral element method for simulating colloidal micro-devices. *Computational Fluid and Solid Mechanics*. 2003;1–2:1413–1416.
55. Yeo K, Maxey MR. Simulation of concentrated suspensions using the force-coupling method. *Journal of Computational Physics*. 2010;229(6):2401–2421.
56. Dance SL, Maxey MR. Particle density stratification in transient sedimentation. *Physical Review E*. 2003;68(3):11.
57. Krieger IM, Dougherty TJ. A mechanism for non-Newtonian flow in suspensions of rigid spheres. *Transactions of the Society of Rheology*. 1959;3(1):137–152.
58. Batchelor GK, Green JT. Determination of bulk stress in a suspension of spherical-particles to order C-2. *Journal of Fluid Mechanics*. 1972;56:401–427.
59. Ladd AJC, Frenkel D. Dissipative hydrodynamic interactions via lattice-gas cellular automata *Physics of Fluids A: Fluid Dynamics*. 1990;2(11):1921–1924.
60. Glowinski R, Pan TW, Hesla TI, Joseph DD. A distributed Lagrange multiplier fictitious domain method for particulate flows. *International Journal of Multiphase Flow*. 1999;25 (5):755–794.
61. Glowinski R, Pan TW, Hesla TI, Joseph DD, Periaux J. A fictitious domain approach to the direct numerical simulation of incompressible viscous flow past moving rigid bodies: Application to particulate flow. *Journal of Computational Physics*. 2001;169(2):363–426.
62. Patankar NA, Singh P, Joseph DD, Glowinski R, Pan TW. A new formulation of the distributed Lagrange multiplier/fictitious domain method for particulate flows. *International Journal of Multiphase Flow*. 2000;26(9):1509–1524.
63. Gallier S, Lemaire E, Lobry L, Peters F. A fictitious domain approach for the simulation of dense suspensions. *Journal of Computational Physics*. 2014;256:367–387.
64. Hao J, Pan TW, Glowinski R, Joseph DD. A fictitious domain/distributed Lagrange multiplier method for the particulate flow of Oldroyd-B fluids: A positive definiteness preserving approach. *Journal of Non-Newtonian Fluid Mechanics*. 2009;156(1–2):95–111.
65. Wachs A. A DEM-DLM/FD method for direct numerical simulation of particulate flows: Sedimentation of polygonal isometric particles in a Newtonian fluid with collisions. *Computers & Fluids*. 2009;38(8):1608–1628.
66. Shi X, Phan-Thien N. Distributed Lagrange multiplier/fictitious domain method in the framework of lattice Boltzmann method for fluid structure interactions. *Journal of Computational Physics*. 2005;206:81–94.
67. Peskin CS. Flow patterns around heart valves–numerical method. *Journal of Computational Physics*. 1972;10(2):252.
68. Peskin CS. Numerical analysis of blood-flow in heart *Journal of Computational Physics*. 1977;25(3):220–252.
69. Fogelson AL, Peskin CS. A fast numerical-method for solving the 3-dimensional Stokes equations in the presence of suspended particles. *Journal of Computational Physics*. 1988;79(1):50–69.
70. Beyer RP, Leveque RJ. Analysis of a one-dimensional model for the immersed boundary method. *SIAM Journal of Numerical Analysis*. 1992;29(2):332–364.
71. Delong S, Usabiaga FB, Delgado-Buscalioni R, Griffith BE, Donev A. Brownian dynamics without Green's functions. *Journal of Chemical Physics*. 2014;140(13):23.
72. Landau LD, Lifshits EM. *Fluid Mechanics*, 2nd ed. Oxford: Pergamon Press; 1987. xiii, 539 p.

73. Ladd AJC, Gang H, Zhu JX, Weitz DA. Temporal and spatial dependence of hydrodynamic correlations–simulation and experiment *Physical Review E.* 1995;52(6):6550–6572.
74. Vowinckel B, Kempe T, Frohlich J. Fluid-particle interaction in turbulent open channel flow with fully-resolved mobile beds. *Advances in Water Resources.* 2014;72:32–44.
75. Feng ZG, Musong SG. Direct numerical simulation of heat and mass transfer of spheres in a fluidized bed. *Powder Technology.* 2014;262:62–70.
76. Zhou Q, Fan LS. A second-order accurate immersed boundary-lattice Boltzmann method for particle-laden flows. *Journal of Computational Physics.* 2014;268:269–301.
77. Kruger T, Kaoui B, Harting J. Interplay of inertia and deformability on rheological properties of a suspension of capsules. *Journal of Fluid Mechanics.* 2014;751:725–745.
78. Deen NG, Peters E, Padding JT, Kuipers JAM. Review of direct numerical simulation of fluid–particle mass, momentum and heat transfer in dense gas-solid flows. *Chemical Engineering Science.* 2014;116:710–724.
79. Tanaka H, Araki T. Simulation method of colloidal suspensions with hydrodynamic interactions: Fluid particle dynamics. *Physical Review Letters.* 2000;85(6):1338–1341.
80. Marth W, Aland S, Voigt A. Margination of white blood cells: A computational approach by a hydrodynamic phase field model. *Journal of Fluid Mechanics.* 2016;790:18.
81. Araki T, Tanaka H. Colloidal aggregation in a nematic liquid crystal: Topological arrest of particles by a single-stroke disclination line. *Physical Review Letters.* 2006;97(12):127801.
82. Furukawa A, Marenduzzo D, Cates ME. Activity-induced clustering in model dumbbell swimmers: The role of hydrodynamic interactions. *Physical Review E.* 2014;90(2):16.
83. Nakayama Y, Yamamoto R. Simulation method to resolve hydrodynamic interactions in colloidal dispersions. *Physical Review E.* 2005;71(3):7.
84. Nakayama Y, Kim K, Yamamoto R. Simulating (electro) hydrodynamic effects in colloidal dispersions: Smoothed profile method. *European Physical Journal E.* 2008;26(4):361–368.
85. Molina JJ, Otomura K, Shiba H, Kobayashi H, Sano M, Yamamoto R. Rheological evaluation of colloidal dispersions using the smoothed profile method: Formulation and applications. *Journal of Fluid Mechanics.* 2016;792:590–619.
86. Sharma N, Patankar NA. Direct numerical simulation of the Brownian motion of particles by using fluctuating hydrodynamic equations. *Journal of Computational Physics.* 2004;201(2):466–486.
87. Ladd AJC. Numerical simulations of particulate suspensions via a discretized Boltzmann-equation. 1. Theoretical foundation. *Journal of Fluid Mechanics.* 1994;271:285–309.
88. Ladd AJC, Verberg R. Lattice-Boltzmann simulations of particle-fluid suspensions. *Journal of Statistical Physics.* 2001;104(5–6):1191–1251.
89. Monaghan JJ. Smoothed particle hydrodynamics *Annual Review of Astronomy and Astrophysics.* 1992;30(1):543–574.
90. Monaghan JJ. Smoothed particle hydrodynamics. *Reports on Progress in Physics.* 2005;68(8):1703–1759.
91. Hoover WG. *Smooth Particle Applied Mechanics: The State of the Art.* Singapore: World Scientific; 2006. xiii, 300 p.
92. Li S, Liu WK. *Meshfree Particle Methods.* Berlin: Springer; 2004. iii, 502 p.
93. Bian X, Litvinov S, Qian R, Ellero M, Adams NA. Multiscale modeling of particle in suspension with smoothed dissipative particle dynamics. *Physics of Fluids.* 2012;24(1):20.

94. Vazquez-Quesada A, Bian X, Ellero M. Three-dimensional simulations of dilute and concentrated suspensions using smoothed particle hydrodynamics. *Computational Particle Mechanics*. 2016;3(2):167–178.
95. Harting J, Frijters S, Ramaioli M, Robinson M, Wolf DE, Luding S. Recent advances in the simulation of particle-laden flows. *European Physical Journal-Special Topics*. 2014;223(11):2253–2267.
96. Español P, Revenga M. Smoothed dissipative particle dynamics. *Physical Review E*. 2003;67(2):12.
97. Bian X, Litvinov S, Ellero M, Wagner NJ. Hydrodynamic shear thickening of particulate suspension under confinement. *Journal of Non-Newtonian Fluid Mechanics*. 2014;213:39–49.
98. Belytschko T, Lu YY, Gu L. Element-free Galerkin methods. *International Journal for Numerical Methods in Engineering*. 1994;37(2):229–256.
99. Liu WK, Jun S, Zhang YF. Reproducing kernel particle methods. *International Journal for Numerical Methods in Fluids*. 1995;20(8–9):1081–1106.
100. Onate E, Idelsohn S, Zienkiewicz OC, Taylor RL. A finite point method in computational mechanics. Applications to convective transport and fluid flow. *International Journal for Numerical Methods in Engineering*. 1996;39(22):3839–3866.
101. Hoogerbrugge PJ, Koelman J. Simulating microscopic hydrodynamic phenomena with dissipative particle dynamics. *Europhysics Letters*. 1992;19(3):155–160.
102. Español P, Warren P. Statistical-mechanics of dissipative particle dynamics. *Europhysics Letters*. 1995;30(4):191–196.
103. Groot RD, Warren PB. Dissipative particle dynamics: Bridging the gap between atomistic and mesoscopic simulation. *Journal of Chemical Physics*. 1997;107(11):4423–4435.
104. Liu MB, Liu GR, Zhou LW, Chang JZ. Dissipative particle dynamics (DPD): An overview and recent developments. *Archives of Computational Methods in Engineering*. 2015;22(4):529–556.
105. Koelman J, Hoogerbrugge PJ. Dynamic simulation of hard-sphere suspensions under shear *Europhysics Letters*. 1993;21(3):363–368.
106. Mai-Duy N, Pan D, Phan-Thien N, Khoo BC. Dissipative particle dynamics modeling of low Reynolds number incompressible flows. *Journal of Rheology*. 2013;57(2):585–604.
107. Huang Y, Marson RL, Larson RG. Inertial migration of a rigid sphere in plane Poiseuille flow as a test of dissipative particle dynamics simulations. *Journal of Chemical Physics*. 2018;149(16):17.
108. Marsh CA, Backx G, Ernst MH. Static and dynamic properties of dissipative particle dynamics. *Physical Review E*. 1997;56(2):1676–1691.
109. Pan WX, Caswell B, Karniadakis GE. Rheology, microstructure and migration in Brownian colloidal suspensions. *Langmuir*. 2010;26(1):133–142.
110. Whittle M, Travis KP. Dynamic simulations of colloids by core-modified dissipative particle dynamics. *Journal of Chemical Physics*. 2010;132(12):16.
111. Jamali S, Boromand A, Wagner N, Maia J. Microstructure and rheology of soft to rigid shear-thickening colloidal suspensions. *Journal of Rheology*. 2015;59(6):1377–1395.
112. Boromand A, Jamali S, Grove B, Maia JM. A generalized frictional and hydrodynamic model of the dynamics and structure of dense colloidal suspensions. *Journal of Rheology*. 2018;62(4):905–918.
113. Royer JR, Blair DL, Hudson SD. Rheological signature of frictional interactions in shear thickening suspensions. *Physical Review Letters*. 2016;116(18):5.

114. Cwalina CD, Wagner NJ. Material properties of the shear-thickened state in concentrated near hard-sphere colloidal dispersions. *Journal of Rheology*. 2014;58(4):949–967.
115. Boek ES, Coveney PV, Lekkerkerker HNW, vanderSchoot P. Simulating the rheology of dense colloidal suspensions using dissipative particle dynamics. *Physical Review E*. 1997;55(3):3124–3133.
116. Martys NS. Study of a dissipative particle dynamics based approach for modeling suspensions. *Journal of Rheology*. 2005;49(2):401–424.
117. Huang YD, Marson RL, Larson RG. Inertial migration of neutrally buoyant prolate and oblate spheroids in plane Poiseuille flow using dissipative particle dynamics simulations. *Computational Materials Science*. 2019;162:178–185.
118. Marson RL, Huang YD, Huang M, Fu TT, Larson RG. Inertio-capillary cross-streamline drift of droplets in Poiseuille flow using dissipative particle dynamics simulations. *Soft Matter*. 2018;14(12):2267–2280.
119. Malevanets A, Kapral R. Mesoscopic model for solvent dynamics. *Journal of Chemical Physics*. 1999;110(17):8605–8613.
120. Ihle T, Kroll DM. Stochastic rotation dynamics. I. Formalism, Galilean invariance, and Green--Kubo relations. *Physical Review E*. 2003;67(6):11.
121. Gompper G, Ihle T, Kroll DM, Winkler RG. Multi-particle collision dynamics: A particle-based mesoscale simulation approach to the hydrodynamics of complex fluids. In: Holm C, Kremer K (eds.) *Advanced Computer Simulation Approaches for Soft Matter Sciences III. Advances in Polymer Science*, vol. 221. Berlin: Springer-Verlag; 2009. pp. 1–87.
122. Zhao TY, Wang XG, Jiang L, Larson RG. Assessment of mesoscopic particle-based methods in microfluidic geometries. *Journal of Chemical Physics*. 2013;139(8):10.
123. Padding JT, Louis AA. Hydrodynamic interactions and Brownian forces in colloidal suspensions: Coarse-graining over time and length scales. *Physical Review E*. 2006;74 (3):29.
124. Ripoll M, Winkler RG, Mussawisade K, Gompper G. Mesoscale hydrodynamics simulations of attractive rod-like colloids in shear flow. *Journal of Physics: Condensed Matter*. 2008;20(40):11.
125. Padding JT, Louis AA. Interplay between hydrodynamic and Brownian fluctuations in sedimenting colloidal suspensions. *Physical Review E*. 2008;77(1):11.
126. Frisch U, Hasslacher B, Pomeau Y. Lattice-gas automata for the Navier–Stokes equation. *Physical Review Letters*. 1986;56(14):1505–1508.
127. Dunweg B, Schiller UD, Ladd AJC. Statistical mechanics of the fluctuating lattice Boltzmann equation. *Physical Review E*. 2007;76(3):10.
128. Dunweg B, Ladd AJC. Lattice Boltzmann simulations of soft matter systems. In Holm C, Kremer K (eds.) *Advanced Computer Simulation Approaches for Soft Matter Sciences III. Advances in Polymer Science*, vol. 221. Berlin: Springer-Verlag; 2009. pp. 89–166.
129. Succi S. *The Lattice Boltzmann Equation for Fluid Dynamics and Beyond*. Oxford: Clarendon Press; 2001. xvi, 288 p.
130. Li Q, Luo KH, Kang QJ, He YL, Chen Q, Liu Q. Lattice Boltzmann methods for multiphase flow and phase-change heat transfer. *Progress in Energy and Combustion Science*. 2016;52:62–105.
131. Lallemand P, Luo LS. Theory of the lattice Boltzmann method: Dispersion, dissipation, isotropy, Galilean invariance, and stability. *Physical Review E*. 2000;61(6):6546–6562.
132. Ladd AJC. Numerical simulations of particulate suspensions via a discretized Boltzmann-equation. 2. Numerical results. *Journal of Fluid Mechanics*. 1994;271:311–339.

133. Adhikari R, Stratford K, Cates ME, Wagner AJ. Fluctuating lattice Boltzmann. *Europhysics Letters*. 2005;71(3):473–479.
134. Ahlrichs P, Dunweg B. Lattice-Boltzmann simulation of polymer-solvent systems. *International Journal of Modern Physics C*. 1998;9(8):1429–1438.
135. Ahlrichs P, Dunweg B. Simulation of a single polymer chain in solution by combining lattice Boltzmann and molecular dynamics. *Journal of Chemical Physics*. 1999;111 (17):8225–8239.
136. Peskin CS. The immersed boundary method. *Acta Numerica*. 2002;11:479–517.
137. Nash RW, Adhikari R, Cates ME. Singular forces and pointlike colloids in lattice Boltzmann hydrodynamics. *Physical Review E*. 2008;77(2):11.
138. Phillips TN, Roberts GW. Lattice Boltzmann models for non-Newtonian flows. *IMA Journal of Applied Mathematics*. 2011;76(5):790–816.
139. Papenkort S, Voigtmann T. Channel flow of a tensorial shear-thinning Maxwell model: Lattice Boltzmann simulations. *Journal of Chemical Physics*. 2014;140(16):13.
140. Ladd AJC. Lattice-Boltzmann methods for suspensions of solid particles. *Molecular Physics*. 2015;113(17–18):2531–2537.
141. Hashemi Z, Rahnama M, Jafari S. Lattice Boltzmann simulation of healthy and defective red blood cell settling in blood plasma. *Journal of Biomechanical Engineering–Transactions of the ASME*. 2016;138(5):10.
142. Sun DK, Wang Y, Dong AP, Sun BD. A three-dimensional quantitative study on the hydrodynamic focusing of particles with the immersed boundary – Lattice Boltzmann method. *International Journal of Heat Mass Transfer*. 2016;94:306–315.
143. Leonardi A, Cabrera M, Wittel FK, Kaitna R, Mendoza M, Wu W, et al. Granular-front formation in free-surface flow of concentrated suspensions. *Physical Review E*. 2015;92 (5):13.

4 Microstructure under Flow

Norman J. Wagner

4.1 Introduction

Previous chapters have highlighted the critical role of microstructure in determining colloidal suspension rheology. Figure 1.1 illustrates how processing affects microstructure, and therefore, processing is often convolved with rheology through microstructure. In this chapter, two experimental methods for determining microstructure under shear flow are briefly summarized to complete the presentation of foundational material for understanding the applications to follow.

An important learning from Chapter 2, expressed in Eq. (2.97), is that the contribution of colloidal interactions to the stress is a direct summation of the forces acting between two particles, summed over the positions of the particles. This can be expressed as an integral over the pair distribution function $g(\mathbf{r};\dot{\gamma})$, as shown in Eq. (2.103), where the vector **r** defines the center-to-center distance and direction of a neighboring particle. The equilibrium structures for hard sphere dispersions are illustrated in Figure 1.5, along with the direct connection to the pressure. It is the shear distortion that is relevant for determining the rheology in Section 2.5.2. Figure 2.8 illustrates such a distorted microstructure, by simulations using methods presented in Chapter 3, along with its relationship to the rheology. Thus, experimental measurement of the shear-distorted microstructure can be used to validate simulation and modeling, as well as in a direct calculation of the stress, as illustrated in chapter 3 of *CSR* [1]. Furthermore, shear distortion of the microstructure is essential to generate deviatoric stress contributions, such as viscosity and normal stress differences.

The contribution of Brownian motion can be treated in an analogous fashion as the interparticle forces because the Brownian motion acts in a statistical manner to homogenize the microstructure. G. K. Batchelor showed how to treat the effects of this stochastic force as an effective interparticle force [2]. The sum of the interparticle forces and the Brownian force contributions to the stress are termed the thermodynamic contribution to the stress, as at equilibrium these two types of forces balance to yield the equilibrium microstructure. Hydrodynamic interactions (HI) have a different symmetry because their contribution to the stress tensor arises from a stress dipole [3], a discussion of which is provided in textbooks such as Russel et al. [4] and in Chapter 2. Because hydrodynamic interactions are directly proportional to the shear rate, and the dissipation of energy in the fluid is due to the presence of particles, this contribution to the stress exists even without distortion of the microstructure.

The hydrodynamic contribution to the stress arises from the distortion of the flow field due to particles, which for very dilute suspensions yields the Einstein contribution to the viscosity, Eq. (1.1), as derived in chapter 2 of *CSR*. As HI contributions are also calculated by integration over the pair distribution function, these too will change with shear distortion of the microstructure. Perhaps most importantly, lubrication hydrodynamic interactions diverge when smooth particles approach, such that driving particles into close proximity can greatly increase the stress, as in shear thickening. This is discussed in some detail in chapter 8 of *CSR*. However, these contributions have a different symmetry because the integrals over the pair distribution function are weighted not by a force, but by a stresslet. This is because the particles are force and torque free in the absence of any external fields. A stresslet is a force dipole, analogous to an electric dipole. This difference in symmetry between a force and a force dipole becomes important when considering what features of the shear-distorted microstructure contribute to the stresses. This is presented in Section 4.2.

4.2 Structure Factors from Scattering

Most undergraduate courses in materials science and physical chemistry introduce some aspects of radiation scattering as a method for materials characterization, such as identifying crystal structure in atomic systems. A basic background is assumed, where concepts such as Bragg's law of diffraction provide guidance for qualitative understanding of the section. Indeed, many applications of scattering in suspension rheology are to identify qualitative or semi-quantitative changes in microstructure due to the flow, and as such, do not require a detailed knowledge of the theory behind scattering from colloidal suspensions. For those interested in mastering the subject, textbooks, including the recent book by Glatter [5], are a resource for further learning. In the following a few illustrations of the use of scattering as a probe of suspension microstructure during flow are presented as an introduction to the experimental methods and their research value.

4.2.1 Suspension Structure under Flow

Light, x-ray, and neutron scattering can all be used to measure the Fourier transform of the pair distribution function, known as the structure factor, $S(\mathbf{q};\dot{\gamma})$. Differences between these sources of radiation used for the scattering measurements lead to differences in the types of samples that can be probed, e.g., optically transparent, sensitive biological colloids, etc. Furthermore, differences in the wavelengths of the radiation sources yield information on different length scales, such that combinations of methods can often be valuable, as will be demonstrated in this section. Returning to the structure factor, this function is in what is termed *recriprocal* space, which corresponds to the Fourier transform of the *real* space pair distribution function $g(\mathbf{r};\dot{\gamma})$. Here $\mathbf{q}$ is the scattering vector that yields the momentum transfer associated with the radiation scattering and it has units of L^{-1}. Indeed, rheo-SANS (small angle

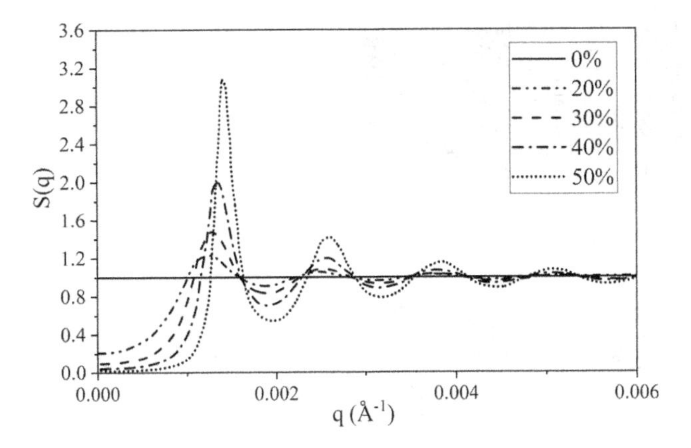

Figure 4.1 Equilibrium structure factor for suspensions of Brownian hard spheres of diameter 500 nm at the volume fractions indicated.

neutron scattering) methods have proven to be especially valuable for developing structure-property relations in complex fluids more generally [6]. The equilibrium structure factor introduced in Section 2.6.1 for a suspension of colloidal hard spheres is shown in Figure 4.1.

In the limit of infinite dilution $S(q) = 1$, indicating that there are no correlations between the particles when dilute. Note that $q = |\mathbf{q}|$. Increasing the particle concentration leads to crowding and the development of a cage of neighbors when $\phi \cong 0.4$ and above. This caging is essential for understanding the formation of attractive and repulsive driven glasses, which are discussed in detail in Chapter 5. This caging at the particle level is reflected in the growth of the first peak, indicating more nearest neighbors. Textbook references for understanding the mechanism of radiation scattering include Glatter [5] and Roe [7]. Bragg's law, $q2\pi/r$, for example, indicates that the q value corresponds to the peak, ~0.0014 Å^{-1} . For these particles that are about 450 nm in diameter this corresponds to the center-to-center separation distance for which particles are nearly at contact. The higher order peaks of decreasing intensity indicate that this structure becomes less and less correlated with distance from the reference particle. Note for a crystal, sharp peaks would be evident for these repeat inverse distances.

Of importance here is that $S(\mathbf{q};\dot{\gamma})$ is the 3D spatial Fourier transform of $g(\mathbf{r};\dot{\gamma})$, such that knowledge of one enables calculation of the other, with the usual caveats of the effects of noise and limited, accessible range of scattering vector along with other, experimental limitations. Two simple rules that arise from the mathematics of Fourier transforms aid in understanding how to interpret scattering. Firstly, Bragg's law illustrates how distances are inversely related to the scattering vector, such that q has units of L^{-1}, or inverse length. Secondly, features are rotated by 90° in reciprocal space relative to real space. This is possible to understand by invoking Bragg's law, as illustrated by the following thought experiment. A rod of length l and diameter d aligned along the flow will generate a scattering pattern that is highly anisotropic, but

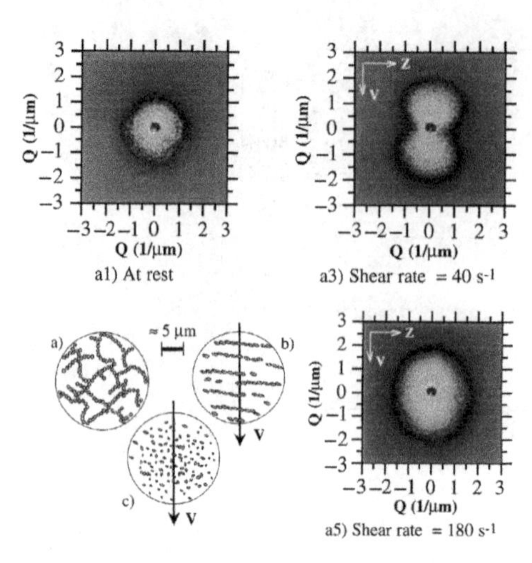

Figure 4.2 Scattering images from a thixotropic clay suspension under flow in the velocity–vorticity (1–3) plane. Clockwise from upper left: At rest, the suspension of anisotropic particles with random orientation yields a spherically symmetric pattern (at rest), while increasing shear leads to a "butterfly pattern" (40 s^{-1}), which becomes more isotropic at even higher shear rates (180 s^{-1}). This corresponds to real space structures shown in the lower left panel, where strong anisotropy in the vorticity direction (b) leads to the butterfly pattern aligned along the flow in reciprocal space. Reprinted with permission from Pignon et al., Light scattering pattern and rheology of sheared thixotropic clay gel. *Physical Review Letters* 1997;79(23):4689–4692 [9] by the American Physical Society, DOI 10.1103/PhysRevLett.79.4689

aligned orthogonal to the flow. This is because the scattering from the long length l is at short scattering vectors $q = 2\pi/l$, while the scattering for the short diameter d extends to longer scattering vectors $q = 2\pi/d$. In this manner, the anisotropy in the scattering from aligned rods can directly yield the aspect ratio [8].

The use of scattering to view particle alignment, or assembly into structures under flow, is illustrated in Figure 4.2, where thixotropic clay particles are probed in the 1–3, or velocity–vorticity direction. Although the clay particles are disc-like, the particles are gelled in an isotropic structure, and hence, the scattering at rest is isotropic. As illustrated in Figure 4.2, further increasing shear breaks up the flocculated network, leaving bands of clay with alignment in the vorticity direction. This results in a so-called butterfly pattern under flow [9,10]. Further shearing at high shear rates disperses the clay randomly and, again, a nearly isotropic scattering pattern emerges, but with higher intensity at larger scattering vector ("Q" which is the same as "q"), indicating smaller scattering objects, i.e., individual clay particles or small aggregates. Direct metrics of quantitative particle alignment can also be achieved simultaneously with rheological measurements using rheo-SANS measurements, as has been demonstrated for precipitated calcium carbonate particle suspensions under shear flow [8], along with other rod-like particle systems [6]. Rheo-SAXS (small angle x-ray scattering) methods probe the relationship between particle orientation and shear thinning rheology [11]. Importantly, colloidal gels have been shown to exhibit butterfly scattering

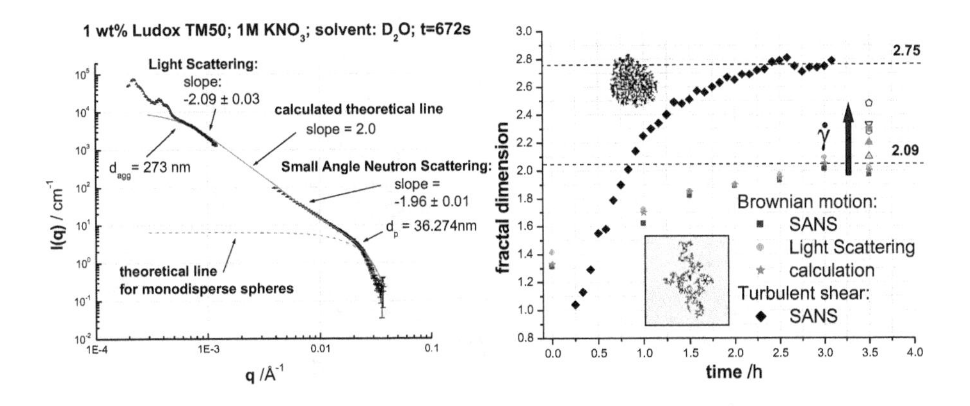

Figure 4.3 Rheo-SANS and Rheo-SALS studies of perikinetic and orthokinetic (under shear) aggregation of a colloidal silica in aqueous solution. The left panel shows a composite of the 1-D scattering at rest at 672 s after addition of salt, which leads to fractal aggregation with mass-fractal dimension around 2.1. The right panel shows how shearing affects this fractal structure. Shearing leads to orthokinetic aggregation and a much more compact structure with fractal dimension around 2.75. Also shown in the right panel is that shearing after 3.5 h of perikinetic aggregation drives further compaction. Adapted from the doctoral dissertation of M. Sommer [15]

in both the 1–3 and 1–2 plane using rheo-SANS and flow-SANS, providing a deeper understanding of how thixotropic gels yield and flow [12,13], as will be discussed in Chapter 5. Wide angle x-ray scattering under shear flow has been used to study rod-like self-assembled surfactant aggregates by novel instrumentation probing the 1–2 plane of flow, which enables a more quantitative understanding of particle alignment and dynamics under shear [14].

A straightforward use of rheo-scattering methods is to track particle aggregation and break-up in thixotropic systems. An example is shown in Figure 4.3, where aqueous suspensions of colloidal silica, Ludox^{TM}, are investigated by both light and neutron scattering at rest and under shear flow [15]. The left panel in Figure 4.3 is typical of scattering from a fractal structure formed at rest by perikinetic aggregation; more about this process of aggregation and the specific physics responsible for this fractal dimension can be found in chapter 6 of *CSR* [1]. Here, the very significant increase in scattering enables easy measurement of the fractal structure by scattering methods. The application of shearing leads to shear-driven or orthokinetic aggregation, which results in a much more compact structure with a higher fractal dimension. Also shown are results for shearing after perikinetic aggregation, again leading to compaction. These results along with those shown in Figure 4.2 illustrate how scattering methods under shear flow can be useful for studying aggregating thixotropic suspensions. Additional rheo-SANS studies indicate shearing can not only break down or build up fractal structures, but can also introduce significant concentration gradients under flow [16]. These methods are used to study some of the industrially relevant systems discussed further in Chapter 9.

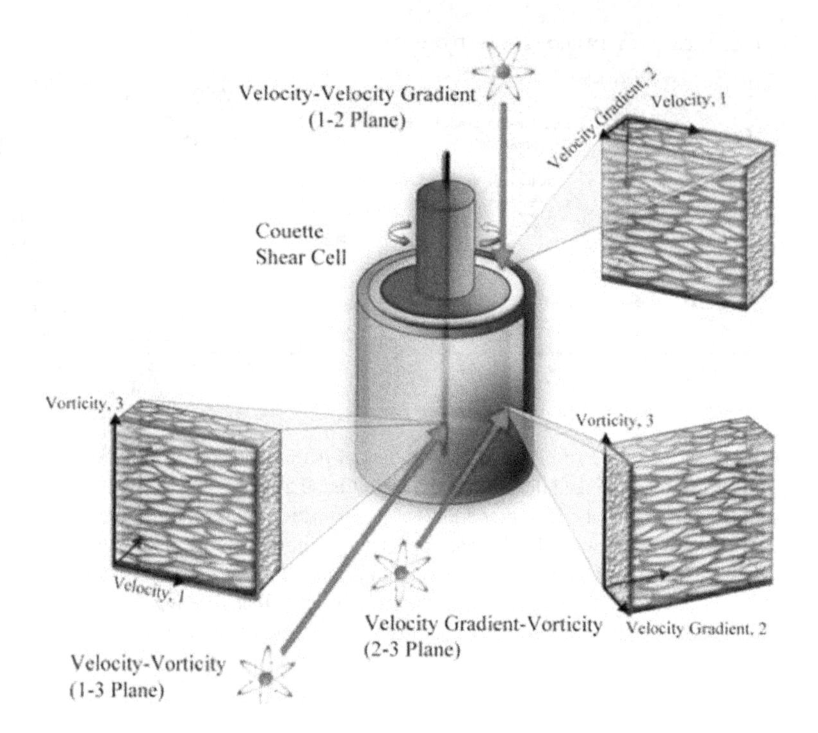

Figure 4.4 Illustration of the three primary projections for scattering measurements in the Couette geometry. Illustration courtesy of Keyi Xu.

A suspension under shear flow has a distorted microstructure in all three directions as defined by the shear flow, i.e., flow, shear gradient, and vorticity directions, typically labeled (1, 2, 3) for convenience. Scattering yields a two-dimensional projection of the structure factor, such that, in analogy to a shadow, any projection along one of these three axes contains only some of this structural information. Again, in analogy to a shadow, multiple projections are required to reconstruct the fully deformed microstructure; however, often much information can be gained by just one projection. Figure 4.4 illustrates the geometry of scattering experiments aligned along these primary axes defined by the flow.

For spherical particles, the scattering is isotropic at rest and the largest structural rearrangement under shear flow is evident in the 1–2 or velocity–shear gradient plane, which is commonly called the plane of shear. Probing this requires special instrumentation capable of projecting the radiation down the gradient direction, through the gap [14,17–19]. Typical results for a shearing colloidal suspension are shown in Figure 4.5, where increasing Péclet number leads to a strong anisotropy in the angular distribution of the nearest neighbors, which is evident as anisotropy in the scattering. Both shear thinning and shear thickening rheology can be associated with aspects of these nonequilibrium microstructures.

These spectra can be compared against theoretical results using the Smoluchowski equation approach presented in Chapter 2, as well as simulation results using the

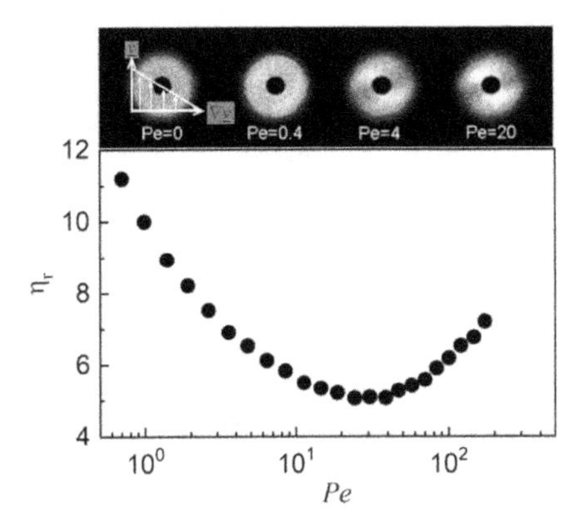

Figure 4.5 Small angle neutron scattering spectra obtained in the 1–2, velocity–flow gradient plane as a function of *Pe* for a 20 vol% suspension of near hard spheres under flow, where the flow curve is provided for reference. Adapted from *CSR*, figure 8.11 with the original data from Kalman [20]

Stokesian Dynamics method covered in Chapter 3. Figure 3.11 of *CSR* presents complementary simulation results from Foss and Brady [21] for comparison. The general shape for shear flow, corrected for angular rotation due to the transformation from real space to reciprocal space, follows that shown in Figure 4.6(b), where the plus/minus signs indicate compression and extension quadrants, respectively [22]. Rheo-SANS measurements of this strong angular anisotropy evident in the 1–2 plane show good agreement with simulation results, as shown in Figure 4.6(c) and (d) [17]. These plots of pair probability from simulation (c) and intensity (d) around the nearest neighbor peak are flat at equilibrium, but show strong anisotropy developing in the relative position of the neighboring particles under strong flow.

Given these geometries, a common method for analyzing scattering data from flowing systems, as well as for examining structures distorted by flow in simulations, is to expand the structure in spherical harmonics, as is often done in molecular orbitals, for example. Following the methods outlined in Wagner and Ackerson [23], the shear-distorted microstructure can be expanded as follows:

$$S(\mathbf{q};Pe) = 1 + \sum_{l,m} B^{+}_{l,m}(q;Pe)(Y_{l,m}(\Omega_k) + (-1)^m Y_{l,-m}(\Omega_k)) \tag{4.1}$$

The Péclet number is defined in Eq. (1.7) as the ratio of the characteristic time for Brownian motion to that for shearing. The scalar coefficients $B^{+}_{l,m}$ are functions of the magnitude of the scattering vector and depend on shear rate. The indices *l*, *m* are positive and $m \leq l$. The angular dependence is captured through the spherical harmonics, $(Y_{l,m}(\Omega_k) + (-1)^m Y_{l,-m}(\Omega_k))$.

While the spherical harmonic expansion is an infinite summation, at equilibrium only the first term $l = 0$, $m = 0$, the spherically symmetric term is nonzero. As discussed in Chapter 2, the first deformation to the microstructure, i.e., the linear response that is responsible for the zero shear viscosity, is given by a contribution with

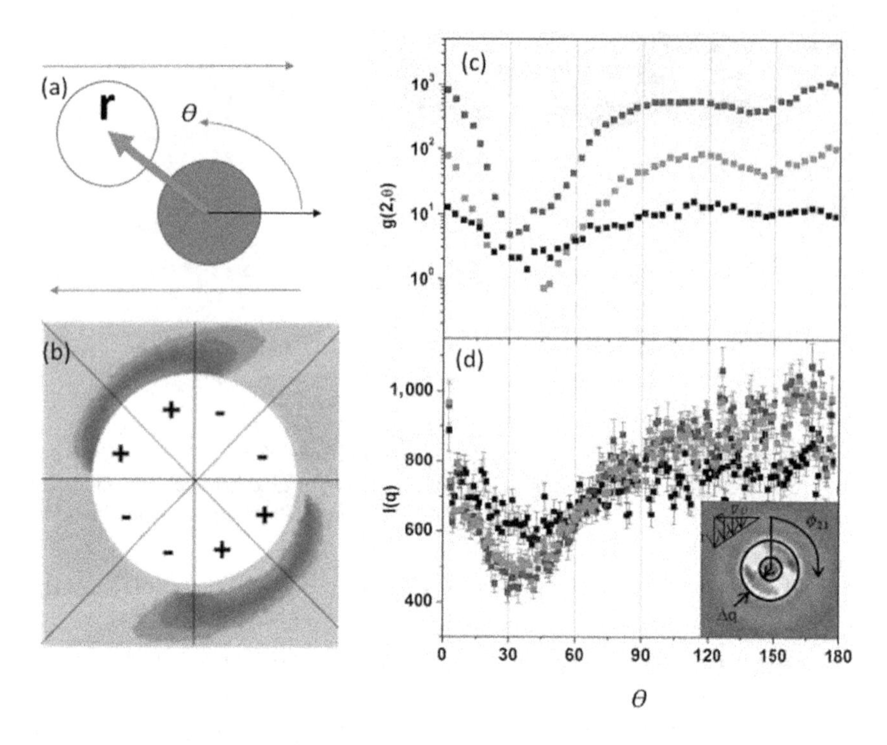

Figure 4.6 (a) Illustration of the definition of the flow field in the plane of flow, with a reference particle (black) and the vector position of a neighboring particle, defined by the center location. (b) Particle probability plot around a reference particle for $Pe = 5$, where the signs illustrate compression (+) and extension (–) of the flow. Reproduced with permission from Bergenholtz et al., The non-Newtonian rheology of dilute colloidal suspensions. *Journal of Fluid Mechanics* 2002;456:239–275 [22]. (c) Simulation results for the pair probability for $Pe = 21$, 25, 1,000, redrawn from Foss and Brady [21] compared to experimental measurements (d) for a near hard sphere dispersion at $Pe = 2.6$, 26, and 53. (c) and (d) are reproduced with permission from Gurnon and Wagner, Microstructure and rheology relationships for shear thickening colloidal dispersions. *Journal of Fluid Mechanics* 2015;769:242–276 [17]

the symmetry $l = 2$, $m = 1$. Shearing for $Pe > 0$ will distort the microstructure further such that higher order terms will become significant. A presentation of results for all three planes of flow for a broad range of Pe for a concentrated, near hard sphere dispersion can be found in Gurnon and Wagner [17]. There it is shown that the $l = 2$, $m = 1$ term is important at low shear rates while higher order terms become significant at higher shear rates as the microstructure becomes highly distorted and deviates further from its spherical, equilibrium pattern.

While particle alignment is critical for materials processing and examples have been given where scattering reveals the conditions under which such alignment can be achieved, other materials processing seeks to organize colloidal particles into shear ordered structures [24]. An example is the use of rheo-SANS to determine conditions under which soft colloids, such as those discussed in Chapters 6 and 7, will organize into different crystal structures depending on the frequency and amplitude of their

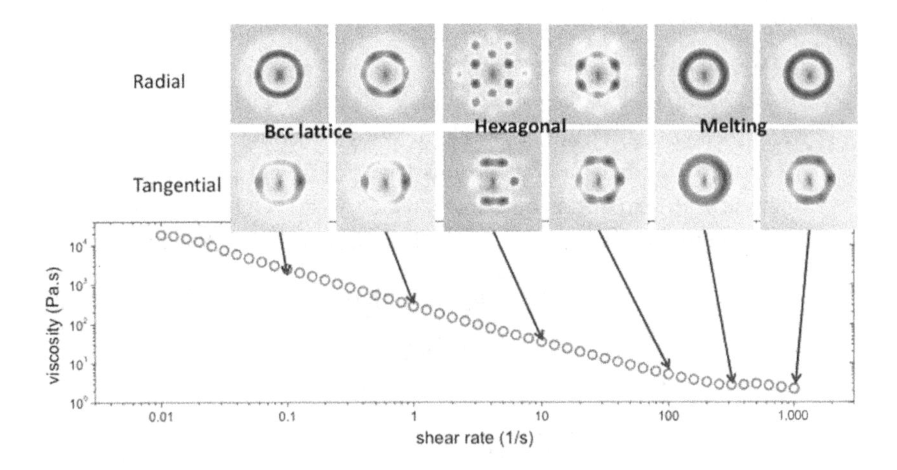

Figure 4.7 Shear viscosity for a concentrated suspension of soft colloids along with rheo-SANS scattering patterns in the 1–3 (radial) and 2–3 (tangential) geometries. Note that the smoothly shear thinning rheology actually reflects shear transformation from a BCC lattice to a flowing hexagonally lattice oriented in the flow direction. Figure courtesy of Carlos López-Barrón, adapted from López-Barrón et al. [26]

shear history in steady and oscillatory shearing [25,26]. An example for soft colloids comprised of self-assembled block-copolymers, so called hairy particles, is shown in Figure 4.7. The original suspension has BCC lattice structure, but is polycrystalline. Such a structure cannot shear uniformly so the crystal transforms into sliding layers of hexagonally ordered colloids. Increasing the shear rate, however, results in a very weak shear thickening that reflects shear melting of the crystal. Microstructural measurements during processing enable creating materials with different degrees of order with controlled orientation [26].

4.2.2 Stresses Derived from the Microstructure

The stresses are calculated by substituting the spherical harmonic expansion in Eq. (4.1) into spatial integrals weighted by the relevant forces. As the forces themselves can be expressed as spherical harmonics, and spherical harmonics are orthogonal, then only structure terms with comparable symmetry survive. Exact details of the mathematical manipulations can be found in Gurnon and Wagner [17] and Wagner and Ackerson [23]. Importantly, all centrosymmetric interparticle forces, such as those discussed in Chapter 1, have $l = 2$, $m = 1$ symmetry. Thus, only one term in this infinite expansion contributes to the thermodynamic contribution to the shear stress, as:

$$\sigma^T(Pe) = -\frac{\rho}{\pi\sqrt{30\pi}}\int_0^\infty \theta^*(q)B_{2,1}^+(q;Pe)q^2dq \tag{4.2}$$

The function $\theta^*(q)$ is the Hankel transform of the force acting between particles [23]. Note that the thermodynamic stress depends on the structure deformation, which is the only place that the shear rate dependence enters. While the initial deformation scales linearly with Pe, such that the stress contribution scales linearly with shear rate, the structure deformation quickly saturates such that the stress contribution eventually becomes constant. Therefore, this structure saturation leads to shear thinning rheology.

As the hydrodynamic interactions have a different symmetry, different harmonics in the structure expansion contribute to the stresses. Only terms of $l = 0, 2, 4$ contribute to the hydrodynamic contribution to the shear stress, such that:

$$\sigma^H(Pe) = Pe^* f\left(B_{0,0}^+(q;Pe), B_{2,0}^+(q;Pe), B_{2,2}^+(q;Pe), B_{4,0}^+(q;Pe), B_{4,2}^+(q;Pe)\right) \quad (4.3)$$

The detailed expressions can be found elsewhere [17,23,27]. Importantly, these scale in proportion to *Pe* so the hydrodynamic contribution to the viscosity remains constant or grows with increasing structure deformation at higher *Pe*, leading to a shear thickening viscosity. Stress-SANS rules can be defined, relating spherical harmonics of the distorted microstructure to the suspension deviatoric stresses [17]. These rules are comparable to the stress–optical rules developed for colloidal suspensions [28–30]. An example for hard spheres is shown in Figure 4.8, where by means of flow-SANS and rheo-SANS it is determined that the suspension stress is dominated by hydrodynamic terms at higher shear rates.

Measurements of the microstructure in all three planes of flow enable calculation of all three material properties associated with the shear stresses and the normal stress differences. Importantly, the thermodynamic contributions depend on ($l = 2$, $m = 0, 2$) harmonics, while now the ($l = 4$, $m = 1, 3$) terms contribute to the hydrodynamic contributions. An example is shown in Figure 4.8(b), where the first normal stress difference by direct measurement is compared to calculations of the stress-SANS rule. Here, the thermodynamic contribution is observed to be small but positive, while the hydrodynamic component is much more significant at high shear rates, and negative,

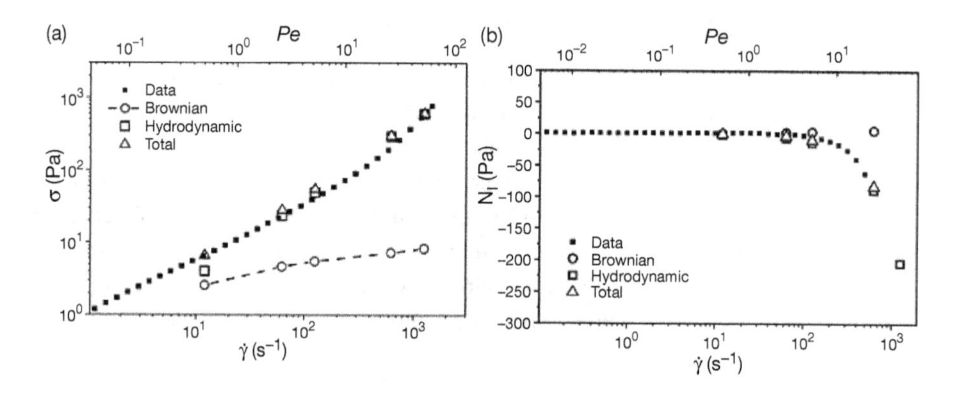

Figure 4.8 Stress-SANS rule for a concentrated, near hard sphere colloidal suspension. The black squares show the measured stress, while the open symbols show the components calculated using the Stress-SANS rule. (a) Shear stress; (b) first normal stress difference. Figures adapted from Gurnon and Wagner [17]

leading to a strong, negative value at high shear rates [17,31]. The anisotropy in the local neighbor distribution, shown in Figure 4.6(c) and (d), along with the dominance of the hydrodynamic interactions is responsible for this negative value of the first normal stress difference.

The examples shown here are for the shear flow of suspensions of near hard sphere colloidal dispersions, where the particles are coated to provide stability and reduce contact friction [32]. Note that for dilatant suspensions and suspensions where contact mechanics are important at high shear rates, significantly different microstructures and rheological properties are predicted as the force chain network develops along both the compression and extension axes [33,34]. Furthermore, predictions have also been made for extensional flows [35], but experimental challenges in probing such flows using scattering for colloidal dispersions remain. Meanwhile, studies of anisotropic particles, where strong alignment leads to strong anisotropy in scattering, are prevalent [6,8,36,37]. Dynamic oscillatory studies under LAOS are also possible, where oscillatory flows can create significant order [25,26]. While the use of scattering methods provides powerful means of probing the microstructure of colloidal dispersions under flow, limitations on instrumentation and resolution make this primarily a research tool. In particular, neutrons and to some extent x-rays can penetrate into opaque samples, while light requires nearly optically clear samples. While neutrons and x-rays are particularly suitable for studying particles of submicron size, the study of model systems of larger particles that are optically matched by direct optical microscopy is popular and is the subject of Section 4.3.

4.3 Direct Observation Using Microscopy

Typical microscopy is diffraction limited such that optical measurements are limited to ~ micron length scales. However, model systems and confocal imaging methods (for a review, see Prasad et al. [38]) enable tracking the location of micron sized particles to within 10 nm in the focal plane (x, y), with lower resolution in the "z" or orthogonal direction, such that particle configurations can be measured and particles can be tracked on the nanoscale. The development of model systems and advances in confocal microscopy have revolutionized our experimental abilities to observe not only the microstructure of colloidal dispersions at equilibrium as well as under a variety of flows, but has also spawned an entire new field of research known as particle microrheology. Chapter 11 of *CSR* included an introduction to the basic concepts of microrheology by Furst, and a recent, extensive monograph by Furst and Squires [39] provides a comprehensive overview of techniques including both passive and active microrheology, particle tracking, laser tweezers, interferometric techniques, and light scattering microrheology. Some examples of these methods will be presented in Chapter 5, and the interested reader is referred to these references for further study. Here, we focus on the measurement of microstructure using modern optical microscopy. Note that further applications of this method to 2D systems has also been reviewed by Vermant in chapter 11 of *CSR*.

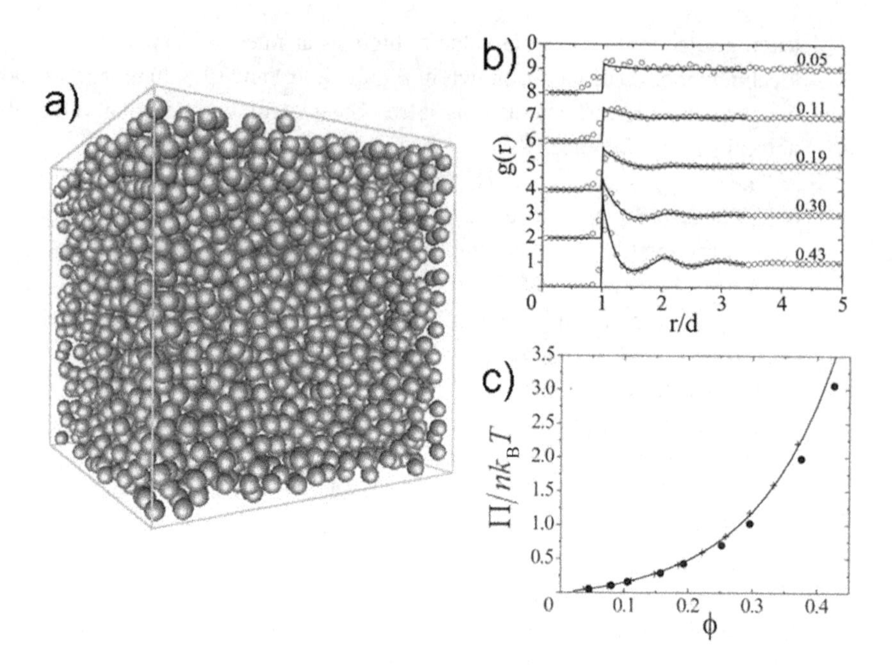

Figure 4.9 Direct imaging of a 25 vol% hard sphere suspension showing (a) reconstruction of the measured colloidal microstructure; (b) radial distribution functions directly calculated from the images compared against theory (lines); and (c) osmotic pressure calculated from the images as compared against theory. Also appears as figure 3.9 in *CSR*. Used with permission from Dullens et al. [40], copyright 2006, National Academy of Sciences, USA

Direct imaging of a colloidal hard sphere dispersion at 25 vol% by Dullens et al. [40] allows for locating enough particles in a sufficiently large volume to enable accurate calculation of the bulk pressure and the radial distribution function. Figure 4.9 is representative of a typical result, where the computer drawing of particles in the volume probed is based on centroid locations identified by optical microscopy. Agreement with theory provides confidence in the ability to image sufficient numbers of particles with the required accuracy and time resolution to quantitatively study equilibrium systems.

With faster methods for scanning, imaging of microstructure under flow becomes possible. An example is shown in Figure 4.10, where Cheng et al. [41] developed a shear apparatus whereby a translation stage oscillates a suspension of 0.96 micron diameter index-matched particles while being viewed by confocal microscopy. As the sample is oscillated, the pair distribution function also oscillates and the shear distortion follows the plate motion. Interestingly, due to the relatively thin sample thickness of only six microns, which is necessary to be accessible for the scan rate and depth of field of the confocal microscope, ordering effects are observed. The order parameter shown in Figure 4.10(f) describes the degree of shear ordering, which is observed to increase during shear thinning and decrease during shear thickening, consistent with the scattering measurements on soft nanoparticle suspensions presented in Figure 4.6.

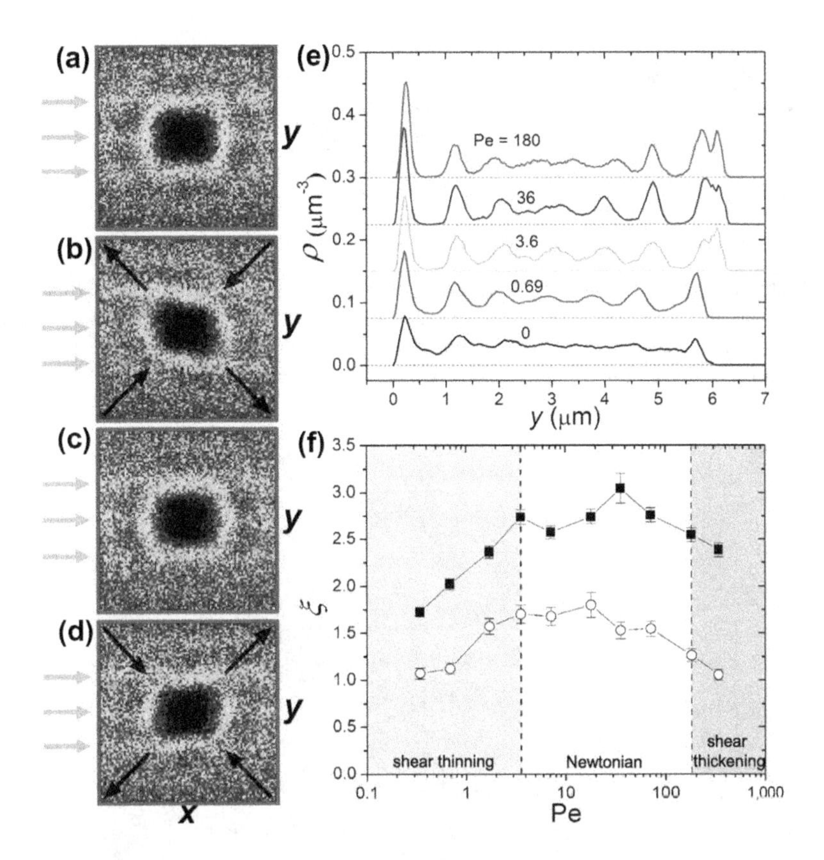

Figure 4.10 (a–d) The pair probability $g(x, y)$ in the plane of shear for $\phi = 0.34$ at $Pe = 0.36$ for oscillatory shearing with a phase of ~$\pi/2$ (a), π (b), $3\pi/2$ (c), and 2π (d). The compression and extension axes are identified by the arrows. (e) A plot of the local density in the shear gradient direction, $\rho(y)$, for the sample. Curves for different Pe are shifted vertically for clarity. (f) Order parameter, ξ, versus Pe for $\phi = 0.34$ (square) and $\phi = 0.48$ (circle) samples. Reproduced from Cheng et al., Imaging the microscopic structure of shear thinning and thickening colloidal suspensions. *Science* 2011;333(6047):1276–1279 [41], used with permission from the AAAS

A major finding of this work is the direct observation of hydroclusters responsible for the increased stress during shear thickening, as discussed in Chapter 1. This is shown in Figure 4.11, where particles that are in close proximity are identified as belonging to a cluster, and clusters of at least six particles are represented by the spheres. The locations of the surrounding particles are plotted as much smaller spheres for clarity of presentation. Importantly, these observations of microscopy under flow are consistent with predictions of Stokesian Dynamics simulations [42,43] and neutron scattering measurements [44]. A limitation of this work is the need to scan in the z-direction sufficiently fast such that accurate positions can be determined despite the relative motion in the sample [45].

While scattering measurements of microstructure are ensemble averages, direct microscopic observations are severely limited in the number of particles that can be

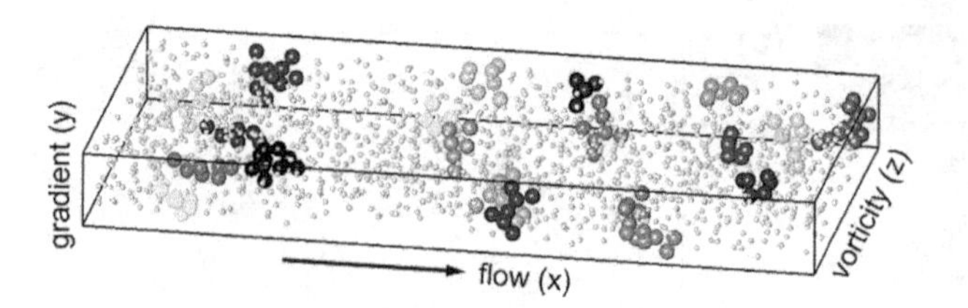

Figure 4.11 Instantaneous real-space configuration of hydroclusters with $N \geq 6$ at a volume fraction of 0.47. Different shading indicates distinct clusters. Other particles not in these larger clusters are smaller for clarity. The boundary box is $31.2 \times 15.4 \times 3.1\ \mu m^3$. Reproduced from Cheng et al., Imaging the microscopic structure of shear thinning and thickening colloidal suspensions. *Science* 2011;333(6047):1276–1279 [41], used with permission from the AAAS

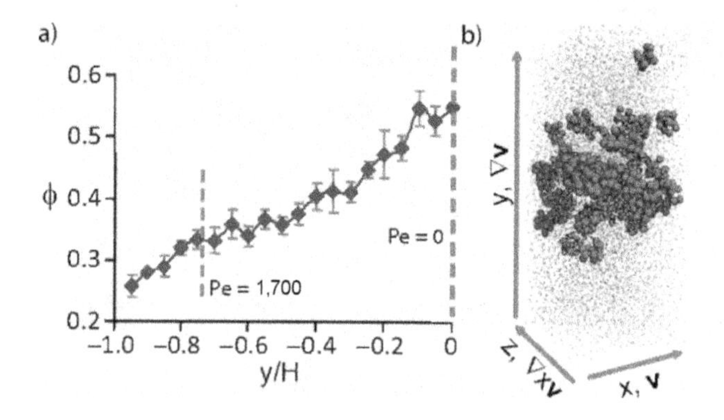

Figure 4.12 (a) Colloidal particle density profile across the microchannel from the wall (y/H = −1) with high *Pe* to the centerline, where *Pe* = 0, showing significant particle migration. (b) Rendering of experimentally determined particle positions as small points and true scale larger spheres indicate particles having FCC or HCP bond order. Reprinted with permission from Gao et al., Direct investigation of anisotropic suspension structure in pressure-driven flow. *Physical Review E* 2010;81(4):041403 [46]. Copyright 2010 by the American Physical Society. DOI: 10.1103/PhysRevE.81.041403

observed, especially for flowing systems. However, the ability to resolve spatial heterogeneities at the particle level, as shown in Figure 4.12, is a distinct advantage. An example for pressure driven flow of a Brownian suspension of micron sized silica particles reported by Gao et al. [46] is shown in Figure 4.12. Flow in the microchannel causes particle migration toward the centerline, which is anticipated for this type of flow and is discussed in section 2.6.1 of *CSR*. Direct imaging also enables identifying groupings of particles with crystalline-like order in the channel, which are also shown. Another important finding in this work is the reported agreement between the observed pair distribution functions under this channel flow with predictions using Stokesian Dynamics simulations [46]. This example also illustrates the opportunity to explore multiple aspects of colloidal microstructure development in more complex flows relevant for particle processing.

4.4 Summary and Outlook

This brief introduction to two methods for measuring microstructure under flow provides both background for results to be presented in the following chapters as well as a motivation for using such techniques to better understand the flow and rheology of colloidal dispersions. Note that alternative methods not covered here, such as dielectric spectroscopy and conductivity [47,48], can also provide microstructural information for some specific samples. Generally, however, scattering techniques provide powerful methods for probing flowing suspensions and developing structure-property relationships, such as the stress-SANS rules. Neutron and x-ray methods do not require optically index-matched suspensions and are especially suited for probing suspensions of nanoparticles to submicron size colloids. Scattering methods under flow are not limited to Couette shear and slit flows [49], as new instrumentation for Poiseuille and extensional flows are being investigated [50] as well as scattering from arbitrary 2D flows using a fluidic four-roll mill [51].

Microscopy is a powerful method for probing micron-sized, index matched suspensions both at rest and under flow. It is especially powerful when coupled with microchannel flows or for examining suspensions either near boundaries, such as in rheo-confocal instruments [52]. Continued advances in optical methods and fluorescence labeling techniques push microscopy investigations further into the nanoscale [53]. The recent offering by rheology manufacturers of turn-key instruments combining confocal microscopy with high-end rheometers, as well as rheo-SALS instruments, along with the prevalence of rheo-SANS and rheo-SAXS instruments at beamlines [18,47,54,55], as well as extensions of these methods to Poiseuille and extensional flow [50], provides opportunities for advanced studies of structure-property relationships for a broad range of models as well as industrial systems [6].

References

1. Mewis J, Wagner NJ. *Colloidal Suspension Rheology*. Cambridge: Cambridge University Press; 2012. 393 p.
2. Batchelor GK. The effect of Brownian motion on the bulk stress in a suspension of spherical particles. *Journal of Fluid Mechanics*. 1977;83(1):97.
3. Batchelor GK. The stress system in a suspension of force-free particles. *Journal of Fluid Mechanics*. 1970;41(3):545–570.
4. Russel WB, Saville DA, Schowalter WR. *Colloidal Dispersions*. Cambridge: Cambridge University Press; 1989.
5. Glatter O. *Scattering Methods and Their Application in Colloid and Interface Science*. Amsterdam: Elsevier; 2018. 392 p.
6. Eberle APR, Porcar L. Flow-SANS and Rheo-SANS applied to soft matter. *Current Opinion in Colloid & Interface Science*. 2012;17(1):33–43.
7. Roe RJ. *Methods of X-ray and Neutron Scattering in Polymer Science*. New York: Oxford University Press; 2000. 331 p.

8. Egres R, Nettesheim F, Wagner NJ. Rheo-SANS investigation of acicular-precipitated calcium carbonate colloidal suspensions through the shear thickening transition. *Journal of Rheology*. 2006;50(5):685–709.
9. Pignon F, Magnin A, Piau J-M. Butterfly light scattering pattern and rheology of a sheared thixotropic clay gel. *Physical Review Letters*. 1997;79(23):4689–4692.
10. Pignon F, Magnin A, Piau JM. Thixotropic behavior of clay dispersions: Combinations of scattering and rheometric techniques. *Journal of Rheology*. 1998;42(6):1349–1373.
11. Philippe AM, Baravian C, Imperor-Clerc M, Silva JD, Paineau E, Bihannic I, et al. Rheo-SAXS investigation of shear-thinning behaviour of very anisometric repulsive disc-like clay suspensions. *Journal of Physics: Condensed Matter*. 2011;23(19):194112.
12. Eberle APR, Martys N, Porcar L, Kline SR, George WL, Kim JM, et al. Shear viscosity and structural scalings in model adhesive hard-sphere gels. *Physical Review E*. 2014;89(5):050302(R).
13. Eberle APR, Wagner NJ, Akgun B, Satija SK. Temperature-dependent nanostructure of an end-tethered octadecane brush in tetradecane and nanoparticle phase behavior. *Langmuir*. 2010;26(5):3003–3007.
14. Caputo FE, Ugaz VM, Burghardt WR, Berret J-F. Transient 1–2 plane small-angle x-ray scattering measurements of micellar orientation in aligning and tumbling nematic surfactant solutions. *Journal of Rheology*. 2002;46(4):927–946.
15. Sommer MM. Mechanical Production of Nanoparticles in Stirred Media Mills [PhD]. Goettingen: Technical University of Erlangen-Nuernberg; 2007.
16. Hipp JB, Richards JJ, Wagner NJ. Structure-property relationships of sheared carbon black suspensions determined by simultaneous rheological and neutron scattering measurements. *Journal of Rheology*. 2019;63(3):423–436.
17. Gurnon AK, Wagner NJ. Microstructure and rheology relationships for shear thickening colloidal dispersions. *Journal of Fluid Mechanics*. 2015;769:242–276.
18. Gurnon AK, Godfrin PD, Wagner NJ, Eberle APR, Butler P, Porcar L. Measuring material microstructure under flow using 1–2 plane flow-small angle neutron scattering. *Jove-Journal of Visualized Experiments*. 2014(84):e51068.
19. Liberatore MW, Nettesheim F, Wagner NJ, Porcar L. Spatially resolved small-angle neutron scattering in the 1–2 plane: A study of shear-induced phase-separating wormlike micelles. *Physical Review E*. 2006;73(2):020504.
20. Kalman D. *Microstructure and Rheology of Concentrated Suspensions of Near Hard-Sphere Colloids*. Newark: University of Delaware; 2010.
21. Foss DR, Brady JF. Structure, diffusion and rheology of Brownian suspensions by Stokesian dynamics simulation. *Journal of Fluid Mechanics*. 2000;407:167–200.
22. Bergenholtz J, Brady JF, Vicic M. The non-Newtonian rheology of dilute colloidal suspensions. *Journal of Fluid Mechanics*. 2002;456:239–275.
23. Wagner NJ, Ackerson BJ. Analysis of nonequilibrium structures of shearing colloidal suspensions. *Journal of Chemical Physics*. 1992;97(2):1473–1483.
24. McMullan JM, Wagner NJ. Directed self-assembly of suspensions by large amplitude oscillatory shear flow. *Journal of Rheology*. 2009;53(3):575–588.
25. López-Barrón CR, Porcar L, Eberle APR, Wagner NJ. Dynamics of melting and recrystallization in a polymeric micellar crystal subjected to large amplitude oscillatory shear flow. *Physical Review Letters*. 2012;108(25):258301.
26. López-Barrón CR, Wagner NJ, Porcar L. Layering, melting, and recrystallization of a close-packed micellar crystal under steady and large-amplitude oscillatory shear flows. *Journal of Rheology*. 2015;59(3):793–820.

27. Maranzano BJ, Wagner NJ. Flow-small angle neutron scattering measurements of colloidal dispersion microstructure evolution through the shear thickening transition. *Journal of Chemical Physics*. 2002;117(22):10291–10302.
28. D'Haene P, Mewis J, Fuller GG. Scattering dichroism measurements of flow-induced structure of a shear thickening suspension. *Journal of Colloid and Interface Science*. 1993;156(2):350–358.
29. Wagner NJ, Fuller GG, Russel WB. The dichroism and birefringence of a hard-sphere suspension under shear. *The Journal of Chemical Physics*. 1988;89(3):1580–1587.
30. Bender JW, Wagner NJ. Optical measurement of the contributions of colloidal forces to the rheology of concentrated suspensions. *Journal of Colloid and Interface Science*. 1995;172(1):171–184.
31. Cwalina CD, Wagner NJ. Material properties of the shear-thickened state in concentrated near hard-sphere colloidal dispersions. *Journal of Rheology*. 2014;58(4):949–967.
32. Lee Y-F, Luo Y, Brown SC, Wagner NJ. Experimental test of a frictional contact model for shear thickening in concentrated colloidal dispersions. *Journal of Rheology*. 2020;64(2):267–282.
33. Morris JF. A review of microstructure in concentrated suspensions and its implications for rheology and bulk flow. *Rheologica Acta*. 2009;48(8):909–923.
34. Morris JF. Shear thickening of concentrated suspensions: Recent developments and relation to other phenomena. *Annual Review of Fluid Mechanics*. 2020;52(1):121–144.
35. Seto R, Giusteri GG, Martiniello A. Microstructure and thickening of dense suspensions under extensional and shear flows. *Journal of Fluid Mechanics*. 2017;825:R3.
36. Jogun SM, Zukoski CF. Rheology and microstructure of dense suspensions of plate-shaped colloidal particles. *Journal of Rheology*. 1999;43(4):847–871.
37. Pignon F, Magnin A, Piau JM, Cabane B, Lindner P, Diat O. A yield stress thixotropic clay suspension: Investigation of structure by light, neutron, and X-ray scattering. *Physical Review E*. 1997;56(3):3281–3289.
38. Prasad V, Semwogerere D, Weeks ER. Confocal microscopy of colloids. *Journal of Physics: Condensed Matter*. 2007;19(11):113102.
39. Furst EM, Squires TM. *Microrheology*. Oxford: Oxford University Press; 2017. 451 p.
40. Dullens RPA, Aarts D, Kegel WK. Direct measurement of the free energy by optical microscopy. *Proceedings of the National Academy of Sciences of the United States of America*. 2006;103(3):529-31.
41. Cheng X, McCoy JH, Israelachvili JN, Cohen I. Imaging the microscopic structure of shear thinning and thickening colloidal suspensions. *Science*. 2011;333(6047):1276–1279.
42. Brady JF, Bossis G. Stokesian dynamics. *Annual Review of Fluid Mechanics*. 1988;20:111–157.
43. Silbert LE, Farr RS, Melrose JR, Ball RC. Stress distributions in flowing aggregated colloidal suspensions. *Journal of Chemical Physics*. 1999;111(10):4780–4789.
44. Kalman DP, Wagner NJ. Microstructure of shear-thickening concentrated suspensions determined by flow-USANS. *Rheologica Acta*. 2009;48(8):897–908.
45. Pandey R, Spannuth M, Conrad JC. Confocal imaging of confined quiescent and flowing colloid-polymer mixtures. *Jove-Journal of Visualized Experiments*. 2014;(87):e51461.
46. Gao C, Kulkarni SD, Morris JF, Gilchrist JF. Direct investigation of anisotropic suspension structure in pressure-driven flow. *Physical Review E*. 2010;81(4):041403.
47. Richards JJ, Gagnon CVL, Krzywon JR, Wagner NJ, Butler PD. Dielectric rheoSANS – simultaneous interrogation of impedance, rheology and small angle neutron scattering of complex fluids. *Jove-Journal of Visualized Experiments*. 2017;(122):e55318.

48. Genz U, Helsen JA, Mewis J. Dielectric spectroscopy of reversibly flocculated dispersions during flow. *Journal of Colloid and Interface Science*. 1994;165(1):212–220.
49. Laun HM, Bung R, Hess S, Loose W, Hess O, Hahn K, et al. Rheological and small-angle neutron-scattering investigation of shear-induced particle structures of concentrated polymer dispersions submitted to plane Poiseuille and Couette-flow. *Journal of Rheology*. 1992;36(4):743.
50. Bharati A, Hudson SD, Weigandt KM. Poiseuille and extensional flow small-angle scattering for developing structure–rheology relationships in soft matter systems. *Current Opinion in Colloid & Interface Science*. 2019;42:137–146.
51. Corona PT, Ruocco N, Weigandt KM, Leal LG, Helgeson ME. Probing flow-induced nanostructure of complex fluids in arbitrary 2D flows using a fluidic four-roll mill (FFoRM). *Scientific Reports*. 2018;8(1):15559.
52. Dutta SK, Mbi A, Arevalo RC, Blair DL. Development of a confocal rheometer for soft and biological materials. *Review of Scientific Instruments*. 2013;84(6):063702.
53. Wang X, Yi H, Gdor I, Hereld M, Scherer NF. Nanoscale resolution 3D snapshot particle tracking by multifocal microscopy. *Nano Letters*. 2019;19(10):6781–6787.
54. Porcar L, Pozzo D, Langenbucher G, Moyer J, Butler PD. Rheo–small-angle neutron scattering at the National Institute of Standards and Technology Center for Neutron Research. *Review of Scientific Instruments*. 2011;82(8):083902.
55. Richards JJ, Wagner NJ, Butler PD. A strain-controlled RheoSANS instrument for the measurement of the microstructural, electrical, and mechanical properties of soft materials. *Review of Scientific Instruments*. 2017;88(10):105115.

5 Rheology of Colloidal Glasses and Gels

George Petekidis and Norman J. Wagner

5.1 Introduction

Many naturally occurring real life colloidal systems, such as blood and natural sediments, as well as manufactured systems, such as foods, pharmaceuticals, cosmetics, coatings, inks, ceramics, and even construction materials, such as asphalts and cements, are either in or go through a gelled or glassy state. The goal of this chapter is to provide fundamental understanding of their rheological behavior and its connection back to microstructure and colloidal properties for the simplest of model systems, that of spherical, homogeneous, nonreacting colloidal suspensions with simple repulsive or attractive interactions. Even these simple systems show an incredibly rich rheological and phase behavior, including phenomena such as dynamical arrest, phase separation, and aging. Therefore, the information provided in this chapter should be helpful in formulating and understanding more practical systems such as those discussed in Chapters 7–9. The information in this chapter builds upon chapter 6 of *CSR*, which broadly surveys suspensions with attractive interactions [1]. Colloidal particles can also be used as model systems [2,3] to study fundamental phenomena in condensed matter physics and soft materials. Examples are phase behavior and metastable out-of-equilibrium states such as glasses and gels [3–6]. Therefore, the contents of this chapter should have broader scientific utility.

Ideal Hard Sphere (HS) colloids are probably the simplest of this class of systems exhibiting a one-dimensional phase diagram with a single parameter: the particle volume fraction, ϕ (see Chapter 1 for a detailed explanation of HS colloidal dispersions), as their interaction potential is temperature independent. As shown in Figure 1.3, at volume fractions below $\phi = 0.494$ the system is liquid, while for monodisperse enough spheres, a liquid-crystal coexistence is found for $0.494 < \phi < 0.545$, above which an FCC crystal phase is the thermodynamic equilibrium phase [4]. In HS suspensions the liquid to crystal phase transition is a first order transition driven solely by entropy, as the local free volume, and thus entropy, is maximized through ordering above volume fractions around $\phi \approx 0.5$. Note that weak polydispersity ($\geq 7-8\%$) can suppress this transition, so it is often not observed [7]. However at even higher volume fractions ($\phi \geq 0.58$) the system is kinetically trapped in a metastable, amorphous state, a glass, where individual particles are locally constrained in a cage formed by their neighbors and their long-time diffusion out of cage is suppressed [4,7,8], as shown in Figure 1.4. As this glass arises by crowding, this is considered a "repulsive" glass as opposed to an "attractive" driven glass [9,10].

Approaching the liquid–solid (crystal or glass) transition, particle dynamics exhibit short-time in-cage (β-relaxation) and a progressively slower long-time, out-of-cage, diffusion (α-relaxation). The average α-relaxation time increases with concentration and diverges at the glass transition, ϕ_g [7,8]. Beyond ϕ_g, out-of-cage diffusion is essentially frozen and the suspension is in a glassy state. Note that ideal hard sphere interactions cannot be easily achieved in real systems such that a short-range repulsive interaction between particles has to be introduced to mitigate attractive van der Waals forces [1,11]. This is achieved either by a steric or an electrostatic stabilization with an interaction range much less than the particle diameter. Colloids exhibiting longer-range soft repulsive potentials, such as due to grafted polymers, are discussed in Chapter 6.

Concentrated suspensions are viscoelastic and exhibit a shear rate dependent viscosity with their zero shear value diverging at the glass transition [1], as discussed for HSs in Chapter 1. In HS colloidal suspensions the elasticity and yield stress are of entropic origin due to particle caging, although local, in-cage, Brownian motion is still possible and cooperative motion at larger length-scales is detected as dynamic heterogeneities [12]. These are connected to hopping processes near the glass transition volume fraction and result in aging and slow creep under weak stresses [13–15].

The addition of attractive interactions leads to a richer phase diagram, with phase separation into a dilute "gas" and a concentrated liquid or crystal phase, reminiscent of atomic systems. Examples are provided in Figures 1.16 and 1.17. In addition to the arrested amorphous glasses interfering with thermodynamic equilibrium, the presence of interparticle attractions causes, depending on attraction characteristics and volume fraction, frustrated out-of-equilibrium states [16]. These range from dilute fractal-like percolating networks [1] and concentrated cluster based gels formed via arrested phase separation [16–18] to more homogeneous attractive glasses at very high volume fractions [9], and "equilibrium" gels [16,19].

For a colloidal glass, tuning the interparticle interactions from repulsive to weakly attractive can lead to melting of the glass to a liquid suspension, followed by re-vitrification [9,10,20,21]. This re-entrant transition is evident as a nonlinear dependence of the shear viscosity on attraction strength in concentrated suspensions [22]. This is because weak attractions can open up space for particles to diffuse out of their cage of neighbors. However, further increasing the strength of interparticle attraction leads to another nonergodic state – the attractive glass – where large scale mobility is suppressed both due to bonding between particles and caging by their neighbors. Decreasing the volume fraction while keeping the interparticle interaction unchanged leads from an attractive glass to a colloidal gel where cages are substituted by clusters that are bonded in a load-bearing network [1,23]. Therefore colloidal suspensions exhibit a transition from an ergodic[1] liquid to nonergodic solids of a purely steric (due

[1] Ergodicity refers to the property that a system can sample all possible configurations through thermal motion given a sufficiently long time. A dynamically arrested state, such as a glass or a gel, is typically not ergodic, and care must be taken when interpreting experimental data on such systems, in part because their properties may be dependent on the sample history.

to excluded volume) character by increasing the volume fraction, or of an enthalpic origin by increasing the attraction strength, or both.

At lower volume fractions colloidal gels typically have a fractal microstructure. With strong enough attractions, particles bond once they touch, creating a rather open gel structure with low fractal dimension (1.7–1.9). These gels often form via diffusion-limited cluster aggregation (DLCA) [1,24,25]. For weaker attractions, comparable to the thermal energy, particles may rearrange breaking and reforming bonds. A reaction-limited cluster aggregation (RLCA) process creates a more compact microstructure with higher fractal dimension (2–2.1) [1]. At intermediate particle volume fractions, gels can form depending on the nature of the interparticle interaction as either equilibrium gels that are homogeneous or heterogeneous gels due to spinodal decomposition [16]. While the former are equilibrium structures, the interplay between thermodynamics and kinetics dictate the evolution of the latter system as nonequilibrium, arrested phase separation. Gravity can still play a role in driving gel collapse, effectively limiting the lowest volume fraction gel attainable [26,27].

The linear viscoelasticity and yielding, nonlinear rheology of HS colloidal glasses and attractive gels and glasses are reviewed in this chapter. Equilibrium rheological properties and particle motion in the suspension are connected to the interparticle interactions and colloidal microstructure. Likewise, yielding and nonlinear flow are connected to microstructural changes under shear. Aging and shear banding phenomena are also briefly discussed. Some key observations are introduced in Section 5.2.

5.2 Landmark Observations

The Hard Sphere (HS) paradigm, introducing colloids as the model soft matter analog of large atoms, has been established by the determination of the phase diagram, presented in Figure 1.3. Figure 5.1 depicts a series of suspensions of nearly HS PMMA particles sterically stabilized in organic solvent (cis-decalin) at different volume fractions from experiments by Pusey and van Megen [4] and Pusey et al. [5]. Here the liquid-crystal coexistence is observed at volume fractions between 0.494 and 0.545, while the amorphous glassy state is seen in the two highest volume fractions. The crystal phases are observed for relatively monodisperse systems; when polydispersity is larger than about 7–8%, the system does not crystallize but rather remains in a super-cooled liquid state [5,28].

Chapter 4 shows how colloidal suspensions can be viewed directly by fluorescent confocal microscopy [29]. Figure 5.2 illustrates some of these phenomena, depicting a typical image of a hard sphere colloidal glass, a colloid-polymer depletion gel, and a 2-D fractal gel. Figure 5.2(a) is an image of a colloidal glass of PMMA HS-like particles at $\phi = 0.6$, showing an amorphous liquid-like order. At lower volume fractions attractive particles form frustrated gel states. A depletion colloidal gel, where attractive interactions are induced by the addition of linear polymer chains, is shown in Figure 5.2(b), for $\phi = 0.44$, attraction strength at contact $U_{\mathrm{dep}} = -16k_{\mathrm{B}}T$ and range

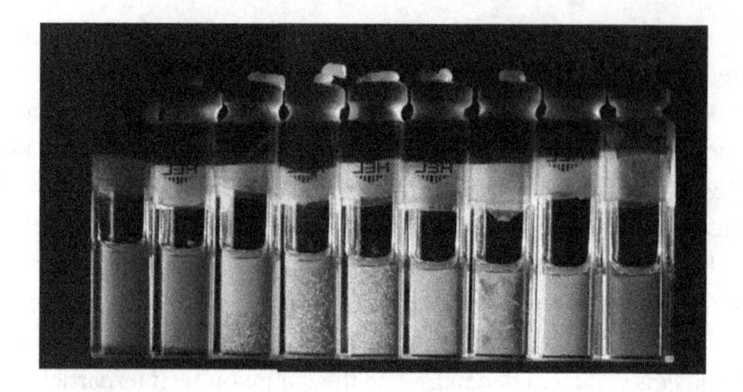

Figure 5.1 HS-like PMMA particles in cis-decalin at increasing volume fractions from left to right, one day after mixing. The first sample is in the liquid regime, the next three are in the liquid-crystal coexistence region, the next two are in the fully crystalline phase and the final two are in the glassy state. Reproduced with permission from Pusey et al., Hard spheres: Crystallization and glass formation. *Philosophical Transactions of the Royal Society A: Mathematical, Physical and Engineering Sciences* 2009;367(1909):4993–5011 [5], DOI 10.1098/rsta.2009.0181

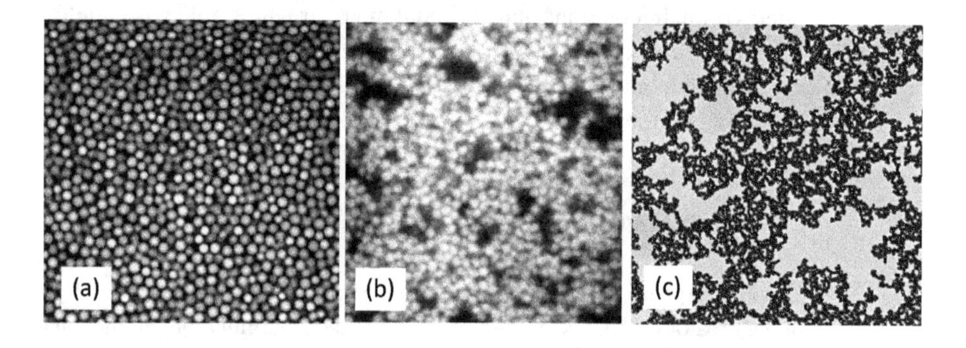

Figure 5.2 Confocal Microscopy Images from: (a) a PMMA HS glass at $\phi = 0.6$; (b) a depletion gel of PMMA particles at intermediate volume fraction with $\phi = 0.44$ and $U_{\text{dep}} = -16\ k_B T$, and a range of attraction $\delta = 0.05$. (c) Bright field microscopy image from a fractal colloidal gel of PS particles as formed in a water-decane interface. (a) and (b) reproduced with permission from Joshi and Petekidis, Yield stress fluids and ageing. *Rheologica Acta* 2018;57(6–7):521–549 [30]. (c) Reproduced with permission from Reynaert et al., Interfacial rheology of stable and weakly aggregated two-dimensional suspensions. *Physical Chemistry Chemical Physics* 2007;9(48):6463–6475 [31] by the Royal Society of Chemistry. Permission conveyed through Copyright Clearance Center, Inc.

of attraction $\delta = 0.05$. At even lower volume fractions fractal gels are formed for high enough attraction strengths, as shown in Figure 5.2(c).

The transition from liquid to gel occurs as the attraction strength is increased, and this can be followed by monitoring the linear viscoelastic moduli. For homogeneous gels, the critical gel temperature can be determined through the classic Winter–Chambon criterion [32], which was first applied to polymer gels. Figure 5.3 shows

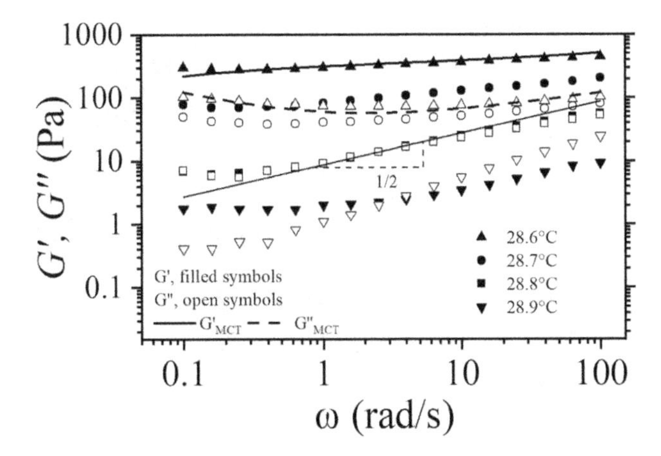

Figure 5.3 Linear viscoelastic moduli for a thermoreversible gel comprised of ~20% silica nanoparticles that show a transition from liquid-like behavior to gel behavior with decreasing temperature [33]. The trend line of $\omega^{1/2}$ identifies the gel point using the Winter–Chambon criterion [32], while the solid and dotted curves are fits of the moduli to the mode coupling model of Mason and Weitz [34].

this thermal gelation transition for suspensions of octadecyl-coated silica nanoparticles at $\phi = 0.28$ in tetradecane, where the critical gel criterion $G' = G'' \sim \omega^{1/2}$ is identified at about 28.8°C (see section 6.6 of *CSR* for further discussion) [33]. As shown, liquid-like behavior is observed above the gel temperature and, upon quenching, soft-solid behavior is observed over the accessible frequency range [34,35]. However, as these are physical gels and not chemically crosslinked, they will always flow and exhibit liquid-like behavior given sufficient time [34].

At even higher volume fractions, in the glass regime, the addition of attractive interactions leads, remarkably, to the melting of the glass, as was predicted by MCT and observed experimentally by dynamic light scattering in colloid-polymer mixtures with depletion attraction [9,10,20]. As depicted in the sketch in Figure 5.4(c), moderate attraction opens up space in the cage of neighbors, facilitating particle escape and melting the glass. Further increase of attraction strength leads to more long-lived particle bonding causing additional frustration of long-time particle dynamics. Therefore with stronger attractions the system freezes again in an attractive glass state, where large scale motion is frustrated both due to bonding and caging. Such particle dynamics have profound effects on the mechanical properties of the system as will be presented in the following section.

5.3 Colloidal Glasses due to Interparticle Repulsion

5.3.1 Steady Shear Rheology

As described in Chapter 1 and shown in Figure 1.2, increasing the particle volume fraction leads to an increasingly complex shear rheology. As shown in Figure 1.3,

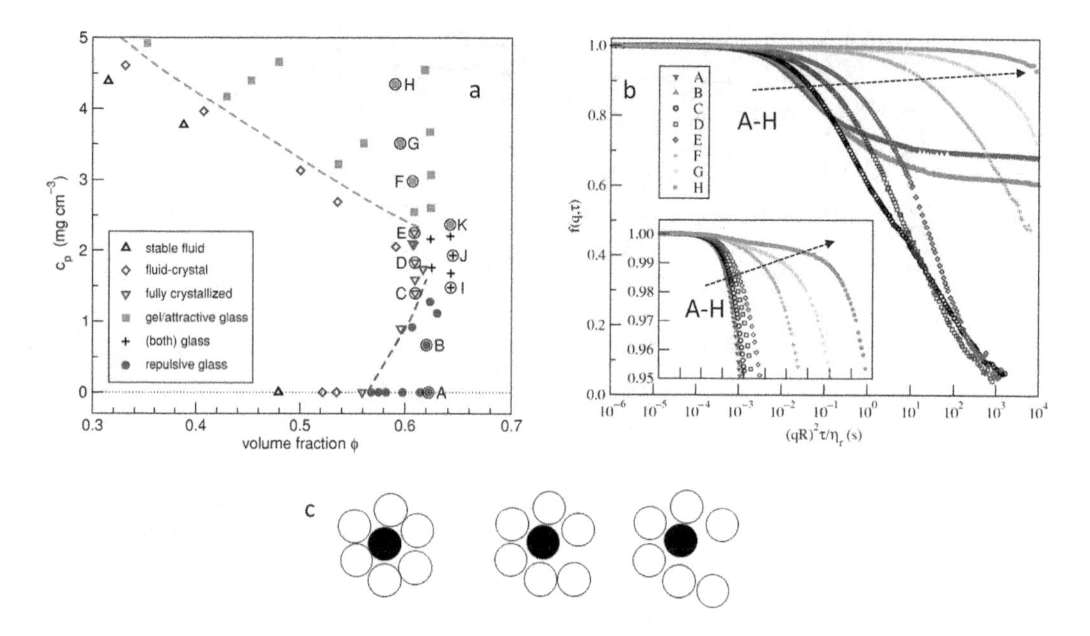

Figure 5.4 (a) Attractive colloids phase diagram with emphasis on high volume fraction where a re-entrant glass–liquid–attractive glass transition is observed (closed circles: repulsive glasses, open triangles: re-entrant liquid, closed squares: attractive glasses); (b): Correlation functions measured by dynamic light scattering at the corresponding points (A–E) shown in part (a); (c) schematic of the cage in a repulsive glass (c, left), re-entrant liquid (c, center) and attractive glass (c, right), adapted from Poon et al. [21]. (a) and (b) Reprinted with permission from Pham et al., Glasses in hard spheres with short-range attraction. *Physical Review E* 2004;69(1):011503 [20] by the American Physical Society

the suspension transitions from a fluid to a glass at packing fractions around 0.58. This rheological response is demonstrated in the flow curves shown in Figure 5.5, where the steady state stress of a flowing suspension is plotted against the shear rate for suspensions of HS-like PMMA particles in dodecane at several volume fractions below and above the glass transition volume fraction, ϕ_g [37]. While in the liquid regime a finite viscosity is measured at low shear rates, above ϕ_g the zero shear viscosity becomes infinite, or in practice too high to be measured [29,38,39]. This divergence has been discussed in literature [40] as well as in Chapter 1. Glass behavior is evident in Figure 5.5 as a constant stress plateau at low shear rates. The stress follows a *Herschel-Bulkley* (HB) model (Table 1.3) $\sigma = \sigma_y + k\dot{\gamma}^n$ with the stress plateau value, σ_y, providing one definition of the yield stress. In the glassy or highly concentrated regime such flow curves have been measured in suspensions of HS-like PMMA particles [37,41], charged stabilized poly(styrene-ethylacrylate) latex [42] and silica particles [1,43], as well as strongly crosslinked core-shell microgel particles [44]. As shown in Figure 5.6(b), the measured exponent takes values from $n = 0.5$ up to the limiting value of $n = 1$, where the response would be *Bingham like* and the sample would acquire a Newtonian type flow with a constant

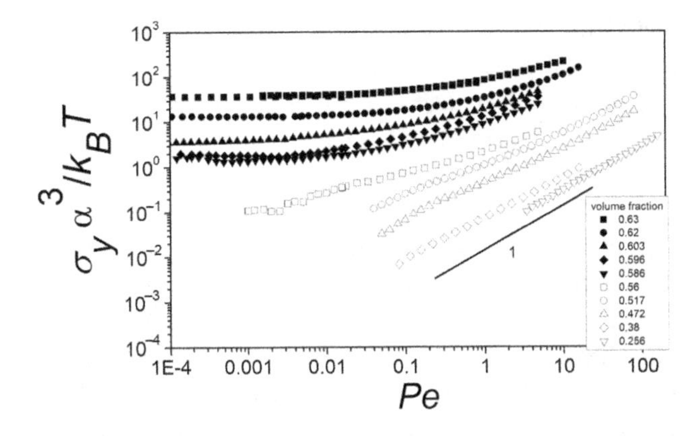

Figure 5.5 Flow curves of hard sphere suspensions and glasses of PMMA particles. Here the stress is normalized by the thermal energy divided by the particle diameter cubed and the shear rate by the free Brownian time, i.e. $Pe = \dot{\gamma}\tau_B$. Adapted from Petekidis et al. [37]

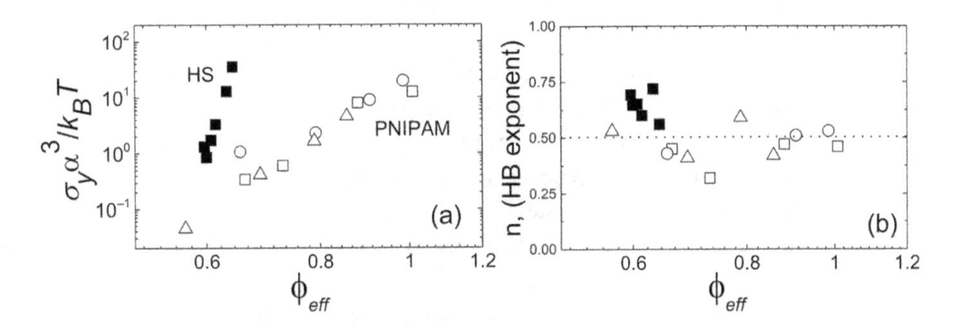

Figure 5.6 (a) Volume fraction dependence of the scaled yield stress for HS-like PMMA glasses and comparison with softer core-shell PS-PNIPAM particles. (b) HB exponent for different nearly hard sphere systems in the glassy regime as a function of volume fraction. Adapted from Koumakis et al. [47]

viscosity at high shear rates. Note however, that for softer particles, such as deformable microgels, emulsions, interpenetrating multiarm stars, the values of n usually cluster around $n = 0.5$, a finding which is discussed in Chapter 6 for soft colloids in the jammed state [45].

The yield stress defined by the low shear rate stress plateau of a flow curve corresponds to a dynamic yield stress, as opposed to a static yield stress at which the sample starts to flow during a transient start-up shear test [1,46]. These different yield stresses are defined in section 9.4.2 of *CSR*. Note also that flow curves in yield stress systems are usually measured from high to low shear rates, i.e., starting from a flowing sample and progressively lowering the shear rate, allowing enough time for the sample to reach steady state at each shear rate, in order to avoid transient and thixotropic effects.

The normalized yield stress deduced from flow curves in colloidal glasses of HS-like PMMA particles of different sizes suspended in organic solvents (cis-decalin or a nonvolatile octadecene/bromonaphtalene mixture) and the corresponding high shear rate shear thinning exponent n are plotted as a function of effective volume fraction in Figure 5.6 [47]. For comparison the corresponding quantities from softer core-shell microgels (PS-PNIPAM) suspended in water are also indicated. For the latter, as discussed extensively in Chapter 6, the effective volume fraction can reach values higher than one, entering well in the jammed state due to their soft character, demonstrated as deformability and/or outer shell interpenetration. This comparison offers two main findings. Firstly, the volume fraction dependence of the yield stress in HS glasses is much stronger than in their softer counterparts since the stress should diverge approaching rcp. The strong volume fraction dependence of the yield stress above ϕ_g and the divergence of the zero shear viscosity upon approaching ϕ_g from below was measured for a variety of nearly hard sphere colloids [1]. The former is also in agreement with the volume fraction dependence of the storage modulus, G'. Secondly, in HS glasses the limiting HB exponents are clearly larger than $n = 0.5$, although the high shear rate Newtonian flow regime, observed at lower volume fractions, is not easily achieved in the glassy state. In comparison, for the softer core-shell microgels shear thinning exponents are around the value of $n = 0.5$ as predicted for soft colloids by the elastohydrodynamic lubrication model [45].

Flow curve measurements in highly concentrated liquids and glasses of colloidal hard spheres are in practice limited by phenomena such as wall slip and/or shear banding at low shear rates [48,49] and by shear thickening in the high shear rate regime [1,50]. Thus the experimental protocols, as well as theoretical descriptions, should be appropriately designed to either address, or if possible suppress, these phenomena. For example, wall slip can be suppressed by the use of rough plates; however, shear banding may then appear. More about these measurement issues and methods for circumventing them can be found in chapter 9 of *CSR*.

5.3.2 Linear Viscoelasticity – Oscillatory Rheology

The development of a solid-like response with increasing volume fraction is also demonstrated in the linear viscoelastic spectrum [1]. The shear moduli are shown in Figure 5.7 for two volume fractions in the glass regime for HS-like PMMA particles with two different radii [47]. Different sizes can be scaled successfully by the thermal energy density, $k_B T/a^3$ when plotted against the scaled frequency, $Pe_\omega = \omega\tau_B$. The experimental data are compared with MCT [34]. A similar response has been observed in various experiments with PMMA particles in the glass regime [15,41,51–53], including shear induced crystals [54], as well as in earlier experiments with silica HS suspensions [43].

The linear viscoelastic spectrum is a consequence of microscopic particle motions and interactions. For concentrated liquids the intermediate scattering function, measured by dynamic light scattering [8], displays a two-step decay at high volume fractions reflecting at short times the localized particle diffusion within a cage, and at long times

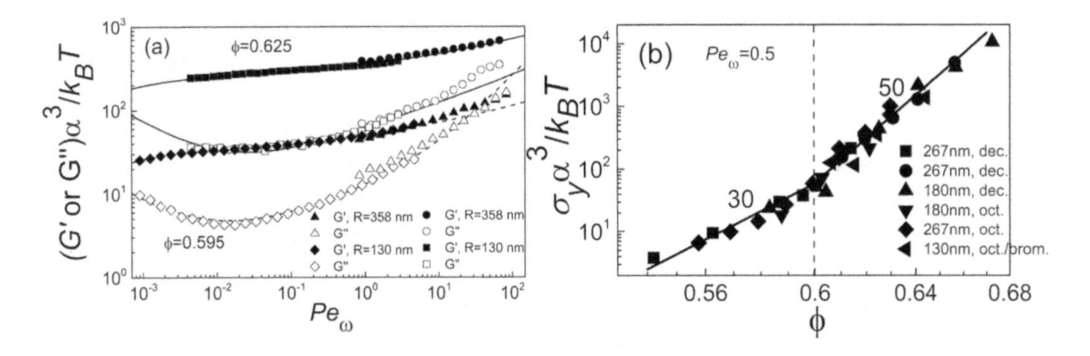

Figure 5.7 (a) Linear viscoelastic moduli of hard spheres of sterically stabilized PMMA particles in the glass regime for two different sizes and comparison with MCT predictions; (b) normalized storage modulus, at $Pe_\omega = 0.5$ as a function of volume fraction in a log–log plot [47]. Solid lines indicate approximate values of the power law slope, m, according to $G' \propto \phi^m$, in two volume fraction regimes. Adapted from Koumakis et al. [47]

the out-of-cage diffusion to longer distances [4,8,9]. The latter is finite below the glass transition as the suspension is an ergodic liquid with a finite zero shear viscosity [1,8]. Above ϕ_g, long-time diffusion becomes zero as the cage is essentially frozen and the sample becomes glassy. Such dynamic arrest leads to a solid-like mechanical response, with the linear viscoelasticity modeled by MCT, as discussed in Chapter 2. The short-time in-cage diffusion, still present in the glassy regime, is manifested in the linear viscoelastic spectrum at high frequencies, as shown in Figure 5.7(a), with G'' becoming larger than G' above a characteristic crossover frequency. In the intermediate frequency regime, where G' is almost constant and higher than G'', a plateau modulus, G'_p can be defined while G'' exhibits a weak minimum. This behavior is linked with a plateau in the particle MSD [7,29,56] reflecting the extent of particle localization due to the excluded volume constraints by its neighbors. On the other hand, the low frequency crossover of G' and G'' followed by a flow regime, with $G'' > G'$ at lower frequencies, is related with an out-of-cage particle motion and the zero shear viscosity [1]. As seen in Figure 5.7(a), in glassy samples such a crossover regime takes place at much lower frequencies outside the experimental window, or is completely absent indicating an ideal glass state.

The characteristic frequency at the minimum of G'' is related with an inverse characteristic time, $\tau_{\min}^{-1}$, for the transition from in-cage (β-relaxation) to out-of-cage (α-relaxation) motion. This transition time exhibits a nonmonotonic dependence on volume fraction for different HS-like PMMA systems [34,47,54], in agreement with MCT predictions and experiments for the short-time diffusion relaxation time measured by DLS [56]. Its peak value is observed at the glass transition volume fraction, ϕ_g, due to an interplay of the increasing particle localization caused by a slowing down of in-cage dynamics, and the cage becoming tighter approaching rcp [34]. Therefore, as the volume fraction is increased the cage in a HS suspension becomes tighter and particles get more localized. This leads to increased viscoelastic moduli, as indicated in Figure 5.7(a). The linear viscoelastic spectra can also be determined from the particle MSD, through the Generalized Stokes–Einstein (GSE) relation yielding the complex shear modulus, utilizing the Kramers–Kronig relations [34], as shown in Chapter 2.

The storage modulus shown Figure 5.7(b) is indicative of suspensions of particles with a steep HS-like potential. Note that the equilibrium elastic modulus can be derived from the interaction potential between particles and the pair correlation function in the framework of the Zwanzig–Mountain formalism [57] as proposed by Buscall [58] and Wagner [59]. This approach provides an approximate relation between the ϕ dependence of the plateau modulus, $G'_p \propto \phi^m$, which is approximately equal to the high frequency modulus, G'_∞, with an interparticle potential $U(r) \propto r^{-n}$, with $m = 1 + n/3$ [60]. This relation has been utilized in several studies to rationalize the volume fraction dependence of the storage modulus, linking it with the interparticle potential for suspensions of PMMA particles [61], microgels [61,62], and charge-stabilized colloids [1,63,64]. From these studies [1] it is clear that the steeper the repulsive interaction, the stronger the volume fraction dependence of the elastic modulus.

A more rigorous model for the volume fraction dependence of G', predicted in the framework of thermally activated barrier hopping in HS glasses, yields an exponential increase, $G' \propto (k_\mathrm{B}T/a^3)\exp(26\phi)$ that can be related to a localization length, r_L, as $G' \propto (k_\mathrm{B}T/a^3)(2a/r_L)^{2.13}$ [14]. This approach, which is discussed in Chapter 2, predicts higher absolute values of G' at low volume fractions but with a weaker volume fraction dependence compared to experimental findings [14,54].

Analogous viscoelastic spectra have been measured in HS colloidal crystals formed by imposing large amplitude oscillatory shear on colloidal glasses. Interestingly, the elastic modulus of the crystal state, with a shear oriented FCC structure, is smaller than the corresponding amorphous glass at the same volume fraction [54]. This is due to the higher close packing volume fraction of the crystal structure ($\phi = 0.74$) as compared to the amorphous glass ($\phi = 0.64$). Indeed, when compared at the same average local free volume, or distance from close packing, the crystal structure, in thermodynamic equilibrium, exhibits a higher elastic modulus than the metastable amorphous glassy state.

5.3.3 Transient Rheology

The application of an external shear stress leads to microscopic structural rearrangements that are macroscopically demonstrated as yielding of the colloidal glass. Eventually a steady state flow is reached with a constant effective viscosity. Such shear induced solid-to-liquid transition has been probed experimentally in different types of nonlinear rheological tests. Simultaneous scattering and/or microscopy experiments, as discussed in Chapter 4, have identified the underlying microscopic mechanisms related with shear induced fluidization and the microstructure and dynamics at the particle level under steady state flow [1,37,41,65–68].

A typical nonlinear test consists of applying a constant shear rate to a sample originally at rest and probing the stress response. During such steady start-up shear test a hard sphere glass exhibits an initial linear increase of the stress which, depending on the volume fraction and shear rate, is followed by a stress overshoot and a steady state plateau [69]. The position of the stress overshoot can be considered as the point where the sample transitions from a viscoelastic solid to viscoelastic liquid, thus providing one estimate of the yield strain and stress. This point can be associated with a static yield strain and stress [1,46], while the dynamic yield strain and stress can be defined at the onset of steady state flow. The latter coincides with the value determined from the low shear rate plateau of the flow curve. Furthermore, an elastic yield strain can be identified prior to the stress overshoot as the end of the linear stress–strain regime, indicating the onset of irrecoverable strain [1].

A set of start-up shear tests showing this behavior for a HS-like PMMA glass at different shear rates is shown in Figure 5.8 [69]. The initial stress increase involves an elastic deformation of the structure, which, together with some viscous energy dissipation, leads to a sublinear stress–strain dependence [69]. The deformation of the cage structure is depicted in Figure 5.8(b), as deduced from Brownian Dynamics (BD) simulations that mimic the experimental HS system under shear. The difference of the pair correlation function under shear from that at rest is followed in the velocity-gradient plane, as a function of strain during the transient start-up test. This demonstrates that particles, on average, approach each other in the compression axis, while increasing their separation along the extension axis. Thus, a deformed cage structure is created, with the higher structural anisotropy occurring at the peak stress, before it partially, but not fully, relaxes when the steady state flow is established.

As seen in Figure 5.8(a) the position of the peak, and thus the static yield strain, is around 10% and is weakly increasing with shear rate, in agreement with findings by

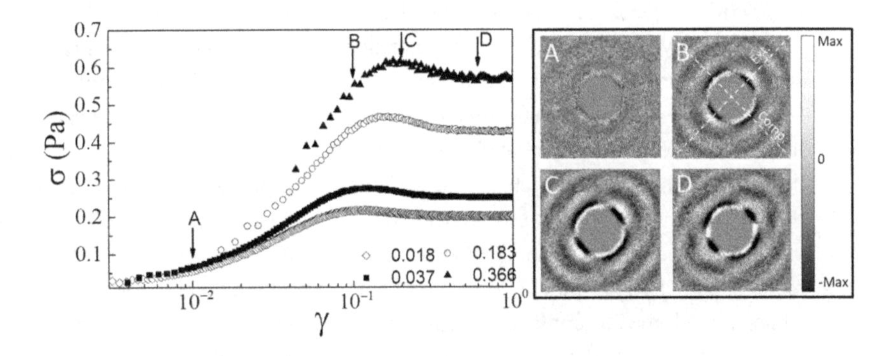

Figure 5.8 (Left) Stress during start-up in hard sphere glasses of sterically stabilized PMMA particles at $\phi = 0.587$ for different scaled *Pe* as indicated. (Right) BD simulations at $Pe = 1$ and $\phi = 0.58$. Projection in the velocity-gradient (*xy*) plane of the difference of the pair distribution function under shear from that at rest at different strains during start-up shear. The four images (a–d) correspond to the structure at 1%, 10%, 20%, and 60% strain. White indicates higher and black lower density compared to the system at rest. The extension and compression axes are indicated by dashed lines. Reprinted with permission from Koumakis et al., Yielding of hard-sphere glasses during start-up shear. *Physical Review Letters* 2012;108 (9):098303 [69(b)] by the American Physical Society

computer simulations [69]. The origin of this behavior lies in the interplay between Brownian motion and shear advection, which allows for higher cage deformation, prior to yielding, as the Brownian motion becomes less important with increasing *Pe*. In contrast, at low *Pe*, Brownian motion assists cage melting which leads to a lower static yield strain. On the other hand, the volume fraction dependence of the yield strain, as deduced from the stress overshoot, exhibits a nonmonotonic trend, as discussed in Section 5.3.5.

Stress overshoots are not limited to the glassy state, but are also observed in concentrated liquids where in-cage localization of particles by their neighbors is only temporary and particles eventually escape through a finite long-time diffusion. Yet, as demonstrated by computer simulations [70], at high enough *Pe* stress overshoots in start-up shear reflect the transient response from an undistorted structure at rest to anisotropic structure under steady shear. In fact, the time dependence of the pair correlation function, $g(r)$, following the evolution of the structural anisotropy, can be directly used in HS suspensions to quantitatively determine the transient stress response [69] as:

$$\sigma_{xy} = -n_0^2 k_B T \alpha^2 (2a) g_0(2a) \iint n_x n_y f \, d\Omega \tag{5.1}$$

with n_x,n_y unit normals, n_0 the particle number density and f the deviation of the pair correlation function under shear from the one at rest according to $g(2a) = g_0(2a)(1 + f(2a))$ [71].

Stress overshoot peaks such as those depicted in Figure 5.8 have been observed in a variety of other soft matter systems such as polymers, worm-like micelles, nanocomposites, colloidal gels, and jammed soft particles, as will be discussed in Chapter 6. Moreover, theoretical work exists based on Soft Glassy Rheology (SGR) [72] and mode coupling theory (MCT) [73,74] and compared with experiments on HS-like particles [68,75,76] and thermosensitive microgels [77]. Computer simulations also provide insights linking microscopic structural effects with mechanical response [78–80]. The overshoot stems from an elastic energy storage mechanism during deformation of a microstructure and a subsequent dissipative energy release related with structural changes and irreversible rearrangements beyond the stress peak, which lead to the fluidization of the system. For HS glasses MCT relates the appearance of the stress peak to a local transient super diffusion and a negative value of a generalized dynamical shear modulus, $G(t)$ [68]. The latter reflects the partial recoil and relaxation of the deformed structure around the yield point as also discussed in Chapter 2. Moreover, confocal microscopy experiments and computer simulations detect such super diffusive dynamics under steady state flow conditions indicating a continuous cage breakage and reformation [69].

Aging Effects

Similarly to other nonlinear rheological tests, the stress overshoot in hard sphere and soft repulsive glasses is affected by aging. Start-up shear experiments on core-shell thermosensitive microgels [81] are shown in Figure 5.9(a). The stress overshoot becomes stronger with increasing waiting time, in agreement with findings for the HS-like PMMA particles [69]. An advantage in using such thermosensitive particles is

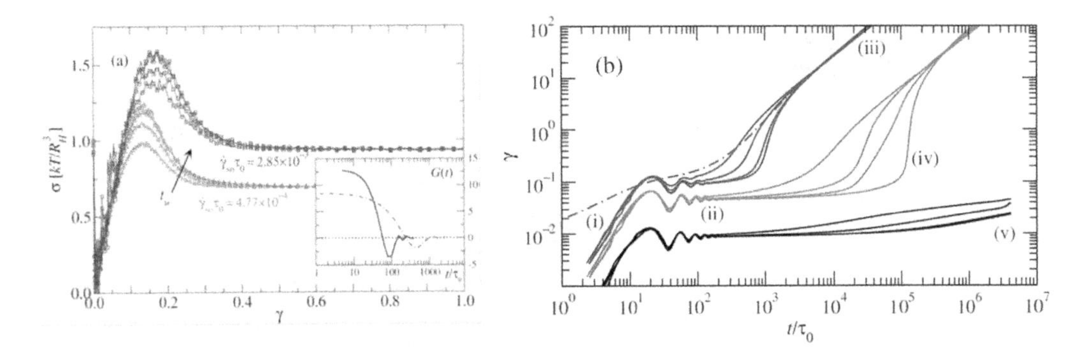

Figure 5.9 (a) Aging effects on stress overshoots: Start-up shear in core-shell PNIPAM particles (softer than sterically stabilized PMMA particles) for two different Pe (=2.85 × 10^{-3} and 4.77 × 10^{-4}) at different waiting times t_w. Inset: Shear modulus, $G(t)$, for Pe = 2.85 × 10^{-3} and t_w = 6000 s (solid line) and $\gamma(t)$ from creep measurements (dash-dot line). (b) Aging effects during creep measurements. Strain, $\gamma(t)$, during creep measurements with PNIPAM particles for different waiting times and stresses. Reprinted with permission from Siebenburger et al., Creep in colloidal glasses. *Physical Review Letters* 2012;108(25):255701 [81] by the American Physical Society

that the liquid to glass transition can be achieved as particles swell by a temperature quench. It should be noted that the strength of the stress overshoot is larger for suspensions of PNIPAM than PMMA particles due to the softer character of the former [47]. Here, the inset also depicts the time dependent elastic modulus, $G(t)$, used within MCT to describe the stress overshoot response, where the negative part of $G(t)$ is due to elastic recoil.

Another method to study yielding of a colloidal glass is to apply a constant stress and measure the strain deformation, i.e., a creep measurement, as shown in Figure 5.9(b). The time-dependent strain $\gamma(t)$ is shown for three different stresses and various waiting times t_w. The sample exhibits an initial fast increase of strain followed by a constant strain plateau for samples below the yield stress, or continued, slow creep. Slow creep is due to the internal rearrangements characteristic of sample aging under constant stress. For this creep, the viscosity decreases sublinearly with time, reminiscent of Andrade type of behavior seen in metallic systems [15,81]. For applied stresses above the yield stress the sample eventually reaches steady state flow.

In these experiments, while the initial linear response and the long-time flow are not affected by sample aging, the transient regime is, as shown in Figure 5.9(b). This is a consequence of the super-diffusive particle dynamics found in computer simulations and experiments, which is apparent in aged samples. Thus, aging in colloidal glasses primarily affects the transient, nonlinear behavior, although, interestingly, linear viscoelasticity is hardly affected [15,69,82].

Creep and strain recovery data for a HS-like glass at $\phi = 0.62$ are shown in Figure 5.10(a) for stresses near and above the yield stress [37]. Similarly to Figure 5.9(b), $\gamma(t)$ increases and reaches an almost constant plateau for stresses below the yield stress (here around 8 Pa), with values up to the yield strain. For higher

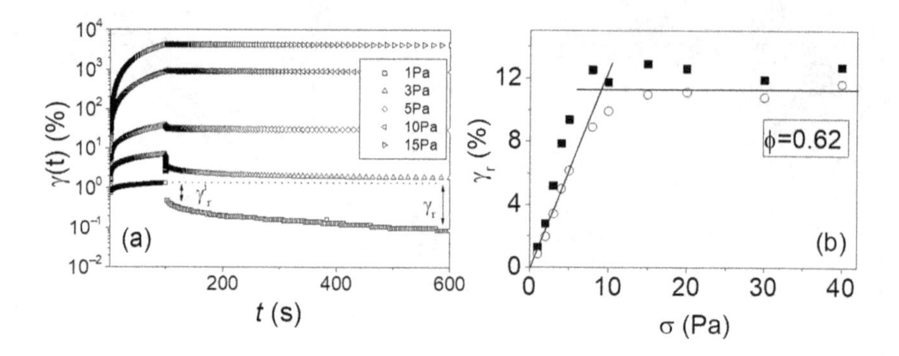

Figure 5.10 (a) The time dependence of the strain, γ(t), during a step stress (creep) and recovery experiment for a PMMA hard sphere glass at $\phi = 0.62$, for several stresses as indicated. The instantaneous and total recovered strain are indicated by vertical arrows. (b) The recovered strain (both instantaneous and final) after the cessation of flow, corresponding to the data of (a) as a function of applied stress. Adapted from Petekidis et al. [37]

stresses the sample reaches steady state flow up to several thousand percent until the stress is switched off. From the solid mechanics point of view the elasticity of the sample should be directly detected by measuring the strain recovered after the applied stress is removed. Such recovered strain is shown in Figure 5.10(b). After stress removal at $t = 100$ s the strain exhibits a sudden drop, which is taken to be the instantaneous recovered strain, γ_r^i. This is followed by a slower relaxation to the final recovered strain, γ_r, reached after several hundreds of seconds. The values of the recovered strain are shown in Figure 5.10(b), where γ_r^i and γ_r initially increase linearly up to a critical stress, which can be identified as a measure of the yield stress. At higher stresses the recovered strain reaches a constant value, identified as the yield strain of the system. The creep and recovery test provides one of the clearest measures of the yield stress and strain, and offers direct evidence of the elastic character of the deformed structure in a flowing hard sphere glass.

5.3.4 Stress Relaxation

Upon shear cessation, the shear-deformed microstructure, shown in Figure 4.6 and 5.8(b), will relax back to equilibrium via particle Brownian motion. The corresponding stress relaxation has been measured experimentally and in computer simulations, as shown in Figure 5.11 [52]. A common feature in experiments, simulations, and MCT, for both HS-like PMMA particles and softer thermosensitive core-shell microgels, is the existence of residual frozen-in stresses in glassy samples (see also Chapter 6). This behavior constitutes the colloidal analog of frozen-in stresses induced in molecular glasses by a fast thermal quench (akin to Prince Rupert's drops [83]).

In concentrated liquids below the glass transition, stresses are found to fully relax slowly with relaxation times similar to those of the quiescent liquid. In glassy systems stresses relax only partly after arresting the flow. The fraction of the stress that does not relax decreases with increasing preshear rate. Such finite residual stresses are a

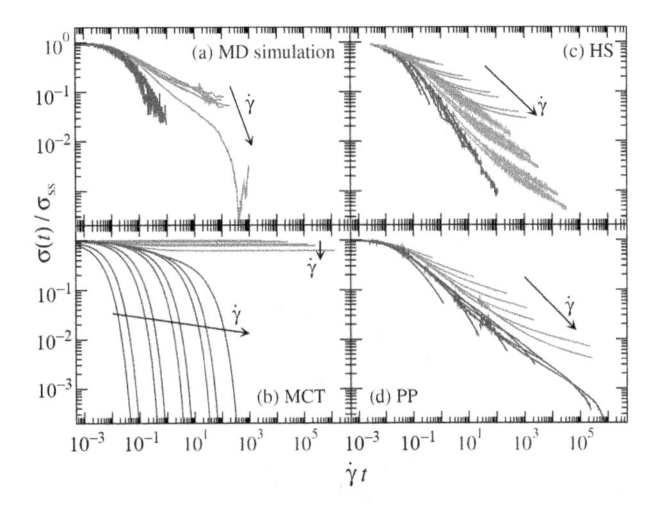

Figure 5.11 Residual stresses after arresting flow for nearly-hard sphere colloidal glasses at different volume fractions in the liquid (concave lines) and glassy (convex lines) states, at several shear rates, plotted a function of time multiplied by the shear rate prior to switch-off: (a) Molecular Dynamics simulations from a binary glass-forming mixture with a Yukawa potential; (b) mode coupling theory (MCT) predictions with an isotropic hard sphere model (see Chapter 2), (c) experiments for suspensions of polydisperse, PMMA particles, and (d) experiments for thermosensitive core-shell PNIPAM microgels. Reprinted with permission from Ballauff et al. Residual stresses in glasses. *Physical Review Letters* 2013;110(21):215701 [52] by the American Physical Society

manifestation of long-lived memory effects in the glass and of its out-of-equilibrium character. Thermodynamic parameters and shear history affect the structure, dynamics, and mechanical properties of the system. Such shear history effects are even more pronounced in attractive nonergodic states, such as the colloidal gels discussed in Section 5.4. Residual stresses have been observed for suspensions of soft colloids, as shown in Chapter 6 [84], and other frustrated soft matter systems with different interactions such as colloidal gels [85], aqueous Laponite suspensions [86], and carbopol microgels [87].

The elastic response of a colloidal glass can also be probed by step-strain tests, when the maximum stress is plotted against the corresponding value of the step strain [53]. This reveals an initial elastic response at small step strains and constant stress values for step strains above the yield strain. Moreover, in agreement with the curves of Figure 5.11, the stress relaxation after a step strain is weaker and slower for strains below the yield strain and faster and more complete for step strains at higher strains, where the sample has been shear melted [53].

5.3.5 Yielding

The yield strain and yield stress in hard sphere glasses depend primarily on the volume fraction, while the absolute value of the yield stress scales with the thermal

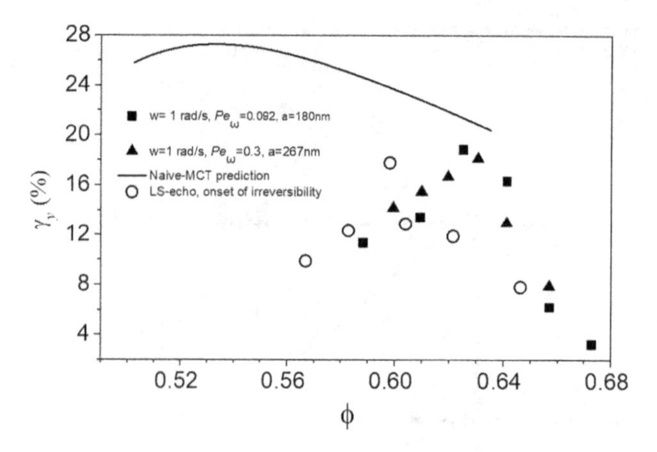

Figure 5.12 Yield strain of HS glasses as a function of volume fraction: Solid symbols are from LAOS sweeps at the crossover strain, adapted from Koumakis et al. [47]. Open circles are experimental data for the onset of irreversible particle rearrangements measured in LS echo [65]. Solid line is theoretical prediction from naïve-MCT and activated hopping model of Kobelev and Schweizer [14].

energy density ($k_B T/a^3$), similar to scalings for the elastic modulus of a Brownian suspension, Eq. (1.8). While the yield stress follows a similar increase with ϕ, a careful investigation indicates that the yield strain, as determined from the crossover of the moduli, exhibits a weak nonmonotonic dependence on ϕ. Experimental data for the volume fraction dependence of the yield strain in HS glasses are shown in Figure 5.12. The yield strain from dynamic strain sweep tests (LAOS) in the low frequency regime for suspensions of two different size HS-like PMMA particles. A nonmonotonic dependence, with a maximum yield strain of about 20% achieved around $\phi = 0.63$, is observed, which indicates that such glasses are very brittle materials. Due to polydispersity, random close packing in these samples is in the range of 0.67–0.69. A similar nonmonotonic behavior is detected in the yield strain determined by the strain recovered after creep, or by the stress overshoot in step rates tests in HS glasses, and this behavior is in qualitative agreement with the onset of irreversible particle rearrangements probed in the same suspensions under LAOS by Light Scattering (LS) Echo experiments [65,66]. Colloidal crystals formed by applying LAOS on HS glasses exhibit similar, although weaker, yield strain maximum at almost the same volume fraction as their glassy counterparts [54]. In contrast, suspensions of soft deformable particles, such as those discussed in Chapter 6, show a continuous increase in the yield strain through close packing and beyond into the jammed state.

A comparison of this data with theoretical predictions of a microscopic theory based on an interplay of mechanically driven particle motions with thermally activated barrier hopping processes [14] is also presented in Figure 5.12. The model predicts a weakly nonmonotonic volume fraction dependence of the absolute yield strain (solid curve) due to a subtly different dependence of the yield stress and the storage modulus on volume fraction. All measures of the yield strain predicted by the model decrease approaching close packing, in qualitative agreement with experiment. Although the exact behavior depends on the shear rate and the age of the sample, the yield strain decreases as the volume fraction increases because the available free volume between particles diminishes approaching rcp. Therefore, there is less room for the cage to be deformed elastically before it breaks at yielding. At the same time, the strength and position of the stress overshoot moves to lower values as volume fraction is decreased towards the liquid state, where the yield strain is expected to vanish. This behavior suggests that a denser system is more brittle under shear deformation, which is consistent with tighter packing of particles in cages of neighbors that are more strongly localized [14].

5.3.6 Shear Localization: Shear Banding and Slip in HS Glasses

In the transient and nonlinear rheological tests discussed above the velocity profile is assumed linear. However, HS glasses can exhibit localized or spatial inhomogeneities in shear, mainly at low shear rates. In the case of smooth walls, HS glasses have been shown to exhibit wall slip below a critical shear rate that depends on volume fraction [48]. A typical second stress plateau appears in flow curves, as shown in Chapter 6 for suspensions of soft deformable particles. Roughening the walls, with asperities similar to the particle size, suppresses wall slip. The slip stress then exceeds the yield stress and the typical HB flow curves such as those shown in Figure 5.5 are recovered. However, when avoiding slip, shear banding is detected for shear rates below a critical *Pe* [49,88]. In the case of HS glasses, shear banding has been proposed to be driven by an instability due to shear concentration coupling and a migration of particles towards the low shear rate region [41], rather than by a nonmonotonic flow curve as can be the case in other soft matter systems [89,90]. Here, the shear banding is due to the strong dependence of the yield stress on concentration. Small fluctuations in the latter lead to the distorted profile. Similar localization effects have been observed in pipe flows with HS-like PMMA particles [91].

5.3.7 Summary and Outlook

Hard sphere glasses formed due to excluded volume caging of particles by their neighbors at high volume fraction display a solid-like response. Their linear viscoelastic response emanates from hindered Brownian motion and scales with the thermal energy density. The elastic modulus increases strongly with volume fraction as localization due to caging becomes tighter. HS glasses flow due to microscopic deformation of the local cage structure and irreversible particle rearrangements associated with out-of-cage diffusion of particles. Cage breaking and reformation occur continuously, but the glass does not flow as a simple liquid. Yielding can be probed either by steady or oscillatory shear, with the corresponding yield stresses and strains being similar, but not necessarily identical.

When applying a constant shear rate, the stress exhibits an overshoot that signals a transition from elastic to viscous response that is dependent on sample age. Yet, the linear viscoelasticity of such glasses is essentially independent of age. Constant stress tests reveal transient super-diffusive response as the sample starts to yield. This also becomes more pronounced with aging. Stress relaxation is, in part, due to elastic recovery of the deformed cage of neighbors; however, there remains a residual, frozen-in, stress resembling that observed in pre-stretched molecular glasses. The yield stress increases with volume fraction, while the yield strain exhibits a nonmonotonic dependence on volume fraction. As random closed packing is approached, the yield strain decreases due to the reduction of the free volume.

At low shear rates hard sphere glasses may show wall slip on smooth surfaces when the stress is lower than the bulk yield stress. Wall slip can be suppressed by using rough walls and additional methods covered in chapter 9 of *CSR*, but then HS glasses may also shear band below a critical shear rate that depends on volume fraction.

Although the linear and nonlinear rheology of HS glasses has been extensively studied by experiments, computer simulations, and theory, several open questions remain. These include the role of dynamic heterogeneities in yielding, as well as the microscopic strain, stress localization, and possible shear banding during transient tests that lead to yielding. Also questions remain about aging phenomena in nonlinear tests, such as stress overshoots and superdiffusive behavior in creep and the specific mechanisms causing them [69]. Finally, the possible role of contract friction in these systems remains relatively unexplored [92].

5.4 Colloidal Gels and Attractive Glasses

We next turn to colloidal particles with interparticle attractions and discuss their linear and nonlinear rheology and the connection to the underlying structure. As discussed in Section 5.1, in the presence of attractive forces colloidal systems can form colloidal gels and attractive glasses depending on the attraction strength and particle volume fraction. This is also discussed in chapter 6 of *CSR* [1].

The rheology of attractive systems is affected by several features of the phase behavior which, depending on the range of and strength of attraction, can involve "gas", liquid, and crystal states [1,16], as well as gels and glasses. The formation of equilibrium clusters due to a delicate balance of short range attractions and long range repulsions [93–95] and the kinetics and structural evolution of fractal flocs at low volume fractions [1] are also affecting the rheological properties of the system. In turn, deformation and flow affect the formation or breakage of such structures [1]. Such phenomena are quite common in both industrial formulations and biological fluids (see also Chapter 9). The interplay of colloidal gelation with an underlying phase separation mechanism impacts onto the structural evolution and flow properties of the gel, invoking a shear history dependence and a strong thixotropy of the system [1,96].

A central question related with the actual definition of a colloidal gel is: What are the necessary conditions for a system of attractive colloidal particles to exhibit a solid-like response, i.e., to acquire mechanical, load bearing, ability? The determination of the gel point can be investigated by different approaches and experimental methods, probing, for example, structurally the formation of a percolated network (percolation point), dynamically the onset of nonergodicity where long-time motion is frustrated (point of dynamic arrest), or detecting by rheology the divergence of zero shear viscosity and the onset of a yield stress (liquid-to-solid transition). Consequently, the question is to what extent and under what conditions these points coincide. Another question is whether there are different routes to gelation and how do they affect the final material properties? The answer to these questions has been the subject of active research into the nature of colloidal gels and their response under shear [1,16,97–100]. The specific characteristics of the bonds, the local micromechanics, topology, and microstructure of the clusters have been investigated in addition to the interplay with the thermodynamic phase separation noted above.

5.4.1 The Mechanisms and the Underlying State Diagram

Experimental evidence and theoretical predictions suggest that two main scenarios are involved in the formation of colloidal gels, related with the position of the percolation and glass lines with respect to the thermodynamic boundaries and the phase separation regime [16,101,102]. Examples are shown in Figure 5.13. Both the percolation line, which defines the formation of a space-spanning connected network [1], and the attractive glass line as predicted by MCT (see Chapter 2), indicate the transition to dynamically arrested states depending on the nature of the system [16,99,103].

In the left part of Figure 5.13, the percolation line lies above the region of phase separation, except at low volume fractions below the critical point, and therefore the system may form "*equilibrium*" (or *homogeneous*) *gels* when quenched to temperatures below the gel line but above the onset of phase separation. As will be shown, dynamical arrest occurs by "rigidity" percolation, which is predicted to be an actual thermodynamic phase transition to an arrested state without macroscopic or microscale phase separation [106]. Such rigid bonds lead to glasses and gels by *rigidity percolation*, as derived by He and Thorpe [106]. An important criterion of rigidity percolation is that the average number of bonds per particle in the network is 2.4 at the transition from liquid to solid, and further, that this is an equilibrium phase transition. This is in contrast to systems without bonds, where packing leads to jamming, which require at least six contacts per particle to achieve Maxwell isotacticity and rigidity [107]. In contrast, other suspensions are known to gel as illustrated in the right part of Figure 5.13, as "*arrested phase separation*" drives gelation. When quenched below the phase separation line, the system will start phase separating and coarsening, but phase separation will be arrested at the glass line at high ϕ, resulting in a gel network of dense glassy clusters [108]. The dense colloidal phase that forms becomes arrested and percolates, such that the suspension is unable to phase separate macroscopically. The local dense phase forming the gel is comprised of a tightly packed, nonequilibrium glass that lies on the attractive glass line, which is predicted to intersect the phase separation line and to extend into the phase separation regime, as shown in Figure 5.13 (right).

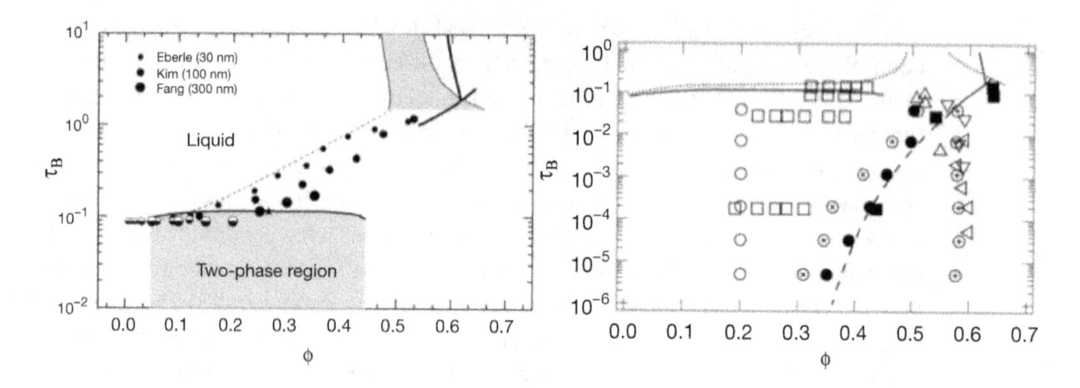

Figure 5.13 Left: Dynamical arrest transition state diagram as Baxter sticky parameter (dimensionless temperature) and volume fraction. Data are for thermoreversible suspensions with different particle diameters. The half-filled symbols are samples that settle due to gravity while filled symbols are stable gels. The dashed line is connectivity percolation. Adapted from [27]. Right: State diagram including data for colloid-polymer mixtures The initial volume fraction and reduced temperature of gel samples are shown for the PMMA-PS mixtures from Whitaker et al. [104] (open circles) and the thermosensitive silica particles from Ramakrishnan et al. [105] (open squares). The density of the clusters, ϕ_g, are determined from simulations and application of a graph theory (closed circles, PMMA particles from [104]; closed squares, octadecyl silica from [105]). The local volume fraction of particle-rich regions are also shown for the PMMA particles from [104] (circle-plus) together with data for similar PMMA particles from Lu et al. [18] (triangles), Fluid–fluid phase coexistence is indicated by the solid black line, as well as the attractive and repulsive glass lines, while an extrapolation through the experimental data is indicated with the dashed line. The AHS phase boundaries with the fluid–solid coexistence lines for the dimensionless range of interaction of 0.1 are given by short dashed lines. Adapted from Whitaker et al. [104]

Interestingly, the two glass lines shown in Figure 5.13 meet at high concentrations and the attractive driven glass line terminates at a singular point (A_3), according to MCT predictions [103]. The attractive glass differs in the nonergodicity factor and the associated localization length from its HS counterpart, with the latter defined by the attraction range, being smaller than the cage localization in the HS glass [103,109]. Because the interparticle potential is attractive in nature in the attractive driven glass, colloidal particles are more constrained and such attractive glasses exhibit a much higher linear elastic modulus than a repulsive glass [53,109]. More importantly though, they show an utterly different nonlinear mechanical response, as will be discussed in this section.

At low volume fractions, gravitational effects due to particle size and density differences with the solvent can significantly shift the dynamical arrest transition to stronger attractions. A detailed study of this effect has been presented by Kim et al. [27], where the effect persists until the packing fraction is into the glassy regime. For example, data for suspensions of a homologous series of particles is shown in the left panel of Figure 5.13. These effects limit our ability to create gels at low volume fractions where settling or creaming, depending on the particle density, is usually observed. The gravitational Péclet number is defined as the time required for

Brownian motion to bring particles of radius a, forming clusters of fractal dimension d_f, together into a network relative to the rate of gravitational settling/creaming of these clusters, as [27]:

$$Pe_g = \frac{4\pi a^4(\Delta\rho)g}{3k_BT}\phi^{\left(\frac{d_f+1}{d_f-3}\right)}. \tag{5.2}$$

Experiments across a broad range of particle sizes and gravitational accelerations (using centrifugation) show that the critical value for gel formation is of order 0.01 [27]. Hence, lowering the volume fraction holding all else constant will eventually increase Pe_g above this critical value and the gel will settle or cream. While investigators may attempt to density match the particles to the suspending medium in order to reduce gravitational effects, the quadratic dependence on particle size suggests these effects will always be relevant for experiments performed on particles that are optically visible, i.e., micrometers in size.

5.4.2 Gel Micromechanics and Local Cluster Structure

Gelation through arrested phase separation is commonly seen in colloidal systems with isotropic attractive potential [102,110]. In the case of dilute suspensions of strongly attractive colloids, an irreversible aggregation process creates fractal clusters growing in space till they form a space spanning, percolating network. These DLCA gels exhibit the hallmarks of spinodal decomposition, with a typical peak in scattering experiments that evolves towards lower scattering angles as the structure coarsens [1,18,25,111]. Depletion gels are the most common model system that form gels via arrested phase separation, both at low and high densities [18,108,112,113]. Depending on the attraction strength and particle volume fraction the gel coarsens in time towards more dense clusters, as particles rotate and rearrange, maximizing the number of nearest neighbors and minimizing the overall free energy. Therefore a main feature of such "nonequilibrium" gels is their coarsening via Brownian motion leading to aging that can be accelerated by shear or gravitational settling or creaming.

The microstructure of a typical suspension with depletion interactions driven by the addition of polymer is shown in Figure 5.14. The depletion potential is presented in Chapter 1 and shown in Figure 1.15. A colloid-polymer mixture of sterically stabilized PMMA particles at a low volume fraction ($\phi = 0.15$) with added polymer chains of varying molecular weight and concentration enables tuning the range and strength of depletion attraction [114]. A fluid state of individual particles is seen for low attractions ($U_{dep} = -0.9k_BT$), which becomes a fluid of clusters when the attraction strength is increased ($U_{dep} = -1.6k_BT$) and eventually for stronger attractions ($U_{dep} = -2.7k_BT$), a percolated gel network. The compact structure of the clusters and the thick strands in the percolated network suggest that these structures are formed through the arrested phase separation mechanism.

As the accuracy of confocal microscopy experiments is insufficient to directly determine the local density inside the gel clusters, Whitaker et al. [104] suggest an

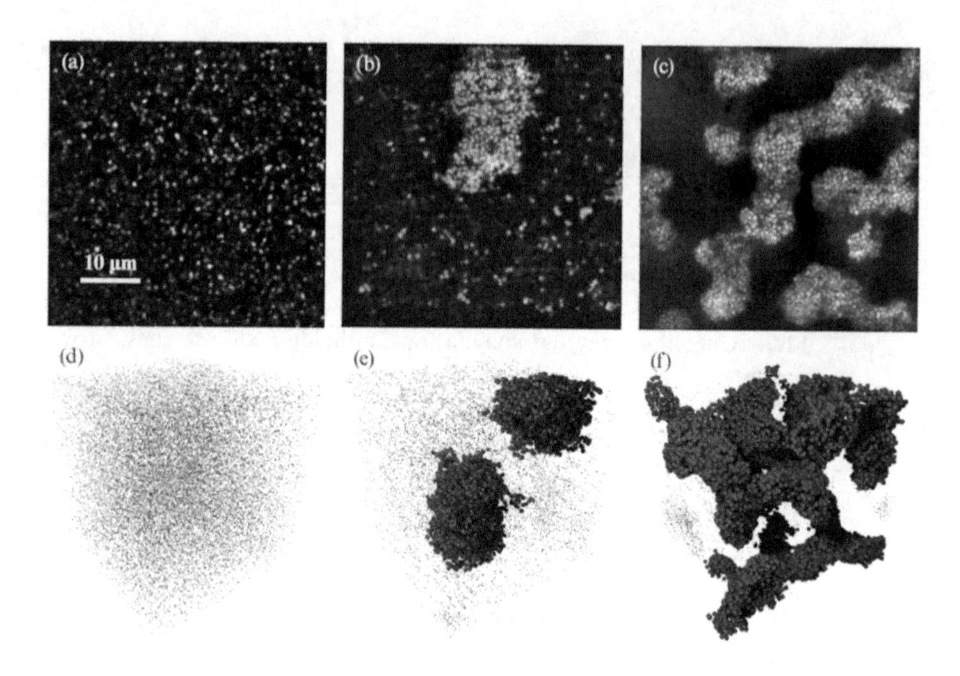

Figure 5.14 (a)–(c) 2D confocal microscope images and (d)–(f) 3D reconstructions of colloid-polymer mixture with $\phi = 0.15$ and linear polystyrene polymer chains as depletant. In the 3D reconstructed data, particles in clusters with more than 500 particles are shown at actual size while the rest are shown at 1/5 actual radius. In (a) and (d) a homogeneous fluid of single particles (with $\delta = 0.17$, $U_{dep} = -0.9k_BT$) is shown; in (b) and (e) a fluid of clusters (with $\delta = 0.15$, $U_{dep} = -1.6k_BT$) and in (c) and (f) a percolated gel (with $\delta = 0.11$, $U_{dep} = -2.7k_BT$). Reprinted with permission from Lu et al., Fluids of clusters in attractive colloids. *Physical Review Letters* 2006;96(2):028306 [114] by the American Physical Society, DOI 10.1103/PhysRevLett.96.028306

alternative analysis. These authors used graph-theory cluster decomposition to identify a structure that fulfills the necessary isostatic condition for rigid, and thus load bearing, microstructure. In this way, the local cluster density is found to decrease with increasing attraction and becomes equal to the density predicted at the glass line, as shown in Figure 5.13. The local cluster density is also deduced in a consistent way from the measured elastic modulus, G' , as measured in experiments and predicted by simulations, using the model by Zaccone et al. [115]. This analysis yields a decrease of the local density within the gel clusters with increasing attraction for both PMMA and silica depletion gels. Moreover, the increase of G' with increasing attraction strength is linked with the creation of more homogeneous gels, as has been reported in several other experiments and simulations of phase separating gels [17,116,117]. This observation suggests the use of different preshear protocols to tune the microstructure and, therefore, the mechanical properties of nonequilibrium colloidal gels formed via arrested phase separation [85,118], as will be discussed in Section 5.4.3.

The formation of homogeneous gels is possible provided that particles in the gel network have bond lifetimes and bond rigidities for a stable mechanical structure. This is not possible for centrosymmetric attractive potentials, where particles are free to rotate and rearrange, even at the limit of vanishing range of attraction (sticky spheres) [102,119]. However, experiments and theoretical studies provide strong evidence for the existence of routes to equilibrium gels. If equilibrium states, the properties of these gels should be independent of processing history. For example, suspensions of thermosensitive particles (octadecyl-coated silica) have been shown to exhibit this type of behavior [19,27,36,120–122]. The interactions of these sticky, thermoreversible particles [19,122] can be mapped to the Baxter potential and gelation is found to take place at the rigidity percolation line and outside the phase separation regime, providing strong evidence for homogeneous gelation.

In addition to suspensions of adhesive, spherical particles, particles with patchy interactions have received much attention, as they form open networks stabilized as the interparticle bond lifetime increases [110,123]. Similarly equilibrium gels may be formed by anisotropic, discotic particles acquiring different charges on their flat surfaces and edges, such as laponites and clays [124], and attractive ellipsoids or rod-like colloids [123,125]. A versatile system that can be tuned to form either phase separating or homogenous gels has been reported [101,126]. This consists of oil microemulsions mixed with telechelic molecules that provide bridging attractions between the droplets, where functionality, bond lifetime, strength of attraction, and volume fraction can be adjusted independently. In this manner the gel line can be shifted without the phase separation being affected. Experiments in different systems [101,126] and theoretical modeling [127] suggest these systems can form homogenous gels in the stable fluid regime of the phase diagram.

Micromechanics

In addition to thermodynamic parameters, local micromechanics also play an important role. This includes the forces between particle surfaces at the nanoscale level, as well as external fields, such as gravity and shear history in the gel formation and mechanical properties. A question that arises then is to what extent contact friction between particle surfaces or other tangential forces due to specific interactions, e.g., grafted polymers, inhomogeneous surface charge distribution, and surface roughness, affect the gelation process, the linear mechanical properties, and the nonlinear yielding response of colloidal gels. To address this, Furst and Pantina [128,129] demonstrated through laser tweezer experiments the existence of local bending moments due to static friction between neighboring particles in strongly flocculated polysterene latex gels. As shown in Figure 5.15, a linear aggregate of particles is subjected to microscopic forces by laser tweezers, as indicated by the arrows, from which the bending rigidity of the aggregate and subsequently of the individual bonds between two colloids can be deduced. At the microscopic level, a localized yielding is manifested as a sudden single particle rearrangement within the linear aggregate at a critical bending moment. The critical bending moment is then linked at the macroscopic level to the yield stress of the gel, indicating the important role of tangential forces on the

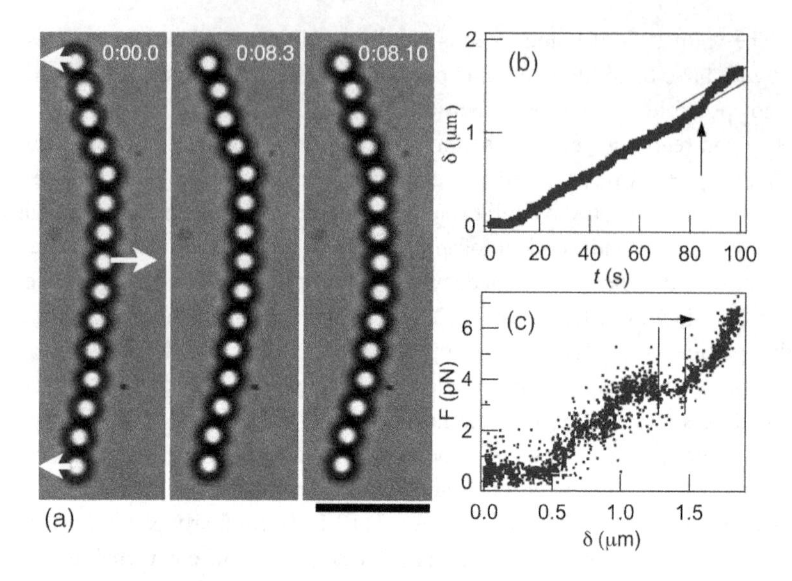

Figure 5.15 (a) Image sequence demonstrating the microscopic process that characterizes the nonlinear mechanics of aggregates. A 15-particle aggregate of 2.8-μm-diameter PS particles in 50 mM $MgCl_2$ is bent using a three-point geometry, in which the middle particle is moved by a translating optical trap. Two stationary optical traps holding the top and bottom most particles exert restoring forces. The corresponding deflection of the aggregate and force are shown in (b) and (c), respectively. Reprinted with permission from Furst and Pantina, Yielding in colloidal gels due to nonlinear microstructure bending mechanics. *Physical Review E* 2007;75(5):050402 [128], by the American Physical Society. DOI 0.1103/PhysRevE.75.050402

nonlinear response of colloidal gels. The details of local interactions are, therefore, expected to affect the phase behavior and gelation kinetics in a way that may define the type of gelation a system follows. Centrosymmetric attractions are expected to lead to gelation through arrested phase separation, whereas patchy and anisotropic, or directional, interactions can lead to homogeneous gelation in the single-phase regime. Hence the existence of additional local tangential interactions may hinder free particle rotation and rearrangements of interacting particles, promoting equilibrium gelation. Such rigid bonds may be why suspensions, such as those comprised of thermosensitive octadecyl-coated silica particles, demonstrate characteristics of homogenous gels.

When the interactions between particles are sufficiently strong, actual contacts may form that are comparatively long-lived. In such a case fractal-like aggregates form that can percolate into stress bearing networks [19]. This is common in various real-life systems, e.g., blood, paints, cement, carbon black, asphalt, and mineral colloidal suspensions, discussed in the chapters to follow. Importantly, these physical gels are often homogeneous on the meso-scale, unlike the heterogeneous, depletion gels. Also, they can be at or near equilibrium, which results in gels and glasses with minimal aging and which are stable against shear [121] and gravity [27]. Importantly, because such gels are homogeneous and have rigidity in their bond angles, they have much higher moduli [130], and, as such, are precursors to materials with exceptional properties. This is because the internal defects in materials made from glasses and homogenous gels are on the order of the particle size, rather than on the comparatively large void mesoscale that dominates phase-separating systems. A close connection can be drawn between physical gelation and chemical gelation in polymers [107].

Local Microstructure

A point of interest is the local microstructure within the strands and clusters of a gel that is necessary to provide load-bearing ability and thus is the required ingredient for a percolating gel network to acquire a solid-like response. The character of the load-bearing units in a gel network and their link with dynamical arrest has been investigated theoretically as well as in experiments and simulations [97,98,131]. For systems where spherical, frictionless particles interact by central-symmetric interactions, mechanical stability and isostaticity require, according to Maxwell [132], constraints with six-fold symmetry, i.e., networks involving particles with at least six bonded neighbors. Confocal microscopy experiments and computer simulations of charged colloid-polymer mixtures forming gels with competing attractive and repulsive interactions at low to intermediate volume fractions $\left(\phi \sim 0.2\right)$ show that dynamical arrest of the gel is linked to directed percolation rather than isotropic percolation [131]. The former differs from the latter in containing network strands that span the whole system without loops or dangling sections.

Tsurusawa et al. [97], using colloid-polymer mixtures and confocal microscopy, showed that solid-like response is acquired only after an isotropic percolation of locally isostatic structures inside the clusters is reached. In dilute suspensions, directed percolation of all particles happens simultaneously with isotropic percolation, whereas at higher concentrations these two timescales deviate from each other. A schematic of these two paths towards gelation are shown in Figure 5.16, illustrating the creation of isostatic structures within the gel network at dilute $(\phi = 0.08)$ and intermediate concentration $(\phi = 0.27)$ [97]. The emergence of a solid-like mechanical response is linked to isotropic percolation of isostatic particles. In dilute concentrations (Figure 5.16(a)) this coincides with directed percolation of all particles, in agreement with findings by Kohl et al. [131]. At higher concentrations percolation of isostatic structures occurs at a later stage following a directed percolation of more floppy particle strands, as shown in Figure 5.16(b).

The central role of isostatic structures has been identified, not only for the linear viscoelasticity at rest, but also for the nonlinear response of gels, such as under step strain tests that lead to yielding and flow [133]. In this case the onset of yielding was related with the breakage of rigid isostatic clusters, identified as those with contact number of bonded neighbors of six, in accordance with local jamming criterion for frictionless particles [134]. Similarly in gels with competing repulsive and attractive interactions, Bernal spiral microstructures are central in providing elasticity in the suspension [95] and define its rheological response.

Suspensions of adhesive hard spheres [19,135,136] create a system spanning network with a minimum average coordination number significantly less than six [122]. Here, as

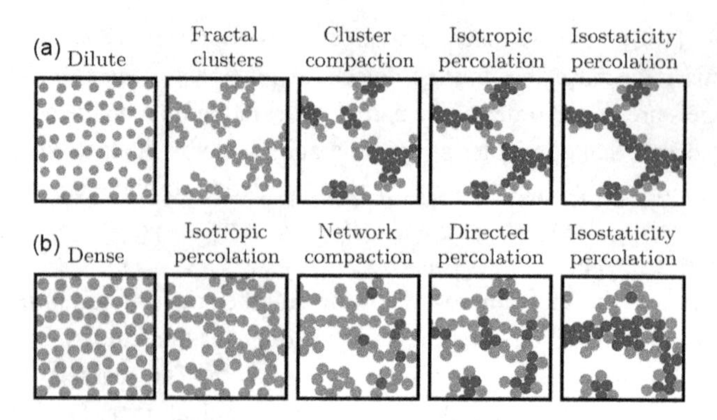

Figure 5.16 Sketch of the two possible paths to mechanically stable gels in depletion systems. (a) Dilute path. (b) Dense path. Isostatic particles are shown in purple; nonisostatic particles are in gray. Reprinted with permission from AAAS from Tsurusawa et al., Science Advances [97], © The Authors, some right reserved; exclusive license American Association for the Advancement of Science. Distributed under a Creative Commons Attribution NonCommercial Licence 4.0 (CC-BY-NC); http://creativecommons.org/licenses/by-nc/4.0/

for the case of particles with directed interactions, the existence of specific local contact interactions, of frictional or other nature, might be the origin of mechanical stability. The criterion for dynamic arrest corresponds to an average coordination number of 2.4, as predicted by *rigidity* percolation [106], which is much less than the isostaticity criterion. This can be viewed in analogy with the lowering of close packing in suspensions of rough frictional particles as compared to smooth particles, which is also central to the discussion of the role of friction and hydrodynamic interactions in shear thickening of concentrated colloidal suspensions [137,138].

In gels generated via arrested phase separation, the structure evolves with aging time. Coarsening of the structure occurs when the attraction strength is of the order of a few k_BT so that thermal fluctuations are able to induce bond restructuring and particle rearrangements within the clusters of the gel network. If, however, the attraction strength is of the order tens of k_BT or more, coarsening becomes extremely slow or is totally suppressed, and structure, dynamics, and mechanical properties are essentially frozen. While the details of gel aging depend on the range and strength of attraction and the colloid volume fraction, experimental and simulation studies show that aging leads to thicker strands or more compact clusters and that both the elastic and viscous moduli increase [117]. Zia et al. [139] showed, via large scale Brownian dynamics simulations, that gel strands are comprised of a glassy interior of immobile particles at volume fractions equal to or higher than 0.58, while mobile particles at the surface of the strands are the ones that mainly drive coarsening through their migration towards deeper energy minima. Similar results have been found at higher volume fractions ($\phi = 0.3$ and 0.44), although, with increasing volume fraction, the gel morphology shifts from a network of strands with larger voids to more homogeneous network of clusters with smaller voids [117]. The onset of mechanical stability has not been probed specifically but these simulations again demonstrate the importance of locally dense, isostatic clusters in the response of the elastic stress discussed earlier in this section.

5.4.3 Rheology of Phase Separating Gels and Attractive Glasses

In this section we focus our discussion on the rheological response of the dynamically arrested states with solid-like response at rest. These include the two types of colloidal gels encountered at low and intermediate particle volume fractions, i.e., homogeneous gels and gels formed via arrested phase separation, as well as attraction driven glasses found at the highest volume fractions.

Linear Response

Approaching the gel point by increasing the attraction strength at a specific volume fraction leads to a divergence of zero shear viscosity and the emergence of a finite elastic modulus. Structural changes involved are particle aggregation, creation of clusters and finally of percolating networks with solid-like properties. In the case of dilute suspensions of strongly attractive particles that proceed according to diffusion limited aggregation [1], theoretical models predict fractal networks with thinner arms as the load bearing units, where particles have on average significantly less than six bonded neighbors. Fractal gel models predict power law increases of G' with volume fraction with exponents that depend on the structural characteristics and the attraction strength [1]. Moreover, the viscosity typically decreases with shear rate (shear thinning), reflecting shear induced network or cluster break-up. Experiments and theoretical modeling of such a rheological response is summarized in CSR [1].

For concentrated depletion, systems theory based on the naïve-MCT discussed in Chapter 2 predicts that the zero frequency elastic modulus should scale as:

$$\frac{G'}{k_B T/\alpha^3} = \frac{2.32}{8}\frac{\phi \alpha^2}{r_L^2}, \tag{5.3}$$

with the localization length, r_L [140,141]. This prediction is compared against experiments in Figure 5.17(b), for a depletion gel with $\phi = 0.4$, where the nonadsorbing polymer concentration is varied to control the strength of attractions. Theoretical predictions are overestimated significantly compared to experiments but when renormalized by the number of particles present in a cluster, N_c, as illustrated in Figure 5.17(a), the trend with experiments is recovered. This rescaling reflects the cluster inhomogeneity present in real systems as opposed to the homogeneous microstructure used as an input in MCT.

A critical comparison with other depletion systems such as PMMA-polystyrene or polybutadiene mixtures in various solvents shows qualitative agreement of the rescaled predicted modulus using, however, scaling factors varying from $N_c = 4$ [17] to $N_c = 1500$ [82]. This indicates the importance of different preparation or preshear protocols, which can cause variations in the structure and mechanical

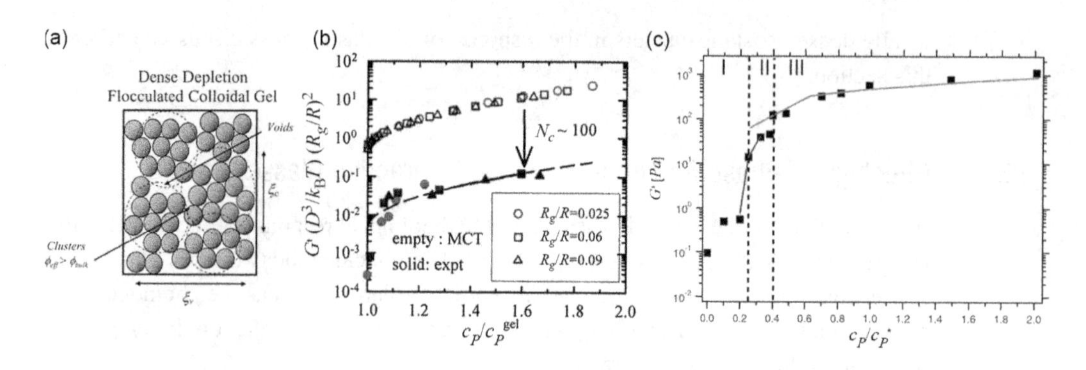

Figure 5.17 (a) Schematic of a dense depletion gel, showing interpenetrating, dense percolated clusters of an average size ξ_c, containing N_c number of particles. Reprinted with permission from Shah et al., Viscoelasticity and rheology of depletion flocculated gels and fluids. *Journal of Chemical Physics* 2003;119(16):8747–8760 [142] by AIP Publishing. (b) Comparison of MCT-PRISM results (empty symbols) for the scaled zero-frequency elastic modulus as a function of polymer depletant concentration, scaled by the gel point concentration for different size ratios, R_g/a = 0.025 (circles), 0.06 (squares), and 0.09 (triangles) with the corresponding experiments from Shah et al. [142] (solid symbols) at $\phi = 0.4$. The dashed line represents the MCT-PRISM prediction for R_g/a = 0.025 divided by a number of particles in a cluster N_c, ranging from 80 to 350. Reprinted with permission from Chen and Schweizer, Microscopic theory of gelation and elasticity in polymer-particle suspensions. *Journal of Chemical Physics* 2004;120(15):7212–7222 [140] by AIP Publishing. (c) Normalized elastic modulus extracted from dynamic frequency experiments in a depletion gel of PMMA particles, at ω=10 rad/s, as a function of polymer concentration at $\phi = 0.4$. Solid lines represent theoretical predictions for the elastic modulus for transient networks for low polymer concentration and gels at higher concentration. Reprinted with permission from Laurati et al., Nonlinear rheology of colloidal gels with intermediate volume fraction. *Journal of Rheology* 2011;55(3):673–706 [116] by The Society of Rheology

properties in arrested phase separating gels. At higher attraction strengths, or depletant concentrations, well inside the gel state, a weak power law increase of G' is observed [17,116], Figure 5.17(c). This can be rationalized with a simple elasticity model by the similar power law dependence of the bond strength on depletant concentration, and the almost constant correlation length found in this regime [17].

A more comprehensive model for intermediate volume fraction gels is presented by Zaccone et al. [115]. These authors calculate the elasticity by taking into account the heterogeneous structure of the gel via a hierarchical scheme defining elastic contributions at the particle and the mesoscopic, or cluster, level. In this model, dense clusters act as the rigid units that provide mechanical stability and propagate the elastic stress throughout the network. The model has been applied successfully to experimental systems forming arrested phase separating gels [17,116,117], without the need of any rescaling. A variation of this model was used by Whitaker et al. [104] to determine the internal density of the load bearing clusters in a self-consistent way with the measured elastic modulus, as discussed with regards to the left panel in Figure 5.13.

Nonlinear Rheology

As discussed with reference to Figure 5.2, phase separating gels have significantly more heterogeneous structure than HS glasses. Together with the characteristics of the interparticle attractions, the microstructure and particle dynamics determine the yielding response and nonlinear mechanical properties. Conceptually the effect of attraction can best be demonstrated by comparing a hard sphere suspension with an attractive glass at the same volume fraction. At such high volume fractions ($\phi > 0.58$) the microstructure of the two systems is similar. Importantly, the addition of attraction introduces an additional length and time scale related with the range of attraction and the bond escape time, respectively. The linear viscoelastic response resembles that of repulsive HS glasses, shown in Figure 5.7(a), with increased moduli due to the addition of interparticle bonding and the increased particle localization [53,143], as predicted by MCT [103]. More spectacular differences, however, are observed when comparing the nonlinear response of repulsive glasses to that of attractive glasses.

Start-up shear experiments for both a repulsive (HS) and an attractive glass at the same volume fraction ($\phi = 0.6$) are shown in Figure 5.18(a). HS glasses exhibit a single stress overshoot, signifying a single yield point and mechanism related with cage deformation and breaking that drives shear induced melting, while in attractive glasses a double stress overshoot is clearly detected [53,82,143]. In LAOS the two yield points may be identified by the two peaks in G'', revealing two strain amplitudes where energy dissipation is maximized, or equivalently by the inflection point of the stress versus strain amplitude curve [53,82]. The double stress overshoot depends on shear rate, volume fraction, and interparticle potential, and indicates a two-step yielding process. Similar two-step yielding is detected in large amplitude oscillatory shear experiments, Figure 5.18(b), and step strain tests when the maximum stress is plotted against the imposed strain [53]. In attractive glasses the first stress overshoot is

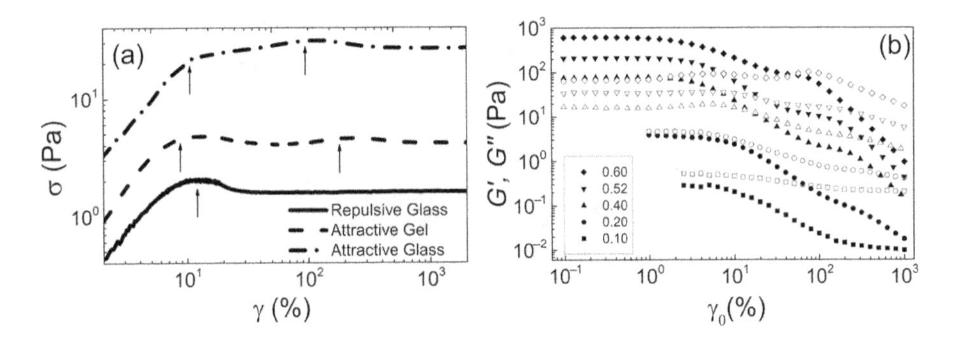

Figure 5.18 Nonlinear rheological tests in a colloid-polymer mixture: (a) Stress overshoots during start-up shear tests at $\dot{\gamma} = 0.5s^{-1}$ in a repulsive glass (solid curve) an attractive colloidal glass (dashed line), both at $\phi = 0.6$, $U_{\text{dep}} = -15\ k_B\text{T}$ and range $\delta = 0.05$, and a colloidal gel (dash-dot line) with the same attraction strength at $\phi = 0.44$. Arrows indicate the position of the stress overshoots. (b) Dynamic strain sweeps at $\omega = 10$ rad/s for five different volume fractions, as indicated, with equal attraction characteristics ($U_{\text{dep}} = -15\ k_B\text{T}$ and $\delta = 0.05$). Adapted from Koumakis and Petekidis [82]

related with breaking and restructuring of bonds, or particles escaping bonds and exchanging neighbors, as the yield strain is comparable and scales with the range of attraction [53,82]. The second overshoot is linked to particle rearrangements at a larger length-scale, namely that of the cage. This second yield point signifies cage breaking, during which particles escape from their cage of neighbors. Note that caging still exists even when attractive bonds are ruptured after the first yield point. Avoiding details, we point out that the second yield strain is indeed higher than the yield strain measured in HS glasses (see Figure 5.18(a)), indicating that the cage can be deformed significantly more in the presence of attractions [53,82,143].

As the volume fraction is decreased at constant attraction strength, the system turns from an attractive glass to a colloidal gel where the larger structural length scale now corresponds to the average cluster size, as seen in Figure 5.2(b), instead of the cage size in the case of repulsive (or attractive) glasses [82]. Both in start-up shear and large amplitude oscillatory shear (LAOS), two-step yielding is observed in colloidal gels as well, as shown in Figure 5.19. As volume fraction decreases, the second yield point moves to higher strains for steady shear start-up tests, while its strength progressively diminishes and becomes very weak for $\phi < 0.2$. Such behavior is attributed to the transition from cage to cluster dominated large scale heterogeneities, which are measured by the size of the clusters, and increase as volume fraction decreases from $\phi = 0.52$ to 0.1. As experiments and computer simulation have demonstrated, the second yield strain is related with large scale, shear-induced rearrangements of clusters. Depending on the shear rate, volume fraction and attraction characteristics this may involve cluster restructuring and reorganization at the particle level, which leads either to a compaction of clusters or its breakage into small pieces or even to individual particles. The first yield point is still related with bond breaking and restructuring within the gel network, which enables deformation and flow of clusters. Similarly to attractive glasses, the first yield strain scales with the range of attraction and is nearly constant at intermediate and high volume fractions ($\phi > 0.3$). The first yield point only moves to higher strain values at lower volume fractions [82]. The

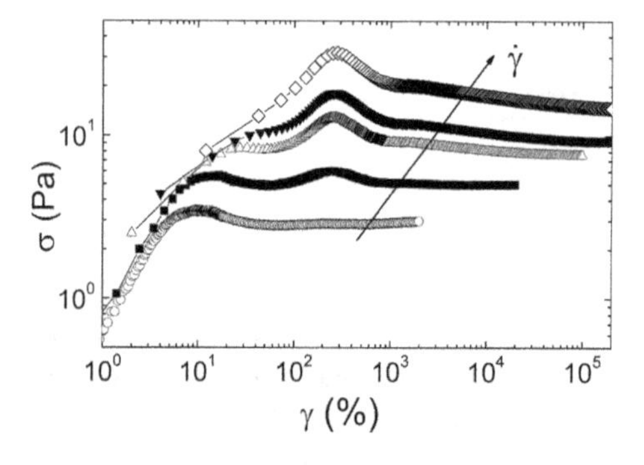

Figure 5.19 Step rate tests for a colloid polymer mixture with $\phi = 0.44$ with $\delta = 0.083$ and $U_{dep} = -15\, k_B T$. Rates are increasing from bottom to top (0.1, 1, 5, 10, 30 s^{-1}). Adapted from Koumakis and Petekidis [82]

structure of the gels formed by arrested phase separation then become more open with larger voids and thinner network strands. This indicates that restructuring and elongation of the clusters or strands takes place before they break to facilitate flow of the gel. In this regime a second yield point is essentially absent as there is enough free volume for flowing clusters or smaller pieces to rearrange and flow. Therefore the mechanisms responsible for two-step yielding in gels are linked with length-scales at the level of the bond and the cluster, rather than the bond and the cage as for attractive glasses.

The shear rate dependence of the two-step yielding is quite revealing with regard to the local mechanisms involved. At low shear rates only the first stress overshoot is detected, as seen in Figure 5.19, while as shear rate is increased further the second overshoot appears and progressively gets stronger. This shear rate dependent behavior is due to a variation of microstructure in the sheared gel network at different shear rates. At low shear rates advection induces bond breakage and rearrangements around the first yield point, leading to more compact clusters that push the system towards phase separation. Cluster densification causes larger free volume and facilitates steady state flow at low shear rates, with a constant bond breaking and reformation, without further reduction of their average size. It should be noted though that under such low shear rates wall slip or shear banding might be present depending on the roughness and interactions at the walls [144]. As shear rate is increased the appearance of the second stress overshoot is caused by a further restructuring of clusters and rupture in smaller pieces, or even to individual particles at high enough shear rates.

This picture is supported by a combination of direct rheo-confocal experiments and computer simulations, as those depicted in Figure 5.20. Here the rheological response of a concentrated depletion gel at different shear rates is shown together with the particle-level microstructure during steady state flow. Experiments were carried out by direct confocal microscopy imaging on a rheometer (rheo-confocal) while Brownian Dynamics (BD) simulations for the same system monitored stresses and microstructure under shear. The flow curves from experiments and BD simulation exhibit a yield stress plateau at low rates and are qualitatively the same, although the stress from BD simulations is lower due to the lack of full hydrodynamic interactions (HI) in the latter. Direct imaging of the structure of this arrested phase separating gel under shear during rheo-confocal experiments is in full agreement with BD simulations showing the creation of large dense clusters and empty voids at low shear rates with a progressive decrease of cluster/void size as shear rate is increased up to the limit of single particles at high rates. The shear rate dependence of the cluster break-up scale quantitatively with a nondimensional number formed as the ratio of viscous to attractive forces [82,118]:

$$Pe_{\mathrm{dep}} \equiv F_{\mathrm{visc}}/F_{\mathrm{dep}} = \frac{6\pi\eta\alpha\dot{\gamma}\alpha}{U_{\mathrm{dep}}(r=2\alpha)/2\delta a} = \frac{12\pi\eta\delta\alpha^3}{U_{\mathrm{dep}}(r=2\alpha)}\dot{\gamma}. \tag{5.4}$$

Indeed experiments and simulations suggest that for $Pe_{\mathrm{dep}} \gg 1$ clusters rupture to individual particles. For systems with different strength and range of attraction, structural measures such as the average cluster/void volumes and number of bonds

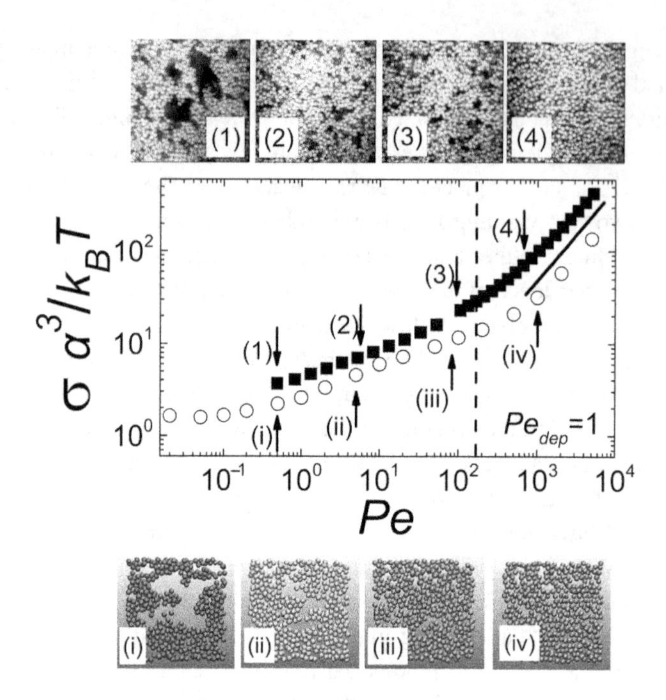

Figure 5.20 Flow curves from experiments (squares) and BD simulations (circles), along with confocal (top) and BD simulation (bottom) images of colloidal gel structures under flow. The experimental system is a colloid-polymer (depletion) gel at $\phi = 0.4$, $\delta = 0.05$ and $U_{dep}(2\alpha) = -16\,k_BT$, while simulation used a depletion potential with $\delta = 0.1$ and $U_{dep}(2\alpha) = -20\,k_BT$. Adapted from Koumakis et al. [118]

scale on a master curve when plotted as a function of Pe_{dep} [85]. This is essentially the same as the Mason number for adhesive hard sphere gels presented in Section 5.4.4.

Although the absence of hydrodynamic interactions in the BD simulations does not affect, at least qualitatively, the steady state flow structure shown in Figure 5.20, there is clear evidence that the presence of hydrodynamic interaction is crucial for the existence of the second yield stress overshoot in dilute and intermediate volume fraction gels. In several studies of colloidal gels with Brownian Dynamics simulations, the second stress overshoot (or yield point in LAOS) is essentially absent, in contrast to experimental findings [82,85,101]. Moreover, Dissipative Particle Dynamics (DPD) simulations, as discussed in Chapter 3, have shown that when full HI are included the second stress overshoot emerges in start-up shear for colloidal gels at $\phi = 0.2$ and various attraction strengths and shear rates [145].

Tuning of Colloidal Gels by Shear

The shear-induced structures of the depletion gels shown in Figure 5.20 evolve differently once shear is switched off, leading eventually to different out-of-equilibrium gel states with varying linear and nonlinear mechanical properties. The general trend established for gels at intermediate volume fractions, such as that shown in

Figures 5.19 and 5.20, is that high shear rates ($Pe_{dep} > 1$) break up gels into individual particles and upon shear cessation, the structure evolves to form a stronger gel with a relatively homogeneous structure. On the other hand, low pre-shear rates create more heterogeneous structures with denser clusters and larger voids which remain stable after shear cessation and exhibit the reduced elasticity of a weaker solid [118]. Interestingly, large amplitude oscillatory shear causes a stronger variation in the linear viscoelastic properties of a depletion gel than steady shear, especially when strain amplitudes intermediate between the first and second yield strain are applied [85]. Furthermore, oscillatory shear at different strain amplitudes affect the transient response in the start-up shear test and the relative strength of the two strength overshoots. This is related with the underlying structures created by pre-shear. The two-step yielding process is thus more pronounced when more heterogeneous gels are created by pre-shear, whereas structures that are more homogenous (and stronger) exhibit weaker two-step yielding [85]. Hence preshear history provides a unique way to tune gels in metastable states with different mechanical properties that are otherwise not accessible. This is a demonstration of the strong thixotropic character of arrested phase separating colloidal gels, in contrast to other dynamically arrested states such as colloidal glasses and equilibrium gels. Similar effects have been observed in a variety of model attractive colloidal gels, such as attractive microgels [146] or industrial systems, where tuning of the mechanical properties is important in a wide range of applications and processing, such as shown in Chapter 9.

Aging

The aging of phase separating gels by coarsening of the suspension microstructure results in an evolution of the mechanical properties. Figure 5.21(a) shows the time dependence of the moduli and of the two yield stresses after a shear rejuvenation of an intermediate volume fraction colloid-polymer gel. Both the elastic modulus and the

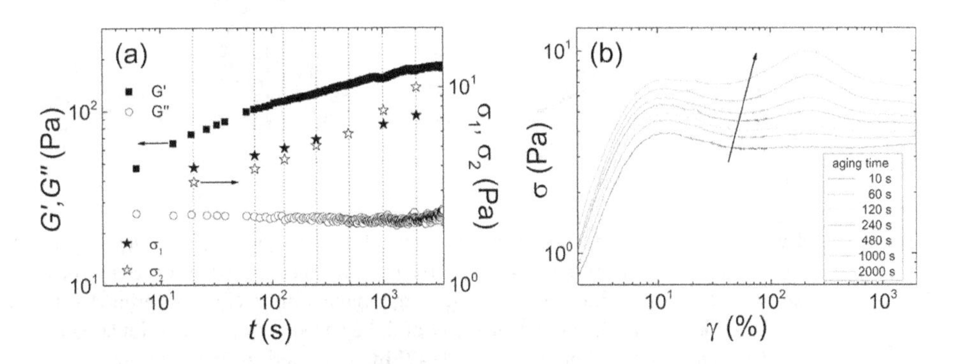

Figure 5.21 Aging of a depletion gel with $\phi = 0.44$, $\delta = 0.083$ and $U_{dep} = -15\, k_BT$. (a) Time evolution of the linear viscoelastic moduli, and the stresses corresponding to the two stress overshoots; (b) the transient stress response at start-up shear for different delay times after rejuvenation. Aging time increases with the direction of the arrow. Adapted from Koumakis and Petekidis [82]

peak stresses increase with time. However, as shown in Figure 5.21(b), the second overshoot is the most affected as it strongly increases with time as the gel coarsens, forming larger and better defined clusters. Hence, the second yield point is most sensitive to shear history.

Yielding under Constant Stress

Similar to observations for colloidal glasses, applying a constant stress well below the yield stress to colloidal gels results in a virtually constant elastic strain with minimal creep. However, in contrast to colloidal glasses, colloidal gels exhibit a long transient creep response for stresses around the yield stress due to their more heterogeneous structure that is amenable to the large scale, stress induced restructuring shown in Figure 5.20. This behavior is manifested by phenomena of delayed yielding and delayed solidification. A colloidal gel under constant stress eventually yields and flows after a transient period, as shown in Figure 5.22(a) and (b) for different stresses applied to a colloidal gel of carbon black particles at low volume fractions.

Delayed dynamical arrest occurs when a colloidal gel under constant stress initially yields and flows but, with time, shear induced structural changes lead to re-solidification. This is illustrated in the right panel of Figure 5.22 by computer simulations of suspensions of particles with moderate attractions, i.e., U_{dep} of a few k_BT. Depending on the applied stresses three regimes were observed: At high stresses,

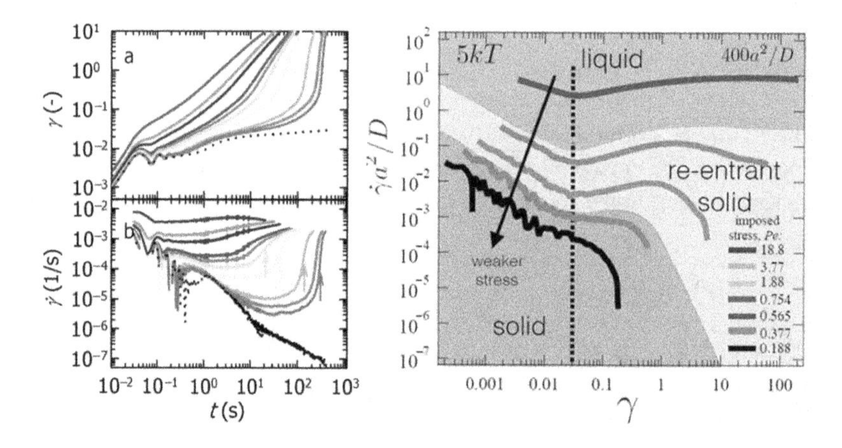

Figure 5.22 Left: (a) Creep response of carbon black gels at 8 wt% for stresses 50–15 Pa, decreasing from left to right and (b) corresponding shear rates during creep for the different stresses. The arrows illustrate the time of the maximum in $d\dot{\gamma}/dt$. Reprinted with permission from Sprakel et al., Stress enhancement in the delayed yielding of colloidal gels. *Physical Review Letters* 2011;106(24):248303 [147] by the American Physical Society, DOI 10.1103/PhysRevLett.106.248303. Right: Computer simulations of a colloidal gel during creep for different stresses and for two strengths of attraction, $5k_BT$ and $6k_BT$. The three regimes in stresses denote a solid or liquid like response, at all times and a delayed re-entrant solid response at intermediate stresses. Reprinted with permission from Landrum et al., Delayed yield in colloidal gels: Creep, flow, and re-entrant solid regimes. *Journal of Rheology* 2016;60 (4):783–807 [148] by the Society of Rheology

well above the yield stress, the sample yields and flows at constant shear rate, near the yield stress the sample shows re-entrant solidification after yielding, while at low stresses the sample remains arrested. Creep experiments in PMMA colloid-polymer gels at similar volume fractions ($\phi = 0.3$) and attraction strengths (δ = 0.083 and $U_{dep} = -6k_BT$) show all these transient responses [149]. These phenomena demonstrate the complex interplay of stress induced restructuring involving breakage and reformation of bonds, clusters, and networks in colloidal gels under constant stress.

5.4.4 Rheology of Homogeneous Gels

Linear Rheology and Dynamics

Studies of the thermoreversible systems consisting of octacedyl-coated silica in tetradecane provide a roadmap for the general behavior of this class of materials [19,27,36,121,122,150,151]. As an example, consider the data presented in Figure 5.3, which shows SAOS sweeps for a physically gelling colloidal dispersion of ~30 nm diameter silica particles over a very narrow temperature range [33]. As temperature is lowered, the storage modulus increases and eventually exceeds the loss modulus in magnitude over the entire frequency range. As the suspension is homogeneous, the gel can be described with the theories derived in Chapter 2. Shown in Figure 5.3 are the fits to the mode coupling theory discussed in Chapter 2, which identifies a relatively long β-relaxation process of order seconds, indicating that the particles are strongly coupled by attractive forces. Note that the Brownian relaxation time, Eq. (1.4), for a particle of 30 nm in this suspending media is of the order of tens of microseconds. Clearly, the local particle motion is greatly hindered here by interparticle attractions, as suggested by the cage schematics in Figure 5.4. The model assumes a homogeneous suspension. Hence, it is possible to apply MCT to these gel structures, even at such intermediate volume fractions, because the samples are not phase separating or heterogeneous. For example, using the naïve MCT model for the elastic modulus of a gel, as discussed in Chapter 2 and Section 5.4.3, the measured elastic modulus corresponds to a localization length of the order of ~3 nm in the fully gelled state, which is comparable to the size of the organic brush layer responsible for the stickiness [33]. In contrast, the phase-separating depletion gels exhibit a much larger localization length and have much weaker moduli due to their heterogeneous structure [105,142].

Also shown in Figure 5.3 is the comparison of the data at 28.8°C to the critical gel criterion proposed by Winter and Chambon [32], which predicts for this particular case $G' = G'' \sim \omega^{1/2}$. This rheological scaling reflects a power-law distribution of relaxation times, which can be related to an underlying hierarchy of structure in the stress-bearing network at the onset of stress-percolation (for a discussion, see section 6.6 in *CSR*). Importantly, experiments, simulation, and theory show that the rheology of homogeneous physical gels can be interpreted using concepts originally derived for chemical polymer gels [107,152].

In Figure 5.23 a consistent data set is compiled of rheological, fiberoptic quasielastic light scattering (FOQELS) measurements of diffusion, and small angle neutron

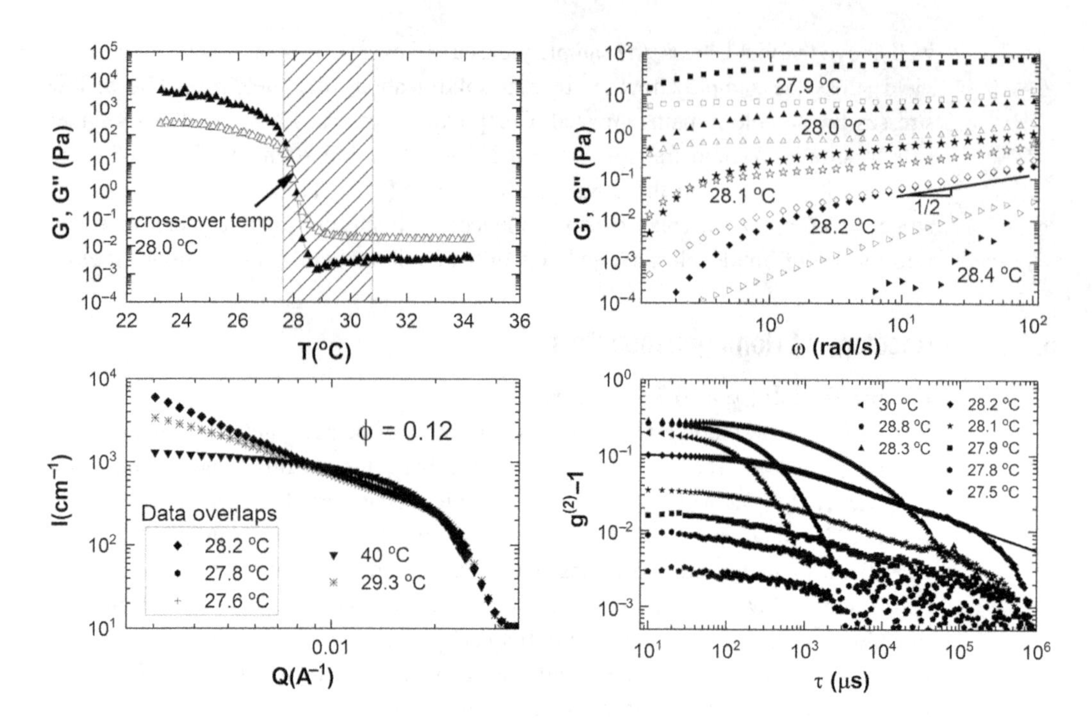

Figure 5.23 Behavior of thermoreversible gels, clockwise from upper right: SAOS as a function of temperature at fixed frequency; SAOS frequency sweeps at temperatures above and below the gel point, with the gel temperature identified; FOQELS correlation functions, where the critical gel temperature shows power law scaling; SANS structure factors above and below the gel temperature. Adapted from Eberle et al. [19]

scattering (SANS) from a similar thermoreversible gel through the gel transition [153]. At high temperatures in the liquid state, the loss modulus dominates. As the sample is cooled, both moduli grow many orders of magnitude and the elastic modulus becomes significantly greater than the loss modulus. The frequency-dependent rheological behavior has characteristic terminal liquid-like behavior at very low frequencies, both at and above a critical gel concentration. Because this is a physical gel and the interparticle "bonds" have a finite lifetime, such gels will flow at long times. Deviations from ideal gel behavior are evident for creep at long times as well as liquid-like response at very low frequencies. Hence, gel-behavior is observed when the α-relaxation time is long relative to the experiment time, i.e., the Deborah number for the process is comparatively large. Practically, gels near the gel transition will flow under gravity, for example, when a jar containing the gel is left upside down for extended periods. Thus, while the definition of critical gelation is well-defined, physical gels creep and flow with time, raising questions concerning the role of the experimental observation time, similarly to considerations surrounding the gel property of a yield stress [154], as well as the discussion of the local structure responsible for load bearing in Section 5.4.2.

Dynamic light scattering results also reflect this gel transition. As shown in Figure 5.23, the same physics that leads to the power law behavior of the shear modulus also leads to a power-law relaxation of the self-intermediate scattering function, defined in Chapter 2, as can be seen in the plot of $g^{(2)}-1$. Therefore, dynamic light scattering can also identify the critical gel state [155,156]. This power law index (–0.65) yields a mass fractal dimension of $d_f \sim 1.8$, consistent with the gelation mechanism of diffusion limited aggregation, which is within uncertainty of the value observed by small angle neutron scattering of $d_f \sim 1.7$ [153]. The neutron scattering from the suspension, discussed in Chapter 4, shows fractal-like scattering intensity in the gel state at low temperature and liquid-like scattering at high temperatures, further confirming the rheologically determined gel transition.

Phase Behavior and Structure

The adhesive hard sphere, or "sticky" potential was introduced in Chapter 1, Figure 1.14, along with the phase behavior of the adhesive hard sphere colloidal suspension, shown in Figures 1.17 and 5.13. The state diagram has been validated by experimental observations of the reversible transition from liquid to percolated gel, spanning a broad range of compositions [152]. Furthermore, unlike the depletion gels, the presence of a contact "bond" leads to a homogeneous gel state that is within the single-phase region. Consequently, thermoreversible gels can be rejuvenated by heating and melting, followed by cooling and reforming, with minimal hysteresis.

The physical bonds between colloidal particles are rigid in the octadecyl-silica/tetradecane suspensions due to co-crystallization of the surface organic coating [153]. The rigidity percolation criterion of He and Thorpe [106] that the average number of bonds per particle is 2.4 has been verified experimentally for the octadecyl-silica suspensions with the aid of computer simulations, and the result of this analysis is shown in Figure 5.24. In this plot the experimentally determined dynamic arrest transition is plotted as the reduced second virial coefficient B_2^* versus volume fraction [122]. This coefficient is related to the Baxter stickiness parameter (Eq. 1.32) by the relationship:

$$B_2^*(T) = (1 - 1/4\tau_B). \tag{5.5}$$

The reduced second virial coefficient is more intuitive perhaps than the Baxter stickiness parameter, as a value of one corresponds to hard sphere excluded volume. This is the high temperature limit for the AHS model. Increasing attractions reduces the value of B_2^*, which becomes net attractive when $B_2^* < 0$, corresponding to a decreasing value of τ_B, or lowering of the temperature.

The state diagram in Figure 5.24 is truncated to emphasize the relationship between the liquid, gel, glass, and two-phase states in the AHS suspension. Importantly, the *predicted* rigidity percolation transition, where the average number of bonds becomes 2.4, is well validated by experiments on the model AHS suspensions from dilute conditions up through the glass [122]. It should be noted that this transition lies slightly below the predicted percolation transition determined for connectivity as well as below the liquid–vapor bimodal at low concentrations [19]. However, difficulties

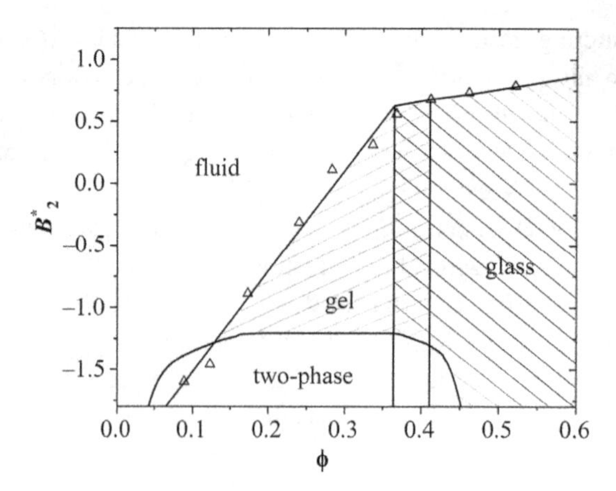

Figure 5.24 State diagram for adhesive hard sphere suspensions in terms of the reduced second virial coefficient, such that the value for hard spheres is $B_2^* = 1$. Measurements (triangles) for the dynamical arrest of an AHS model system are compared against the predictions of rigidity percolation (line), while the hatched regions represent gel and glass regions. Adapted from Valadez-Perez et al. [122]

arise at very low volume fractions as the large fractal clusters that form are subject to gravitational settling, as discussed in relation to Eq. (5.2), which limits experiments in gravity. Rigidity percolation does not distinguish between the conventionally defined "gel" and "glass" because the criterion for dynamical arrest from the liquid state is identical for all volume fractions. In other words, there is no formal distinction between the gel and glass within this theoretical framework. However, an apparent change in slope of the rigidity percolation line around $\phi = 0.4$ corresponds to a fundamental transition in the dynamical properties of the colloidal liquid. As depicted in Figure 1.4, below $\phi = 0.4$ the colloids can diffuse without serious hindrance. Yet, increasing particle concentration leads to the concept of cage formation, whereby particle motion becomes strongly coupled to that of the nearest neighbors, as discussed in Sections 5.2 and 5.3. The onset of caging in liquids more generally is reviewed in this context by Zaccarelli et al. [112] and the concept is well supported by simulation and experiment. While this change in behavior is not a sharp transition, it strongly limits the applicability of mode coupling theory, which relies on the existence of cages of neighbors as a starting assumption.

The concept of caging enables understanding why the rigidity percolation transition has the behavior shown in Figure 5.24. At high volume fractions the particles are already strongly constrained by the cage constituted by the neighboring particles such that very weak interactions are sufficient to reach the criterion of an average bond number of 2.4. In fact, the attractive driven glass (ADG) virial coefficient is still positive, although less than the hard sphere value of 1.0. Reducing the particle concentration dilutes the cage of neighbors such that it is no longer identifiable. In that case particle crowding no longer assists the formation of a solid network and a stronger attraction is required to hold the system together. Further dilution requires

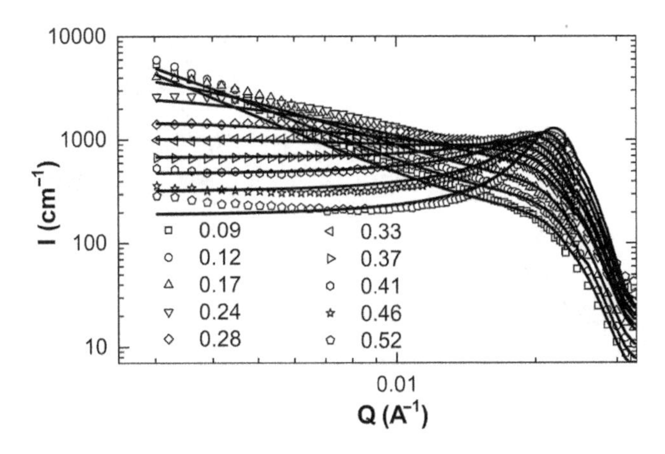

Figure 5.25 Small angle neutron scattering intensities for states at the rigidity percolation transition for the volume fractions and temperatures corresponding to the data points shown in Figure 5.24. Adapted from Eberle et al. [19]

stronger bonds to hold the network together. These bonds are thermal and thus, labile, as this is a physical gel. Consequently, the concept of a time scale is important and the bond lifetime helps to define the timescale over which one can effectively consider solid-like behavior in the bulk. Colloidal suspensions designed to bridge the behavior of thermoreversible homogeneous systems with those of the heterogeneous gels have been reported and the differences studied [101,126,157].

The microstructure of gels and glasses is captured in confocal images in Figure 5.2, and additional evidence for this continuous evolution in microstructure is shown in Figure 5.25. Neutron scattering intensities from the octadecyl-silica suspensions corresponding to the states along the rigidity percolation transition are shown as symbols in Figure 5.24, along with fits to the sticky hard sphere model [19]. As described in Chapter 4, these can be directly interpreted as follows. At the lowest volume fractions the constant slope of the scattering intensity at low scattering vectors corresponds to a fractal-like structure, consistent with DLCA. With increasing volume fraction it is no longer possible to generate fractals of any significant size before filling the space. Hence, the scattering at low scattering vectors decreases and the range of the power-law behavior becomes more limited. The increase in intensity at larger scattering vectors is proportional to the increasing particle concentration, such that the significant decrease at low scattering vectors signals a transition from an open fractal network to a dense, liquid-like state. This is evident at the highest volume fractions, in the attractive driven glass state, where a correlation peak develops at a characteristic distance corresponding to the nearest neighbor location. This correlation peak moves to larger scattering vectors and sharpens with increasing volume fraction, which simply reflects the packing effects associated with excluded volume interactions. Note that the transition from fractal-like gel to glass-like microstructure is smooth and there are no sharp transitions. This continuous evolution in behavior of the scattering from suspensions along the transition line further supports the unifying

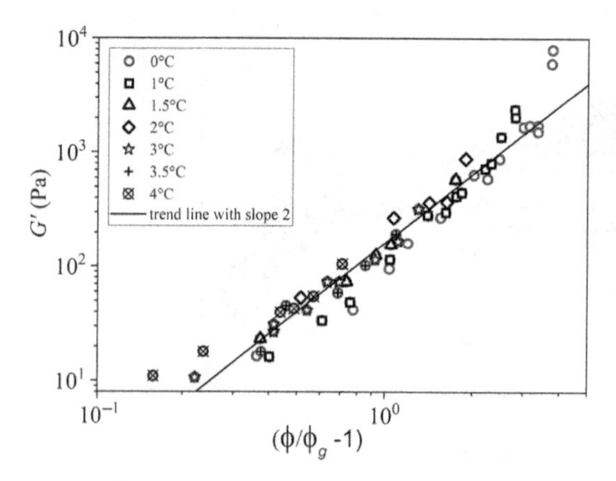

Figure 5.26 Scaling of the elastic moduli for AHS colloidal gels with volume fraction plotted according to percolation scaling, where the relevant variable is the distance to the gel concentration, shown for varying temperatures. Adapted from Rueb and Zukoski [158]

concept of rigidity percolation where homogenous gelation and attraction driven glass formation are predicted to be the same equilibrium rigidity transition from the liquid state. Percolation of stress-bearing particle aggregates drives dynamical arrest into a material with solid-like properties on the time-scale of observation, i.e., homogeneous physical gelation or vitrification.

The elastic moduli in the gel or glass state, as reported by Rueb and Zukoski [158], are observed to follow predictions of percolation theory with a power law exponent of 2, as shown in Figure 5.26 and discussed in detail in section 6.6.1 of *CSR*. In this study the samples were held at temperature and presheared to rejuvenate them prior to measurement at each temperature. Rigidity percolation predicts that the elasticity, and hence the yield stress, should vanish at the rigidity transition, and, further, that they depend on the average contact number with power law of 1.5 ± 0.2 [106], but this has not been tested by experiment. The hydrodynamic component of the zero shear viscosity has been predicted to scale linearly with the mean cluster size in the liquid as rigidity percolation is approached, as deduced from simulations by Wang et al. [107]. The authors predict that the zero shear viscosity should be proportional to the cube of this cluster size for long-lived bonds, but these predictions have not been tested experimentally.

Nonlinear Rheology

At the gel point the sample transitions to a behavior without a well-defined zero shear viscosity, as shown by the flow curves for suspensions of thermoreversible particles in Figure 5.27(a). Temperatures below the gel transition lead to yield stresses, which increase with increasing quench depth, shown in Figure 5.27(b). Creep tests provide one of many methods to measure the yield stress, see section 9.4.2 of *CSR*. The initial response and gel-ringing, as described in section 7.4.4 of *CSR*, is followed by a plateau where the sample does not strain further in response to the stress.

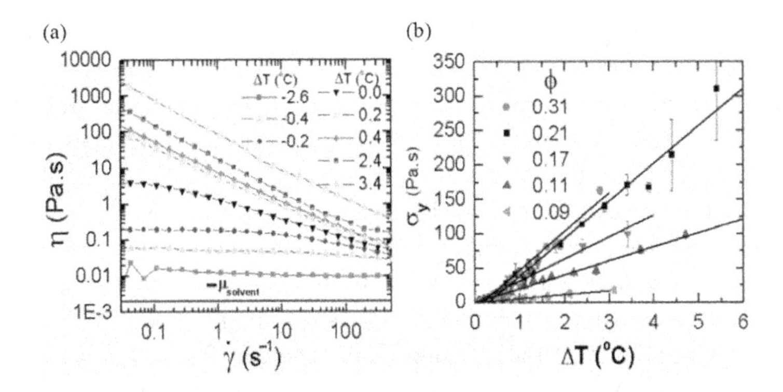

Figure 5.27 (a) Flow curves for an AHS gel as a function of the temperature increment above (negative) or below (positive) the gel transition. (b) Yield stress measurements on the same particles at varying volume fraction and at the temperature difference from the gel transition. Reprinted with permission from Eberle et al., Shear viscosity and structural scalings in model adhesive hard-sphere gels. *Physical Review E* 2014;89(5):050302(R) [121] by the American Physical Society, DOI 10.1103/PhysRevE.89.050302

Performing the test at ever increasing shear stresses leads to an abrupt yielding event and flow. As with all measurements of physical gels and glasses, the results essentially depend on the observation time [33].

A unifying concept for the shear rheology of these systems is to reduce the flow curves using a ratio of shear force to the dominant interparticle force. For gels the structure is dominated by the attractions driving gelation and, so, the force of attraction is what must be overcome to rupture the gel and induce yielding and flow. The force of attraction is the derivative of the interaction potential, Eq. (1.26). This can be used to form a Péclet number for depletion gels Pe_{dep}, as shown in Eq. (5.4) and discussed therein. Similarly, we can define a dimensionless group that is also referred to as a Mason number:

$$Mn = \frac{\text{viscous force}}{\text{interparticle force}} = 6\pi\eta a^2\dot{\gamma} \Big/ \frac{d\Phi}{dr}\Big|_{\max}. \tag{5.6}$$

While the overall strength of attraction can be readily measured, the interparticle force of attraction is not generally known. However, the yield stress scales in the following manner with the force of attraction [159]:

$$\sigma_y \propto \frac{\phi^2}{a^2}\frac{d\Phi}{dr}\Big|_{\max} \tag{5.7}$$

Substitution in Eq. (5.6) yields:

$$Mn = C\phi^2 \frac{\eta\dot{\gamma}}{\sigma_y}, \tag{5.8}$$

where the constant C captures the proportionality in Eq. (5.7). As a further refinement, the numerator of Eq. (5.8) can be replaced by the stress, recognizing that the viscous stress acting on the particles in the gel is more closely captured by the actual stress than the medium viscosity times shear rate. This becomes then the inverse of the Bingham number, which compares the yield stress to the applied stress, as $Bi = \sigma_y/\sigma$ and is used classically to understand the rheology of yield-stress fluids such as these gels and attraction driven glasses.

A validation test by Eberle et al. [121] used this concept to reduce the experimental data for the gel state shown in Figure 5.27, which spans over five orders of magnitude of viscosity, as well as corresponding DPD simulations for which the actual strength of attraction is known, to a universal curve, as shown in Figure 5.28. The success of this reduction shows that large variations in the strength of attraction can be accounted for through the yield stress measurements, where M varies over more than eight orders of magnitude in Figure 5.28. It was also verified that the structure, as determined by small angle neutron scattering (not shown here), also followed the same scaling, providing evidence for a simple structure-property relationship for these gels [121].

Experimental studies show that the flow curves can be reduced by plotting σ/σ_y versus a reduced shear rate [150]. Furthermore, studies of the microstructure of an AHS colloidal gel during both steady shearing and LAOS show microstructural breakage: "... of the fractal gel network at low Deborah (*De*) or Weissenberg (*Wi*) numbers and the flow of compacted clusters at intermediate *De* or *Wi*, followed by fluidization at very high *Wi*" ([150], p. 1306) in agreement with findings for depletion gels [85,118], as discussed in Section 5.4.3 and shown in Figure 5.20. A direct connection between the evolving structural stress and the microstructure has been established using Rheo-SANS methods discussed in Chapter 4. These show that a gel with a relatively open microstructure under the appropriate conditions of shearing can

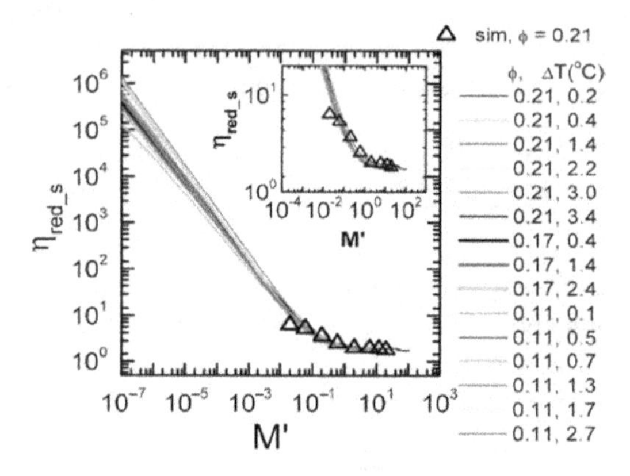

Figure 5.28 Reduced steady shear viscosity, with data in part from Figure 5.27 versus the dimensionless flow strength gauged against the strength of attraction for various volume fractions and temperatures below the gel transition. Solid lines are Carreau–Yasuda fits to the experimental viscosity, and open triangles are DPD simulations for a fixed concentration $\phi = 0.21$ and three different potentials. Reprinted with permission from Eberle et al., Shear viscosity and structural scalings in model adhesive hard-sphere gels. *Physical Review E* 2014;89 (5):050302(R) [121] by the American Physical Society, DOI 10.1103/PhysRevE.89.050302

be compacted to yield a material with different properties upon flow cessation and re-gelation [150,158]. This has important consequences for materials processing, and is equally important for shear rejuvenation of both types of gels and glasses, as discussed in Section 5.4.3 on tuning properties [85,118].

Gel Aging and Shear Rejuvenation

Quenching an AHS suspension below the dynamical arrest transition temperature leads to rapid property development followed by slow aging. An example is shown in Figure 5.29, where a $\phi = 0.28$ AHS suspension is brought by two different routes to 28.6°C, which is slightly below the gel transition temperature of 28.8°C. Slow aging is evident once the target temperature is reached. Importantly, the rheological properties become independent of thermal history, consistent with the concept of an equilibrium gel. There are some predictions that this aging should scale with time as $t^{0.4}$ [33]. The aging has been shown to be due to structure evolution, where the microstructure evolves toward its thermodynamic equilibrium under very slow dynamics below the dynamical arrest transition. A quantitative connection between the mechanical properties during these transitions, and the microstructure and local particle dynamics has been established by multiple experimental methods and investigators [33,160].

Shearing while quenching has interesting consequences, as illustrated in Figure 5.30 [33], based on similar, earlier observations by Rueb and Zukoski on a

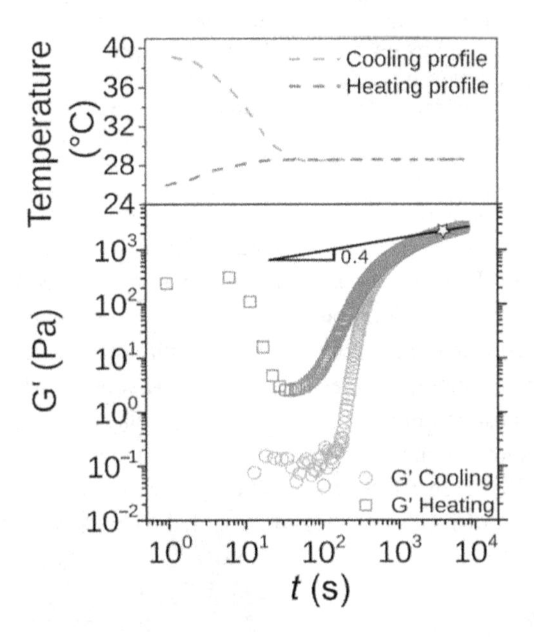

Figure 5.29 Evolution of the elastic moduli during a thermal quench for a gel originally at 40°C for the cooling profile. In the "heating profile" the gel is first rapidly quenched to 25°C and immediately heated to 28.6°C The gel temperature for this sample is 28.8°C. Reprinted with permission from Gordon et al., The rheology and microstructure of an aging thermoreversible colloidal gel. *Journal of Rheology* 2017;61(1):23–34 [33] by The Society of Rheology

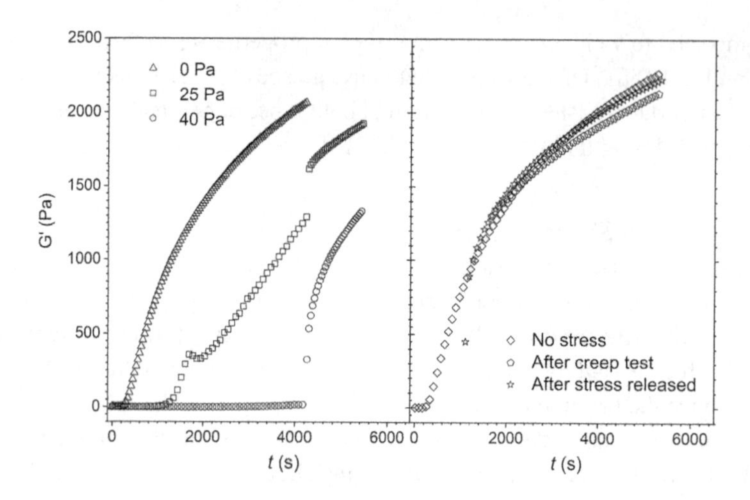

Figure 5.30 Left: Moduli development under a temperature quench. SAOS data (1 Hz) shows the quiescent gel formation (0 Pa). For the LAOS tests, the sample was aged to 4250 s, which corresponds to the reference state of the unstressed gels, under oscillatory stresses of 25 and 40 Pa (all at 1 Hz). The stress was then released and gel formation was monitored as in the SAOS test. Right: Gel formation after long-time aging experiments. Datasets shown are (1) "no stress" case where the sample ages under SAOS, (2) "after creep test" case where the gel was subjected to a 20 minute 65 Pa creep test followed by a 20 minute recovery period prior to aging under SAOS and, (3) "after stress released" case where 40 Pa stress was applied to the gel during its formation, and subsequent aging was measured under SAOS after the stress was removed. Time was shifted backward by 4985 and 3150 s for the "after creep test" and "after stress released" experiments, respectively, to show the same gel evolution is observed, irrespective of the preshearing condition. Adapted from Gordon et al. [33]

similar system [158]. In this experiment, the elastic moduli are monitored as the gel forms during a temperature quench below the gel state. This is compared to behavior under two shear stress levels. Shearing suppresses gel formation, but once stopped, the elastic properties rapidly reform, an evolution similar to that for the quiescent gel. This is shown in the right panel, where the moduli evolution with time for suspensions at 0 Pa and 40 Pa are compared, along with results for a sample similarly sheared at 65 Pa, which is above the yield stress. For this system, shearing delays the microstructure evolution, but upon cessation, as shown, the suspensions continue to evolve along a similar mechanical path [33]. Such effects are commonly observed in many industrial systems, such as presented in Chapter 9, where mixing maintains flowability.

5.4.5 Summary and Outlook

Arrested phase separation is a route to colloidal gelation observed widely in experimental systems such as depletion gels. The mechanism leads to network structures with solid-like response with variability in heterogeneity depending on volume fraction, attraction strength, and preshear history. For depletion gels and attractive glasses two additional length scales, i.e., bond and cluster or cage and the resulting heterogeneity, define the rheology. As a consequence, two-step yielding is evident both in steady and oscillatory shear, and is found to be strongly dependent on preshear history and shear rate. Studies of the relaxation spectrum have been proposed to distinguish gel from glass behavior [161]. A related manifestation of the thixotropic character of arrested phase separation gels is the tunability of their structure and mechanical properties by preshearing, with oscillatory preshear found to be more efficient in affecting both the structure and the rheology. As a consequence of the underlying phase separation mechanism in these colloidal gels, preshear at low shear rates, or intermediate strain amplitudes in LAOS, cause cluster compaction leading to very heterogeneous structures, which upon shear cessation lead to weaker solids. Meanwhile, stronger preshear at large rates or strain amplitudes creates more homogenous and stronger gels upon flow cessation [85,118].

In contrast, physical gelation with rigid interparticle contacts provides an important route to homogeneous gel formation in the single-phase region of the phase diagram. Experiments, simulation, and theory provide a unifying framework in terms of rigidity percolation, which has characteristics of an equilibrium phase transition. For short range attractions the details of the interactions are not important and dynamical arrest can be identified in terms of a stickiness parameter or a second virial coefficient. A continuous transition in behavior from gel to attractive driven glass is observed, where increasing volume fraction reduces the strength of attraction required for dynamical arrest due to excluded volume and caging effects. The details of the potential of interactions become important when considering the flow behavior. However, here again a unifying concept is to use experimental measurements of the yield stress in the form of an inverse Bingham number as a metric for the maximum force of adhesion in order to express how the gel yields and flows. Recent simulations that account for the rigidity of the interparticle bonds and associated thixotropy modeling have described the nonlinear flow behavior of these systems, connecting local microstructure to rheological response [162].

Physical gelation poses experimental challenges, as there is an inherent interparticle interaction or bond lifetime, such that physical gels and glasses creep and flow over comparatively long observation times. Consequently, aging and shear rejuvenation are observed and can be understood in terms of the microstructural rearrangements that become asymptotically slower the deeper the quench into the dynamical arrested state. Similarities have been drawn to glass formation in polymers [33]. Importantly, these homogeneous gels achieve well-defined states that can be rejuvenated by temperature changes and/or shearing. They provide a roadmap to other, irreversible gels that form gels with particle–particle contacts, including paints [163] and many of the practical colloidal materials discussed in detail in Chapter 9. Bridging these two very different systems, homogenous rigidity percolation versus dynamically arrested gels through spinodal decomposition gels is a topic of current research [101].

Particle shape plays an important role in determining the structure and properties of gels and glasses. Recent work extending the state diagram to the next level of complexity, adhesive hard rod Brownian suspensions, suggests a nonlinear dependence on aspect ratio for low aspect ratios [125]. Semi-empirical predictions have been made for a variety of shapes, suggesting both the repulsive glass and dynamic arrest transition due to attractions will be strongly dependent on aspect ratio and shape [164]. Additional challenges have been identified by Wang and Swan [165], who developed simulations of patchy particles with anisotropic interactions to study the influence of surface heterogeneities for patchy particles as a route to creating physical gels. As discussed above, such effects lead to homogeneous gelation in contrast to phase separation as shown for smooth spheres. The degree of surface heterogeneity affects the percolation transition such that the details of the surface heterogeneities are now important in locating the transition. The work presented in this chapter provides a basis for understanding dynamical arrest in suspensions of such more complex systems, including those to be discussed in the following chapters.

Chapter Notation

d_f fractal dimension [–]
N_c number of particles in a cluster [–]
r_L localization length [m]

Greek Symbols

γ_r total recovered strain [–]
γ_r^i initial recovered strain [–]
γ_{ur} unrecovered strain [–]
δ range of attraction [m]

References

1. Mewis J, Wagner NJ. *Colloidal Suspension Rheology*. Cambridge: Cambridge University Press; 2012. 393 p.
2. Likos CN. Soft matter with soft particles. *Soft Matter*. 2006;2(6):478–498.
3. Poon W. Colloids as big atoms. *Science*. 2004;304(5672):830–831.
4. Pusey PN, van Megen W. Phase behaviour of concentrated suspensions of nearly hard colloidal spheres. *Nature*. 1986;320(6060):340–342.
5. Pusey PN, Zaccarelli E, Valeriani C, Sanz E, Poon WCK, Cates ME. Hard spheres: Crystallization and glass formation. *Philosophical Transactions of the Royal Society A: Mathematical Physical and Engineering Sciences*. 2009;367(1909):4993–5011.
6. Weeks ER, Crocker JC, Levitt AC, Schofield A, Weitz DA. Three dimensional imaging of structural relaxation near the colloidal glass transtion. *Science*. 2000;287(5453):627.

7. Pusey PN. Liquids, freezing and the glass transition. In: Lesvesque D, Hansen JP, Zinn-Justin J (eds.) *Les Houches Session 51*. Amsterdam: North-Holland; 1991, pp. 3–28.
8. Pusey PN, van Megen W. Observation of a glass transition of spherical colloidal particles. *Physical Review Letters*. 1987;59(18):2083.
9. Pham KN, Puertas AM, Bergenholtz J, Egelhaaf SU, Moussaïd A, Pusey PN, et al. Multiple glassy states in a simple model system. *Science*. 2002;296(5565):104–106.
10. Eckert T, Bartsch E. Re-entrant glass transition in a colloid-polymer mixture with depletion attractions. *Physical Review Letters*. 2002;89(12):125701.
11. Russel WB, Saville DA, Schowalter WR. *Colloidal Dispersions*. Cambridge: Cambridge University Press; 1989.
12. Kegel WK, van Blaaderen A. Direct observation of dynamical hetergeneities in colloidal hard sphere suspensions. *Science*. 2000;287(5451):290.
13. Brambilla G, El Masri D, Pierno M, Berthier L, Cipelletti L, Petekidis G, et al. Probing the equilibrium dynamics of colloidal hard spheres above the mode-coupling glass transition. *Physical Review Letters*. 2009;102(8):085703.
14. Kobelev V, Schweizer KS. Strain softening, yielding, and shear thining in glassy colloidal suspensions. *Phys. Rev. E*. 2005;71(2 Pt 1):021401.
15. Ballesta P, Petekidis G. Creep and aging of hard-sphere glasses under constant stress. *Physical Review E*. 2016;93(4):042613.
16. Zaccarelli E. Colloidal gels: Equilibrium and non-equilibrium routes. *Journal of Physics: Condensed Matter*. 2007;19(32):323101.
17. Laurati M, Petekidis G, Koumakis N, Cardinaux F, Schofield AB, Brader JM, et al. Structure, dynamics, and rheology of colloid-polymer mixtures: From liquids to gels. *Journal of Chemical Physics*. 2009;130(13):134907.
18. Lu PJ, Zaccarelli E, Ciulla F, Schofield AB, Sciortino F, Weitz DA. Gelation of particles with short-range attraction. *Nature*. 2008;453(7194):499–503.
19. Eberle APR, Wagner NJ, Castaneda-Priego R. Dynamical arrest transition in nanoparticle dispersions with short-range interactions. *Physical Review Letters*. 2011;106(10):105704.
20. Pham KN, Egelhaaf SU, Pusey PN, Poon WCK. Glasses in hard spheres with short-range attraction. *Physical Review E*. 2004;69(1):011503-1.
21. Poon WCK, Pham KN, Egelhaaf SU, Pusey PN. "Unsticking" a colloidal glass, and sticking it again. *Journal of Physics Condensed Matter*. 2003;15(1):S269–S275.
22. Krishnamurthy LN, Wagner NJ. Letter to the editor: Comment on "Effect of attractions on shear thickening in dense suspensions". *Journal of Rheology*. 48, 1321 (2004). *Journal of Rheology*. 2005;49(3):799–803.
23. Zaccarelli E, Poon WCK. Colloidal glasses and gels: The interplay of bonding and caging. *Proceedings of the National Academy of Sciences of the United States of America*. 2009;106(36):15203–15208.
24. Lin MY, Lindsay HM, Weitz DA, Ball RC, Klein R, Meakin P. Universality in colloid aggregation. *Nature*. 1989;339(6223):360–362.
25. Carpineti M, Giglio M. Spinodal-type dynamics in fractal aggregation of colloidal clusters. *Physical Review Letters*. 1992;68(22):3327–3330.
26. Poon WCK, Haw MD. Mesoscopic structure formation in colloidal aggregation and gelation. *Advances in Colloid and Interface Science*. 1997;73:71–126.
27. Kim JM, Fang J, Eberle APR, Castaneda-Priego R, Wagner NJ. Gel transition in adhesive hard-sphere colloidal dispersions: The role of gravitational effects. *Physical Review Letters*. 2013;110(20):208302.

28. Phan S-E, Russel WB, Cheng ZD, Zhu JX, Chaikin PM, Dunsmuir JH, et al. Phase transition, equation of state, and limiting shear viscosities of hard sphere dispersions. *Physical Review E*. 1996;54(6):6633–6645.
29. Hunter GL, Weeks ER. The physics of the colloidal glass transition. *Reports on Progress in Physics*. 2012;75(6):066501.
30. Joshi YM, Petekidis G. Yield stress fluids and ageing. *Rheologica Acta*. 2018;57 (6–7):521–549.
31. Reynaert S, Moldenaers P, Vermant J. Interfacial rheology of stable and weakly aggregated two-dimensional suspensions. *Physical Chemistry Chemical Physics*. 2007;9(48):6463–6475.
32. Winter HH, Chambon F. Analysis of linear viscoelasticity of a cross-linking polymer at the gel point. *Journal of Rheology*. 1986;30(2):367–382.
33. Gordon MB, Kloxin CJ, Wagner NJ. The rheology and microstructure of an aging thermoreversible colloidal gel. *Journal of Rheology*. 2017;61(1):23–34.
34. Mason TG, Weitz DA. Linear viscoelasticity of colloidal hard-sphere suspensions near the glass-transition. *Physical Review Letters*. 1995;75(14):2770–2773.
35. Helgeson ME, Wagner NJ, Vlassopoulos D. Viscoelasticity and shear melting of colloidal star polymer glasses. *Journal of Rheology*. 2007;51(2):297–316.
36. Eberle APR, Castaneda-Priego R, Kim JM, Wagner NJ. Dynamical arrest, percolation, gelation, and glass formation in model nanoparticle dispersions with thermoreversible adhesive interactions. *Langmuir*. 2012;28(3):1866–1878.
37. Petekidis G, Vlassopoulos D, Pusey PN. Yielding and flow of sheared colloidal glasses. *Journal of Physics Condensed Matter*. 2004;16(38):S3955–S3963.
38. Segre PN, Meeker SP, Pusey PN, Poon WCK. Viscoity and structureal relaxation in suspensions of hard-sphere colloids. *Physical Review Letters*. 1995;75(5):958–961.
39. Pusey PN, Segre PN, Behrend OP, Meeker SP, Poon WCK. Dynamics of concentrated colloidal suspensions. *Physica A*. 1997;235(1–2):1–8.
40. Russel WB, Wagner NJ, Mewis J. Divergence in the low shear viscosity for Brownian hard-sphere dispersions: At random close packing or the glass transition? *Journal of Rheology*. 2013;57(6):1555–1567.
41. Besseling R, Isa L, Ballesta P, Petekidis G, Cates ME, Poon WCK. Shear banding and flow-concentration coupling in colloidal glasses. *Physical Review Letters*. 2010;105 (26):268301.
42. Laun HM. Rheological properties of aqueous polymer dispersions. *Angewandte Makromolekulare Chemie*. 1984;123(1):335–359.
43. Jones DAR, Leary B, Boger DV. The rheology of a concentrated colloidal suspensions of hard spheres. *Journal of Colloid and Interface Science*. 1991;147(2):479–495.
44. Fuchs M, Ballauff M. Flow curves of dense colloidal dispersions: Schematic model analysis of the shear-dependent viscosity near the colloidal glass transition. *Journal of Chemical Physics*. 2005;122(9):094707.
45. Seth JR, Mohan L, Locatelli-Champagne C, Cloitre M, Bonnecaze RT. A micromechanical model to predict the flow of soft particle glasses. *Nature Materials*. 2011;10(11):838–843.
46. Bonnecaze RT, Brady JF. Dynamic simulation of an electrorheological fluid. *Journal of Chemical Physics*. 1992;96(3):2183–2202.
47. Koumakis N, Pamvouxoglou A, Poulos A, and Petekdis G. Direct comparison of the rheology of hard and soft particle glasses. *Soft Matter*. 2012;8(12):4271–4284.
48. Ballesta P, Besseling R, Isa L, Petekidis G, Poon WCK. Slip and flow of hard-sphere colloidal glasses. *Physical Review Letters*. 2008;101(25):258301.

49. Ballesta P, Petekidis G, Isa L, Poon WCK, Besseling R. Wall slip and flow of concentrated hard-sphere colloidal suspensions. *Journal of Rheology*. 2012;56(5):1005–1037.
50. Wagner NJ, Brady JF. Shear thickening in colloidal dispersions. *Physics Today*. 2009;62 (10):27–32.
51. Di Cola E, Moussaid A, Sztucki M, Narayanan T, Zaccarelli E. Correlation between structure and rheology of a model colloidal glass. *Journal of Chemical Physics*. 2009;131 (14):144903.
52. Ballauff M, Brader JM, Egelhaaf SU, Fuchs M, Horbach J, Koumakis N, et al. Residual stresses in glasses. *Physical Review Letters*. 2013;110(21):215701.
53. Pham KN, Petekidis G, Vlassopoulos D, Egelhaaf SU, Poon WCK, Pusey PN. Yielding behavior of repulsion- and attraction-dominated colloidal glasses. *Journal of Rheology*. 2008;52(2):649–676.
54. Koumakis N, Schofield AB, Petekidis G. Effects of shear induced crystallization on the rheology and ageing of hard sphere glasses. *Soft Matter*. 2008;4(10):2008–2018.
55. van Megen W, Mortensen TC, Williams SR, Muller J. Measurement of the self-intermediate scattering function of suspensions of hard spherical particles near the glass transition. *Physical Review E*. 1998;58(5):6073–6085.
56. Van Megen W, Underwood SM. Glass transition in colloidal hard spheres: Measurement and mode-coupling-theory analysis of the coherent intermediate scattering function. *Physical Review E*. 1994;49(5):4206–4220.
57. Zwanzig R, Mountain RD. High-frequency moduli of simple fluids. *Journal of Chemical Physics*. 1965;43(12):4464.
58. Buscall R. Effect of long-range repulsive forces on the viscosity of concenttrated lattices – Comparison of experimental data with an effective hard-sphere model. *Journal of the Chemical Society, Faraday Transactions*. 1991;87(6):1365–1370.
59. Wagner NJ. The high-frequency shear modulus of colloidal suspensions and the effects of hydrodynamic interactions *Journal of Colloid and Interface Science*. 1993;161 (1):169–181.
60. Paulin SE, Ackerson BJ, Wolfe MS. Equilibrium and shear induced nonequilibrium phase behavior of PMMA microgel spheres. *Journal of Colloid and Interface Science*. 1996;178 (1):251–262.
61. Le Grand A, Petekidis G. Effects of particle softness on the rheology and yielding of colloidal glasses. *Rheologica Acta*. 2008;47(5–6):579–590.
62. Senff H, Richtering W. Temperature sensitive microgel suspensions: Colloidal phase behavior and rheology of soft spheres. *Journal of Chemical Physics*. 1999;111 (4):1705–1711.
63. Chow MK, Zukoski CF. Nonequilibrium behavior of dense suspensions of uniform particles - Volume fraction and size dependence of rheology and microstructure. *Journal of Rheology*. 1995;39(1):33–59.
64. Raynaud L, Ernst B, Verge C, Mewis J. Rheology of aqueous latices with adsorbed stabilizer layers. *Journal of Colloid and Interface Science*. 1996;181(1):11–19.
65. Petekidis G, Moussaid A, Pusey PN. Rearrangements in hard-sphere glasses under oscillatory shear strain. *Physical Review E*. 2002;66(5):051402.
66. Petekidis G, Vlassopoulos D, Pusey PN. Yielding and flow of colloidal glasses. *Faraday Discussions*. 2003;123:287–302.
67. Schall P, Weitz DA, Spaepen F. Structural rearrangements that govern flow in colloidal glasses. *Science*. 2007;318(5858):1895–1899.

68. Amann CP, Denisov D, Dang MT, Struth B, Schall P, Fuchs M. Shear-induced breaking of cages in colloidal glasses: Scattering experiments and mode coupling theory. *The Journal of Chemical Physics*. 2015;143(3):034505.
69. (a) Koumakis N, Laurati M, Jacob AR, Mutch KJ, Abdellali A, Schofield AB, et al. Start-up shear of concentrated colloidal hard spheres: Stresses, dynamics, and structure. *Journal of Rheology*. 2016;60(4):603–623.
(b) Koumakis N, Laurati M, Egelhaaf SU, Brady JF, Petekidis G. Yielding of Hard-Sphere Glasses during Start-Up Shear. *Physical Review Letters*. 2012;108(9):098303.
70. Marenne S, Morris JF, Foss DR, Brady JF. Unsteady shear flows of colloidal hard-sphere suspensions by dynamic simulation. *Journal of Rheology*. 2017;61(3):477–501.
71. Brady JF. The rheological behavior of concentrated colloidal dispersions. *Journal of Chemical Physics*. 1993;99(1):567–581.
72. Sollich P. Rheological constitutive equation for a model of soft glassy materials. *Physical Review E*. 1998;58(1):738–759.
73. Brader JM, Cates ME, Fuchs M. First-principles constitutive equation for suspension rheology. *Physical Review Letters*. 2008;101(13):138301.
74. Priya M, Voigtmann T. Nonlinear rheology of dense colloidal systems with short-ranged attraction: A mode-coupling theory analysis. *Journal of Rheology*. 2014;58 (5):1163–1187.
75. Laurati M, Mutch KJ, Koumakis N, Zausch J, Amann CP, Schofield AB, et al. Transient dynamics in dense colloidal suspensions under shear: Shear rate dependence. *Journal of Physics: Condensed Matter*. 2012;24(46):464104.
76. Mutch KJ, Laurati M, Amann CP, Fuchs M, Egelhaaf SU. Time-dependent flow in arrested states – transient behaviour. *European Physical Journal–Special Topics*. 2013;222(11):2803–2817.
77. Amann CM, Siebenbürger M, Ballauff M, Fuchs M. Nonlinear rheology of glass-forming colloidal dispersions: transient stress–strain relations from anisotropic mode coupling theory and thermosensitive microgels. *Journal of Physics: Condensed Matter*. 2015;27(19):194121.
78. Padding JT, Boek ES, Briels WJ. Dynamics and rheology of wormlike micelles emerging from particulate computer simulations. *The Journal of Chemical Physics*. 2008;129 (7):074903-11.
79. Rottler J, Robbins MO. Shear yielding of amorphous glassy solids: Effect of temperature and strain rate. *Physical Review E*. 2003;68(1):011507.
80. Zausch J, Horbach J, Laurati M, Egelhaaf SU, Brader JM, Voigtmann T, et al. From equilibrium to steady state: The transient dynamics of colloidal liquids under shear. *Journal of Physics: Condensed Matter*. 2008;20(40):404210.
81. Siebenburger M, Ballauff M, Voigtmann T. Creep in colloidal glasses. *Physical Review Letters*. 2012;108(25):255701.
82. Koumakis N, Petekidis G. Two step yielding in attractive colloids: Transition from gels to attractive glasses. *Soft Matter*. 2011;7(6):2456–2470.
83. Schirber M. Focus: Controlling persistent stress in glass. *Physics*. 2013;6(60): comment on: Ballauff M, Brader JM, Egelhaaf SU, Fuchs M, Horbach J, Koumakis N, et al. Residual stresses in glasses. *Physical Review Letters*. 110(21):215701.
84. Mohan L, Bonnecaze RT, Cloitre M. Microscopic origin of internal stresses in jammed soft particle suspensions. *Physical Review Letters*. 2013;111(26):268301.
85. Moghimi E, Jacob AR, Koumakis N, Petekidis G. Colloidal gels tuned by oscillatory shear. *Soft Matter*. 2017;13(12):2371–2383.

86. Negi AS, Osuji CO. Physical aging and relaxation of residual stresses in a colloidal glass following flow cessation. *Journal of Rheology*. 2010;54(5):943–958.
87. Lidon P, Villa L, Manneville S. Power-law creep and residual stresses in a carbopol gel. *Rheologica Acta*. 2017;56(3):307–323.
88. Chikkadi V, Miedema DM, Dang MT, Nienhuis B, Schall P. Shear banding of colloidal glasses: Observation of a dynamic first-order transition. *Physical Review Letters*. 2014;113(20):208301.
89. Fielding SM. Shear banding in soft glassy materials. *Reports on Progress in Physics*. 2014;77(10):102601.
90. Fielding SM, Cates ME, Sollich P. Shear banding, aging and noise dynamics in soft glassy materials. *Soft Matter*. 2009;5(12):2378–2382.
91. Besseling R, Isa L, Weeks ER, Poon WCK. Quantitative imaging of colloidal flows. *Advances in Colloid and Interface Science*. 2009;146(1–2):1–17.
92. Morris JF. Shear thickening of concentrated suspensions: Recent developments and relation to other phenomena. *Annual Review of Fluid Mechanics*. 2020;52(1):121–144.
93. Stradner A, Sedgwick H, Cardinaux F, Poon WCK, Egelhaaf SU, Schurtenberger P. Equilibrium cluster formation in concentrated protein solutions and colloids. *Nature*. 2004;432(7016):492–495.
94. Stradner A, Cardinaux F, Egelhaaf SU, Schurtenberger A. Do equilibrium clusters exist in concentrated lysozyme solutions? *Proceedings of the National Academy of Sciences of the United States of America*. 2008;105(44):E75.
95. Campbell AI, Anderson VJ, van Duijneveldt JS, Bartlett P. Dynamical arrest in attractive colloids: The effect of long-range repulsion. *Physical Review Letters*. 2005;94(20):208301.
96. Mewis J. Thixotropy – general review *Journal of Non-Newtonian Fluid Mechanics*. 1979;6(1):1–20.
97. Tsurusawa H, Leocmach M, Russo J, Tanaka H. Direct link between mechanical stability in gels and percolation of isostatic particles. *Science Advances*. 2019;5(5):eaav6090.
98. Royall CP, Williams SR, Ohtsuka T, Tanaka H. Direct observation of a local structural mechanism for dynamic arrest. *Nature Materials*. 2008;7(7):556–561.
99. Sciortino F, Buldyrev SV, De Michele C, Foffi G, Ghofraniha N, La Nave E, et al. Routes to colloidal gel formation. *Computer Physics Communications*. 2005;169(1):166–171.
100. Trappe V, Sandkuhler P. Colloidal gels—Low-density disordered solid-like states. *Current Opinion in Colloid & Interface Science*. 2004;8(6):494–500.
101. Helgeson ME, Gao YX, Moran SE, Lee J, Godfrin M, Tripathi A, et al. Homogeneous percolation versus arrested phase separation in attractively-driven nanoemulsion colloidal gels. *Soft Matter*. 2014;10(17):3122–3133.
102. Foffi G, De Michele C, Sciortino F, Tartaglia P. Arrested phase separation in a short-ranged attractive colloidal system: A numerical study. *Journal of Chemical Physics*. 2005;122(22):224903.
103. Dawson K, Foffi G, Fuchs M, Gotze W, Sciortino F, Sperl M, et al. Higher-order glass-transition singularities in colloidal systems with attractive interactions. *Physical Review E*. 2001;63(1):011401.
104. Whitaker KA, Varga Z, Hsiao LC, Solomon MJ, Swan JW, Furst EM. Colloidal gel elasticity arises from the packing of locally glassy clusters. *Nature Communications*. 2019;10(1):2237.
105. Ramakrishnan S, Chen YL, Schweizer KS, Zukoski CF. Elasticity and clustering in concentrated depletion gels. *Physical Review E*. 2004;70(4):040401.

106. He H, Thorpe MF. Elastic properties of glasses. *Physical Review Letters*. 1985;54 (19):2107–2110.
107. Wang G, Fiore AM, Swan JW. On the viscosity of adhesive hard sphere dispersions: Critical scaling and the role of rigid contacts. *Journal of Rheology*. 2019;63(2):229–245.
108. Verhaegh NAM, Asnaghi D, Lekkerkerker HNW, Giglio M, Cipelletti L. Transient gelation by spinodal decomposition in colloid-polymer mixtures. *Physica A*. 1997;242 (1–2):104–118.
109. Zaccarelli E, Foffi G, Dawson KA, Sciortino F, Tartaglia P. Mechanical properties of a model of attractive colloidal solutions. *Physical Review E*. 2001;63(3):031501.
110. Sciortino F, Zaccarelli E. Reversible gels of patchy particles. *Current Opinion in Solid State & Materials Science*. 2011;15(6):246–253.
111. Cipelletti L, Manley S, Ball RC, Weitz DA. Universal aging features in the restructuring of fractal colloidal gels. *Physical Review Letters*. 2000;84(10):2275–2278.
112. Zaccarelli E, Lu PJ, Ciulla F, Weitz DA, Sciortino F. Gelation as arrested phase separation in short-ranged attractive colloid-polymer mixtures. *Journal of Physics-Condensed Matter*. 2008;20(49):494242.
113. Poon WCK, Selfe JS, Robertson MB, Ilett SM, Pirie AD, Pusey PN. An experimental study of a model colloid-polymer mixture. *Journal of Physics II*. 1993;3(7):1075–1086.
114. Lu PJ, Conrad JC, Wyss HM, Schofield AB, Weitz DA. Fluids of clusters in attractive colloids. *Physical Review Letters*. 2006;96(2):028306.
115. Zaccone A, Wu H, Del Gado E. Elasticity of arrested short-ranged attractive colloids: Homogeneous and heterogeneous glasses. *Physical Review Letters*. 2009;103 (20):208301.
116. Laurati M, Egelhaaf SU, Petekidis G. Nonlinear rheology of colloidal gels with intermediate volume fraction. *Journal of Rheology*. 2011;55(3):673–706.
117. Johnson LC, Zia RN, Moghimi E, Petekidis G. Influence of structure on the linear response rheology of colloidal gels. *Journal of Rheology*. 2019;63(4):583–608.
118. Koumakis N, Moghimi E, Besseling R, Poon WCK, Brady JF, Petekidis G. Tuning colloidal gels by shear. *Soft Matter*. 2015;11(23):4640–4648.
119. Foffi G, De Michele C, Sciortino F, Tartaglia P. Scaling of dynamics with the range of interaction in short-range attractive colloids. *Physical Review Letters*. 2005;94(7):078301.
120. Ramakrishnan S, Gopalakrishnan V, Zukoski CF. Clustering and mechanics in dense depletion and thermal gels. *Langmuir*. 2005;21(22):9917–9925.
121. Eberle APR, Martys N, Porcar L, Kline SR, George WL, Kim JM, et al. Shear viscosity and structural scalings in model adhesive hard-sphere gels. *Physical Review E*. 2014;89 (5):050302(R).
122. Valadez-Perez NE, Liu Y, Eberle APR, Wagner NJ, Castaneda-Priego R. Dynamical arrest in adhesive hard-sphere dispersions driven by rigidity percolation. *Physical Review E*. 2013;88(6):060302(R).
123. Sciortino F, Zaccarelli E. Equilibrium gels of limited valence colloids. *Current Opinion in Colloid & Interface Science*. 2017;30:90–96.
124. Angelini R, Zaccarelli E, Marques FAD, Sztucki M, Fluerasu A, Ruocco G, et al. Glass-glass transition during aging of a colloidal clay. *Nature Communications*. 2014;5(1):4049.
125. Murphy RP, Hong KL, Wagner NJ. Thermoreversible gels composed of colloidal silica rods with short range attractions. *Langmuir*. 2016;32(33):8424–8435.
126. Michel E, Filali M, Aznar R, Porte G, Appell J. Percolation in a model transient network: Rheology and dynamic light scattering. *Langmuir*. 2000;16(23):8702–8711.

127. Zilman A, Kieffer J, Molino F, Porte G, Safran SA. Entropic phase separation in polymer-microemulsion networks. *Physical Review Letters*. 2003;91(1):015901.
128. Furst EM, Pantina JP. Yielding in colloidal gels due to nonlinear microstructure bending mechanics. *Physical Review E*. 2007;75(5):050402.
129. Pantina JP, Furst EM. Elasticity and critical bending moment of model colloidal aggregates. *Physical Review Letters*. 2005;94(13):138301.
130. van Doorn JM, Sprakel J, Kodger TE. Temperature-triggered colloidal gelation through well-defined grafted polymeric surfaces. *Gels*. 2017;3(2):21.
131. Kohl M, Capellmann RF, Laurati M, Egelhaaf SU, Schmiedeberg M. Directed percolation identified as equilibrium pre-transition towards non-equilibrium arrested gel states. *Nature Communications*. 2016;7(1):11817.
132. Maxwell JCL. On the calculation of the equilibrium and stiffness of frames. *The London, Edinburgh, and Dublin Philosophical Magazine and Journal of Science*. 1864;27 (182):294–299.
133. Hsiao LC, Newman RS, Glotzer SC, Solomon MJ. Role of isostaticity and load-bearing microstructure in the elasticity of yielded colloidal gels. *Proceedings of the National Academy of Sciences of the United States of America*. 2012;109(40):16029–16034.
134. van Hecke M. Jamming of soft particles: Geometry, mechanics, scaling and isostaticity. *Journal of Physics-Condensed Matter*. 2010;22(3):033101.
135. Verduin H, Dhont JKG. Phase diagram of a model adhesive hard-sphere dispersion. *Journal of Colloid and Interface Science*. 1995;172(2):425–437.
136. Grant MC, Russel WB. Volume-fraction dependence of elastic-moduli and transition-temperatures for colloidal silica-gels. *Physical Review E*. 1993;47(4):2606–2614.
137. Brown E, Jaeger HM. Shear thickening in concentrated suspensions: Phenomenology, mechanisms and relations to jamming. *Reports on Progress in Physics*. 2014;77 (4):046602.
138. Jamali S, Brady JF. Alternative frictional model for discontinuous shear thickening of dense suspensions: Hydrodynamics. *Physical Review Letters*. 2019;123(13):138002.
139. Zia RN, Landrum BJ, Russel WB. A micro-mechanical study of coarsening and rheology of colloidal gels: Cage building, cage hopping, and Smoluchowski's ratchet. *Journal of Rheology*. 2014;58(5):1121–1157.
140. Chen YL, Schweizer KS. Microscopic theory of gelation and elasticity in polymer-particle suspensions. *Journal of Chemical Physics*. 2004;120(15):7212–7222.
141. Shah SA, Chen YL, Ramakrishnan S, Schweizer KS, Zukoski CF. Microstructure of dense colloid-polymer suspensions and gels. *Journal of Physics: Condensed Matter*. 2003;15 (27):4751–4778.
142. Shah SA, Chen YL, Schweizer KS, Zukoski CF. Phase behavior and concentration fluctuations in suspensions of hard spheres and nearly ideal polymers. *Journal of Chemical Physics*. 2003;118(7):3350–3361.
143. Pham KN, Petekidis G, Vlassopoulos D, Egelhaaf SU, Pusey PN, Poon WCK. Yielding of colloidal glasses. *EPL (Europhysics Letters)*. 2006;75(4):624.
144. Ballesta P, Koumakis N, Besseling R, Poon WCK, Petekidis G. Slip of gels in colloid-polymer mixtures under shear. *Soft Matter*. 2013;9(12):3237–3245.
145. Boromand A, Jamali S, Maia JM. Structural fingerprints of yielding mechanisms in attractive colloidal gels. *Soft Matter*. 2017;13(2):458–473.
146. Shao Z, Negi AS, Osuji CO. Role of interparticle attraction in the yielding response of microgel suspensions. *Soft Matter*. 2013;9(22):5492–5500.

147. Sprakel J, Lindstrom SB, Kodger TE, Weitz DA. Stress enhancement in the delayed yielding of colloidal gels. *Physical Review Letters*. 2011;106(24):248303.
148. Landrum BJ, Russel WB, Zia RN. Delayed yield in colloidal gels: Creep, flow, and re-entrant solid regimes. *Journal of Rheology*. 2016;60(4):783–807.
149. Moghimi E. Microscopic Dynamics and Rheology of Colloidal Gels [PhD thesis]. Crete: Univeristy of Crete; 2016.
150. Kim JM, Eberle APR, Gurnon AK, Porcar L, Wagner NJ. The microstructure and rheology of a model, thixotropic nanoparticle gel under steady shear and large amplitude oscillatory shear (LAOS). *Journal of Rheology*. 2014;58(5):1301–1328.
151. Royall CP, Williams SR, Tanaka H. Vitrification and gelation in sticky spheres. *Journal of Chemical Physics*. 2018;148(4):044501.
152. Eberle APR, Wagner NJ, Castanada-Priego R. Dynamical arrest transition in nanoparticle dispersions with short-range interactions. *Physical Review Letters*. 2011;106(10):105704.
153. Eberle APR, Wagner NJ, Akgun B, Satija SK. Temperature-dependent nanostructure of an end-tethered octadecane brush in tetradecane and nanoparticle phase behavior. *Langmuir*. 2010;26(5):3003–3007.
154. Barnes HA. The yieldstress—A review of “panta rhei”—Everything flows? *Journal of Non-Newtonian Fluid Mechanics*. 1999;81(1–2):133–178.
155. Martin JE, Adolf D, Wilcoxon JP. Viscoleasticiy of near-critical gels. *Physical Review Letters*. 1988;61(22):2620–2623.
156. Martin JE, Wilcoxon J, Odinek J. Decay of density-fluctuations in gels. *Physical Review A*. 1991;43(2):858–872.
157. Hurtado PI, Berthier L, Kob W. Heterogeneous diffusion in a reversible gel. *Physical Review Letters*. 2007;98(13):135503.
158. Rueb CJ, Zukoski CF. Viscoelastic properties of colloidal gels. *Journal of Rheology*. 1997;41(2):197–218.
159. Mewis J, Wagner NJ. Thixotropy. *Advances in Colloid and Interface Science*. 2009;147–148:214–227.
160. Guo HY, Ramakrishnan S, Harden JL, Leheny RL. Connecting nanoscale motion and rheology of gel-forming colloidal suspensions. *Physical Review E*. 2010;81(5):050401(R).
161. Winter HH. Glass transition as the rheological inverse of gelation. *Macromolecules*. 2013;46(6):2425–2432.
162. Park JD, Ahn KH, Wagner NJ. Structure-rheology relationship for a homogeneous colloidal gel under shear startup. *Journal of Rheology*. 2017;61(1):117–137.
163. Elliott SL, Butera RJ, Hanus LH, Wagner NJ. Fundamentals of aggregation in concentrated dispersions: Fiber-optic quasielastic light scattering and linear viscoelastic measurements. *Faraday Discussions*. 2003;123:369–383.
164. Tripathy M, Schweizer KS. Activated dynamics in dense fluids of attractive nonspherical particles. I. Kinetic crossover, dynamic free energies, and the physical nature of glasses and gels. *Physical Review E*. 2011;83(4):10.
165. Wang G, Swan J. Surface heterogeneity affects percolation and gelation of colloids: dynamic simulations with random patchy spheres. *Soft Matter*. 2019;15(25):5094–5108.

6 Suspensions of Soft Colloidal Particles

Dimitris Vlassopoulos and Michel Cloitre

6.1 Introduction

Softness imparts a plethora of rheological phenomena and transitions in colloidal suspensions. It offers unprecedented opportunities for tailoring the rheology with significant implications in applications. With hard sphere suspensions serving as the reference system, examined in *CSR* and Chapter 5, we guide the reader through the key elements of softness and their consequences on the behavior of colloidal suspensions. We discuss softness in terms of pair interaction potential and particle elasticity and describe briefly different types of soft colloids. Whereas we focus on spherical systems, at different instances the role of shape is mentioned. The presentation is based on two classes of well-defined soft colloids, microgels and hairy particles, which encompass the two main features of softness at high fractions, i.e., shape/volume adjustment and mutual interpenetrability. The state behavior of one-component systems in a solvent background is then discussed. An inherent subtle issue here is the volume fraction, which, unlike that of hard spheres, cannot be determined unambiguously due to the imparted volume change of soft particles at high packing fractions. Often, an effective volume fraction is used. Further, we consider the linear viscoelastic properties, focusing on how softness affects the dependence of the shear viscosity and plateau modulus of glasses on the volume fraction. Rheology has emerged as a prime diagnostic tool for solid–liquid–solid transitions and for distinguishing metastable states such as gels and glasses. A wealth of nonlinear rheological features are also presented, i.e., shear thinning, yielding and flow of glassy and jammed suspensions, wall slip, and shear banding instabilities. Finally, entropic and enthalpic mixtures involving soft colloids exhibit an extraordinary (DV) richness of dynamic states and offer a unique means to tailor flow properties of suspensions and composites.

6.2 Landmark Observations

There is a large body of literature addressing the effects of softness on the rheological properties of colloids. Experiments have shown that shear viscosity depends very sensitively on the departure from hard sphere behavior. As seen in Figure 6.1, both the low and high shear viscosities, η_0 and η_∞, respectively, exhibit different dependence

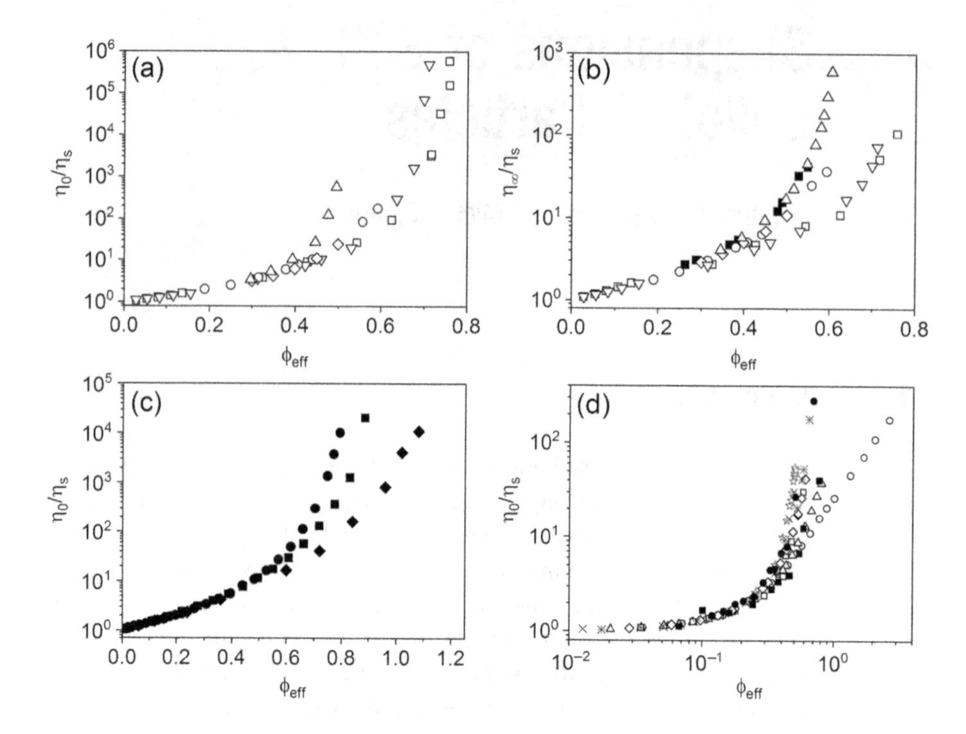

Figure 6.1 Top: Relative zero shear (a) and high shear viscosity (b) of sterically stabilized particle suspensions as a function of the effective volume fraction (data from [1]). The particles are: (i) PMMA latex particles stabilized with a poly(12-hydroxystearic acid) layer in various solvents. Each symbol refers to a particular particle radius to layer thickness ratio: 61 (■); 30 (Δ); 5 (▽, □, different solvents); (ii) crosslinked polystyrene particles with hydrodynamic radius in the range 75–215 nm in different solvents (◊); (iii) sterically stabilized silica particles in cyclohexane with a hydrodynamic radius of 78 nm (○). Bottom: Relative zero shear viscosity versus effective volume fraction for (c) acrylic microgels in water with different crosslink densities: N_x = 28 (●), 70 (■), and 140 (♦) (data from [2]) and (d) hairy particles (data from [3,4]): (i) polybutadiene stars of different functionalities f and arm degree of polymerization N_a (chemical, based on synthesis) in different solvents, f-N_a: 128–130 (◊); 128–1480 (□); 64-130 (Δ); 32-1480 (○); (ii) polyisoprene-polystyrene block copolymer micelles in decane: f = 1470, N_a = 20,675 (■); f = 1710, N_a = 5630 (●) and PMMA hard spheres in decalin (✳,★) [4].

on the effective volume fraction for different particle softness. Here, the effective volume fraction ϕ_{eff} (discussed in *CSR* and later in section 6.5.1) is defined from the hydrodynamic volume of the whole particle at infinite dilution. The classic set of data shown in Figure 6.1(a and b), taken from Mewis et al. [1], refers to sterically stabilized PMMA spherical particles, having different core-to-layer size ratios. Clearly, the data do not collapse onto a master curve at high fractions beyond the Einstein–Sutherland limit. Instead, softness leads to a reduction of the relative viscosity at the same effective volume fraction, so that the viscosities shift to the right with respect to the hard sphere behavior. This is true for a wide range of soft particles, as also shown in figure 4.27 of *CSR*. We extend this discussion and provide a fundamental understanding of this behavior, which allows for a fine-tuning of the viscosity.

The viscosity can be tuned by systematic variation of the particle softness, which is achieved by changing the molecular parameters affecting it. For two soft particle paradigms discussed in detail in this section, microgels and star polymers, the respective parameters are the degree of crosslinking expressed in terms of the average number of monomers between two crosslinks (N_x), and the number of arms or branching functionality (f). Figure 6.1(c and d) depicts the zero shear viscosities plotted against the effective volume fraction. Data are shown for different levels of softness, modulated by crosslink density and functionality for microgels and stars, respectively. We will see that an exact description of the viscosity divergence, beyond empirical modeling, remains elusive.

Early approaches have considered the role of softness on the rheology of dilute suspensions through particle deformability [5]. In concentrated suspensions, the capacity of soft particles to change their shape also affects the rheological properties. Figure 6.2(a) shows the morphology of micron-sized emulsions at a volume fraction of about 0.8. The droplets form facets at contact in order to balance the compression forces that brings then into contact. Because of faceting, the jamming regime of Brownian particles differs from the entropic glass in the sense that the properties are dictated by particle contacts rather than thermal motion [6,7]. Figure 6.2(b) shows the same morphology in a layer of particles densely grafted with DNA strands [8]. In addition to faceting and contact formation, there exist other modes of deformation like osmotic deswelling, compression and interpenetration [9–14]. A complete visual observation of deformation, interpenetration, and compression of highly concentrated poly(N-isopropylacrylamide) (PNIPAm) microgel suspensions was recently obtained by means of two-color super-resolution microscopy [11]; a representative example is depicted in Figure 6.2(c).

The plateau modulus in the glassy and jamming regimes is affected by the softness of the particles and so is its (effective) volume fraction dependence, which is weaker for softer particles [4,15,16]. This is discussed in Section 6.5. Another manifestation of softness concerns the rheological properties of mixtures of colloidal particles with a nonabsorbing polymer. In the presence of a linear polymer, a hard sphere glass melts, but with increasing polymer concentration eventually a re-entrant attractive glass is formed, which has a much larger modulus (see figure 6.12 of *CSR* and Chapter 5). A star polymer/linear polymer mixture has a completely different behavior. The star polymers, which are made of polymer chains joined together through one end, are taken here as model soft spheres (Section 6.3). The star polymer/linear polymer mixture forms a re-entrant gel due to the combination of star compression and depletion attraction (the latter as in hard spheres), both mediated by the osmotic pressure exerted by the polymeric additives [17]. Figure 6.3(a) shows that the re-entrance is accompanied by a substantial modification of the linear viscoelastic properties and a drop of the plateau modulus at low frequency. This observation is made quantitative in Figure 6.3(b), where the plateau storage modulus of the two mixtures is plotted as a function of the

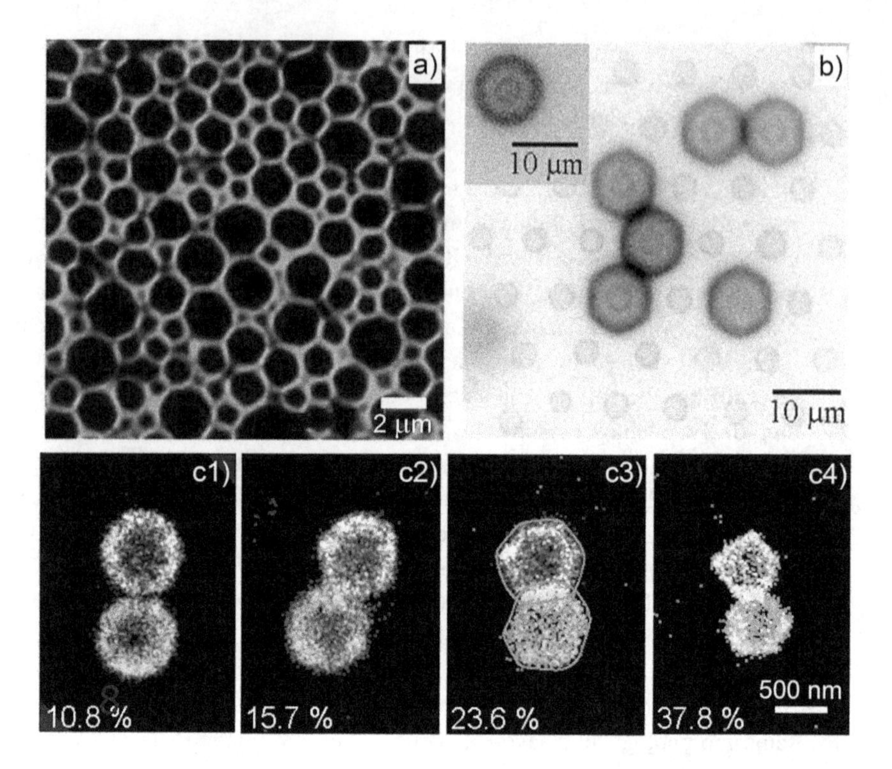

Figure 6.2 Imaging soft colloids. (a) Confocal fluorescence microscopy of concentrated oil-in-water emulsion at a volume fraction of about 0.8. Reprinted with permission from Seth et al., A micromechanical model to predict the flow of soft particle glasses. *Nature Materials* 2011;10 (11):838–843 [7] by Macmillan Publishers Ltd, DOI 10.1007/s00397-017-1002-7. (b) Confocal fluorescence microscopy of DNA-coated soft colloids brought into contact by means of magnetically mediated quasi two-dimensional compression experiments. The image clearly shows the facets formed due to the resistance of the charged coronas to mutual interpenetration, even at pressures approaching the megapascal range. The number of arms is 4.67×10^5 and each arm has 10^4 base pairs. Reprinted with permission from Zhang et al., Direct visualization of conformation and dense packing of DNA-based soft colloids. *Physical Review Letters* 2014;113 (26):268303 [8] by the American Physical Society, DOI 10.1103/PhysRevLett.113.268303. (c) Two-color super-resolution microscopy of poly(N-isopropylacrylamide) microgel suspensions at different concentrations (expressed in wt%); the sequence of images shows that two neighboring microgels successively undergo deformation (c1), deformation and interpenetration (c2 and c3), and compression (c4). c2 is at the onset of interpenetration; contour lines are shown in c3. Reprinted with permission from AAAS from Science Advances [11]. © The Authors, some right reserved; exclusive license American Association for the Advancement of Science. Distributed under a Creative Commons Attribution NonCommercial License 4.0 (CC-BY-NC); http://creativecommons.org/licenses/by-nc/4.0/

concentration of the nonadsorbing polymer. The modulus is normalized with its value in the absence of polymer, when the colloids are in the repulsive glassy state (*RG*), whereas the linear polymer concentration (c_L) is normalized with its overlapping value c_L^* so that it represents an effective volume fraction. We clearly see that the modulus of the hard sphere/polymer mixture increases whereas that of the star

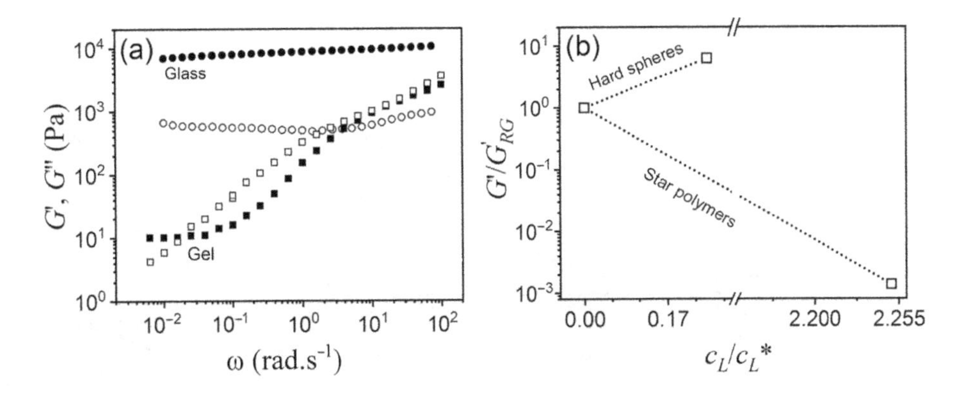

Figure 6.3 (a) Storage (filled symbols) and loss (open symbols) moduli of a star polymer colloidal glass (circles) and re-entrant gel (squares) formed upon addition of linear homopolymers. (b) Illustration of changing plateau modulus (normalized to that of the repulsive glass, G'_{RG}) of colloid-polymer mixtures with increasing concentration of linear polymer (normalized to its overlap concentration, c_L^*) for two mixtures involving hard spheres and soft spheres (star polymers). Data are from Truzzolillo et al. [17]

polymer/linear polymer mixture decreases significantly. This topic is further elaborated in Section 6.8.

Among the many nonlinear shear rheology features of colloidal materials, one can identify yielding, shear thinning, wall slip, shear banding, and shear-induced transitions as the most intriguing. With the exception of shear thickening, which diminishes and eventually disappears as softness increases (see chapter 8 of *CSR* for further discussion), all other phenomena are observed in soft colloids and often have a different physical origin when compared to hard spheres. They will be presented in Section 6.6. In the following, we briefly discuss a rheological feature of great significance to soft colloids – the yield stress. A large number of commercial and commodity products constitute colloidal pastes with yielding properties, from Carbopol to drilling muds and from yoghurt to shaving cream. They have been studied extensively and, despite the fact that a connection with microstructure in these complicated systems is in general not possible, it has been shown that it is useful to distinguish between simple and thixotropic yield stress fluids [18–20]. Simple yield stress fluids share similar static and dynamic yield stresses, irrespective of the flow history. At the macroscopic level, they yield and flow in an homogeneous way although they exhibit transient shear banding, as found with Carbopol measured in a Couette geometry having a small gap and rough walls [21]. Thixotropic yield stress fluids akin to gels are strongly history-dependent and are prone to shear banding at low shear rates [18].

Beyond this tentative classification, the evolution of the flow curves with concentration is intimately linked to the internal microstructure of the particles. This is illustrated in Figure 6.4, which compares the flow curves of microgel and star polymers suspensions at different concentrations in the jamming and glassy regimes.

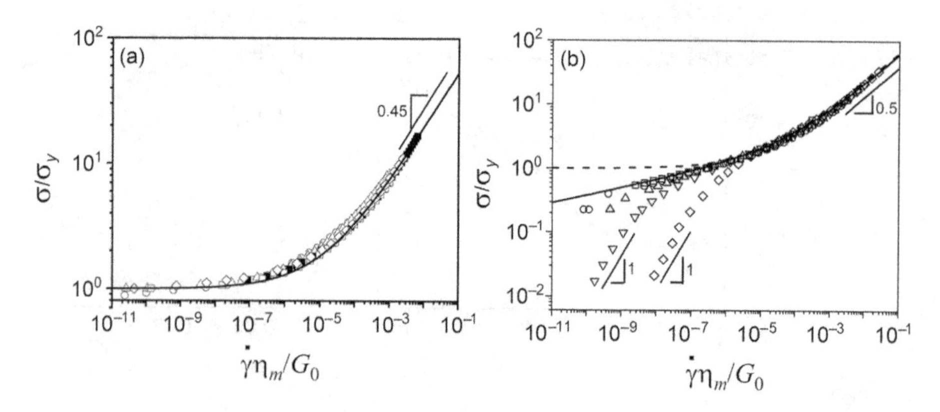

Figure 6.4 (a) Universal flow curve for concentrated microgel suspensions in the jammed glass regime. The line represents the best fit to the Herschel–Bulkley equation; the solid segment depicts the slope in the high shear rate limit. The crosslink density is $N_x = 140$; the concentrations are $c = 20$ (◊), 30 (○, ■), 40 (Δ); 60 (○) mg/g; the solvent viscosity: $\eta_m = 1$ mPa.s (open symbols) and 10.2 mPa.s (full symbol). (b) Universal flow curve for colloidal star ($f = 391$, $N_a = 450$) glasses at increasing concentrations from right to left: $c = 109$ (◊), 121 (▽), 147 (Δ), 199 (○) mg/mL. The dashed curve depicts the generic Herschel–Bulkley equation shown in (a), whereas the solid line is a generalized Herschel–Bulkley equation. The solid segments depict the slopes in the high and low shear rate regimes. Data in (a) and (b) are from Cloitre et al. [22] and Erwin et al. [23], respectively

The microgels are crosslinked polymeric networks swollen by a solvent. The star polymers resemble soft hairy particles with a very small core compared to the grafted hairs (see Section 6.3). Interestingly, normalizing the stress by its yield stress value (determined by oscillatory strain amplitude sweep tests, for example) and the shear rate by a characteristic time, η_m/G_0, determined by the ratio of the solvent viscosity to the plateau modulus of the suspensions, collapses the data measured for different concentrations and solvent viscosities [22]. There is however a clear difference between microgel and star polymer suspensions. In log–log coordinates, the flow curves of microgel suspensions in the jammed regime have a characteristic shape consisting of a plateau at low shear rates and a power-law variation with an exponent close to 0.5 at high shear rates. This is also observed for many other soft particle suspensions like concentrated emulsions [7]. At a quantitative level these flow curves are well described by the Herschel–Bulkley equation (see Section 6.6). On the other hand, whereas the flow curves of star polymer glassy suspensions also exhibit a square-root power law variation at higher shear rates, in the lower shear rates regime the stress departs from a horizontal plateau before finally exhibiting a terminal regime where it is proportional to shear rate. The observation of a terminal regime in glassy suspensions of star polymer suspensions has been linked to the existence of large arm fluctuations and a subsequent facilitated center-of-mass motion, which are responsible for alpha relaxation processes at experimentally accessible frequencies [23]. These modes generally do not exist in microgel suspensions, hence making the case of hairy particles singular.

Note that the experimental detection or not of alpha relaxation in stars depends on their functionality and arm length, which in turn tune their softness [9,23].

The robustness of the scaling used in Figure 6.4 highlights the role of the bulk elasticity with regards to the way that soft particle suspensions deform and flow. Another effect of elasticity in the behavior of concentrated soft particle suspensions concerns the existence of slow dynamics. Upon shear flow cessation, internal stresses of elastic origin remain trapped in the suspensions and slowly relax [24]. These aspects will be discussed in Section 6.8, along with the associated aging behavior, and a critical comparison of systems with different softness will be made, where possible.

6.3 Classification of Soft Colloids

6.3.1 Spherical Particles of Varying Internal Microstructure

Soft particles form a broad range of materials encompassing very different chemical composition and architecture [9,25]. Examples are schematically illustrated in Figure 6.5.

Charged Particles

Solid particles acquire some degree of softness when they are stabilized against aggregation. Ionic groups, arising from the preferential dissolution of chemical species or the ionization of anionic or cationic organic species, leave their counterions in solution; these mobile ions form electrical double-layers through which the particles exert repulsive interactions. At low ionic strength, double-layers expand and inter-particle repulsion acts over long distances, leading to the formation of crystals or glasses. The double-layers surrounding the particles are sensitive to compression and/or distort easily under an external force, making these systems soft and deformable. The interested reader is referred to chapter 4 of *CSR* for further details.

Liquid Dispersions: Emulsions, Vesicles, and Capsules

Emulsions consist of a mixture of two immiscible fluids, one of which, generally oil, is dispersed as small droplets in the continuous phase of the other fluid, generally water. The droplets are stabilized by a surfactant (see Figure 6.5). The preparation, properties, and phase behavior of emulsions have been described in great length. Well-defined monodisperse emulsions can be prepared using specific shearing procedures or microfluidic techniques. In concentrated solutions, amphiphilic molecules are known to self-assemble into a variety of spatially organized flexible structures. They include anisotropic lamellar phases, which under steady shear flow can form multi-lamellar vesicles. The size of these objects results from a balance between the viscous stress associated with the shearing motion and the elastic stress of the membranes. Their structure is metastable, just like for emulsions, but relaxes extremely slowly over a few days to several months, depending on the lamellar composition.

Other particles such as simple vesicles and capsules are made of an internal fluid surrounded by a thin membrane [26,27]. Vesicles have a membrane that imposes fixed volume and surface–area conditions. Capsules are referred to as particles with an elastic membrane whose area is not strictly being conserved. Red blood cells share properties with vesicles and capsules, the membrane having viscoelastic and area-preserving properties [27] (see Chapter 8).

Microgel Particles

Microgel particles constitute an important class of soft particles, both for fundamental science and applications. They consist of a crosslinked polymeric network swollen by a solvent. Because of this architecture, microgels are partially impenetrable, just like hard colloids, but at the same time inherently soft and deformable like polymers. They are central components of advanced functional colloidal materials, with promising applications for bioencapsulation and controlled targeted drug release, metal ion adsorption, photonic materials, and rheology control. A rich literature describes how the synthesis, composition, and architecture of microgels (monomer composition, crosslink density, particle size, surface charge, and functional groups) can be customized to meet the requirements of applications [6,28]. The size of the particles covers several orders of magnitude, from 10 nm to 1 μm or more. Smaller sizes can be achieved with the so-called single-chain nanoparticles [29]. Recent microfluidic techniques can produce well-characterized microgels with sizes in the range 10–100 μm. Neutral or polyelectrolyte water-soluble microgels offer interesting opportunities for novel and environmentally safe applications. The spatial distribution of crosslinks inside microgels determines the local microstructure of the particles, which can be homogeneous, core-shell, or even hairy. Microgel particles can swell and deswell under the action of external stimuli or at

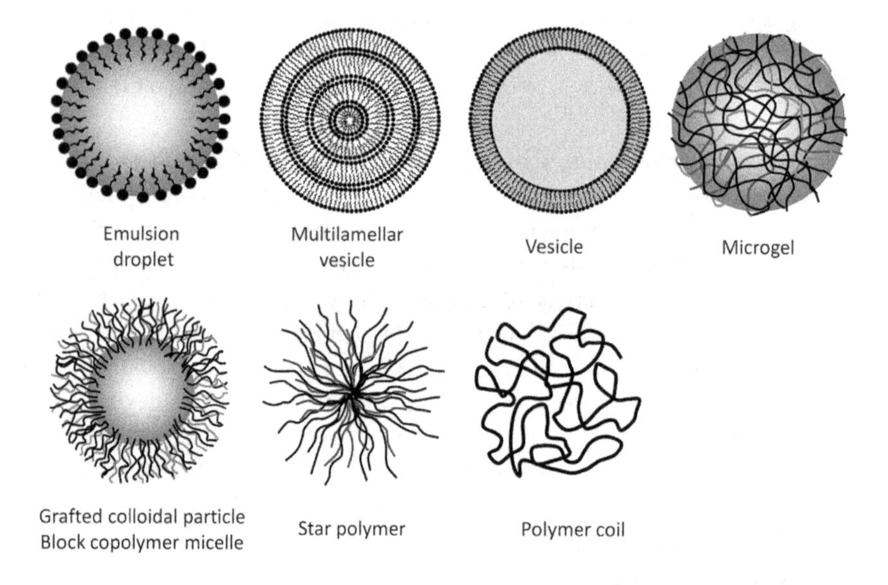

Figure 6.5 Schematic illustrating several types of spherical soft colloidal particles with increasing softness towards the bottom right. The illustration is not to scale.

high concentrations or in the presence of nonadsorbing additives (due to osmotic pressure), whereas weakly crosslinked microgels with dangling hairs can interpenetrate at high concentrations. This has been demonstrated for poly(N-isopropylacrylamide) (PNIPAm) particles by implementing different experimental techniques [30].

Elastomer particles constitute a special case. They are intrinsically soft and deformable, typically of submicron size. They are often used as reinforcing agents to improve the impact resistance of polymer matrices [31]. The rubber of choice is often a styrene-butadiene copolymer. Core-shell particles consisting of an elastomeric core of polybutadiene or *n*-butyl-acrylate and a thin rigid shell of polymethylmethacrylate have been developed for technical applications [32].

Hairy Colloids: Grafted Colloidal Particles, Stars, Micelles

Grafted colloidal particles consist of a core surrounded by a polymeric corona. Examples include star polymers, block copolymer micelles, and polymer-grafted particles. These systems exhibit a high degree of complexity with the three main signatures of softness, i.e., deformability, swelling/deswelling, and interpenetration. They find many applications in drug delivery, reinforcement, processing aids, compatibilizers in polymeric mixtures or Pickering emulsions, and in complex fluids by tailoring their flow properties. Details on their physical properties can be found in Vlassopoulos and Fytas [25].

Multiarm star polymers have recently emerged as archetype grafted colloids, with properties intermediate between those of polymers and hard spheres. They consist of f polymer chains grafted onto a solid core, which plays the role of a topological constraint. When the branching functionality f is large, stars are spherical objects, and for $f \to \infty$ the hard sphere limit is reached. The particular topology of stars results in a variation in monomer density distribution in the particle [33–35]. Typical particle sizes cover the range of 10–100 nm. Several reviews describe the synthesis, structure, and dynamics of star polymers both in melt and in solution [25].

Block copolymers in selective solvents exhibit a remarkable capacity to self-assemble into micellar structures. Their morphology and shape depend on the chain architecture, the block composition, and the affinity of the solvent for the different blocks. The solvophobic blocks constitute the core of the micelles, while the soluble blocks form a soft and deformable corona. Copolymer micelles are thus partially impenetrable, just like colloids, but at the same time inherently soft and deformable like polymers. The phase behavior, structure (often influenced by the dynamic exchange of arms [36,37]), dynamics and applications, both in aqueous and organic solvents, have been the subject of considerable studies [38].

Grafting or adsorbing polymer chains onto solid particles provides efficient steric stabilization against aggregation. The adsorbed or grafted polymer layer offers steric repulsion and, if sufficiently dense and thick, can counterbalance van der Waals attraction. The polymeric nature of the layer softens the interparticle interactions and makes the particles intrinsically deformable. As for the other hairy particles, swelling effects and interpenetration are possible. On the other hand, they can be intrinsically metastable (if the core is glassy) or unstable (if the core dissolves in the solvent when the temperature is altered). In the latter case, chemical crosslinking of the core represents an efficient way to obtain stable micelles [39,40].

6.3.2 Nonspherical Particles

Characteristic examples of nonspherical particles are illustrated in Figure 6.6. Depending on block fraction and interactions, block copolymer micelles can be nonspherical (cylindrical, disk-like) with well-defined shape, size, and grafting distribution [25,41], which can be stabilized via core crosslinking [39]. Grafted nonspherical particles such as clays [42], ellipsoids [43], faceted- or bowl-shaped microgels [44], rods of fd-virus with tunable interactions [45,46] serve as prototype soft anisotropic colloids. The key effect of shape anisotropy is the orientational correlations which may give rise to liquid crystalline order and also glass or gel, with rich texture dynamics [47,48]. For example, using a PNIPAm coat allows tuning the strength of attraction by temperature, leading to thermoreversible nematic gels, whose viscoelastic properties can be tailored by ionic strength as well [45] and scaled onto universal master curves for both fluid-like and solid-like phases [49]. Other examples of nonspherical soft colloidal systems are cellulose nanofiber particles with varying aspect ratios, various self-assembled structures of amphiphiles, telechelic star polymers forming patchy assemblies [50–52], and bottlebrushes or dendronized polymers [53], which represent anisotropic molecular objects. The field is broad and rapidly evolving, so the purpose of this section is simply to alert the reader.

6.3.3 Particle Elasticity

It is convenient to characterize the elasticity of colloidal particles by the effective elastic free energy F_{el} [9]. This parameter accounts only for the elasticity of a single particle of given size but does not consider strain or energy localization at particle contacts. Polymer coils and infinitely hard spheres represent the extreme situations where

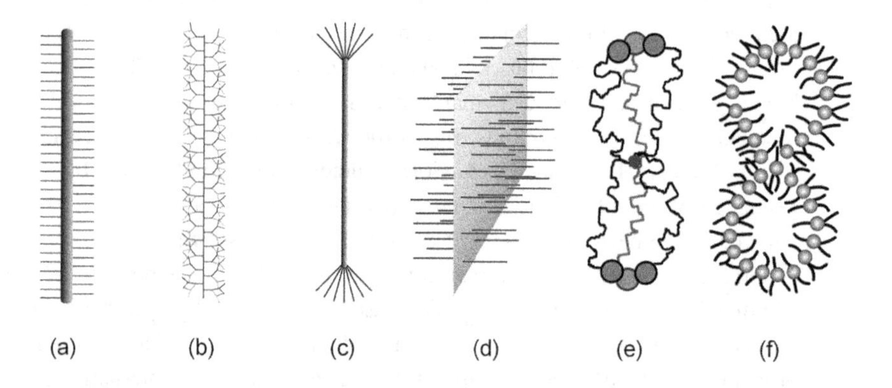

Figure 6.6 Schematic illustration of different types of anisotropic soft colloidal particles (out of scale): (a) grafted rod; (b) dendronized polymer; (c) rod grafted at two ends; (d) grafted plate; (e) patchy structure from associating polymers; and (f) self-assembled amphiphilic molecules.

$F_{el} \cong k_BT$ and $F_{el} = \infty$, respectively. The colloids presented in Section 6.3.1, although very different in size, composition, and structure, lay between these two limits. The elasticity of emulsion or nanoemulsion droplets originates from the interfacial tension of the oil–water interface. When a strain is exerted on the droplets, their shape is changed and the area of the interface increases, storing energy that is released when the droplets recover their initial shape. The free energy associated with small deformations is the surface energy $F_{el} \propto \gamma_s R^2$, where γ_s is the interfacial tension and R the radius of the droplet. Multilamellar vesicles exhibit strong analogies with emulsion particles. The lamellar phase constituting the vesicles is characterized by two elastic moduli, accounting for the compression ($\bar{B}$) and bending (K) of the layers [54]. The combination $(K\bar{B})^{1/2}$ plays the role of an effective surface tension when the lamellae undergo small deformations, leading to an elastic free energy of the form $F_{el} \propto (K\bar{B})^{1/2}R^2$.

The free energy of spherical elastic particles of radius R and Young modulus E is $F_{el} \propto ER^3$. The elasticity of elastomer particles or microgels is associated with the change of free energy of the strands between entanglements or crosslinks during deformation; hence it is of entropic origin. A reasonable estimate of the modulus of elastomeric particles is $E \cong k_BT/N_xV_0$, where N_x is the number of statistical units between entanglements and V_0 is the volume of a statistical unit. The elasticity of microgels depends on many parameters, such as the crosslink density and the degree of ionization or the ionic strength for polyelectrolyte microgels. Useful estimates can be obtained from predictions available for macroscopic gels. It is important to note, however, that the elastic moduli of polymer networks or polymer-colloid particles are strongly dependent on the conditions of synthesis and preparation which are responsible for inhomogeneities of composition and crosslink density, like dangling ends. Micromanipulation techniques have been used to directly characterize the swelling behavior and the elastic properties of individual micron-sized microgels [55].

The elasticity of star polymer and copolymer micelles is also of entropic origin because of the polymeric nature of the particles. In solution, the structure of a star polymer is intrinsically heterogeneous with a dense rigid core, surrounded by a melt-like region impenetrable to solvent and an outer corona, which in the framework of the Daoud–Cotton model can be described as a dense packing of blobs [33]. For good solvent conditions, this region involves an inner ideal part which is only penetrable to solvent and an outer excluded volume part where mutual interpenetration can take place in nondilute conditions. At a distance r from the core there are f blobs at the periphery, each with a free energy k_BT and a characteristic size $\xi(r) \propto rf^{-1/2}$ (in good solvent). The free energy associated with small deformations of the outer corona is thus $F_{el} \propto fk_BT$, which clearly demonstrates the entropic origin of the elasticity of star polymers.

6.3.4 Solvent (Suspending Medium) Free Colloids

Star Polymers and Ordered Block Copolymer Micelles in the Melt State

Densely branched macromolecules are characterized by inhomogeneous intramolecular distribution in monomer density, which gives rise to a complex rheological response [41,56]. This is nicely illustrated in the case of multiarm star polymer [57]. Given their density profile, these stars can be considered as core-shell soft particles in the melt, as in solution, but under ideal conditions [33]. Their structure and dynamic response have been studied extensively. The stress relaxation exhibits hallmarks of both polymers and colloids, due to the interpenetrating arms and excluded volume effects at the macromolecular scale, respectively, the latter leading to a solid-like response for large branching functionality and low chain length [58–60]. Ordered asymmetric block copolymer melts can form spherical micelles which are similar to stars, except that their core is often larger (this depends on the block asymmetry and enthalpic interactions). Their rheology exhibits polymeric and colloidal features and bears similarities to that of colloidal crystals [61]. However, as these micelles are formed due to the enthalpic interactions between blocks, they are temperature sensitive and in this respect more complicated and limited in use. Note that microgels in the melt state have been studied and found to exhibit solid-like response when the crosslink density is small [62].

NOHMs: Nanoparticle Organic Hybrid Materials

More recently, hard spheres grafted with polymers were found to suspend in the melt, bearing similarities with stars, though having a substantially larger core [63,64]. The problem was first studied theoretically [65,66] and later experimentally in terms of optimum graft size and density. Such soft colloids in the melt state can be thought of as solvent-free colloids [63,67,68]. They are called nanoparticle organic hybrid materials (NOHMs). Similarly, melts of polymerically grafted microgel-like cores represent solvent-free colloids [69]. NOHMs are chemically stable and have a large core when compared to star polymers. In the latter case, the core has a typical size amounting to about 10% or less of the overall star [57,58].

6.4 Soft Particle Interactions and State Diagrams

6.4.1 Repulsive Interactions

Soft particles are characterized by a huge diversity of (coarse grained) pair-interaction potentials. Yukawa type potentials have been used successfully in a range of systems with soft steric interactions. Softness is also reflected in the internal microstructure of particles and there are two basic types of potentials accounting for this. The Hertzian potential [70] is appropriate for particles like microgels, emulsions, vesicles, and cells, which deform at contact on increasing the volume fraction. The repulsion between two neighboring particles of initial radii R_i and R_j and center-to-center distance r is:

$$V_{\mathrm{Hertz}}(r) = \begin{cases} \varepsilon\left(1 - \dfrac{r}{R_i + R_j}\right)^{5/2} & \text{if} \quad r \le R_i + R_j \\ 0 & \text{if} \quad r > R_i + R_j \end{cases} \tag{6.1}$$

where the prefactor ε depends on the elastic properties of the particles through the contact modulus $E^* = E/2(1 - \nu^2)$ (E is the Young modulus and ν the Poisson ratio) and the particle radii. The star or logarithmic Yukawa potential [71] applies to particles like star polymers, block copolymer micelles, and other long hairy particles, which undergo both shape/volume changes and weak interpenetration on increasing volume fraction. The potential between two star polymers of functionality f and effective diameter d_s has a Yukawa form at long distances and a logarithmic divergence at short distances [71,72]:

$$\frac{V_{\text{star}}(r)}{k_B T} = \begin{cases} (5/18)f^{3/2}\left[-\ln(r/d_s) + \left(1+\sqrt{f}/2\right)^{-1}\right] & \text{if } r \leq d_s \\ (5/18)f^{3/2}\left(1+\sqrt{f}/2\right)^{-1}(d_s/r)\exp\left[-\sqrt{f}(r-d_s)/2d_s\right] & \text{if } r > d_s \end{cases} \tag{6.2}$$

The validity of this potential has been confirmed experimentally [72,73]. The arms should be sufficiently long for the star to be in the excluded volume regime [33]. However, if they are too long the contribution of the effective entanglements due to interpenetration should be accounted for by considering the associated transient forces [25,74]. Polymer coated particles with a relatively sizable rigid core can be treated in the context of interacting polymeric brushes [75]. Further details can be found in *CSR*.

6.4.2 From Repulsive to Attractive Interactions

Changing the solvent quality, i.e., mediating interactions via the suspending medium, is a way to reduce repulsions and eventually promote attractions. This can also be achieved by adding nonadsorbing polymers, as discussed in Chapter 5. The former situation can be accounted for via a modification of the repulsive potential of hairy particles, which allows it (for example) to become less repulsive on reducing the temperature [76]. Considering a star in a solvent that induces weak excluded volume interactions, by accounting for the free energy of polymers grafted on rigid particles, the pair potential is given by:

$$\frac{V(r)}{k_B T} = V_0\left[-\ln y - \frac{9}{5}(1-y) + \frac{1}{3}(1-y^3) - \frac{1}{30}(1-y^6)\right] \tag{6.3}$$

where $y = (r-2b)/2$, b is the radius of the rigid core, $V_0 = \pi^2 L^3 f/48Na_m^2 b$, a_m is the monomer length, and L is the polymer layer thickness. This expression is valid for $2b < r < 2b + L$. It tends to infinity for $r < 2b$ and is zero for $r > 2b + L$. This potential has been shown to successfully describe the ordering of stars approaching the theta state, as revealed by small angle neutron scattering measurements [76]. When combined with the van der Waals attraction of the cores it can give realistic estimates in the vicinity of the theta temperature.

Entropic soft attraction can lead to particle aggregation, which can be understood in the framework of an effective pair interaction potential between two colloids in a polymer solution, $V_{\text{eff}}(r)$. For the case of star polymers, $V_{\text{eff}}(r)$ accounts for the contributions of the original star pair potential V_{star}, Eq. (6.2), the polymer pair potential V_p, and a star-polymer cross-interaction term V_{sp} [71]. Replacing the linear chain depletant by hard sphere (HS) colloid changes $V_{\text{eff}}(r)$ with respect to the contributions V_{HS} and $V_{\text{star-HS}}$ [77,78].

6.4.3 Defining the Volume Fraction: Effective versus Actual Volume Fraction

The main message from chapter 4 of *CSR* is that mapping the volume fraction of particles onto hard spheres does not work in general for soft colloids. The microstructural features originating from softness influence the volume fraction of soft particle suspensions and consequently the description of their behavior, especially at high concentrations. Soft particles without dangling ends, such as emulsion droplets or some microgels, can adjust their shape and volume to steric constraints by forming facets at contact. Hairy particles can interdigitate in addition to deforming. Therefore, the volume fraction can be tuned by means of external stimuli such as pressure, flow, pH, and temperature [6].

In the literature, an effective volume fraction based on the single particle hydrodynamic radius is generally agreed upon as a universal reference parameter [79]. This particle size can be measured accurately by means of dynamic light scattering at varying temperature or pH or solvent. For particles whose total molar mass is known, such as star polymers, the effective volume fraction is equivalent to the ratio of concentration over its overlap value, $\phi_{\text{eff}} = c/c^*$ An alternative approach is to go back to the idea of mapping viscosity data to an effective hard sphere volume fraction, in the concentration regime where this works (see also *CSR*). The approach is based on the implicit assumption that there exists a linear relationship between the effective volume fraction and the actual concentration: $\phi_{\text{eff}} = k_c c$. By assuming that the Batchelor–Green equation:

$$\eta_r = \eta_0/\eta_m = 1 + 2.5\phi_{\text{eff}} + 7.6\phi_{\text{eff}}^2 \tag{6.4}$$

holds at low concentrations, and by fitting the concentration-dependent relative shear viscosity data $\eta_r(c)$ to a second order polynomial one can determine the coefficient k_c and hence the effective volume fraction [2,80]. Often, for soft spheres only the Einstein–Sutherland linear term in Eq. (6.4) is used.

The effective volume fraction is valid at low concentrations before the osmotic pressure of the neighboring particles and/or external stimuli become important. In this respect, an effective volume fraction exceeding unity can reflect particle overlap and/or deswelling. On the practical side, the change of size of a soft colloidal particle, hence the actual volume fraction, can be estimated for some classes of particles, but there are several approximations that render the process uncertain and nonuniversal. A method consists in mapping the high concentration limit of the viscosity onto the maximum packing volume fraction. Other attempts involve analysis of osmotic shrinkage or deswelling [2,10–13,81,82]. For polyelectrolyte microgels where osmotic deswelling is significant, the actual volume fraction ϕ is found to depend on the effective volume fraction at infinite dilution, ϕ_∞ through the expression [2]:

$$\frac{\phi}{\phi_\infty} = \left[1 - \frac{\Gamma}{1-\Gamma}\,\frac{\phi}{1-\phi}\right]^{3/2} \tag{6.5}$$

where Γ is the fraction of counterions which are not trapped inside the microgel, i.e., free to move in the suspension. This expression is valid up to a certain volume fraction $\phi < 1 - \Gamma$, beyond which the counterion concentrations inside and outside the microgel are equal and deswelling becomes negligible. Further details can be found elsewhere [2,81,82]. Other models for the volume change of nonionic particles account for the osmotic pressure by considering the elasticity of the network-like particle, which is balanced by that of the bath, described by an equation of state [10,12].

The osmotic compression of star polymers has been considered in the context of mixtures with smaller linear homopolymers, which act as depletants (see Section 6.8.1) [83]. The analysis of the size of an isolated soft particle in a bath of depletants is based on the minimization of its free energy. The radius R of a star macromolecule in the presence of linear chains can be evaluated by employing the Flory argument [71,84] and accounting for the osmotic pressure contribution of the chains that depends on their effective volume fraction, defined as the fractional distance from overlap, $\phi_L = c_L/c_L^*$. There are three contributions to the free energy of a single star with unperturbed radius R: the free energy associated with elasticity, F_{el}, the enthalpy originating from interactions due to excluded volume, F_{int}, and the osmotic contribution, F_{osm}, due to the effective expulsion of the linear chains from a cavity region of size $R_{\mathrm{cav}} \approx R$ because of the insertion of a star in a bath of linear chains [71,84]. The total free energy can be written as:

$$F(R) = F_{\mathrm{el}} + F_{\mathrm{int}} + F_{\mathrm{osm}} = k_B T \frac{3fR^2}{2N_a {a_m}^2} + k_B T \frac{\upsilon N_a^2 f^2 {a_m}^3}{2R^3} + \frac{4\pi}{3} R^3 \Pi(\phi_L), \tag{6.6}$$

where a_m is the length scale of monomers, N_a the degree of polymerization of one arm, υ the dimensionless excluded volume parameter, and Π the osmotic pressure exerted by the depletant. Details of the analysis for a single star and for a concentrated solution of star polymers in the presence of depletants can be found elsewhere [84,85]. The penetrability of star polymer molecules makes the transition of the polymer concentration from the average value in the solution to its value in the star interior rather gradual, so that the interfacial cost is not considered [84]. Whereas this analysis applies to good solvent conditions, it can be extended to situations where miscibility issues arise between nanoparticle and solvent and/or chemically different polymers, a problem which has been systematically studied for microgels [86,87].

In conclusion, the main message in this section is that the volume fraction is a delicate issue and its use requires great care. The effective hydrodynamic volume fraction based on the hydrodynamic size is useful to scale transport properties, such as zero shear viscosity or self-diffusion of different soft particles over a range of sizes and concentrations. The actual value of the volume fraction is more accurate over a wider range of concentrations but in general it does not cover the whole range deep into the jamming regime.

6.4.4 State Diagrams for Archetype Soft Colloids

Starting from the state diagram of hard spheres, covered in detail in *CSR*, we now discuss briefly the respective behavior of soft colloids. The state diagrams of soft colloidal suspensions exhibit apparent similarities with that of hard sphere suspensions. Soft colloidal suspensions undergo the same sequence of transitions with increasing volume fraction as Brownian hard spheres: fluid to crystal to entropic repulsive glass and jamming (Figure 6.7). However softness substantially affects the existence and locations of the crystal, glass, and jamming transitions. In particular, there is a clear tendency for the fluid region to be enhanced with softness. This is obvious for polymer coils which are amorphous fluids throughout the whole concentration range.

For relatively monodisperse microgels, the onset of fluid-crystal coexistence occurs at the same volume fraction of 0.494 as in hard spheres. Polydisperse microgels do not crystallize but form an entropic glassy and further a jammed glassy state. However, the glass and jamming transitions occur at higher volume fractions than in hard spheres, with the actual values depending on details of particle microstructure [80,88]. An important question concerns whether an entropic glass and a jammed glass region can be distinguished [89]. This is the case for concentrated emulsions [90] or polyelectrolyte microgels [82,91] that are relatively stiff, i.e., the crosslink density N_x is small, but the results for softer systems are ambiguous [92]. Star polymers crystallize over a very small range of volume fractions at larger effective volume fractions than microgels but the small crystal region may be completely masked by a large glassy region. The suppression of crystallization in suspensions of long hairy particles is related to the fact that arms can interpenetrate above the overlap concentration c^* and fluctuate so that the root-mean-squared

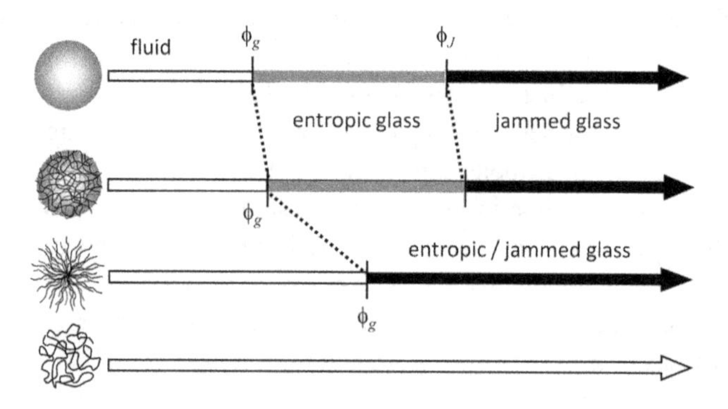

Figure 6.7 Schematic one-dimensional state diagrams of athermal suspensions with varying softness as a function of their effective volume fraction (in the direction of the arrows), from below: polymer coil, star polymer, microgel, hard sphere. For simplicity, we show only the glass (ϕ_g) and jamming (ϕ_J) transitions.

displacement of a particle exceeds 10% of its radius which is the limit for maintaining a crystal, as determined by the Lindeman criterion [93]. The actual existence and nature of a jamming transition in star suspensions are still open questions. It should be noted that block copolymer micelles with possible dynamic arm exchange, or densely grafted hard nanoparticles with larger cores, do crystallize readily [36,94,95]. At very long times, star glasses will eventually crystallize [96]. Thermal annealing can promote crystallization [97].

6.4.5 Shear-Induced Crystallization

Steady or oscillatory shear flow can induce a variety of crystal phases in concentrated soft colloids. Various phases can be obtained (fcc, bcc, layered structures) depending on the flow conditions: Weissenberg number, strain amplitude. Shear-induced structural transitions in polymeric micelles have stimulated a lot of experimental rheo-SANS investigations with spherical [98–103] and slightly nonspherical soft particles [104]. Shear induced crystallization, melting, and rearrangements have also been investigated at high concentrations in systems as different as PMMA microgels [105], thermosensitive microgels [106], emulsions [107–109], and core-shell particles [110]. Using Particle Dynamic Simulations, it has been shown that crystallization in athermal suspensions is associated with specific rheological responses that have been described in terms of shear-activated mechanisms [111,112]. The shear-induced crystalline phases can sometimes sustain large deformations without melting, as for example in the case of block copolymer micelles with large cores [113]. These effects have also been reported for flexible opals in the absence of a suspending medium [64,114]. Shear can also promote the acceleration of the nucleation process and eventual ordering of particles in the supercooled state [115]. Its coupling with significant softness, as for example in the case of star polymers, can also induce order-to-order transition (bcc- to fcc-dominated) without intermediate melting [103].

Story 6.1 The Origins of Soft Particle Rheology

The first attempt to describe a departure from the Einstein–Sutherland equation for a dilute suspension of rigid spheres is due to Taylor [116], who considered a dilute emulsion of small Newtonian drops in a Newtonian suspending medium, in the absence of inertia and hydrodynamic interactions. Specifically, he accounted for the effect of interfacial tension and the relative role of viscous and interfacial forces on the viscosity of the emulsion. Viscoelasticity was first introduced by Fröhlich and Sack [117], who investigated the rheology of a suspension of elastic spheres in a Newtonian liquid. Later, Cerf [118] worked on the same problem. The results of these studies indicate that, due to the slight deformability of the spheres, the suspension exhibited time-dependent elastic recovery under the influence of an imposed stress. To develop constitutive equations, the use of appropriate time derivatives was rendered necessary. The latter task was undertaken by Oldroyd

Story 6.1 *(cont.)*

[5,119,120], who extended the work of Taylor and of Fröhlich and Sack. He showed that an emulsion exhibits elastic properties arising from the interfacial tension. Kerner [121] calculated the elastic modulus of composites consisting of Hookean spheres embedded in a Hookean matrix at different concentrations. The stress response of dilute suspensions of deformable particles in time-varying shear flows was addressed by Frankel and Acrivos [122] and Goddard and Miller [123]. The coupling of particle deformation with flow gave rise to nonlinear rheological effects in the absence of inertia. Choi and Schowalter [124] treated the linear and nonlinear rheology of more concentrated suspensions of Newtonian or viscoelastic drops in Newtonian media, and studied the drop deformation and orientation with varying capillary number. Palierne [125] derived expressions for the linear viscoelastic moduli of viscoelastic spherical emulsions in viscoelastic media for any concentration and size polydispersity, which depend primarily on the ratio of interfacial tension to droplet size. That work serves as the reference for understanding the linear rheology of immiscible polymer blends. It has been used extensively to analyze experimental data and/or to extract the interfacial tension [126], and several improvements or extensions have been proposed over the years [127].

The above brief review summarizes the evolution of the field, which is based on continuum treatment of the particle-matrix system. Despite the enormous impact of this body of work on non-Newtonian fluid mechanics, the particle microstructure is not taken into account; hence, relating moduli (and nonlinear rheology) to interaction potential in a direct way is not possible. In this chapter we adopt a presentation based on linking macroscopic properties to particle microstructure.

The main teaching for this important early work is the general trends associated with the influence of particle softness on rheology through its deformability and the role of shear rate and of interfacial tension.

6.5 Linear Viscoelasticity and Diffusion Dynamics

6.5.1 Shear Viscosity and Self-diffusion

The zero shear viscosity, η_0, which reflects the thermodynamic contribution to the stress, increases with the volume fraction of the suspensions and eventually diverges in a way that depends on the softness. In an attempt towards a generic description, it is useful to plot the relative zero shear viscosity of various systems as function of the effective hydrodynamic volume fraction [22,23,25]. This is depicted in Figure 6.8, which includes data sets for different classes of spherical particles: hard spheres [128], star polymers of different functionality [23], stable block copolymer micelles [129], and microgels [2]. The plot is generic in the sense that at low volume fractions the data

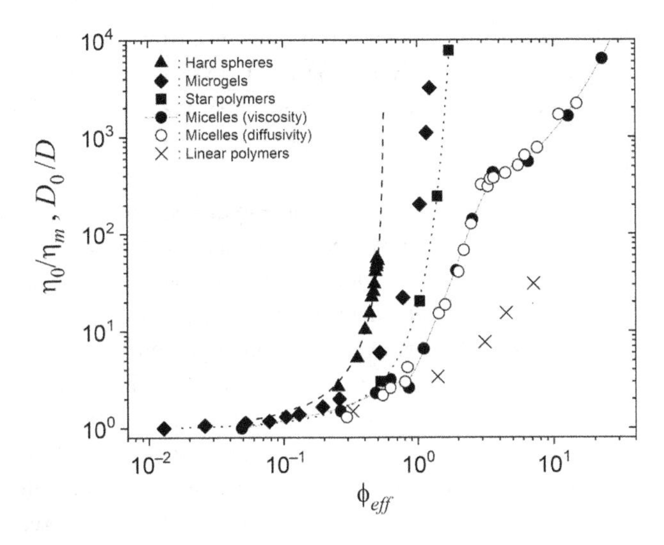

Figure 6.8 Representative experimental data of relative zero shear viscosity and inverse long-time self-diffusion coefficient (normalized to that in a dilute solution) versus the effective volume fraction for various soft spherical colloidal suspensions. For reference, the two extreme cases of hard spheres and linear polymers are also shown. The dashed line through the hard sphere data represents the Quemada model (*CSR*) whereas the dotted lines through the star and microgel data are drawn to guide the eye. Data are taken from other studies [9,23,25,128,129]

collapse, whereas they separate at large volume fractions depending on their softness. Figure 6.8 reveals that softness reduces the relative viscosity and shifts the maximum packing fraction to higher values. Moreover, sufficiently soft particles will osmotically shrink as the volume fraction increases; hence their viscosity will experience a very modest increase (if at all) before eventually diverging at much higher fractions. This is clearly demonstrated in Figure 6.8 for the case of spherical block copolymer micelles with chemically crosslinked soft cores. These particular micelles deswell at high fractions, hence reducing the actual volume fraction and shifting the increase of viscosity to higher fractions [129]. Additional data are shown in Figure 6.1. Swelling and deswelling can be triggered externally by varying the temperature or the ionic strength [130].

For polyelectrolyte microgel particles it has been shown that the divergence of the viscosity with the effective volume fraction depends on the crosslinking density (Figure 6.1(c)) but there is little departure from the hard sphere situation once the volume fraction is appropriately corrected for particle shrinkage [2,131].

Interestingly, available long-time self-diffusion data at different volume fractions, normalized to the single-particle diffusion, are in excellent agreement with the relative viscosity as shown in Figure 6.8. This suggests that the Stokes–Einstein–Sutherland relation is valid even at large volume fractions, in agreement with hard sphere behavior. This has been confirmed up to the glass transition in the case of star-like block copolymer micelles [132]. If one now considers the divergence of the viscosity in order to define an effective glass transition volume fraction, one may plot the normalized viscosity of the suspension against its distance from the glass transition in the context of Angell's fragility analysis [133]. For PNIPAm microgels of varying softness these plots indicate that softer colloids make stronger glasses [134]. This is another situation where accounting for the change of particle size with concentration is important [12,13].

The limiting high shear viscosity, η_∞, reflects the hydrodynamic contribution to the stress. Whereas it is associated with nonlinear response, covered in Section 6.6, it is instructive to discuss it here in relation with the limiting low shear viscosity, η_0. The available data in the literature are not exhaustive because at low concentrations shear thinning is weak and one needs to attain very large and often prohibitive shear rates in order to get the high shear limit, which is very close to the low shear one. Moreover at high concentrations, especially in the jammed regime, instabilities often obscure measurements. Whenever adequate measurements are possible, the flow curve can be fitted to an empirical model (Carreau–Yasuda or Cross model) that allows determining η_0 and η_∞ (see Section 6.6.1). Figure 6.9(a) depicts the relative zero and high shear viscosities for a microgel suspension as a function of mass concentration, spanning the entire range from dilute to jammed regimes [82]. At low concentrations the data collapse, whereas with increasing concentration the zero shear viscosity diverges, in perfect agreement with the normalized longest relaxation time obtained from dynamic light scattering. In Figure 6.9(b), relative high shear viscosity data are plotted against the effective volume fraction determined from a fit of the low concentration data to the Batchelor–Green Eq. (6.4) [135]. Data are presented for two core-shell particles consisting of polystyrene latex cores with adsorbed poly(vinyl alcohol)

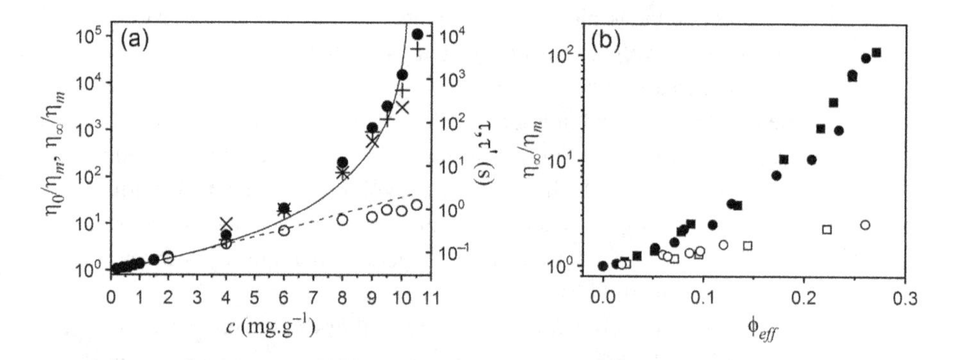

Figure 6.9 (a) Relative low- (●) and high shear (○) viscosity data for microgel suspensions with $N_x = 128$ monomers between two crosslinks and hydrodynamic radius $R_h = 305$ nm at infinite dilution, plotted against mass concentration. The longest relaxation time $\tau'(+)$ from rheology and the structural relaxation time obtained from dynamic light scattering $\tau'(\times)$ are also plotted. Lines are drawn to guide the eye. (b) Relative high shear viscosity data for polystyrene particles with and without adsorbed poly(vinyl alcohol) layers, plotted against the hydrodynamic effective volume fraction of the core. The bare particles have a radius of 113 nm (□) and 175 nm (○). The coated analogues have core radius-to-layer thickness size ratios of 2.53 (■) and 2.75 (●). Data in (a) and (b) are from Pellet and Cloitre [82] and Neuhaeusler and Richtering [135], respectively

layers having different layer thicknesses and core-to-shell size ratios. The soft particle suspensions are non-Newtonian at volume fractions above 0.15 and exhibit a huge increase of the zero shear viscosity. Note that if the effective volume fraction were calculated based on the core size, values larger than the close packing fraction for hard spheres would be obtained, suggesting that the layers are compressed. By contrast, in the same volume fraction range (0.15–0.3, see Figure 6.9b), the bare core suspensions exhibit Newtonian behavior, i.e., identical zero- and high shear viscosities, as well as a weak dependence on volume fraction.

6.5.2 Viscoelastic Relaxation Spectrum and Plateau Modulus

As the concentration increases, the liquid-to-glass transition is signaled by a dramatic evolution of the linear viscoelastic spectrum. The storage modulus G' becomes frequency-independent and larger than the loss modulus G'', which exhibits a shallow minimum. Whereas this is typical for colloidal suspensions in general, subtle differences reveal the role of softness. As an example, in Figure 6.10 we show the linear viscoelastic spectra of three glassy suspensions: star polymers, block copolymer micelles, and microgels. The storage modulus of block copolymers and microgels is independent of frequency. The cage model can be invoked to understand the solid-like behavior of soft glasses as in hard sphere glasses (see also Chapter 5 and *CSR*). For stars, the storage modulus exhibits a weak frequency dependence and the moduli cross at a low frequency, indicating viscoelastic relaxation associated with alpha relaxation. Here, cage escape proceeds by arm disengagement [136]. The accessible terminal relaxation in star polymer solutions sets them apart from other soft glassy materials [23]. This is a consequence of their hairy microstructure which gives rise to internal relaxation modes. It is likely that a similar behavior can be observed in other brush-like or core-shell particles depending on the characteristics of the outer corona. To fully decode the response of these systems and reach terminal relaxation, it is thus necessary to extend the linear viscoelastic

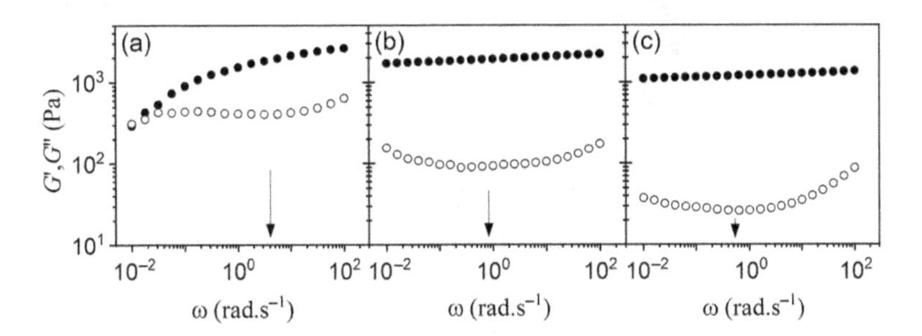

Figure 6.10 Typical frequency-dependent storage (filled circles) and loss moduli (open circles) of three soft colloidal glasses: (a) star polymers [23], (b) block copolymer micelles [139], and (c) microgels [82]. The estimated effective volume fractions are 1.1, 1.1, and 3.5, respectively. The frequencies indicated by arrows are attributed to the end of the in-cage rattling motion or beta relaxation.

spectrum to very low frequencies [60]. In general, this is not achievable by simply changing the temperature and using time-temperature superposition, which is possibly a consequence of the particle pair interactions. Other techniques at constant temperature have been used and proven successful; they include stress relaxation, creep, and time-concentration superposition [136–138].

In stars, hairs can interdigitate and provide an additional contribution to energy dissipation. This effect can be evaluated in Figure 6.10 by comparing the ratios between the storage modulus and the loss modulus, G_0'/G_0'', at the frequencies where the loss modulus exhibits a minimum. Note that the relative concentrations were selected to yield nearly identical storage moduli G_0'. The ratio is larger than 10 for microgel and micellar suspensions, whereas it is only about 3 for the star solution, signaling an enhanced dissipation. These particular micelles consist of branched arms, which significantly reduce interdigitation [139]; hence, they are akin to microgels but with different particle elasticity. To achieve values of the storage modulus similar to that of stars, microgels must have larger effective volume fractions. The Mode Coupling Theory (see Chapter 2) can reasonably well describe the linear viscoelastic spectrum (plateau modulus and one relaxation mode) for soft particle glasses by using the same parameters as for hard spheres [136]. This is discussed in Chapter 5 for hard spheres and in Section 6.6 for soft spheres.

Particle softness is also reflected in the dependence of the plateau modulus on the effective volume fraction, which becomes stronger as the hard sphere limit is approached [9,15,16,140]. The glass-to-jamming transition can be identified by a clear change in the concentration dependence of the plateau modulus, as seen in Figure 6.11 for star polymers and microgels [23,82]. Given that the scaling of the modulus with $k_B T/R^3$ is expected to hold in the entropic glass regime but not in the

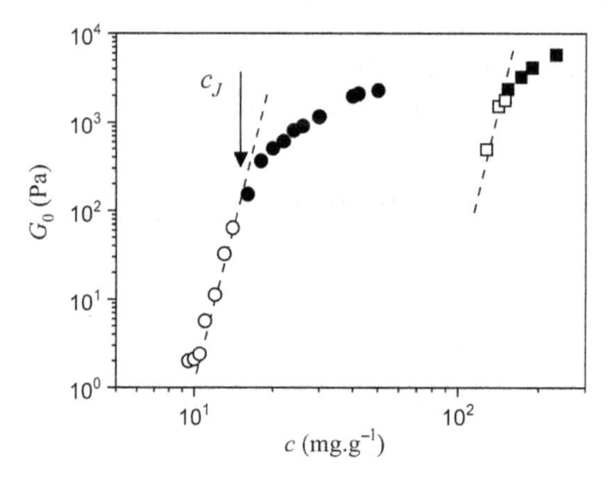

Figure 6.11 Concentration dependence of the plateau modulus of microgels, from Pellet and Cloitre [82] (○, ●), and star polymers, from Erwin et al. [23] (□, ■). For microgels, c_J marks the transition from the entropic glass (○) to the jammed glass regime (●). The dashed lines are drawn to guide the eye. For stars, the respective transition is not yet established, even though the data are suggestive. The dashed lines are drawn to guide the eye.

jammed glass regime, and considering that the actual volume fraction remains an ambiguous quantity for soft spheres, the data are presented in terms of actual modulus versus mass fraction. In this representation, the data indicate a linear dependence of the modulus on concentration in the jammed glass regime, which is weaker than in the entropic glass regime [82]. Such a linear dependence of the modulus has also been observed is other jammed microgels [14,91] and emulsions [90]. For the particular star solution shown, the moduli at low concentration seem to follow the same dependence as microgels in the entropic glass regime [23]. Whereas these findings appear to be in agreement with microscopic theoretical predictions suggesting a soft jamming threshold [89,141], the jammed glass regime in star glasses is not yet fully understood.

Jammed soft suspensions exhibit quantitative differences in their viscoelastic properties in comparison with entropic glasses because of the presence of repulsive elastic interactions which exceed the thermal forces. This has been studied for microgel suspensions [22,82,91,92], concentrated emulsions [90,142] ,and star polymers [141]. The plateau modulus is found to scale with the contact modulus E^* of the particle, see Eq. (6.1), and varies linearly with the distance to the jamming transition: $G_0 \propto E^*(\phi - \phi_J)$. At high frequencies, the loss moduli of many soft systems like emulsions and microgels increase as $G'' \propto \omega^{0.5}$ [142]. By treating microgel particles as elastic non-Brownian spheres dispersed in a suspending medium at high volume fractions, interacting through a Hertzian potential and with elastohydrodynamic lubrication forces, it is possible to predict their pair correlation function, high frequency moduli, and the osmotic modulus by combining particle-scale simulations and mean-field theory [140,143]. The predictions capture the main experimental features without adjustable parameters when the moduli are scaled by the contact modulus of the particles and the frequency by the characteristic time η_m/E^*. Note that the characteristic time used to scale the shear rate in Figure 6.4 uses the suspension plateau modulus G_0 instead of E^* (see also Section 6.6).

6.5.3 Temperature-Induced Effects

The quality of the suspending medium can be tailored by changing the temperature and can have a dramatic effect on the rheology of soft suspensions because of the imparted volume change. A characteristic example is discussed in Kapnistos et al. [144], where concentrated solutions of star polymers and block copolymer micelles in a solvent of intermediate quality (between theta and good) were shown to undergo a counterintuitive reversible liquid-to-solid transition upon heating (see Figure 6.12). The physical origin of this unexpected behavior is the swelling of the particles on heating and the corresponding increase of their effective volume fraction. On the basis of cluster formation, observed by means of dynamic light scattering, it has been argued that the induced solid is a gel. Note, however, that this is a dense system with repulsive interactions, so it is akin to a colloidal glass. The phenomenon is observed only if the arms are sufficiently long (in the excluded volume region [33]) to experience non-negligible swelling on heating. In more recent work with solvent-free colloids bearing similarities to star polymer solutions, a similar phenomenon has been

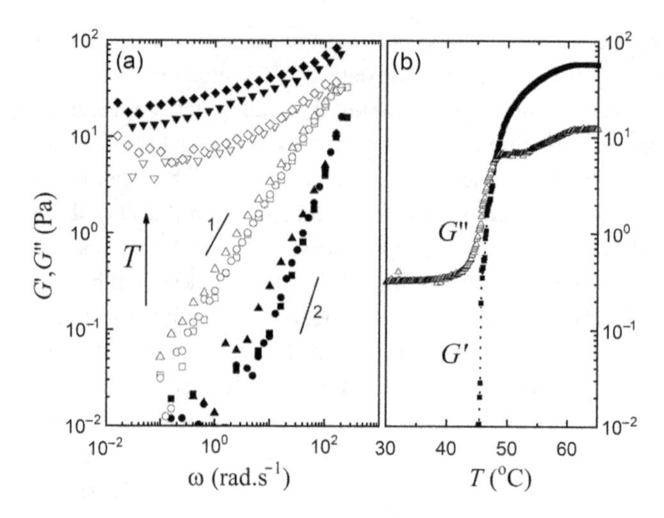

Figure 6.12 (a) Linear storage (solid symbols) and loss moduli (open symbols) versus frequency for a concentrated solution of a colloidal polybutadiene star (with 128 arms and 80,000 g/mol per arm) in decane (4.6% wt or about 13.5c*); ■: 20°C; ● ○: 25°C; ▲△: 30°C; ▼▽: 33°C; ◆◇: 35°C. (b) Dynamic temperature ramp data (G':■; G'':□) for a similar star system (128 arms with 56,000 g/mol per arm, at 5 wt% or about 11c* in decane) at an angular frequency of 1 rad/s and a heating rate of 1°C/min. The data are from Kapnistos et al. [144]

reported, which was called by the authors "thermal jamming". It was attributed to a change of conformation of the interpenetrating tethered chains in order to fill the interparticle space uniformly. The linear viscoelastic spectrum of the jammed solid was captured by the soft glassy rheology model, as discussed in Section 6.6 [145].

More complicated phenomena induced by temperature changes have been reported. Block copolymer micelles in selective solvents, resembling star polymers, undergo thermoreversible gelation on cooling in the vicinity of the order–disorder transition temperature due to enhanced attractions associated with reduced solvent quality [146]. In other situations the soft particle size decreases on heating and a lower critical solution temperature (LCST) is ultimately reached, beyond which phase separation takes place. This is the case for PNIPAm microgels which exhibit LCST in water at about 32°C. Concentrated PNIPAm suspensions reversibly vitrify on cooling, remain viscoelastic liquids at ambient temperatures, and gel on heating beyond the LCST. The structural and rheological signatures are distinct, whereas the extent and kinetics of the transitions as well as the strength of glasses and gels depend on details of the particle microstructure, especially the crosslink density [147,148].

The thermosensitive properties of PNIPAm microgels have been exploited in order to assess the aging response of colloidal glasses subject to both preshear and concentration perturbations induced by temperature jumps in the vicinity of the glass transition [149,150]. The use of sequential creep experiments allows probing the aging of the colloidal glass which is found to exhibit different kinetics compared to a molecular glass, indicating that the structural changes induced by the preshear are different from those induced by the temperature perturbation. The enhancement of the structural and rheological times at the approach to glass transition is also different. Hence, these systems are promising paradigms for exploring potential universalities in glass transition behavior.

6.6 Flow Properties of Soft Particle Suspensions

6.6.1 Phenomenology of Yielding and Flow

Colloidal suspensions of soft particles are known to exhibit remarkable nonlinear shear rheology. At high volume fractions, they behave as typical yield stress fluids that respond either as an elastic solid when the applied stress is lower than a stress termed the yield stress, or as a viscoplastic fluid when a stress greater than the yield value of the material is applied. A comprehensive discussion of the general phenomenology of yielding and flow in these materials is necessary in order to distinguish suspensions with purely repulsive interactions from suspensions where particles experience some degree of attraction and thixotropy plays an important role (see also *CSR*). In both cases, the steady state flow curves represent a remarkable dynamical fingerprint of the systems under investigation.

Let us first consider suspensions with repulsive interactions such as microgel suspensions [22,82,92,151,152], concentrated emulsions [153], star polymer suspensions [23], and multilamellar vesicles [154], where the solid-like behavior arises from the repulsive interactions between the swollen microgels, droplets, star polymers, and vesicles, respectively. Typical flow curves of polyelectrolyte microgel suspensions measured over a wide range of concentration are presented in Figure 6.13 [82]. At low

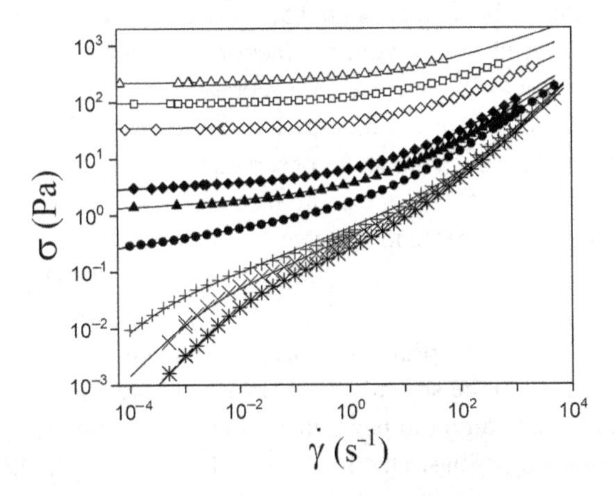

Figure 6.13 Flow curves of microgel suspensions with increasing volume fraction (from bottom to top) in the viscous suspension regime (crosses, $c < 12$ mg/g), entropic glass regime (filled symbols, 12 mg/g $< c <$ 16 mg/g), and jammed regime (open symbols, $16 < c < 50$ mg/g). Continuous lines are fits of the experimental data to the Carreau–Yasuda Eq. (6.7) for viscous suspensions, generalized Herschel–Bulkley Eq. (6.8) for entropic glasses, and Herschel–Bulkley Eq. (6.9) for jammed glasses. The data are from Pellet and Cloitre [82]

volume fractions, the flow curves exhibit Newtonian branches at low and (very) high shear rates, separated by a power-law region at intermediate rates, a shape which is well-represented by the Carreau–Yasuda equation:

$$\sigma(\dot{\gamma}) = \dot{\gamma}\left(\eta_\infty + \frac{\eta_0 - \eta_\infty}{\left(1 + (\dot{\gamma}\tau)^\alpha\right)^\beta}\right), \tag{6.7}$$

where τ is the longest relaxation time and α and β are numerical exponents. As the volume fraction increases, the low shear Newtonian region shifts to very low shear rates due to the huge slowing-down of the dynamics, up to the point where it drops outside the experimental window. The stress continuously decreases at low shear rates but does not vanish, signaling the emergence of a yield stress. When the concentration is further increased, the suspension goes through the glass and jamming transitions with clear qualitative difference between the entropic glass and the jammed glass regimes. In the former, the stress approaches the yield stress value σ_y' as a slowly decreasing plateau well described by a power law. This variation is followed by a second steeper power law variation at high shear rates. The global flow curve is well-represented by a generalized Herschel–Bulkley equation:

$$\sigma = \sigma_y' + k'\dot{\gamma}^p + k\dot{\gamma}^n, \tag{6.8}$$

where the exponents p and n both vary with volume fraction. In the jammed glass regime, the approach of the yield stress σ_y in the low shear regime is characterized by a true plateau ($p \cong 0$) and the stress follows the Herschel–Bulkley equation:

$$\sigma = \sigma_y + k\dot{\gamma}^n. \tag{6.9}$$

The shear thinning exponent n is often close to 0.5, in contrast to the values found in the glass regime which depend on the volume fraction [7,22,82,92,152–154]. The characteristic features shown in Figure 6.13 are representative of the behavior of many entropic and jammed glasses, although the succession of the glass and jamming regimes is not always systematically observed when the softness is varied (see also Section 6.5). This is especially apparent in Carbopol suspensions where the flow properties are sensitive to pH and concentration variations of the polymer mixture, resulting in substantial variations of the Herschel–Bulkley exponent [155,156].

Attractive interactions lead to the formation of a space spanning network, or even a gel, which is able to sustain a finite stress but rearranges and breaks when the stress exceeds a threshold value. In addition to being thixotropic, these materials exhibit an interesting flow phenomenology, illustrated in Figure 6.14(a), where the flow curves of a pure emulsion and the same emulsion loaded with clay particles responsible for attractive interactions are compared [19]. At high shear rates the network of droplets is totally disrupted and the flow curve is superimposed on the Herschel–Bulkley flow curve of the pure emulsion. However, as the shear rate is progressively decreased, the attractive suspension abruptly stops flowing, which results in a sharp discontinuity in the flow curve at a critical rate $\dot{\gamma}_c$ and stress σ_c. Due to the thixotropic nature of the

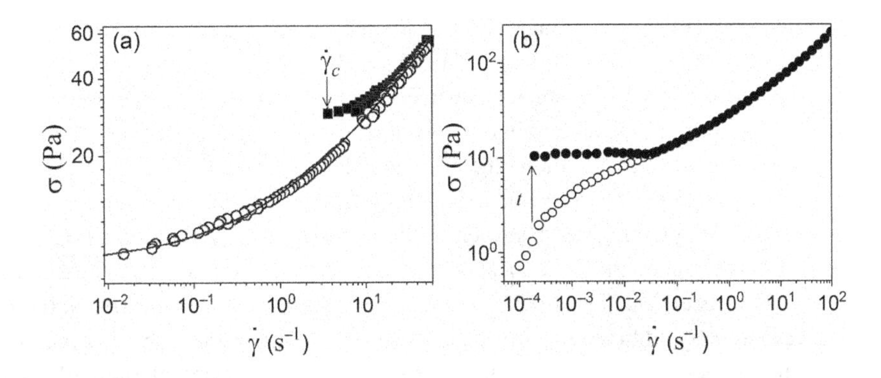

Figure 6.14 (a) Flow curves of a repulsive emulsion (open symbols) and a thixotropic emulsion (filled symbols). The line is the best fit of the Herschel–Bulkley equation to the data. After Ragouilliaux et al. [19]; (b) Flow curves of star polymer solutions after preparation (open circles) and after long aging (filled circles), indicated by the elapsed time t with arrow. Data are from Rogers et al. [157]

attractive emulsion, the behavior observed at and below σ_c is strongly time-dependent. Usually, the mechanical history is parameterized by the time elapsed after complete fluidization upon imposing a strong flow (see Section 6.7.1). Immediately after preshear, an apparent flow is detected for stresses lower than σ_c, but the shear rate progressively vanishes until a discontinuity finally appears at σ_c. In Figure 6.14(b), a different apparent flow curve is observed under controlled shear rate for a glassy star solution, which is weakly thixotropic and also solidifies after pre-shear in the course of time [157]. Here, the weak thixotropy is associated with aging and mutual interpenetration of the stars with long arms (see also Sections 6.4 and 6.7). The flow curves measured at high shear rates immediately after pre-shear and after long aging are superimposed, indicating that they are representative of the fully fluidized state of the material. However, at low shear rates the flow curve of the aged suspension reaches a true plateau. In Section 6.7.6, we shall see that this low shear plateau is associated with nonhomogeneous, shear banded flow.

Different approaches to determine experimentally the yield stress are discussed in *CSR* and Chapter 5. Recently, Large Amplitude Oscillatory Shear (LAOS) has emerged as a powerful and popular tool for inducing and studying rheological transitions associated with structural modifications, such as yielding, melting, or crystallization in colloidal glasses. There is ample literature on the topic, still evolving as reviewed in Hyun et al. [158] and Rogers [159]. The presence of attractions, which is often manifested in colloid-polymer mixtures as gels or attractive glasses, has been linked to the so-called two-step yielding process [160] (details are discussed in Chapter 5). Softness can also lead to two-step yielding, albeit in repulsive glasses, as has been demonstrated in the case of star polymers [136]. The interpenetration of arms and local stretching at high frequencies is thought to be responsible for a second yielding at high strain values, in addition to cage break-up that leads to partial opening at lower yield strains. This is different from the coupling of Brownian motion and hydrodynamics, which may give rise to double yielding at intermediate frequencies in repulsive hard sphere colloidal glasses [161].

Whereas LAOS has been extensively used, it suffers from several limitations. First it is important to keep in mind that the thus-extracted yield stress and strain depend on the frequency. In addition, the fact that the imposed flow history is not constant restricts interpretation to phenomenological approaches. One example is the link of oscillatory yielding to a sequence of processes and the introduction of the cage modulus concept, i.e., the local derivative of stress with respect to strain at zero stress [162]. Recent advances have opened the route for a better understanding of LAOS response by linking the rheological signal to the particle pair distribution function during the oscillation. This has been achieved for microgel pastes by combining rheology and simulations based on the micromechanical model discussed in Section 6.6.4 [163] and for triblock copolymer micelles by inducing crystallization by LAOS and studying the structural evolution in-situ via small angle neutron scattering [113].

6.6.2 Shear Thinning Behavior of Liquid Suspensions

In Brownian hard sphere suspensions, shear thinning occurs when the thermodynamic limit is exceeded and a shear-induced out-of-equilibrium structure builds-up. The onset of shear thinning is controlled by a critical Péclet number (see Chapter 1 and *CSR*). In soft particle suspensions, particle deformation brings an additional contribution to shear thinning [27], for instance, polymeric particles such as star polymers or block copolymer micelles are highly deformable under shear flow. For very dilute suspensions, it is useful to use the Weissenberg–Zimm number as the control parameter: $Wi_Z = \dot{\gamma}\tau_Z$, where τ_Z is the longest relaxion time of a single particle, i.e., the Zimm time. In rheo-SANS experiments or Molecular Dynamics simulations (see Chapter 3) it is observed that a single particle (vesicle or micelle or star) extends in the flow direction when $Wi_Z > 1$ (up to a factor of 4 from its equilibrium size for $Wi_Z = 100$), and shrinks in the other directions (less than factor of 3 for $Wi_Z = 100$) [27,164,165]. At higher concentrations, a weaker deformation occurs at high shear rates.

It is interesting to compare the shear thinning behavior of star and linear polymer solutions [27]. In the latter case, the strong deformation of the macromolecules leads to pronounced shear thinning: the viscosity decreases with the applied shear rate as a power law, $\eta \sim \eta_0\dot{\gamma}^{-n}$ with the exponent n ranging from 0.4 to 0.85, depending on the concentration, polydispersity, and branching. Available experimental and simulation data for different star polymers show that the shear thinning exponents are in the range between 0.4 and 0.5 with little variation when the particle softness is varied. The fact that multiarm star polymers in good solvents, with long arms and high functionalities, are much less shear thinning than entangled polymers conforms to their relatively weak interpenetration [23] due to strong excluded volume interactions between the outer blobs. There is a strong correlation between the material topology, which is dominated by excluded volume or by entanglements, and the dissipation rate under steady shear flow. In this respect, the role of softness in shear thinning is quite subtle.

Particulate colloids like droplets, vesicles, microgels, and capsules also have a complex dynamical behavior under flow, which strongly impacts their shear thinning rheology in dilute regimes. In general, different types of motion are possible, i.e., tank-treading, tumbling, and trembling, depending on the shear rate, elastic shear modulus, initial particle shape, and internal microstructure [27,166]. Particles such as microgels, which can be represented by initially spherical elastic particles, constitute the simplest situation [167]. Their stable steady state flow is characterized by the balance between the viscous force exerted by the fluid and the elastic force on the particle. As a result, the rheological properties of even very dilute suspensions of such soft particles generally exhibit shear thinning behavior at large shear rates. Each class of soft particle systems exhibits specific dynamical phase diagrams and rheology, which have to be investigated independently [27].

6.6.3 Yielding and Flow of Repulsive Entropic Glasses

The glass transition marks the divergence of the zero shear viscosity as discussed in Section 6.5. At the glass transition the dynamics of the suspension becomes nonergodic at rest. The entropic glass regime has been extensively studied in concentrated suspensions of thermosensitive core-shell particles with a soft PNIPAm corona [151]. In the entropic glass regime, the flow curves are conveniently represented in dimensionless coordinates as: $\sigma k_B T/R^3 = f(Pe)$ where $k_B T/R^3$ represents the characteristic scale of the thermal stress (see Chapter 1 and *CSR*). These suspensions constitute excellent model systems for comparing their glassy rheology against a generalized Mode Coupling Theory (MCT) for flowing suspensions [151], see also Chapter 2. This theory explains the rheology of glassy suspensions based on the competition between two opposing effects: on the one hand the local caging of particles, which is responsible for the slow structural relaxation, and on the other hand shear advection of fluctuations which accelerates relaxations [168,169] (see also Chapter 2). Flow leads to a glass-to-fluid transition and a unique stationary state can be reached even if the volume fraction of the particles is larger than the glass transition. This extension of MCT accounts for the existence of the yield stress and provides quantitative predictions of the flow curves at low shear rates. MCT is also capable of describing many other rheological responses including the linear viscoelastic behavior of concentrated dispersions [136,151,170], shear thinning [171], creep [172], and residual stresses [173]. Therefore, in the entropic regime, the effects of softness and deformability can be effectively lumped into the structure factor, at least for low shear rates.

It is interesting to note that some PNIPAm microgel suspensions and concentrated emulsions switch directly from the viscous regime to a nonergodic regime, identified as the jammed state, without apparently crossing this entropic glass regime [89,152,153]. The behavior of star polymers, characterized by an ultrasoft interaction potential (Eq. 6.2), is still more puzzling [23]. The low shear plateau is preceded by a terminal regime at very low rates, indicating that there is an apparent yield stress at intermediate time scales but that the stress ultimately relaxes at low frequencies. This specific behavior has been associated to the extreme softness of star polymers which are able to deform, disengage their arms, and exchange their position though thermally activated processes. It is believed to signify ultrasoft particle behavior in the glassy state whose alpha relaxation is detectable (see also Section 6.2).

6.6.4 Yielding and Flow of Repulsive Jammed Glasses

The jamming transition is a rigidity transition distinct in nature from the glass transition. In the jammed regime, thermal motion essentially plays a negligible role even for otherwise Brownian particles, and most of the properties are associated with a network of elastic contacts [174]. This explains why materials as different as microgel suspensions [22,82,152], concentrated emulsions and foams [153], dense packings of multilamellar vesicles [154], and many other complex materials [175] share generic yielding properties. The yielding process is gradual and characterized by heterogeneous dynamics [176]. In the jammed regime, the flow curves collapse onto master curves with the same Herschel–Bulkley power law exponent (0.5), as discussed in Section 6.2. In that case the shear rate is rescaled by a characteristic time that involves the viscosity of the suspending medium and the plateau modulus, i.e., η_m/G_0, and the stress by the yield stress that can be expressed as the product of the plateau modulus and the yield strain: $\sigma_y = G_0\gamma_y$ [6,7,22,177] (see also Section 6.2). Near the jamming transition the scaling variables are function of $\phi - \phi_J$, the distance from the volume fraction to the jamming point [92,152,153]. On the theoretical side, the rheology of jammed materials is a very active field of research with a wide diversity of approaches, which are summarized here and reviewed in Voigtmann [178].

The Soft Glassy Rheology Model

The phenomenological Soft Glassy Rheology (SGR) model successfully predicts the linear and nonlinear rheology of nonergodic soft materials by considering the activated hopping dynamics of mesoscopic elements in a free-energy landscape. The density of energy barriers at rest has an exponential form: $\rho(E) \propto \exp(-E/x)$ [179,180]. The parameter x plays the role of an effective temperature. The shear rate is assumed to be spatially homogeneous, but each element is characterized by a local strain ℓ and stress σ_ℓ, which are related by an elastic constant k so that $\sigma_\ell = k\ell$. The state of the material is described by the probability of finding an element at time t with a yield energy E and strain ℓ, $P(E, \ell, t)$. The elastic energy locally stored inside the material decreases the energy of the barrier by a quantity $k\ell^2/2$, causing rearrangements and a redistribution of local stresses at a rate $\Gamma_0 \exp\left(-(E - k\ell^2/2)/x\right)$. The evolution of the macroscopic stress is governed by the set of equations:

$$\sigma(t) = k\int \ell P(E, \ell, t)d\ell dE \tag{6.10a}$$

$$\frac{\partial P(E, \ell, t)}{\partial t} = -\dot{\gamma}\frac{\partial P(E, \ell, t)}{\partial \ell} - \Gamma_0 P(E, \ell, t)e^{-(E-k\ell^2/2)/x} + \Gamma(t)\rho(E)\delta(\ell) \tag{6.10b}$$

$$\Gamma(t) = \Gamma_0\int P(E, \ell, t)e^{-(E-k\ell^2/2)/x}d\ell dE \tag{6.10c}$$

Eq. (6.10a) of the macroscopic stress involves an integral of the local stress over the probability distribution $P(E, l, t)$. This probability function evolves in time according to the differential Eq. (6.10b), which involves advection of the elements between hops, interaction-activated yielding at the local rate $\Gamma_0 \exp\left(-\left(E - k\ell^2/2\right)/x\right)$, and the reformation of elements after yielding (δ is the Dirac function). The term corresponding to the latter process involves the total yielding rate given by Eq. (6.10c). By varying the effective temperature parameter, different rheological signatures are obtained: fluid-like for $x > 2$; glassy for $x < 1$; power law fluid for $1 < x < 2$. The SGR model has been remarkably successful in reproducing the macroscopic yield and flow dynamics of jammed materials in steady, oscillatory, and transient flows [180,181] as well as shear banding [182], aging, and thixotropy [183–185].

Fluidity Models

Fluidity models start from a generalization of the Maxwell model for steady shear, where the stress evolves under an imposed shear rate $\dot{\gamma}$ according to [186,187]:

$$\frac{d\sigma}{dt} = G_0\dot{\gamma} - D(t)\sigma \tag{6.11a}$$

$$\frac{dD}{dt} = rD^{\alpha} - \nu D^{\alpha+\beta} + f(\sigma, D\dot{\gamma}) \tag{6.11b}$$

Eq. (6.11a) is essentially a Maxwell equation where the coefficient $D(t)$, termed fluidity, represents the relaxation time of the stress, which is finite in the fluid phase and tends to infinity in the jammed phase. The fluidity can alternatively be connected to the rate of plastic rearrangements or to a local viscosity. The exact form of the fluidity results from the competition between aging, which decreases the fluidity as time passes, and flow-induced rearrangements, which increase the fluidity. The time evolution of the fluidity is given by the dynamical Eq. (6.11b). The constant ν is positive; the parameter r depends on $\phi - \phi_c$, ϕ_c being the volume fraction where the suspensions solidify; α and β are two constants. The function $f(\sigma, D, \dot{\gamma})$ expresses the fact that the applied shear rate induces rearrangements in the materials and for simplicity it is assumed to be a power law function. Fluidity models are able to reproduce the linear and nonlinear rheology of soft particle suspensions [188], aging properties, and the response of thixotropic materials [187]. By taking into account the spatial variation of the fluidity, it is also possible to account for the occurrence of various spatial heterogeneities like shear banding [157,182,189]. When a diffusive term is added to the dynamical equation of the fluidity, the stress evolution in each point is related to the stress fluctuations occurring over the entire system and the model becomes nonlocal. Nonlocal models have been used to explain the specific rheology of concentrated emulsions in confined geometries when the confining walls trigger or inhibit rearrangements, leading to specific fluidity conditions at the boundaries, which in turn modify the bulk flow profiles [190].

Elastoplastic Models

In the SGR and fluidity models, rearrangements take place uniformly over the sheared materials. Actually in the athermal limit, nonaffine deformations occur through individual rearrangements which are localized in time and space. This is the premise of different models like the shear transformation zone models [191] and elastoplastic models [192–194]. Elastoplastic models are a prolific class of models that consider that the destabilization of a mesoscopic region is responsible for the emergence of long-ranged stress and strain fluctuations in the surrounding, which in turn cause a cascade of rearrangements via long range elastic interactions. The state of the material is characterized by the probability $P_i(\sigma, t)$ that a mesoscopic element i has a stress σ at time t. A general form of the differential equation governing the time evolution of $P_i(\sigma, t)$ is given by the kinetic elastoplastic model [194]:

$$\frac{\partial P_i(\sigma,t)}{\partial t} = -G_0\dot{\gamma}_i \frac{\partial P_i(\sigma,t)}{\partial \sigma} - \frac{H(|\sigma| - \sigma_c)}{\tau} P_i(\sigma,t) + D_i \frac{\partial^2 P_i(\sigma,t)}{\partial \sigma^2} + \Gamma_i(t)\delta(\sigma). \tag{6.12}$$

The terms in the right hand side of Eq. (6.12) correspond, respectively, to the elastic deformation caused by the local shear rate $\dot{\gamma}_i$, the local stress relaxation on a time scale τ after rearrangements (H is the Heaviside function and σ_c is a local yield stress), the long-range redistribution of the stress in the material (D_i is the stress diffusion coefficient), and the reformation of elements after yielding ($\Gamma_i(t)$ is the rate of plastic events). A nonlocal relationship relates the stress diffusion and the rate of plastic events. These models reproduce many features found in experiments and simulations [192–195]: the jamming transition, the existence of a yield stress, the shape of the flow curves, and the spatial cooperativity of confined flows. Elastoplastic models are also widely used to probe the relation between the statistics of local rearrangements and the athermal rheology of yield stress fluids [196,197]. Recent developments suggest that the nontrivial value of the Herschel–Bulkley exponent that characterizes the high shear rate behavior of flow curves is linked to the statistics of rearrangements and possible avalanches [198,199].

Micromechanical Models

Whereas the previous models adopt a purely statistical approach, irrespective of the nature of the interactions and material properties, micromechanical models provide a more direct link to the microstructure of the suspension. They are based on two main ingredients, the interparticle and hydrodynamic forces that exist at the particle scale, and the dynamic structure under flow. Most of these models are essentially 2D, which raises the question of whether they are representative of experiments [200–202]. A recent 3D model analyzes the flow of jammed suspensions of soft incompressible spheres by treating explicitly the flow dynamics at the contacts between the compressed particles [6,7,177]. The flow field is split into a far-field contribution created by the applied shear rate and a near-field contribution at the contact level. The total interaction between two particles involves a Hertzian repulsive force associated with the elastic contact between particles, which is coupled to a frictional drag force due to elastohydrodynamic lubrication inside the film of suspending medium separating the two particles. Using a combination of Particle Dynamics Simulations (see Chapter 3), theory and experiments, the model provides constitutive equations for the macroscopic rheological properties.

Figure 6.15(a) represents the pair correlation of a fully equilibrated packing at rest and under flow. During flow there is a clear angular distortion of the microstructure. First, there is accumulation of neighboring particles in the compressive upstream region and depletion in the extensional region. Second, particles are on average more compressed than at equilibrium. This establishes a connection between the yield stress and the elastic force barrier that particles have to overcome in order to slide on past each other [7]. The average force barrier has a Herschel–Bulkley form with the same shear thinning exponent of 0.5 as the macroscopic stress. This result makes a connection between particle scale properties and macroscopic rheology. Figure 6.15(b–d) depicts

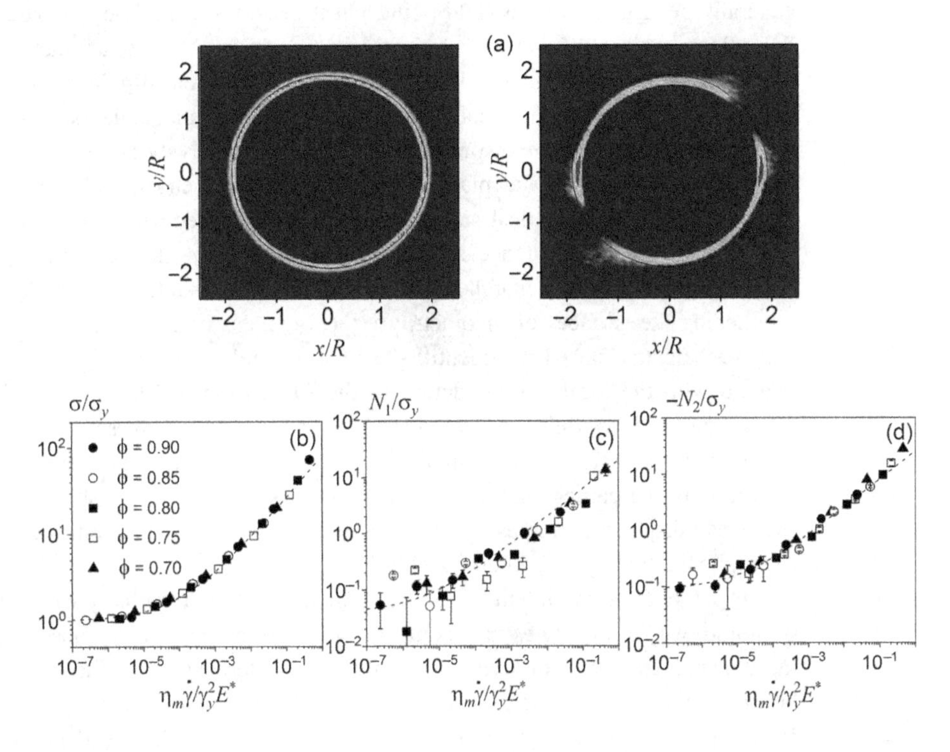

Figure 6.15 (a) Two-dimensional pair correlation function in the velocity–velocity gradient plane, computed for a jammed microgel suspension at rest (left) and under shear flow (right). The black line represents the mean center-to-center distance at rest (courtesy of R. T. Bonnecaze and L. Mohan). Bottom: Universal constitutive behavior for the shear stress (b), first (c), and second (d) normal stress differences from simulations of soft elastic spheres in steady shear at different volume fractions. The dashed lines are the best fits to the data using a micromechanical constitutive equation. Each symbol refers to a volume fraction: 0.70 (▲), 0.75 (□), 0.80 (■), 0.85 (○), 0.90 (●). The data are from Seth et al. [7]

the universal variations obtained for the shear and first and second normal stress differences as functions of the shear rate. The quantities σ, N_1, and N_2 are normalized with the yield stress and the shear rate is normalized with the ratio of the viscosity of the suspending medium over the particle's elastic modulus and square of the yield strain. They are well fitted by the Herschel–Bulkley equation. Recent particle dynamics simulations address the universal scaling of shear and normal forces and the Herschel–Bulkley exponent for soft deformable spheres in the glassy state [177].

6.6.5 Wall Slip

Wall slip represents an extreme situation of strain localization where most of the deformation occurs near the confining walls, while the bulk of the material behaves more or less like a solid body or exhibits negligible deformation. Below the yield point, concentrated suspensions of soft particles such as microgels and emulsions are prone to slip rather than flow homogeneously when sheared near smooth surfaces. Slip generally persists above the yield point, but it becomes negligible compared to bulk flow. Because the slip layer has a finite thickness of a few tens of nanometers or more, slip in concentrated dispersions is often classified as apparent slip, in contrast with the true slip of polymeric materials which takes place at molecular scales. The slip properties of soft particle suspensions have been extensively studied in rheometric shear flows using cone and plate or parallel plate geometries, and pressure driven flows [203]. On the practical side, the use of different geometries provides an easy way to examine the existence of slip: in the presence of slip, the shear properties depend on the gap in a parallel plate geometry (see chapter 9 of *CSR*), whereas roughening the surfaces of rheometric fixtures reduces or even eliminates slip. Two slip mechanisms have been identified and a general theory has been proposed in relation with the local microstructure of the suspension and the nature of particle-surface interactions. Below the yield stress, soft particles are pressed against the wall in Hertzian contacts by the osmotic pressure of the bulk suspension. Due to the proximity of the particle and the wall, the particles are sensitive to short-range forces such as dispersive forces, steric hindrance, electrostatic contributions, and hydrophobic–hydrophilic forces [204,205].

For purely repulsive interactions, the boundaries are preferentially wetted by a film of liquid which lubricates the contacts with the particles and causes slip. This mechanism, termed hydrodynamic slip, is characterized by a linear relationship between the slip velocity and a wall stress relationship $V_s = V^* (\sigma/\sigma_y)$. Hydrodynamic slip with a similar slip law has also been described in hard sphere suspensions [206]. Results for a dense emulsion are depicted in Figure 6.16.

For attractive particle-wall interactions, a mechanism specific to soft particles, which is based on elastohydrodynamic lubrication (EHL), is responsible for the formation of the lubricating layer that causes slip. In EHL, the deformation of the particles is coupled to the flow through the pressure field in the lubricating film so that the flat contacts existing between the particles and the surface are asymmetrically deformed. This breaks the reversibility of the Stokes equation and generates a lift force

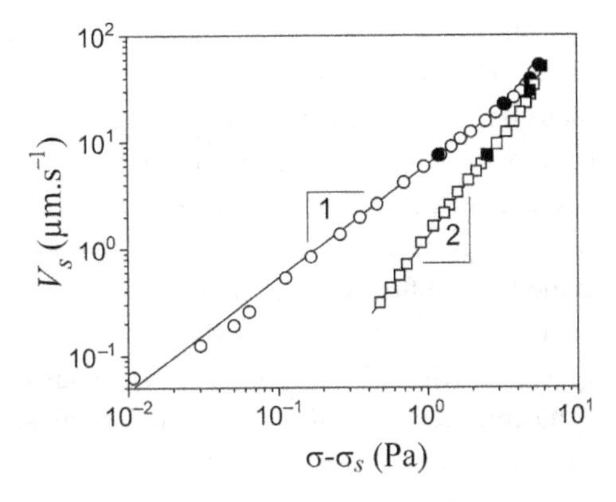

Figure 6.16 Variations of the slip velocity (from rheometry) of a concentrated emulsion sheared along a weakly adhering polymer surface (□, ■) and a repulsive glass surface (○, ●) against the excess stress $\sigma - \sigma_s$ where σ_s is the slip yield stress. Open symbols represent data deduced from rheological measurements; full symbols refer to data measured using particle-tracking velocimetry. All measurements are performed below the yield stress. Data are from Seth et al. [205]

pushing the particles away from the moving surfaces. The balance between the lift force, which depends on the slip velocity, and the bulk osmotic forces sets the thickness of the lubricated layer. The resulting slip law has a quadratic form, $V_s = V^* (\sigma/\sigma_y)^2$, as illustrated in Figure 6.16. In practice, EHL is only possible when the lift force counterbalances the net attractive forces, which requires that the stress exceeds a slip yield stress σ_s. When the attractive forces are large, the slip yield stress may exceed the bulk yield stress and slip is suppressed. It is also important to note that the characteristic velocity $V^* = G_0 R/\eta_m$ only depends on the plateau modulus of the suspension, the initial particle radius, and the viscosity of the suspending medium. Since the modulus is ultimately a function of the effective modulus of the particle and the volume fraction, we see that there is a direct relation between the softness of the particles and the amplitude of slip.

6.6.6 Shear Banding

Two distinct shear banding types of behavior have been observed in concentrated colloidal suspensions [207]. In the absence of wall slip, hard sphere glasses exhibit a particular type of shear banding due to mass transport induced by spatial gradients in the shear rate, the so-called shear gradient concentration coupling instability [208,209]. In essence, a similar coupling of stresses to inhomogeneous concentration profiles has been implemented in order to predict shear banding in entangled polymer solutions [210,211]. The increased concentration of colloids close to one of the shear cell walls results in a local yield stress which is larger than the actual stress, giving rise to an immobile band. This scenario requires that the yield stress is a strongly increasing function of the concentration, which is possible for hard sphere glasses with short-ranged repulsive interactions where the free volume is drastically reduced as the volume fraction increases. This shear banding instability is generally not observed in concentrated suspensions of soft particles with purely repulsive interactions like microgels, emulsions, or vesicles, which flow homogeneously. Aged star polymer solutions with ultrasoft interactions and mutual interpenetration constitute a notable exception, here shear banding is observed in controlled strain experiments at low shear rates in the plateau region of the flow curves, which differs from true yield-stress behavior. Often in such situations, one band is virtually immobile and its extend depends on aging, see also Figure 6.14(b) [157]. A proposed mechanism, known as gradient banding, calls for a mechanical instability associated with strong shear thinning which causes a sudden drop of the stress [137,212,213]. This scenario is known to produce banding in many complex fluids, such as wormlike micellar solutions, lyotropic lamellar phases, polymer solutions, polycrystalline colloids, and dispersions of platelets and rods [214]. However, the link of substantial viscosity thinning to microstructure is still unclear in soft suspensions. In stars, the local interpenetration of the outer blobs may be the clue. A model based of transient forces resulting from perturbations of overlapping polymers when they flow also predicts flow instabilities [215]. In principle, this model predicts shear gradient banding due to interpenetration (or entanglements), in agreement with Cromer and Villet [210]. Nonspherical effective particles such as glasses of sterically stabilized colloidal rods exhibit heterogeneous flow behavior due to the dynamic arrest of chiral-nematic orientational textures and their constituent rods [207].

Shear banding is found experimentally in a variety of yield stress materials such as gels, cement pastes, and emulsions [18,19,216], even in entangled solutions of associating polymers [217]. Again, it occurs at low shear rates and is associated with a stress plateau in the flow curves, whereas both steady state and transient banding have been reported. These systems are described as thixotropic yield stress materials, which points to the role of attractive interactions in the occurrence of shear banding. The structure of these materials is not always well-defined and the exact role of the attractive forces is still not well understood. On a more practical side, a simple phenomenological model has proposed that the onset of shear banding in jammed systems is set by the ratio between a characteristic relaxation time of the system and the restructuring time [218].

6.7 Slow Dynamics

6.7.1 Rheological Aging

Below the yield stress, the properties of concentrated suspensions of soft particles continuously vary in time and strongly depend on sample preparation and rheological history. While these observations have been attributed for a long time to experimental artifacts, there is now clear evidence that they exhibit all the hallmarks of aging phenomena commonly encountered in glassy systems. We refer the interested reader to Joshi and Petekidis [219] for a specialized review. Aging has been extensively studied in materials as diverse as suspensions of microgel particles [181,220,221], laponite suspensions [222–228], multilamellar vesicles [229,230], polymer grafted silica particles [187,231], charge-stabilized polystyrene particles [232], and Carbopol suspensions [233]. A general protocol for investigating rheological aging involves sample loading, rejuvenation, waiting time, and determination of the rheological properties. Rejuvenation is intended to place the material in a reproducible state and is achieved either mechanically or thermally. Mechanical rejuvenation, often called preshear, consists of applying a high shear stress or shear rate well above the yield point. Flow cessation marks the end of rejuvenation and sets the time origin $t = 0$, after which the material is kept at rest for a waiting time interval t_w, also called the age, before further measurements are performed. In thermal rejuvenation, a temperature jump causes a decrease of the volume fraction [149,150] or a change of structure, which drives the suspension into the liquid state. A second thermal quench causes the solidification of the material and sets the time origin.

Many jammed suspensions made of soft particles evolve in a way that is reminiscent of the physical aging observed in glassy polymers [234] or spin glasses [235]. In thermal glasses, a metastable state is reached when the material is brought below the glass transition temperature, and rejuvenation happens when the temperature is increased. In suspensions, rheological rejuvenation is achieved by applying a large stress and metastability appears upon flow cessation. The viscoelastic moduli depend on the time elapsed after flow cessation; the storage modulus $G'(t)$ increases in time while the loss modulus $G''(t)$ decreases, both quantities being well described by logarithmic functions, see Figure 6.17(a). A similar result is shown in Figure 6.17 (b) for the case of multilamellar vesicles, which are thermally rejuvenated and probed by stress relaxation experiments. Since the paste is still slowly evolving, mechanical equilibrium is not reached and the properties are a function of the waiting time t_w. Conventional rheological tests such as creep or start-up flow measurements have also been used to investigate aging. Figure 6.17(c) shows the creep response of a microgel paste subjected to a stress smaller than the yield stress applied at time t_w. The strain following the application of the stress depends on the time the material has been kept at rest. The longer the waiting time before applying the probe stress, the slower the overall response, the smaller the initial elastic jump, and the slower its creep. Plotting the strain response as a function of the variable $(t - t_w)/t_w^{\mu}$, with μ being a positive exponent, collapses all data onto a single master curve. Aging has other manifestations like those shown in Figure 6.17(d), which illustrates start-up experiments performed on suspensions of polymer-grafted silica particles. The initial stress overshoot, which is interpreted as a static yield stress, results from the competition between the initial elastic response and plastic deformation and becomes more pronounced as the waiting time increases because the material becomes stiffer.

The scaling variable $(t - t_w)/t_w^{\mu}$ was first proposed in the context of polymeric glasses [235]. It indicates that the system does not have any intrinsic characteristic time, but a variable one instead, which depends on the waiting time. A phenomenological interpretation is that, during aging, different spatial configurations with a broad

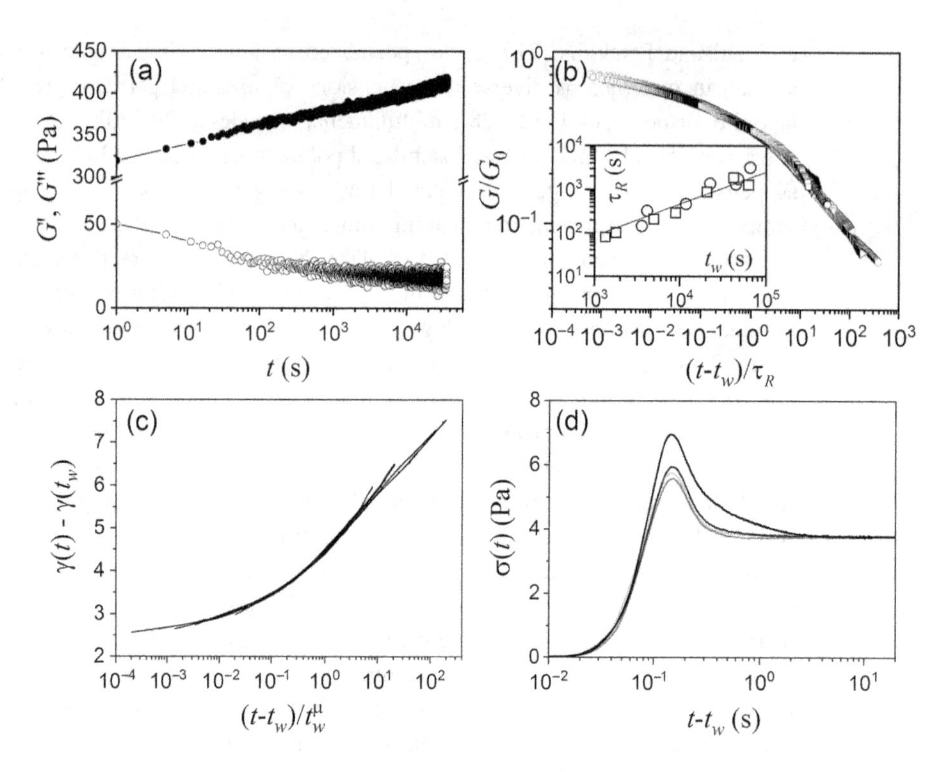

Figure 6.17 Typical rheological responses associated with aging. (a) Characteristic time evolution of the linear viscoelastic moduli (G' increases and G'' decreases) of a jammed microgel suspension after flow cessation. (b) Scaling of the linear relaxation moduli $G(t, t_w)$ of a lamellar gel measured by applying a constant strain at values of t_w ranging from 1.0 to 48.2 hours; the modulus is normalized by its value at $t = t_w$; the time is scaled by the characteristic time τ_R which increases with t_w as a power law: $\tau_R \propto t_w^{\mu}$ with $\mu = 0.78$ (inset). (c) Scaling of the strain response to a stress applied at values of t_w ranging from 15 to 10^4 s, for microgel suspensions. (d) Start-up flow experiments performed at values of t_w ranging from 10 to 10^4 s (from bottom to top) for silica particle suspensions grafted with polyethylene oxide. Data are from Cloitre et al. [220] (a, c), Ramos and Cipelletti [229] (b), and Derec et al. [187] (d)

distribution of relaxation times coexist in the material [220]. At time t_w, only those configurations with relaxation times smaller than t_w have relaxed, others have not but will eventually do later on. The value of the exponent μ is related to the value of the probe stress applied at $t > t_w$. Very often the effect of any applied stress is to partially rejuvenate the material so that μ decreases from 1 (full aging) to 0 (no aging) when the probe stress amplitude increases [220]. However, μ can be larger than 1, in particular when the material is allowed to age during the waiting period $0 < t < t_w$ under a moderate shear stress [232]. This situation is termed overaging and occurs because the stress applied during the waiting time overpopulates the long-time tail of the distribution of relaxation times. It is important to emphasize that the scaling variable $(t - t_w)/t_w^{\mu}$ may not be valid at very long times when the duration of a measurement exceeds the waiting time t_w, because then the material evolves while it is studied. Other forms based on the concept of effective time have been proposed to study the long-time behavior of aging suspensions [226,234].

This description has proved quite successful to rationalize rheological aging in jammed soft suspensions. On the theoretical side, it is nicely supported by predictions from the trap model for glasses [235], the soft glassy rheology model [183], or the fluidity model [187,188]. However, in spite of these remarkable advances, several issues remain open to investigations. A first class of questions concerns the apparent similarity existing between the aging properties of different colloidal systems and the nonequilibrium dynamics of molecular glasses. Detailed experiments on PNIPAm microgel suspensions in the glassy state have recently revealed that there exist qualitative differences between colloidal and molecular glasses [149,150]. The reasons for such differences remain to be elucidated. Secondly, it is unclear whether the slow dynamics of concentrated dispersions is universal. The role played by the particle microstructure is particularly challenging and not fully understood yet. In arrested suspensions of star polymers, it has been found that terminal relaxation can be reached at low frequencies and shear rates, which has been attributed to the ability of the internal modes of the stars to induce cage distortions and provoke cage escape [23]. Aging in these materials takes a different form [236]. It is associated with a slowing-down of terminal relaxation with the waiting time, and the mechanical response of the suspensions then scales with the total time t elapsed after rejuvenation instead of the waiting time t_w as in other particulate suspensions. In other star polymer suspensions with different functionalities it has been observed that aging takes place in two stages: a fast stage where steady state is reached, which characterizes the soft state of the material, and a slow stage during which the material ages, which characterizes the solid state of the material [137,237]. The pathway between the so-called soft and solid states can take different forms including shear banded states depending on the nature of the mechanical solicitation that is applied [137].

6.7.2 Microscopic Signatures of Aging

The local dynamics of out-of-equilibrium jammed suspensions has been characterized through the intermediate scattering function $f(q,t)$ measured using a variety of scattering techniques spanning different length scales, including dynamic light scattering [229,238], diffusing wave spectroscopy [232], and X-ray photon correlation spectroscopy [227]. In general, aging is associated with a generic two-step decay of the dynamic structure factor. A first decay at short times is due to relaxation of the scatterers by Brownian motion. This initial decay terminates in a plateau at the intermediate time domain where particle displacements are arrested. A second, ultra-slow decay leads to the complete relaxation of $f(q,t)$, which is due to rearrangements at length scales as large as several microns or tens of microns. For many systems, the final relaxation of the dynamic structure factor has the same very peculiar form shown in Figure 6.18 [229,230,238]:

$$f(q,t,t_w) \propto \exp\{-[(t-t_w)/\tau_L]^p\}, \tag{6.13}$$

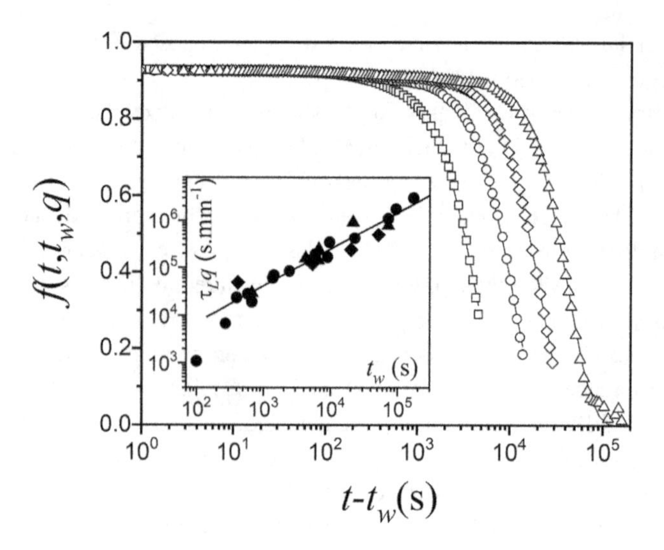

Figure 6.18 Dynamic structure factors of a lamellar gel at different ages (from right to left): $t_w = 7$ min, 1.5 h, 5.6 h, and 14.3 h ($q = 11.2\ \mu m^{-1}$). Solid lines are best fits to a stretched exponential function. The inset shows the age dependence of $q\tau_L$, where τ_L is obtained from DLS measurements at different values of the scattering wavevector: $q = 6$ (▲), 11.2 (♦), and 24.2 (●) μm^{-1}; the straight line is a power law fit with an exponent of 0.77. Data are from Ramos and Cipelletti [229]

where $p \cong 1.5$ and $\tau_L \propto t_w^{\mu}$, with μ being an exponent close to 1. Two important results are to be highlighted here: the final relaxation process exhibits a steeper than exponential decay, which is termed "compressed exponential", and the corresponding, characteristic time slows down with increasing waiting time. Although the relaxation times, which characterize rheological aging and local relaxations, are similar in shape, they differ by orders of magnitude since the two techniques probe microscopic motions over different length scales. Finally, the characteristic time τ_L also exhibits a remarkable scaling with the spatial wavevector q, suggesting that the average particle displacement grows linearly with time and moves ballistically [227,228,230].

The physical origin of these properties has been attributed to the build-up of internal stresses that drive local rearrangements. Recent simulations with solid networks (akin to colloidal gels) have revealed that stress heterogeneities frozen-in upon solidification can partially relax through elastically driven fluctuations. The latter are intermittent because of strong correlations that persist for a long time exceeding the timescale of observation in experiments or simulations, leading to faster than exponential decay [239]. There is ample experimental evidence confirming this suggested mechanism. In multilamellar vesicles, small temperature fluctuations induce intermittent local stresses due to expansion and contraction in the sample, which cause both reversible and irreversible rearrangements whose amplitude decreases with time, leading to an exponential slowing down of the dynamics with material age [240]. In jammed suspensions of soft particles, which are mechanically rejuvenated, cage defects originating from the initial preshear associated with rejuvenation seem to be the driving force for internal rearrangements [24]. In Laponite suspensions, it has been proposed that internal stresses responsible for local rearrangements are due to an increased repulsion between the particles (see also Section 6.7.3). At larger length scales and longer times, aging is associated with the existence of a growing dynamical heterogeneity whose typical length increases with waiting time [241]. This remains an open field of research.

6.7.3 Internal or Residual Stress after Flow Cessation

The internal stress problem characterizes amorphous materials, which are shear-rejuvenated in the liquid state where they are malleable and quenched to the solid state. Upon solidification, the structure does not instantaneously equilibrate and some of the stress existing in the liquid state can be trapped for long periods of time: this is the internal or residual stress [24,242]. In applications, internal stresses can adversely affect product performance but, if wisely controlled, they can be used to create advanced materials with improved properties. At a more fundamental level, the build-up of internal stress upon flow cessation has been invoked to explain the spontaneous slow dynamics observed in a variety of glassy and biological materials and has been connected with the aging phenomena discussed in Section 6.7.2. Whereas hard sphere glasses exhibit qualitatively similar behavior [173], the uniqueness of soft jammed materials is that they interact via elastic contacts at volume fractions exceeding random close packing. The qualitative trend of the two-step relaxation is shared by both hard and soft colloidal glasses, albeit with significant differences in the microscopic mechanisms involved and the physical signatures.

Figure 6.19 shows the stress relaxation following shear rejuvenation for a jammed microgel suspension. The relaxation is a two-step process comprising a fast decay

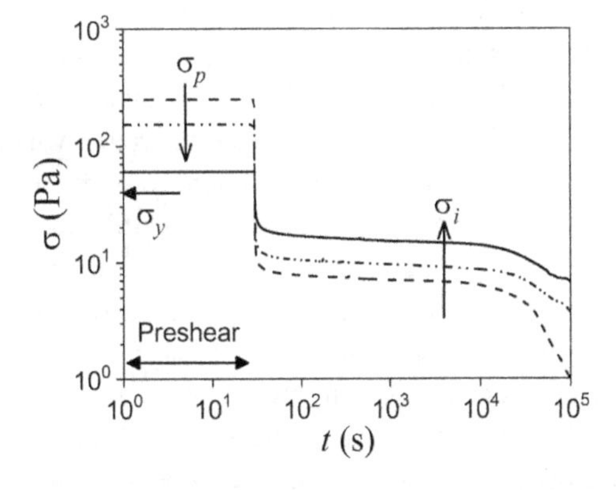

Figure 6.19 Time evolution after flow cessation of the residual stresses trapped in microgel suspensions. The residual stress σ_i is larger when the preshear stress σ_p is smaller. Data are from Mohan et al. [242]

immediately following flow cessation, with a nearly constant plateau at intermediate times, and a final decay at long times. The micromechanical model discussed in Section 6.6.4 provides useful information on the physical processes involved during stress relaxation [24,242]. The initial short-time relaxation corresponds to fast elastic recoil, during which the center-to-center distance between particles returns to its static value. It is not controlled by an intrinsic time like in conventional viscoelastic materials but by a characteristic time which involves the shear rate applied during rejuvenation and the storage modulus of the material (see also Section 6.7.1). Concomitantly, particles undergo ballistic motion over short distances with a velocity that decreases with the inverse of the time elapsed after flow cessation. This scenario distinguishes jammed suspensions of soft particles from entropic glasses near the glass transition where subdiffusive motion takes place [173]. After the initial relaxation, the particles become trapped in local cages formed by their neighbors and their subsequent motion takes place through local contact rearrangements. The particles move with a velocity which is inversely proportional to the time but independent of the pre-shearing conditions. This slow process corresponds to the return of the structure factor to isotropic state, after it has been distorted during pre-shear.

6.8 Mixtures and Osmotic Interactions

6.8.1 Soft Colloid-Polymer Mixtures

Athermal Systems

Soft interactions impart quantitative and qualitative differences in the behavior of colloid-polymer mixtures with or without Newtonian solvent [243]. In the classic mixture involving hard spheres, which has been discussed at length in Chapter 5, osmotic forces are responsible for attractive depletion interactions, inducing aggregation of particles at low volume fractions and a two-step transition from a repulsive glass to a liquid, to a re-entrant attractive glass at high volume fractions. To better appreciate the role of softness, let us consider the simplest case of three-component mixtures comprising a background solvent, a soft colloid, and a noninteracting linear polymer additive. When the colloid is a highly crosslinked microgel, which behaves very much like a hard sphere, the repulsive glass-to-liquid-to-re-entrant attractive glass or depletion flocculation route is again observed [244–246]. Softness can be tuned by adjusting the degree of crosslinking, which shifts the re-entrance boundary to larger polymer concentrations [247]. In general, the osmotic forces, which are inherent to the physics of entropic mixtures, act via two channels in soft colloids: volume reduction through deswelling and depletion. Slightly crosslinked (soft) microgels, which respond to external fields by both volume and shape adjustments (Section 6.4.3), deswell in the presence of small linear polymers. This occurs beyond a certain threshold pressure, i.e., a concentration of linear polymer set by the density of ions inside the microgel [86]. Depending on the relative size (and ionic strength in the case of charged particles) of polymer and microgel, some initial penetration of polymer into the microgel may occur, increasing its osmotic pressure and size before subsequent osmotic deswelling takes place at higher concentrations [87]. Such interplay of polymer chain penetration and volume change has been investigated for charge-free PMMA microgels of different crosslinking densities, in the presence of much smaller linear PMMA chains. The linear chains can penetrate the particles with low crosslink density, hence form homogeneous blends, whereas in the case of high crosslink density this is impossible and particle clusters are formed due to depletion [248].

Hairy particles respond to osmotic forces in a qualitatively similar manner. Taking the paradigm of star polymers, the osmotic pressure due to added linear homopolymers will shrink them, causing a reduction of volume fraction, and eventually lead to depletion interactions and phase separation [83,249–254]. The osmotic compression of the stars in this situation is discussed in Section 6.4.3. Note that in these situations the ratio of additive (here polymer) size to the overall particle size (core and corona) is the control parameter. The grafting density and size of the graft (here star arm) are accounted for in the interaction potential and star compression [249,254].

The transitions between states can be unambiguously diagnosed by means of rheology. An example is shown in Figure 6.20 where the linear viscoelastic moduli

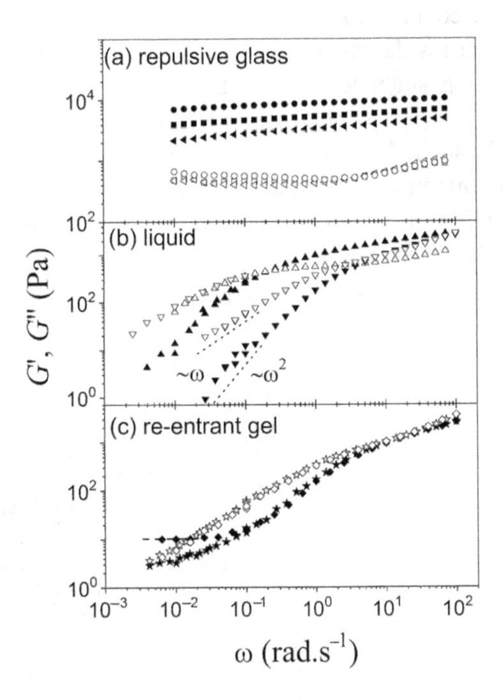

Figure 6.20 Linear viscoelastic spectra of a star-linear polymer mixture (polymer/star size ratio is 0.05) at constant fraction of stars and increasing fraction of added linear polymers (from top to bottom). The mixture evolves from a repulsive glass (a), with $\phi_L = 0$ (●,○), 0.21 (■,□) and 0.43 (◀,◁), to a liquid (b) with $\phi_L = 0.88$ (▲,△) and 1.78 (▼,▽), and to a re-entrant gel (c) with $\phi_L = 2.18$ (★☆) and 2.24 (♦,◊). Filled symbols refer to G' and open symbols to G''. Data are from Truzzolillo et al. [17]

are plotted against angular frequency for mixtures of glassy stars at constant fraction and much smaller linear polymers with increasing fraction [17]. The initial repulsive glass weakens upon addition of linear polymers, as evidenced by the decrease of the storage modulus G'. The glass melts and the liquid eventually becomes a re-entrant solid. Hence, the combination of osmotic compression and depletion attraction of the stars reduces the volume fraction, resulting in a breaking of the entropic cage and eventually formation of a soft solid akin to a depletion gel. Due to the range of the soft potential, the liquid pocket, which exists at intermediate volume fractions of linear chains, is larger for soft particles than for hard particles [17,247]. Another consequence of softness is that the range of polymer-to-colloid size ratio is much larger and reaches larger values than for the hard sphere analog [247,253]. Figure 6.21 illustrates qualitatively the different states in such a star-linear polymer mixture in terms of polymer versus star fraction, Figure 6.21(a), as well as polymer fraction versus polymer-to-colloid size ratio, Figure 6.21(b), with the initial star fraction being in the glassy regime. The rich nonlinear rheology of the different states reflects several of the hallmarks discussed in Section 6.6, and the interested reader is referred to Vlassopoulos and Cloitre [9]. For large polymer-to-star size ratios (of about 1) the above picture changes and on increasing linear polymer concentration in a glassy star suspension, a transition from confined-to-bulk dynamics is observed [255]. In the former case the star–star distance is shorter compared to the average distance between polymer chains, whereas the opposite holds in the latter case.

The impact of softness on the behavior of colloid-polymer mixtures has many implications. For example, the star polymer paradigm has been used to investigate the structure and dynamics of mixtures of polymer grafted nanoparticles and linear polymers in good solvents [256] or block copolymers and linear homopolymers in selective solvents [257]. In the latter case in particular, the homopolymers not only melt the crystalline phases formed by the block copolymer micelles but can also

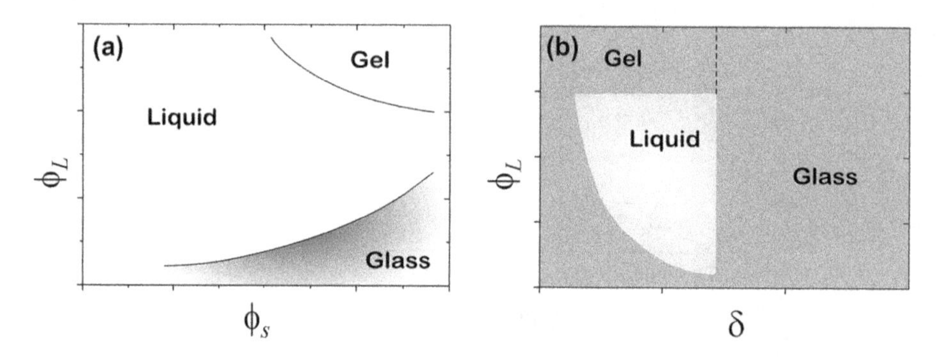

Figure 6.21 Schematic state diagram of soft colloid (star)-linear polymer mixtures, adapted from Vlassopoulos and Cloitre [9] and Truzzolillo et al. 17], with permission. (a) Representation in terms of volume fraction of linear additive, ϕ_L,versus volume fraction of star, ϕ_s, at constant (small) polymer–star size ratio. (b) For soft colloids in the glassy regime at constant fraction, the state behavior of the mixture varies with the volume fraction of the linear additive ϕ_L and the polymer/colloid size ratio δ.

influence their aggregation number, hence providing a control parameter for self-assembly [40,257].

Temperature Effects: Mixtures with Enthalpic Interactions

Besides the classic situation of depletion-induced attraction in colloid-polymer mixtures, effective attraction between two colloidal particles can be achieved by means of enthalpy minimization in a temperature-dependent solvent medium for the polymer. Such an avenue has been exploited for both hard and soft colloidal systems [258,259]. The addition of linear homopolymers to nondilute solutions of colloidal star polymers in nonideal solvents, which improve in quality on heating, can influence their liquid-to-solid transition dramatically [259]. For a given polymer-solvent system, the extent of the effect depends on the polymer-to-star size ratio. In general, if the polymers are too small, they may penetrate the stars and swell them (the linear polymer solution acts effectively as a solvent of better quality for the stars), while additional swelling on heating further promotes this effect. The increased effective volume fraction of the swollen stars facilitates their thermal gelation (the critical temperature for this transition drops compared to that in the absence of small polymers). On the other hand, when the size ratio prohibits penetration of linear polymers into stars, the gelation transition is determined by the balance of three factors: osmotic shrinkage, enthalpic swelling, and depletion [259]. Similar phenomena have been observed in mixtures of microgels involving thermosensitive PNIPAm in water [245,260]. Below its LCST, PNIPAm is in a good-solvent environment [261] and exerts an osmotic pressure on the microgel particles which deswell. Above LCST, the polymer is collapsed and behaves as a small hard particle dispersed among the microgel particles which re-swell. These changes are reflected on the structure and rheology of the mixtures and the formation of binary gels. An intriguing thermoreversible melting of depletion gels from mixtures of star and linear polymers in a good solvent has been reported [17,262]. It is associated with the changing internal conformation of the stars upon heating.

6.8.2 Binary Colloidal Mixtures Involving Soft Particles

Soft–Soft Colloidal Mixtures

By analogy with binary hard colloidal mixtures (see Chapter 5), the entropy of mixing soft colloids of different sizes is at the origin of their phase and state behavior. Mixing stars with different functionalities, hence different softness, and sizes leads to the formation of different types of arrested states. For small size ratios, a single component is kinetically frozen and a single glass is obtained. With increasing size ratio, both particle populations become trapped in a double-glass regime, as in hard sphere mixtures. By adding more and more small stars to the double-glass, an asymmetric glass is eventually formed, which is characterized by anisotropic cages with very rich rheology [263–265]. Figure 6.22(a) illustrates the resulting state diagram.

Mixtures involving block copolymer micelles with frozen and "living" cores and short arms have been studied, and a balance between crystallization and vitrification, resulting from an optimum adjustment of the number of arms, has been reported [37]. Mixing microgels of different softness and size at high fractions has also been shown to have a huge impact on the properties of the resulting composite. Soft and small microgels exhibit a jamming transition at volume fractions larger than for stiffer ones. Concomitantly the dependence of the plateau modulus on the particle fraction is much stronger for stiffer particles. Hence, mixtures of soft and stiff particles enable modulating the elasticity and jamming transition and offer a means to tailor the rheological properties of soft composites [266].

Soft–Hard Colloidal Mixtures

This class of mixtures with hard particles acting as depleted or depletants offers new interesting possibilities to tailor rheology since the combination of particle interactions (soft–soft, hard–hard, and soft–hard) is unique. We examine here both cases, with emphasis on the role of hard spheres as depletants. Let us start with the classic hard colloid-polymer mixture, and replace linear polymers by small soft particles. Mixtures of small star polymers of low functionality (up to 32) and large hard colloids have been found to phase separate for size ratios not exceeding 0.5 and colloid volume fractions below 0.5 [269]. When small microgel particles are added to a liquid suspension of large hard spheres, depletion gels are formed exhibiting similarities with, but also differences from hard colloid/polymer or binary hard colloidal mixtures [270]. If the microgel is thermoreversible like PNIPAm, additional tuning of the system is possible through selective attractions. For example, in the temperature range 25–40°C, PNIPAm can adsorb to the surface of polystyrene latex (hard sphere), and consequently attraction takes place via bridging. Depending on the concentration of microgels in the mixture, for a similar microgel to hard sphere size ratio (of about 0.15), different transitions can take place from fluid to depletion or bridging gel, from weak to strong gel, from gel to attractive glass, and from repulsive to attractive glass, exhibiting similarities with attractive glasses or gels based on hard spheres, such as two-step yielding, but in general richer state behavior, being more versatile [271–274]. Experiments with binary mixtures of core-shell particles comprising a polystyrene hard core with a PNIPAm shell of same size but opposite charges have revealed that, depending on their relative proportion, one may obtain stable clusters, gelation, and asymmetric (with respect to the mixing ratio) phase separation [275]. By increasing the complexity and mixing the same polystyrene-core/PNIPAm shell microgel-like particles and sulfonated polystyrene hard spheres (stabilized via long-range electrostatic repulsions) at different nominal volume fractions and strengths of attraction (mediated by temperature) produces a very rich state diagram. Different glasses and gel-to-defective gel-to-glass transitions are detected by means of rheological measurements, as well as an intermediate, ergodic liquid at large microgel fractions [276].

Let us now consider the use of hard spheres as depletants to large soft spheres. This has been explored for colloidal star polymers [267,277,278]. At low star volume fractions the colloidal liquid phase separates upon adding hard spheres due to depletion. At higher volume fractions in the star glass regime, gradual addition of hard spheres leads to different transitions associated with the osmotic pressure exerted on the stars: ergodic domain, due to depletion, arrested phase separation, again due to depletion, double glass due to deswelling, and depletion. This complex behavior is

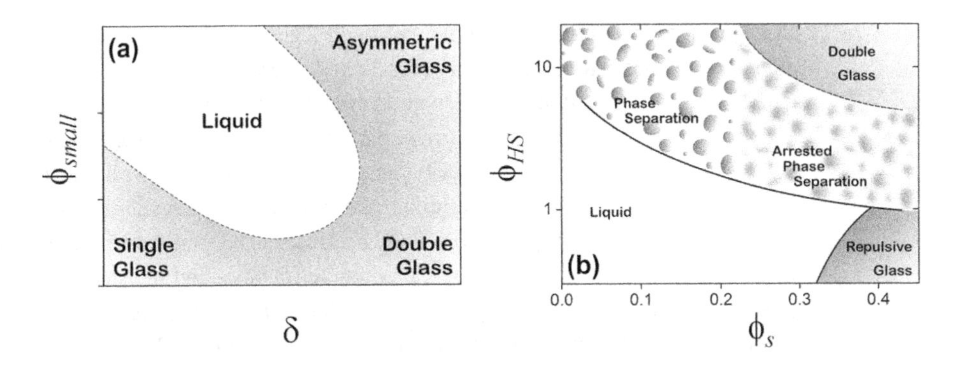

Figure 6.22 (a) Indicative illustration of the state diagram of binary soft–soft colloidal mixtures of stars, representing different situations when the volume fraction of the small star additive, ϕ_{small}, is varied as function of size ratio δ, with the initial fraction of the large star being in the glassy regime and kept constant. Adapted from Mayer et al. [267]. (b) Schematic illustration of the state diagram of a soft–hard colloidal mixture representing different states when the volume fractions of hard sphere additive, ϕ_{HS}, and soft sphere, ϕ_S, are varied. Adapted from Marzi et al. [268]

schematically illustrated in Figure 6.22(b). It has been determined experimentally by rheology and light scattering and predicted by use of MCT and simulations, with remarkable agreement. Such a rich behavior is associated with a very interesting, yet complex nonlinear rheology which, like its linear counterpart, can be tuned at the level of particle interactions [278]. In brief, the double glass exhibits much larger yield and residual stresses when compared to the repulsive glass and arrested phase separation [24] regimes, with the yield strain being of the order of 0.1, the same for all states here as well as all colloidal glasses of pastes [9,207,279] (see also Chapter 5). The arrested phase separation is characterized by a two-step yielding, whereas intracycle stress overshoots during LAOS are evidenced in the repulsive glass regime and vanish as hard colloids are added into the mixtures, conforming to the behavior of soft glassy materials formed by interpenetrable particles [278].

6.8.3 Colloidal Mixtures in the Absence of Solvent Background

Blends of grafted spheres and polymers bear similarities to nanocomposites. It has been suggested that adding linear homopolymers to polymer-grafted nanoparticles originally in the fluid state can result in the formation of anisotropic domains comprising trapped particles [280,281]. This is reminiscent of the asymmetric caging of binary star mixtures, as discussed in Section 6.8.2 [266]. Systematic investigations of polymer-grafted nanoparticles in linear polymer matrices have revealed that by increasing the polymer-to-particle size ratio it is possible to reduce the strength of depletion and promote dispersion of the particles, with important consequences for their dynamic response [282–284]. More recently, blends of crosslinked PMMA microgels with linear PMMA of much smaller size were investigated [248]. In the case of slightly crosslinked microgels the linear chains penetrate and swell the microgels and homogeneous dispersions are obtained. However, highly crosslinked microgels behave as impenetrable hard spheres and aggregate due to the depletion effects induced by linear chains. The concept of depletion-mediated attraction of nanoparticles applies also in the absence of solvent background [9]. Finally, blending polymer-grafted nanoparticles with linear polymers of different chemistry allows for fine-tuning the structure of the resulting composites via a delicate balance of entropic and enthalpic effects. The pertinent parameters are the polymer/nanoparticle size ratio, the nanoparticle fraction, and the strength of enthalpic interactions. The most intriguing outcome of this is the mixing, i.e., the dispersion state of the nanoparticles, promoted by the enthalpic interactions [285]. Systematic experimental studies using SAXS have mapped the whole parameter space and explored, in part, the rheological consequences [286].

6.9 Summary

Softness, expressed as particle deformability and long-range particle interactions, imparts significant changes to the rheology of suspensions, as compared with their hard sphere analogs. The main reason is the ability of soft particles to adjust their shape and volume, as well as mutually interpenetrate if they are grafted, at high concentrations. This has consequences on their volume fraction, elastic interactions, dynamic response, and rheology. It is therefore nontrivial to determine the volume fraction. The conventional approach is to neglect osmotic particle deswelling and consider the effective volume fraction based on the size of an isolated particle at the given conditions (temperature, pH) or, equivalently, on the mapping onto the behavior of hard spheres. In this respect, the zero shear viscosity has a weaker dependence on, and diverges at higher values of the effective volume fraction. Similarly, the plateau modulus in the glassy and jamming regimes also exhibits a weaker dependence on volume fraction. Nevertheless, it is possible to account for volume change in soft particles and estimate their actual volume fraction. Shear thinning is associated with deformation of both particle and the overall microstructure. The flow curves exhibit a plateau or weak power-law of stress versus shear rate at low values of the latter and can be described phenomenologically by a Herschel–Bulkley type of behavior. Depending on particle microstructure, particle–particle and particle-surface interactions, this plateau can be associated with yield stress, shear banding, or wall slip. Aging and internal stresses characterize soft colloidal glasses and pastes and, whereas some features are (qualitatively) common with hard sphere glasses (e.g., residual stress), the microstructural details of particles impart deviations such as different scaling properties of the plateau modulus with aging or volume fraction. A wide range of models based on the use of microstructural or statistical approaches or phenomenological micromechanical evidence offer a valuable predictive toolbox for the rich rheology of this class of systems. Finally, the flow properties of soft entropic and jammed colloidal glasses can be tailored by means of osmotic interactions due to the addition of nonadsorbing linear or star polymers or hard spheres. The main difference from the respective mixtures based on hard spheres is that, besides depletion, the osmotic pressure due to additives also compresses the soft particles. Typical manifestations are the wide-ranging soft interactions and the fact that, contrary to hard colloid-linear polymer mixtures, no re-entrant attractive glass is observed in the soft mixtures, but a re-entrant gel. Replacing the linear polymers by other additives reveals a plethora of interesting structural and rheological properties reflecting the different interactions. Enthalpic interactions, primarily manifested by temperature-mediated change of solvent quality, also lead to changes in the effective volume fraction, hence the occurrence of rheological phenomena such as heating-induced jamming.

Chapter Notation

a_m	length scale of monomer [m]
$\bar{B}$	compression modulus [Pa]
c_L	concentration of linear chains [kg/m^3]
c^*	overlap concentration of polymers [kg/m^3]
c_L^*	overlap concentration of linear chains [kg/m^3]
D	long-time self-diffusion coefficient [m^2/s]
D_0	single particle self-diffusion coefficient [m^2/s]
D	fluidity in Eq. (6.11a) [s^{-1}]
D_i	stress diffusion coefficient in Eq. (6.12) [Pa^2/s]
d_s	effective diameter of a star polymer [nm]
F_{el}	effective elastic free energy of a star macromolecule [J]
F_{int}	enthalpic contribution to the free energy of a star macromolecule [J]
F_{osm}	osmotic contribution to the free energy of a star macromolecule [J]
f	branching functionality, number of arms, of star polymers [–]
$f(q,t)$	intermediate scattering function related to dynamic structure factor [–]
G_0	plateau shear modulus [Pa]
K	bending modulus [Pa]
ℓ	local strain [–]
N_a	degree of polymerization of the star polymer arms [–]
N_x	average number of monomers between two crosslinks [–]
$P_i(\sigma, t)$	probability that a mesoscopic element i has a stress σ at time t [–]
q	wave vector [μm^{-1}]
r	parameter in Eq. (6.11) [s$^{-\alpha-2}$]
t_w	waiting time [s]
V	interparticle potential [J]
V_{eff}	effective pair potential in a polymer solution [J]
V_{Hertz}	Hertzian potential [J]
V_{HS}	hard sphere pair potential [J]
V	volume of a statistical unit [m^3]
V_p	polymer pair potential [J]

V_{sp} cross-interaction star-polymer potential [J]
V_{star} Interaction potential for star polymers [J]
V_s slip velocity [m/s]
V^* characteristic slip velocity [m/s]
Wi_Z Weissenberg–Zimm number [–]

Greek Symbols

α numerical factor, Eq. (6.7) [–]
β numerical factor, Eq. (6.7) [–]
γ_s surface tension [N/m]
γ_y yield strain [–]
Γ fraction of counterions that are not trapped in the microgel particles [–]
$\Gamma_i(t)$ rate of plastic events in Eq. (6.12) [s^{-1}]
δ polymer/colloid size ratio [–]
ε prefactor in Eq. (6.1) [–]
ν Poisson ratio [–]
ξ size of a blob [m]
$\rho(E)$ density of energy barriers at rest in the SGR model [–]
σ_c local yield stress [Pa]
σ_ℓ local stress [Pa]
σ_p preshear stress [Pa]
σ_y' dynamic yield stress in the glassy region [Pa]
σ_y dynamic yield stress [Pa]
υ excluded volume parameter [–]
ϕ_g volume fraction at the glass transition [–]
ϕ_J volume fraction at jamming [–]
ϕ_L fractional distance from overlap concentration [–]
ϕ_s volume fraction of star polymer additive [–]
τ characteristic time [s]
τ_L characteristic time for aging in light scattering experiments [s]
τ_R characteristic time for aging in rheology experiments [s]
τ_z longest relaxation time (Zimm) for a single particle [s]
τ_0 longest relaxation time [s]

References

1. Mewis J, Frith WJ, Strivens TA, Russel WB. The rheology of suspensions containing polymerically stabilized particles. *AIChE Journal*. 1989;35(3):415–422.
2. Borrega R, Cloitre M, Betremieux I, Ernst B, Leibler L. Concentration dependence of the low-shear viscosity of polyelectrolyte micronetworks: From hard sphere to soft microgels. *Europhysics Letters*. 1999:47(6):799–835.

3. Roovers J. Concentration dependence of the relative viscosity of star polymers. *Macromolecules*. 1994;27(19):5359–5364.
4. Vlassopoulos D, Fytas G, Pispas S, Hadjichristidis N. Spherical polymer brushes viewed as soft colloidal particles: zero-shear viscosity. *Physica B*. 2001;296(1–3):184–189.
5. Oldroyd JG. The elastic and viscous properties of emulsions and suspensions. *Proceedings of the Royal Society of London A*. 1953;218(1132):122–132.
6. Cloitre M, Bonnecaze RT. Micromechanics of soft particle glasses. *Advances in Polymer Science*. 2010;236:117–162.
7. Seth JR, Mohan L, Locatelli-Champagne C, Cloitre M, Bonnecaze RT. A micromechanical model to predict the flow of soft particle glasses. *Nature Materials*. 2011;10:838–843.
8. Zhang J, Lettinga PM, Dhont JKG, Stiakakis E. Direct visualization of conformation and dense packing of DNA-based soft colloids. *Physical Review Letters*. 2014;113 (26):268303.
9. Vlassopoulos D, Cloitre M. Tunable rheology of dense soft deformable colloids. *Current Opinion in Colloid & Interface Science*. 2014;19(6):561–574.
10. Riest J, Athanasopoulou L, Egorov SA, Likos CN, Ziherl P. Elasticity of polymeric nanocolloidal particles. *Scientific Reports*. 2015;5:15854.
11. Conley GM, Aebischer P, Nöjd S, Schurtenberger P, Scheffold F. Jamming and overpacking fuzzy microgels: Deformation, interpenetration, and compression. *Science Advances*. 2017;3(10):e1700969.
12. van der Scheer P, van de Laar T, van der Gucht J, Vlassopoulos D, Sprakel J. Fragility and strength in nanoparticle glasses. *ACS Nano*. 2017;11(7):6755–6763.
13. Bouhid dc Aguiar I, van de Laar T, Meireles M, Bouchoux A, Sprakel J, Schroën K. Deswelling and deformation of microgels in concentrated packings. *Scientific Reports*. 2017;7:10223.
14. Conley GM, Zhang C, Aebischer P, Harden JL, Scheffold F. Relationship between rheology and structure of interpenetrating, deforming and compressing microgels. *Nature Communications*. 2019;10:2436 .
15. Koumakis N, Pamvouxoglou A, Poulosa AS, Petekidis G. Direct comparison of the rheology of model hard and soft particle glasses. *Soft Matter*. 2012;8(15):4271–4284.
16. Le Grand F, Petekidis G. Effects of particle softness on the rheology and yielding of colloidal glasses. *Rheologica Acta*. 2008;47(5–6):579–590.
17. Truzzolillo D, Vlassopoulos D, Munam A, Gauthier M. Depletion gels from dense soft colloids: Rheology and thermoreversible melting. *Journal of Rheology*. 2014;58 (5),1441–1462.
18. Coussot P, Raynaud JS, Bertrand F, Moucheront P, Guilbaud JP, Huynh HT, et al. Coexistence of liquid and solid phases in flowing soft-glassy materials. *Physical Review Letters*. 2002;88(21):218301.
19. Ragouilliaux A, Ovarlez G, Shahidzadeh-Bonn N, Herzhaft X, Palermo T, Coussot P. Transition from a simple yield-stress fluid to a thixotropic material. *Physical Review E*. 2007;76(5):051408.
20. Möller P, Fall A, Chikkadi V, Derk D, Bonn D. An attempt to categorize yield stress fluid behaviour. *Philosophical Transactions of the Royal Society A*. 2009;367:5139–5155.
21. Divoux T, Tamarii D, Barentin C, Manneville S. Transient shear banding in a simple yield stress fluid. *Physical Review Letters*. 2010;104(20):208301.

22. Cloitre M, Borrega R, Monti F, Leibler L. Glassy dynamics and flow properties of soft colloidal pastes. *Physical Review Letters*. 2003;90(6):068303.
23. Erwin BM, Cloitre M, Gauthier M, Vlassopoulos D. Dynamics and rheology of colloidal star polymers. *Soft Matter*. 2010;6(12):2825–2833.
24. Mohan L, Bonnecaze RT, Cloitre M. Microscopic origin of internal stresses in jammed soft particle suspensions. *Physical Review Letters*. 2013;111(26):268301.
25. Vlassopoulos D, Fytas G. From polymers to colloids: Engineering the dynamic properties of hairy particles. *Advances in Polymer Science*. 2010;236:1–54.
26. Abreu D, Levant M, Steinberg V, Seifert U. Fluid vesicles in flow. *Advances in Colloid and Interface Science*. 2014;208:29–41.
27. Winkler RG, Fedosov DA, Gompper G. Dynamical and rheological properties of soft colloid suspensions. *Current Opinion in Colloid & Interface Science*. 2014;19(6):594–610.
28. Fernandez-Nieves A, Wyss H, Mattsson J, Weitz DA. (eds.) *Microgel Suspensions: Fundamentals and Applications*. Weinheim: Wiley-VCH; 2011.
29. Perez-Baena I, Moreno AJ, Colmenero J, Pomposo JA. Single-chain nanoparticles vs. star, hyperbranched and dendrimeric polymers: Effect of the nanoscopic architecture on the flow properties of diluted solutions. *Soft Matter*. 2014;10(47):9454–9459.
30. Mohanty PS, Noejd S, van Gruijthuijsen K, Crassous JJ, Obiols-Rabasa M, Schweins R, et al. Interpenetration of polymeric microgels at ultrahigh densities. *Scientific Reports*. 2017;7:1487.
31. Bucknall CB, Paul DR. *Polymer Blends: Formulation and Performance*. New York: Wiley; 2000.
32. Qian JY, Pearson RA, Dimone VL, El-Aasser MS. Synthesis and application of core-shell particles as toughening agents for epoxies. *Journal of Applied Polymer Science*. 1995;58 (2):439–448.
33. Daoud M, Cotton JP. Star shaped polymers: A model for the conformation and its concentration dependence. *Journal de Physique France*. 1982;43(3):531–538.
34. Birshtein TM, Zhulina EB. Conformations of star-branched macromolecules. *Polymer*. 1984;25(10):1453–1461.
35. Grest GS, Fetters LJ, Huang JS, Richter D. Star polymers: Experiment, theory and simulation. *Advances in Chemical Physics*. 1996;94:67–163.
36. Lund R, Willner L, Stellbrink J, Lindner P, Richter D. Logarithmic chain-exchange kinetics of diblock copolymer micelles. *Physical Review Letters*. 2006;96(6):068302.
37. Paud F, Nicolai T, Nicol E, Benyaha F, Brotton G. Dynamic arm exchange facilitates crystallization and jamming of starlike polymers by spontaneous fine-tuning of the number of arms. *Physical Review Letters*. 2013;110(2):028302.
38. Hamley IW. *Block Copolymers in Solution: Fundamentals and Applications*. Chichester: Wiley; 2005.
39. van Ruymbeke E, Pamvouxoglou A, Vlassopoulos D, Petekidis G, Mountrichas G, Pispas S. Stable responsive diblock copolymer micelles for rheology control. *Soft Matter*. 2010;6 (5):881–891.
40. Ruan Y, Gao L, Yao D, Zhang K, Zhang B, Chen Y, et al. Polymer-grafted nanoparticles with precisely controlled structures. *ACS Macro Letters*. 2015;4(10):1067–1071.
41. Halperin A, Tirrell M, Lodge TP. Tethered chains in polymer microstructures. *Advances in Polymer Science*. 1992;100:31–71.
42. Anyfantakis M, Bourlinos A, Vlassopoulos D, Fytas G, Giannelis E, Kumar SK. Solvent-mediated pathways to gelation and phase separation in suspensions of grafted nanoparticles. *Soft Matter*. 2012;5(21):5246–5265.

43. Zhang Z, Pfleiderer P, Schofield AB, Clasen C, Vermant J. Synthesis and directed self-assembly of patterned anisotropic polymeric particles. *Journal of the American Chemical Society*. 2011;133(3):392–395.
44. Crassous JJ, Mihut AM, Månsson LK, Schurtenberger P. Anisotropic responsive microgels with tuneable shape and interactions. *Nanoscale*. 2015;7(38):15971–15982.
45. Reddy NK, Zhanga Z, Lettinga MP, Dhont JKG, Vermant J. Probing structure in colloidal gels of thermoreversible rodlike virus particles: Rheology and scattering. *Journal of Rheology*. 2012;56(5):1153–1164.
46. Grelet E, Rana R. From soft to hard rod behavior in liquid crystalline suspensions of sterically stabilized colloidal filamentous particles. *Soft Matter*. 2016;12(20):4621–4627.
47. Kang K, Dhont JKG. Glass transition in suspensions of charged rods: Structural arrest and texture dynamics. *Physical Review Letters*. 2013;110(1):015901.
48. Solomon MJ, Spicer PT. Microstructural regimes of colloidal rod suspensions, gels and glasses. *Soft Matter* 2010;6(7):1391–1400.
49. Huang F, Rotstein R, Fraden S, Kasza KE, Flynn NT. Phase behavior and rheology of attractive rod-like particles. *Soft Matter*. 2009;5(14):2766–2771.
50. Verso FL, Likos CN. End-functionalized polymers: Versatile building blocks for soft materials. *Polymer*. 2008;49(6):1425–1434.
51. Bianchi E, Blaak R, Likos CN. Patchy colloids: State of the art and perspectives. *Physical Chemistry Chemical Physics*. 2011;13(14):6397–6410.
52. Moghimi E, Chubak I, Statt A, Howard MP, Founta D, Polymeropoulos G, et al. Self-organization and flow of low-functionality telechelic star polymers with varying attraction. *ACS Macro Letters*. 2019;8(7):766–772.
53. Schlüter AD, Halperin A, Kröger M, Vlassopoulos D, Wegner G, Zhang B. Dendronized polymers: Molecular objects between conventional linear polymers and colloidal particles. *ACS Macro Letters*. 2014;3(10):991–998.
54. De Gennes PG, Prost J. *Physics of Liquid Crystals*. Oxford: Oxford University Press; 1994.
55. Eichenbaum GM, Kiser PF, Dobrynin AV, Simon SA, Needham D. Investigation of the swelling response and loading of ionic microgels with drugs and proteins: The dependence on cross-link density. *Macromolecules*. 1999;32(15):4867–4878.
56. Vlassopoulos D. Molecular topology and rheology: Beyond the tube model. *Rheologica Acta*. 2016;55(8):613–632.
57. Roovers J, Zhou LL, Toporowski PM, van der Zwan M, Iatrou H, Hadjichristidis N. Regular star polymers with 64 and 128 arms: Models for polymeric micelles. *Macromolecules*. 1993;26(16):4324–4331.
58. Kapnistos M, Semenov AN, Vlassopoulos D, Roovers J. Viscoelastic response of hyperstar polymers in the linear regime. *Journal of Chemical Physics*. 1999;111(4):1753–1759.
59. Vlassopoulos D, Fytas G, Pakula T, Roovers J. Multiarm star polymer dynamics. *Journal of Physics: Condensed Matter*. 2001;13(41):R855–R876.
60. Gury L, Gauthier M, Cloitre M, Vlassopoulos D. Colloidal jamming in multiarm star polymer melts. *Macromolecules*. 2019;52(12):4617–4623.
61. Sebastian JM, Lai C, Graessley WW, Register RA. Steady-shear rheology of block copolymer melts and concentrated solutions: Disordering stress in body-centered-cubic systems. *Macromolecules*. 2002;35(7):2707–2713.
62. Antonietti M, Pakula T, Bremser W. Rheology of small spherical polystyrene microgels: A direct proof of a new transport mechanism in bulk polymers besides reptation. *Macromolecules* 1995;28(12):4227–4233.

63. Bourlinos AB, Herrera R, Chalkias N, Jiang DD, Zhang Q, Archer LA, et al. Surface-functionalized nanoparticles with liquid-like behavior. *Advanced Materials.* 2005;17 (2):234–237.
64. Pursiainen OLJ, Baumberg JJ, Winkler H, Viel B, Spahn P, Ruhl T. Shear-induced organization in flexible polymer opals. *Advanced Materials.* 2008;20(8):1484–1487.
65. Hasegawa R, Aoki Y, Doi M. Optimum graft density for dispersing particles in polymer melts. *Macromolecules.* 1996;29(20):6656–6662.
66. Ferreira PG, Ajdari A, Leibler L. Scaling law for entropic effects at interfaces between grafted layers and polymer melts. *Macromolecules.* 1998;31(12):3994–4003.
67. Chremos A, Panagiotopoulos AZ, Koch DL. Dynamics of solvent-free grafted nanoparticles. *Journal of Chemical Physics.* 2012;136(4):044902.
68. Kim SA, Mangal R, Archer LA. Relaxation dynamics of nanoparticle-tethered polymer chains. *Macromolecules.* 2015;48(17):6280–6293.
69. Snijkers F, Cho HY, Nese A, Matyjaszewski K, Pyckhout-Hintzen W, Vlassopoulos D. Effects of core microstructure on structure and dynamics of star polymer melts: From polymeric to colloidal response. *Macromolecules.* 2014;47(15):5347–5356.
70. Landau LD, Lifshitz EM. *Theory of Elasticity*, 3rd ed. Amsterdam: Elsevier; 1986.
71. Likos CN. Effective interactions in soft condensed matter physics. *Physics Reports.* 2001;348(4–5):267–439.
72. Likos CN, Loewen H, Watzlawek M, Abbas B, Jucknischke O, Allgaier J, et al. Star polymers viewed as ultrasoft colloidal particles. *Physical Review Letters.* 1998;80 (20):4450–4453.
73. Laurati M, Stellbrink J, Lund R, Willner L, Richter D. Starlike micelles with starlike interactions: A quantitative evaluation of structure factors and phase diagram. *Physical Review Letters.* 2005;94(19):195504.
74. Briels WJ. Transient forces in flowing soft matter. *Soft Matter.* 2009;5(22):4401–4411.
75. Likos CN, Vaynberg KA, Löwen H, Wagner NJ. Colloidal stabilization by adsorbed gelatin. *Langmuir.* 2000;16(9):4100–4108.
76. Likos CN, Loewen H, Poppe A, Willner L, Roovers J, Cubitt B, et al. Ordering phenomena of star polymer solutions approaching the Θ state. *Physical Review E.* 1998;58 (5):6299–6307.
77. Marzi D, Likos CN, Capone B. Coarse graining of star-polymer-colloid nanocomposites. *Journal of Chemical Physics.* 2012;137(1):014902.
78. Mahynski NA, Panagiotopoulos AZ. Phase behavior of athermal colloid-star polymer mixtures. *Journal of Chemical Physics.* 2013;139(2):024907.
79. Poon WCK, Weeks ER, Royall CP. On measuring colloidal volume fractions. *Soft Matter.* 2012;8 (1):21–30.
80. Senff H, Richtering W. Temperature sensitive microgel suspensions: Colloidal phase behavior and rheology of soft spheres. *Journal of Chemical Physics.* 1999;111(4):705–711.
81. Romeo G, Imperiali L, Kim J-W, Fernandez-Nieves A, Weitz DA. Origin of de-swelling and dynamics of dense ionic microgel suspensions. *Journal of Chemical Physics.* 2012;136(12):124905.
82. Pellet C, Cloitre M. The glass and jamming transitions of soft polyelectrolyte microgel suspensions. *Soft Matter.* 2016;12(16):3710–3720.
83. Stiakakis E, Petekidis G, Vlassopoulos D, Likos CN, Iatrou H, Hadjichristidis N, et al. Depletion and cluster formation in soft colloid-polymer mixtures. *Europhysics Letters.* 2005;72(4):664–670.

84. Wilk A, Huissmann S, Stiakakis E, Kohlbrecher J, Vlassopoulos D, Likos CN, et al. Osmotic shrinkage in star/linear polymer mixtures. *The European Physical Journal E.* 2010;32(2):127–134.
85. Truzzolillo D, Vlassopoulos D, Gauthier M. Osmotic interactions, rheology, and arrested phase separation of star-linear polymer mixtures. *Macromolecules.* 2011;44(12):5043–5052.
86. Fernández-Nieves A, Fernández-Barbero A, Vincent B, de las Nieves FJ. Osmotic de-swelling of ionic microgel particles. *Journal of Chemical Physics.* 2003;119 (19):10383–10388.
87. Routh AF, Fernandez-Nieves A, Bradley M, Vincent B. Effect of added free polymer on the swelling of neutral microgel particles: A thermodynamic approach. *Journal of Physical Chemistry B.* 2006;110(25):12721–12727.
88. Fernandez-Nieves A, Lyon LA. The polymer/colloid duality of microgel suspensions. *Annual Review of Physical Chemistry.* 2012;63:25–43.
89. Ikeda A, Berthier L, Sollich P. Disentangling glass and jamming physics in the rheology of soft materials. *Soft Matter.* 2013;9(32):7669–7683.
90. Scheffold F, Cardinaux F, Mason TG. Linear and nonlinear rheology of dense emulsions across the glass and the jamming regimes. *Journal of Physics: Condensed Matter.* 2013;25 (50):502101.
91. Ghosh A, Chaudhary G, Kang JG, Braun PW, Ewoldt RH, Schweizer KS. Linear and nonlinear rheology and structural relaxation in dense glassy and jammed soft repulsive pNIPAM microgel suspensions. *Soft Matter.* 2019;15(5):1038–1052.
92. Basu A, Xu Y, Still T, Arratia PE, Zhang Z, Nordstrom KN, et al. Rheology of soft colloids across the onset of rigidity: Scaling behavior, thermal, and non-thermal responses. *Soft Matter.* 2010;10(17):3027–3035.
93. Witten TA, Pincus PA. Colloid stabilization by long grafted polymers. *Europhysics Letters.* 1986;19(10):2509–2513.
94. McConnell GA, Gast AP, Huang JC, Smith SD. Disorder–order transitions in soft-sphere polymer micelles. *Physical Review Letters.* 1993;71(13):2102–2105.
95. Mortensen K. Structural studies of aqueous solutions of PEO–PPO–PEO triblock copolymers, their micellar aggregates and mesophases; a small-angle neutron scattering study. *Journal of Physics: Condensed Matter.* 1996;8(25A):103–124.
96. Stiakakis E, Wilk A, Kohlbrecher J, Vlassopoulos D, Petekidis G. Slow dynamics, aging and crystallization of multiarm star glasses. *Physical Review E.* 2010;81(2):0205402 (R).
97. Rissanou AN, Yiannourakou M, Economou IG, Bitsanis IA. Temperature-induced crystallization in concentrated suspensions of multiarm star polymers: A molecular dynamics study. *Journal of Chemical Physics.* 2006;124(4):044905.
98. McConnell GA, Lin MY, Gast AP. Long range order in polymeric micelles under steady shear. *Macromolecules.* 1995;28(20):6754–6764.
99. Molino FR, Berret J-F, Porte G, Diat O, Lindner P. Identification of flow mechanisms for a soft crystal. *The European Physical Journal B.* 1998;3(1):59–72.
100. Mortensen K, Theunissen E, Kleppinger R, Almdal K, Reynaers H. Shear-induced morphologies of cubic ordered block copolymer micellar networks studied by in-situ small-angle neutron scattering and rheology. *Macromolecules.* 2002;35(20):7773–7781.
101. Hamley IW. The effect of shear on block copolymer solutions. *Current Opinion in Colloid & Interface Science.* 2000;5;342–50.
102. Jiang J, Burger C, Li C, Li J, Lin MY, Colby RH, et al. Shear-induced layered structure of polymeric micelles by SANS. *Macromolecules.* 2007;40(11):4016–4022.

103. Ruiz-Franco J, Marakis N, Gnan N, Kohlbrecher J, Gauthier M, Lettinga MP, et al. Crystal-to-crystal transition in star colloids under shear. *Physical Review Letters*. 2018;120(7):078003.
104. Chu F, Heptner N, Lu Y, Siebenbürger M, Lindner P, Dzubiella J, et al. Colloidal plastic crystals in a shear field. *Langmuir*. 2015;31(22):5992–6000.
105. Paulin SE, Ackerson BJ, Wolfe MS. Equilibrium and shear induced nonequilibrium phase behavior of PMMA microgel spheres. *Journal of Colloid and Interface Science*. 1996;178 (1):251–262.
106. Stieger M, Lindner P, Richtering W. Structure formation in thermoresponsive microgel suspensions under shear flow. *Journal of Physics: Condensed Matter*. 2004;16(36): S3861–3872.
107. Freiberger N, Medebach M, Glatter O. Melting behavior of shear-induced crystals in dense emulsions as investigated by time-resolved light scattering. *The Journal of Physical Chemistry B*. 2008;112(40):12635–12643.
108. Huang J-R, Mason TG. Shear oscillation light scattering of droplet deformation and reconfiguration in concentrated emulsions. *Europhysics Letters*. 2008;83(2):28004.
109. Huang J-R, Mason TG. Deformation, restructuring, and un-jamming of concentrated droplets in large-amplitude oscillatory shear flows. *Soft Matter*. 2009;5(2):2208–2214.
110. Snoswell DRE, Finlayson CE, Zhao O, Baumberg JJ. Real-time measurements of crystallization processes in viscoelastic polymeric photonic crystals. *Physical Review E*. 2015;92 (5):052315.
111. Khabaz F, Liu, T, Cloitre M, Bonnecaze RT. Shear-induced ordering and crystallization of jammed suspensions of soft particles glasses. *Physical Review Fluids*. 2017;2(9):093301.
112. Khabaz F, Cloitre M, Bonnecaze RT. Structural state diagram of concentrated suspensions of jammed soft particles in oscillatory shear flow. *Physical Review Fluids*. 2018;3(3):033301.
113. López-Barrón CR, Porcar L, Eberle APR, Wagner NJ. Dynamics of melting and recrystallization in a polymeric micellar crystal subjected to large amplitude oscillatory shear flow. *Physical Review Letters*. 2012;108(25):258301.
114. Zhao Q, Finlayson CE, Snoswel DRE, Haines A, Schäfer C, Spahn P, et al. Large-scale ordering of nanoparticles using viscoelastic shear processing. *Nature Communications*. 2016;7:11661.
115. Nikoubashman A, Kahl G, Likos CN. Flow quantization and nonequilibrium nucleation of soft crystals. *Soft Matter*. 2012;8(15):4121–4131.
116. Taylor GI. The viscosity of a fluid containing small drops of another fluid. *Proceedings of the Royal Society of London A*. 1932;138(834):41–48.
117. Fröhlich H, Sack R. Theory of the rheological properties of dispersions. *Proceedings of the Royal Society of London A*. 1946;185(1003):415–430.
118. Cerf R. Recherches théoriques et expérimentales sur l'effet Maxwell des solutions de macromolécules déformables. *Journal de Chimie Physique*. 1951;48:59–84.
119. Oldroyd JG. The effect of interfacial stabilizing films on the elastic and viscous properties of emulsions. *Proceedings of the Royal Society of London A*. 1955;232(1191):567–577.
120. Oldroyd JG. On the formulation of rheological equations of state. *Proceedings of the Royal Society of London A*. 1950;200(1063):523–541.
121. Kerner EH. The elastic and thermoelastic properties of composite media. *Proceedings of the Physical Society. Section B*. 1956;69(8):808–813.
122. Frankel NA, Acrivos A. The constitutive equation for a dilute emulsion. *Journal of Fluid Mechanics*. 1970;44(1):65–78.

123. Goddard JD, Miller C. Nonlinear effects in the rheology of dilute suspensions. *Journal of Fluid Mechanics*. 1967;28(4):657–663.
124. Choi SJ, Schowalter WR. Rheological properties of nondilute suspensions of deformable particles. *Physics of Fluids*. 1975;18(4):420–427.
125. Palierne JF. Linear rheology of viscoelastic emulsions with interfacial tension. *Rheologica Acta*. 1990;29(3):204–14; erratum *Rheologica Acta*. 1991;30(5):497.
126. Vinckier I, Moldenaers P, Mewis J. Relationship between rheology and morphology of model blends in steady shear flow. *Journal of Rheology*. 1996;40(4):613–631.
127. Bousmina M. Rheology of polymer blends: Linear model for viscoelastic emulsions. *Rheologica Acta*. 1999;38(1):73–83.
128. Segrè PN, Meeker SP, Pusey PN, Poon WCK. Viscosity and structural relaxation in suspensions of hard-sphere colloids. *Physical Review Letters*. 1995;75(5):958–961.
129. Loppinet B, Fytas G, Vlassopoulos D, Likos CN, Meier G, Liu GJ. Dynamics of dense suspensions of star-like micelles with responsive fixed cores. *Macromolecular Chemistry and Physics*. 2005;206(1):163–172.
130. Dahbi L, Alexander M, Trappe V, Dhont JKG, Schurtenberger P. Rheology and structural arrest of casein suspensions. *Journal of Colloid and Interface Science*. 2010;34 (2):564–570.
131. Cloitre M, Borrega R, Monti F, Leibler L. Structure and flow behaviour of polyelectrolyte microgels: from suspensions to glasses. *Comptes Rendus Physique*. 2003;4(2):221–230.
132. Gupta S, Stellbrink J, Zaccarelli E, Likos CN, Camargo M, Holmqvist P, et al. Validity of the Stokes–Einstein relation in soft colloids up to the glass transition. *Physical Review Letters*. 2015;115(12):28302.
133. Angell CA. Formation of glasses from liquids and biopolymers. *Science*. 1995;267 (5206):1924–1935.
134. Mattsson J, Wyss HM, Fernandez-Nieves A, Miyazaki K, Hu Z, Reichman DR, et al. Soft colloids make strong glasses. *Nature*. 2009;462:83–86.
135. Neuhaeusler S, Richtering W. Rheology and diffusion in concentrated sterically stabilized polymer dispersions. *Colloids and Surfaces A: Physicochemical and Engineering Aspects*. 1995;96(1):39–51.
136. Helgeson ME, Wagner NJ, Vlassopoulos D. Viscoelasticity and shear melting of colloidal star polymer glasses. *Journal of Rheology*. 2007;51(2):297–316.
137. Erwin BM, Vlassopoulos D, Cloitre M. Rheological fingerprinting of an aging soft colloidal glass. *Journal of Rheology*. 2010;54(4):915–939.
138. Wen YH, Schaefer JL, Archer LA. Dynamics and rheology of soft colloidal glasses. *ACS Macro Letters*. 2015;4(1):119–123.
139. Merlet-Lacroix N, Cloitre M. Swelling and rheology of thermoresponsive gradient copolymer micelles. *Soft Matter*. 2010;6(5):984–993.
140. Seth JR, Cloitre M, Bonnecaze RT. Elastic properties of soft particle pastes. *Journal of Rheology*. 2006;50(3):353–376.
141. Yang J, Schweizer KS. Tunable dynamic fragility and elasticity in dense suspensions of many-arm-star polymer colloids. *Europhysics Letters*. 2010;90(6):66001.
142. Mason TG, Lacasse M-D, Grest GS, Levine D, Bibette J, Weitz DA. Osmotic pressure and viscoelastic shear moduli of concentrated emulsions. *Physical Review E*. 1997:56 (3):3150–3166.
143. Mohan L, Bonnecaze RT. Short-ranged pair distribution function for concentrated suspensions of soft particles. *Soft Matter*. 2012;8(15):4216–4222.

144. Kapnistos M, Vlassopoulos D, Fytas G, Mortensen K, Fleischer G, Roovers J. Reversible thermal gelation in soft spheres. *Physical Review Letters*. 2000;85(19);4072–4075.
145. Agarwal P, Srivastava S, Archer LA. Thermal jamming of a colloidal glass. *Physical Review Letters*. 2011;107(26):268302.
146. Sato T, Watanabe H, Osaki K. Thermoreversible physical gelation of block copolymers in a selective solvent. *Macromolecules*. 2005;33(5):1686–1691.
147. Zhao Y, Cao Y, Yang Y, Wu C. Rheological study of the sol-gel transition of hybrid gels. *Macromolecules*. 2003;36(3):855–859.
148. Crassous JJ, Casal-Dujat L, Medebach M, Obiols-Rabasa M, Vincent R, Reinhold F, et al. Structure and dynamics of soft repulsive colloidal suspensions in the vicinity of the glass transition. *Langmuir*. 2013;29(33):10346–10359.
149. Peng X, McKenna GB. Comparison of the physical aging behavior of a colloidal glass after shear melting and concentration jumps. *Physical Review E*. 2014;90(1):050301(R).
150. Peng X, McKenna GB. Physical aging and structural recovery in a colloidal glass subjected to volume-fraction jump conditions. *Physical Review E*. 2016;93(4):042603.
151. Siebenbürger M, Fuchs M, Winter H, Ballauff M. Viscoelasticity and shear flow of concentrated, noncrystallizing colloidal suspensions: Comparison with mode-coupling theory. *Journal of Rheology*. 2009;53(3):707–726.
152. Nordstrom KN, Arratia PE, Verneuil E, Basu A, Zhang Z, Yodh AG, et al. Microfluidic rheology of soft colloids above and below jamming. *Physical Review Letters*. 2010;105 (17):175701.
153. Paredes J, Michels MAJ, Bonn D. Rheology across the zero-temperature jamming transition. *Physical Review Letters*. 2013;111(1):015701.
154. Fujii S, Richtering W. Size and viscoelasticity of spatially confined multilamellar vesicles. *The European Physical Journal E*. 2006;19(2):139–148.
155. Gutowski IA, Lee D, de Bruyn JR, Frisken BJ. Scaling and mesostructure of Carbopol dispersions. *Rheologica Acta*. 2012;51(5):441–450.
156. Shafiei M, Balhoff M, Hayman NW. Chemical and microstructural controls on viscoplasticity in carbopol hydrogel. *Polymer*. 2018;139:44–51.
157. Rogers SA, Vlassopoulos D, Callaghan PT. Aging, yielding, and shear banding in soft colloidal glasses. *Physical Review Letters*. 2008;100(12):128304.
158. Hyun K, Wilhelm M, Klein CO, Cho KS, Nam JG, Ahn KH, et al. A review of nonlinear oscillatory shear tests: Analysis and application of large amplitude oscillatory shear (LAOS). *Progress in Polymer Science*. 2011;36(12):1697–1753.
159. Rogers S. Large amplitude oscillatory shear: Simple to describe, hard to interpret. *Physics Today*. 2018;71(7):34–40.
160. Koumakis N, Petekidis G. Two step yielding in attractive colloids: Transition from gels to attractive glasses. *Soft Matter*. 2011;7(6):456–470.
161. Koumakis N, Brady JF, Petekidis G. Complex oscillatory yielding of model hard-sphere glasses. *Physical Review Letters*. 2013;110(17):178301.
162. Rogers S, Erwin BM, Vlassopoulos D, Cloitre M. A sequence of physical processes determined and quantified in LAOS: Application to a yield stress fluid. *Journal of Rheology*. 2011;55(2):435–458.
163. Mohan L, Pellet C, Cloitre M, Bonnecaze RT. Local mobility and microstructure in periodically sheared soft particle glasses and their connection to macroscopic rheology. *Journal of Rheology*. 2013;57(3):1023–1046.

164. Dhont JKG, Lettinga MP, Dogic Z, Lenstra TAJ, Wang H, Rathgeber S, et al. Shear-banding and microstructure of colloids in shear flow. *Faraday Discussions*. 2003;123:157–172.
165. Kundu SK, Gupta S, Stellbrink J, Willner L, Richter D. Relating structure and flow of soft colloids. *The European Physical Journal Special Topics*. 2013;222(11):2757–2772.
166. Misbah C. Vacillating breathing and tumbling of vesicles under shear flow. *Physical Review Letters*. 2006;96(2):028104.
167. Gao T, Hu HH, Castañeda PP. Shape dynamics and rheology of soft elastic particles in a shear flow. *Physical Review Letters*. 2012;108(5):058302.
168. Fuchs M, Cates ME. Theory of nonlinear rheology and yielding of dense colloidal suspensions. *Physical Review Letters*. 2002;89(24):248304.
169. Fuchs M, Cates ME. A mode coupling theory for Brownian particles in homogeneous steady shear flow. *Journal of Rheology*. 2009;53(4):957–1000.
170. Crassous JJ, Regisser R, Ballauff M, Willenbacher N. Characterization of the viscoelastic behavior of complex fluids using the piezoelastic axial vibrator. *Journal of Rheology*. 2005;49(4):851–863.
171. Kobelev V, Schweizer KS. Dynamic yielding, shear thinning, and stress rheology of polymer-particle suspensions and gels. *Journal of Chemical Physics*. 2005;123 (16):164903.
172. Siebenbürger M, Ballauff M, Voigtmann T. Creep in colloidal glasses. *Physical Review Letters*. 2012;108(25):255701.
173. Ballauff M, Brader JM, Egelhaaf SU, Fuchs M, Horbach J, Koumakis N, et al. Residual stresses in glasses. *Physical Review Letters*. 2013;110(21):215701.
174. van Hecke M. Jamming of soft particles: Geometry, mechanics, scaling and isostaticity. *Journal of Physics: Condensed Matter*. 2010;22(3):033101.
175. Bonn D, Paredes J, Denn MM, Berthier L, Divoux T, Manneville S. Yield stress materials in soft condensed matter. *Reviews of Modern Physics*. 2017;89(3):035005.
176. Knowlton ED, Pine DJ, Cipelletti L. A microscopic view of the yielding transition in concentrated emulsions. *Soft Matter*. 2014;10(36):6931–6940.
177. Liu T, Khabaz F, Bonnecaze RT, Cloitre M. On the universality of flow properties of soft-particle glasses. *Soft Matter*. 2018;14(34):7064–7074.
178. Voigtmann T. Nonlinear glassy rheology. *Current Opinion in Colloid & Interface Science* 2014;19(6):549–560.
179. Sollich P, Lequeux F, Hébraud P, Cates ME. Rheology of soft glassy materials. *Physical Review Letters*. 1997;78(10):2020–2023.
180. Sollich P. Rheological constitutive equation for a model of soft glassy materials. *Physical Review E*. 1998;58(1):738–759.
181. Purnomo EH, van den Ende D, Mellema J, Mugele F. Linear viscoelastic properties of aging suspensions. *Europhysics Letters*. 2006;76(1):74–80.
182. Fielding SM. Shear banding in soft glassy materials. *Reports on Progress in Physics*. 2014;77(10):102601.
183. Purnomo EH, van den Ende D, Mellema J, Mugele F. Rheological properties of aging thermosensitive suspensions. *Physical Review E*. 2007;76(2):021404.
184. Fielding SM, Sollich P, Cates ME. Aging and rheology in soft materials. *Journal of Rheology*. 2000;44(2):323–369.
185. Fielding SM, Cates ME, Sollich P. Shear banding, aging and noise dynamics in soft glassy materials. *Soft Matter*. 2009;5(12):2378–2382.

186. Derec C, Ajdari A, Lequeux F. Rheology and aging: A simple approach. *The European Physical Journal E.* 2001;4(3):355–361.
187. Derec C, Ducouret G, Ajdari A, Lequeux F. Aging and nonlinear rheology in suspensions of polyethylene oxide–protected silica particles. *Physical Review E.* 2003;67(6):061403.
188. Carrier V, Petekidis G. Nonlinear rheology of colloidal glasses of soft thermosensitive microgel particles. *Journal of Rheology.* 2009;53(2):245–273.
189. Picard G, Ajdari A, Bocquet L, Lequeux F. Simple model for heterogeneous flows of yield stress fluids. *Physical Review E.* 2002;66(5):051501.
190. Goyon J, Colin A, Ovarlez G, Ajdari A, Bocquet L. Spatial cooperativity in soft glassy flows. *Nature.* 2008;454:84–87.
191. Langer JS. Shear-transformation-zone theory of yielding in athermal amorphous materials. *Physical Review E.* 2015;92(1):012318.
192. Hébraud P, Lequeux F. Mode-coupling theory for the pasty rheology of soft glassy materials. *Physical Review Letters.* 1998;81(14):2934–2937.
193. Picard G, Ajdari A, Lequeux F, Bocquet L. Slow flows of yield stress fluids: Complex spatiotemporal behavior within a simple elastoplastic model. *Physical Review E.* 2005;71 (1):010501.
194. Bocquet L, Colin A, Ajdari A. Kinetic theory of plastic flow in soft glassy materials. *Physical Review Letters.* 2009;103(3):036001.
195. Nicolas A, Barrat J-L. Spatial cooperativity in microchannel flows of soft jammed materials: A mesoscopic approach. *Physical Review Letters.* 2013;110(13):138304.
196. Puosi F, Olivier J, Martens K. Probing relevant ingredients in mean-field approaches for the athermal rheology of yield stress materials. *Soft Matter* 2015;11(38):7639–7647.
197. Liu C, Ferrero EE, Puosi F, Barrat J-L, Martens K. Driving rate dependence of avalanche statistics and shapes at the yielding transition. *Physical Review Letters.* 2016;116 (6):065501.
198. Agoritsas E, Martens K. Non-trivial rheological exponents in sheared yield stress fluids. *Soft Matter.* 2017;13(26):4653–4660.
199. Lin J, Wyart M. Microscopic processes controlling the Herschel–Bulkley exponent. *Physical Review Fluids.* 2018;97(1):012603.
200. Durian DJ. Bubble-scale model of foam mechanics: Melting, nonlinear behavior, and avalanches. *Physical Review E.* 1997;55(2);1739–1751.
201. Tighe BP, Woldhuis E, Remmers JJC, van Saarloos W, van Hecke M. Model for the scaling of stresses and fluctuations in flows near jamming. *Physical Review Letters.* 2010;105(8):088303.
202. Mansard V, Colin A, Chaudhuri P, Bocquet L. A molecular dynamics study of non-local effects in the flow of soft jammed particles. *Soft Matter.* 2013;9(31):7489–7500.
203. Cloitre M, Bonnecaze RT. A review on wall slip in high solid dispersions. *Rheologica Acta.* 2017;56(3):283–305.
204. Seth JR, Cloitre M, Bonnecaze RT. Influence of short-range forces on wall-slip in microgel pastes. *Journal of Rheology.* 2008;52(5):1241–1268.
205. Seth JR, Locatelli-Champagne C, Monti F, Bonnecaze RT, Cloitre M. How do soft particle glasses yield and flow near solid surfaces. *Soft Matter.* 2012;8(1):140–148.
206. Ballesta P, Petekidis G, Isa L, Poon WCK, Besseling R. Wall slip and flow of concentrated hard-sphere colloidal suspensions. *Journal of Rheology.* 2012;56(5):1005–1037.
207. Dhont JKG, Kang K, Kriegs H, Danko O, Marakis J, Vlassopoulos D. Nonuniform flow in soft glasses of colloidal rods. *Physical Review Fluids.* 2017;2(4):043301.

208. Besseling R, Isa L, Ballesta P, Petekidis G, Cates ME, Poon WCK. Shear banding and flow-concentration coupling in colloidal glasses. *Physical Review Letters*. 2010;105 (26):268301.
209. Jin H, Kang K, Ahn K-H, Dhont JKG. Flow instability due to coupling of shear-gradients with concentration: Non-uniform flow of (hard-sphere) glasses. *Soft Matter*. 2014;10 (47):9470–9485.
210. Cromer M, Villet MC, Fredrickson GH, Leal LG. Shear banding in polymer solutions. *Physics of Fluids*. 2013;25(5):051703.
211. Cromer M, Fredrickson GH, Leal LG. A study of shear banding in polymer solutions. *Physics of Fluids*. 2014;26(6):063101.
212. Dhont JKG, Briels WJ. Gradient and vorticity banding. *Rheologica Acta*. 2008;47 (3):257–281.
213. Olmsted PD. Perspectives on shear banding in complex fluids. *Rheologica Acta*. 2008;47 (3):283–300.
214. Divoux, T, Fardin MA, Manneville S, Lerouge S. Shear banding of complex fluids. *Annual Review of Fluid Mechanics*. 2016;48:81–103.
215. Briels WJ, Vlassopoulos D, Kang K, Dhont JKG. Constitutive equations for the flow behavior of entangled polymeric systems: Application to star polymers. *Journal of Chemical Physics* 2011;134(12):124901.
216. Ovarlez G, Rodts S, Chateau X, Coussot P. Phenomenology and physical origin of shear localization and shear banding in complex fluids. *Rheologica Acta*. 2009;48(8):831–844.
217. Tang H, Kochetkova T, Kriegs H, Dhont JKG, Lettinga MP. Shear-banding in entangled xanthan solutions: Tunable transition from sharp to broad shear-band interfaces. *Soft Matter*. 2018;14(5):826–836.
218. Coussot P, Ovarlez G. Physical origin of shear-banding in jammed systems. *The European Physical Journal E*. 2010;33(3):183–188.
219. Joshi YM, Petekidis G. Yield stress fluids and ageing. *Rheologica Acta*. 2018;57(6–7):521–549.
220. Cloitre M, Borrega R, Leibler L. Aging and rejuvenation in microgel pastes. *Physical Review Letters*. 2000;95(22):4819–4822.
221. Di X, Win KZ, McKenna GB, Narita T, Lequeux F, Pullela SR, et al. Signatures of structural recovery in colloidal glasses. *Physical Review Letters*. 2011;106(9):095701.
222. Bonn D, Tanaka H, Wegdam G, Kellay H, Meunier J. Aging of a colloidal "Wigner" glass. *Europhysics Letters*. 1999;45(1):52–57.
223. Knaebel A, Bellour M, Munch J-P, Viasnoff V, Lequeux F, Harden JL. Aging behavior of Laponite clay particle suspensions. *Europhysics Letters*. 2000;52(1):73–79.
224. Abou A, Bonn D, Meunier J. Aging dynamics in a colloidal glass of laponite. *Physical Review E*. 2001;64(2):021911.
225. Bonn D, Tanase S, Abou B, Tanaka H, Meunier J. Laponite: Aging and shear rejuvenation of a colloidal glass. *Physical Review Letters*. 2002;89(1):015701.
226. Shahin A, Joshi YM. Prediction of long and short time rheological behavior in soft glassy materials. *Physical Review Letters*. 2011;106(3):038302.
227. Bandyopadhyay R, Liang D, Yardimci H, Sessoms DA, Borthwick MA, Mochrie SGJ, et al. Evolution of particle-scale dynamics in an aging clay suspension. *Physical Review Letters*. 2004;93(22):228302.
228. Srivastava S, Archer LA, Narayanan S. Structure and transport anomalies in soft colloids. *Physical Review Letters*. 2013;110(14):148302.

229. Ramos L, Cipelletti L. Ultraslow dynamics and stress relaxation in the aging of a soft glassy system. *Physical Review Letters*. 2001;87(24):245503.
230. Ramos L, Cipelletti L. Intrinsic aging and effective viscosity in the slow dynamics of a soft glass with tunable elasticity. *Physical Review Letters*. 2005;94(15):158301.
231. Derec C, Ajdari A, Ducouret G, Lequeux F. Rheological characterization of aging in a concentrated colloidal suspension. *Comptes rendus de l'Académie des Sciences IV*. 2000;1 (8):1115–1119.
232. Viasnoff V, Lequeux F. Rejuvenation and overaging in a colloidal glass under shear. *Physical Review Letters*. 2002;89(6):065701.
233. Agarwal M, Joshi YM. Signatures of physical aging and thixotropy in aqueous dispersion of carbopol. *Physics of Fluids*. 2019;31(6):063107.
234. Struik LCE. *Physical Aging in Amorphous Polymers and Other Materials*. Amsterdam: Elsevier; 1978.
235. Bouchaud JP, Cugliandolo LF, Kurchan J, Mézard M. Out of equilibrium dynamics in spin-glasses and other glassy systems. In Young AP (ed.) *Spin Glasses and Random Fields*. Singapore: World Scientific; 1998, pp. 161–223.
236. Erwin BM, Vlassopoulos D, Gauthier M, Cloitre M. Unique slow dynamics and aging phenomena in soft glassy suspensions of multiarm star polymers. *Physical Review E*. 2011;83(6):061402.
237. Christopoulou C, Petekidis G, Erwin BM, Cloitre M, Vlassopoulos D. Ageing and yield behaviour in model soft colloidal glasses. *Philosophical Transactions of the Royal Society A*. 2009;367(1909):5051–5071.
238. Cipelletti L, Manley S, Ramos L, Weitz DA. Universal aging features in the restructuring of fractal colloidal gels. *Physical Review Letters*. 2000;84(10):2275–2278.
239. Bouzid M, Colombo J, Vieira Barbos L, Del Gado E. Elastically driven intermittent microscopic dynamics in soft solids. *Nature Communications*. 2017;8:15846.
240. Mazoyer S, Cipelletti L, Ramos L. Origin of the slow dynamics and the aging of a soft glass. *Physical Review Letters*. 2006;97(23):238301.
241. Jabbari-Farouji S, Zargar R, Wegdam GH, Bonn D. Dynamical heterogeneity in aging colloidal glasses of Laponite. *Soft Matter*. 2012;8(20):5507–5512.
242. Mohan L, Cloitre M, Bonnecaze RT. Build-up and two-step relaxation of internal stresses in jammed suspensions. *Journal of Rheology*. 2015;50:63–84.
243. Likos CN. Soft matter with soft particles. *Soft Matter*. 2006;2(6):478–498.
244. Eckert T, Bartsch E. Re-entrant glass transition in a colloid-polymer mixture with depletion attractions. *Physical Review Letters*. 2002;89(12):125701.
245. Monti F, Fu SY, Iliopoulos I, Cloitre M. Doubly responsive polymer-microgel composites: Rheology and structure. *Langmuir*. 2008;24(20):11474–11482.
246. Willenbacher N, Vesaratchanon JS, Thorwarth O, Bartsch E. An alternative route to highly concentrated, freely flowing colloidal dispersions. *Soft Matter*. 2011;7(12):5777–5788.
247. Wiemann M, Willenbacher N, Bartsch E. Effect of cross-link density on re-entrant melting of microgel colloids. *Colloids and Surfaces A: Physicochemical and Engineering Aspects* 2012;413:78–83.
248. Schneider M, Michels R, Pipich V, Goerigk G, Sauer V, Heim H-P, et al. Morphology of blends with cross-linked PMMA microgels and linear PMMA chains. *Macromolecules*. 2013;46(22):9091–9103.
249. Stiakakis E, Vlassopoulos D Likos CN, Roovers J, Meier G. Polymer-mediated melting in ultrasoft colloidal gels. *Physical Review Letters*. 2002;89(20):208302.

250. Likos CN, Mayer C, Stiakakis E, Petekidis G. Clustering of soft colloids due to polymer additives. *Journal of Physics: Condensed Matter*. 2005;17:S3363–3369.
251. Camargo M, Likos CN. Phase separation in star-linear polymer mixtures. *Journal of Chemical Physics*. 2009;130(20):204904.
252. Camargo M, Likos CN. Unusual features of depletion interactions in soft polymer-based colloids mixed with linear homopolymers. *Physical Review Letters*. 2010;104 (7):078301.
253. Lonetti B, Camargo M, Stellbrink J, Likos CN, Zaccarelli E, Willner L, et al. Ultrasoft colloid-polymer mixtures: Structure and phase diagram. *Physical Review Letters*. 2011;106 (22):228301.
254. Camargo M, Egorov SA, Likos CN. Cluster formation in star-linear polymer mixtures: Equilibrium and dynamical properties. *Soft Matter*. 2012;8(15):4177–4184.
255. Parisi D, Truzzolillo D, Deepak FD, Gauthier M, Vlassopoulos D. Transition from confined to bulk dynamics. *Macromolecules* 2019;52(15):5872–5883.
256. Kandar AK, Basu JK, Narayanan S, Sandy A. Anomalous structural and dynamical phase transitions of soft colloidal binary mixtures. *Soft Matter*. 2012;8(39):10055–10060.
257. Abbas S, Lodge TP. Depletion interactions: Effects of added homopolymer on ordered phases formed by spherical block copolymer micelles. *Macromolecules*. 2008;41 (22):8895–8902.
258. Feng L, Laderman B, Sacanna S, Chaikin P. Re-entrant solidification in polymer–colloid mixtures as a consequence of competing entropic and enthalpic attractions. *Nature Materials*. 2015;14:61–65.
259. Stiakakis E, Vlassopoulos D, Roovers J. Thermal jamming in colloidal star-linear polymer mixtures. *Langmuir*. 2013;19(17):6645–6649.
260. Immink JN, Maris E, Crassous JJ, Stenhammar J, Schurtenberger P. Reversible formation of thermoresponsive binary particle gels with tunable structural and mechanical properties. *ACS Nano*. 2019;13(3):3292–3300.
261. Bischofberger I, Calzolari DCE, De Los Rios P, Jelezarov I, Trappe V. Hydrophobic hydration of poly-N-isopropyl acrylamide: A matter of the mean energetic state of water. *Scientific Reports*. 2014;4:4377.
262. Truzzolillo D, Vlassopoulos D, Gauthier M, Munam A. Thermal melting in depletion gels of hairy nanoparticles. *Soft Matter*. 2013;9(38):9088–9093.
263. Zaccarelli E, Mayer C, Asteriadi A, Likos CN, Sciortino F, Roovers J, et al. Tailoring the flow of soft glasses by soft additives. *Physical Review Letters*. 2005;95(26):268301.
264. Mayer C, Sciortino F, Likos CN, Tartaglia P, Löwen H, Zaccarelli E. Multiple glass transitions in star polymer mixtures: Insights from theory and simulations. *Macromolecules*. 2009;42(1):423–434.
265. Stiakakis E, Erwin BM, Vlassopoulos D, Cloitre M, Munam A, Gauthier M, et al. Probing glassy states in binary mixtures of soft interpenetrable colloids. *Journal of Physics: Condensed Matter*. 2011;23(23):234116.
266. Di Lorenzo F, Seiffert S. Counter-effect of Brownian and elastic forces on the liquid-to-solid transition of microgel suspensions. *Soft Matter*. 2015;11(26):5235–5245.
267. Mayer C, Zaccarelli E, Stiakakis E, Likos CN, Sciortino F, Munam A, et al. Asymmetric caging in soft colloidal mixtures. *Nature Materials*. 2008;7(10):780–784.
268. Marzi D, Capone B, Marakis J, Merola MC, Truzzolillo D, Cipelletti L, et al. Depletion, melting and reentrant solidification in mixtures of soft and hard colloids. *Soft Matter*. 2015;11(42):8296–8312.

269. Dzubiella J, Jusufi A, Likos CN, von Ferber C, Löwen H, Stellbrink J, et al. Phase separation of star-polymer-colloid mixtures. *Physical Review E.* 2001;64(1):010401(R).
270. Bayliss K, van Duijneveldt JS, Faers MA, Vermeer AWP. Comparing colloidal phase separation induced by linear polymer and by microgel particles. *Soft Matter.* 2011;7 (21):10345–10352.
271. Zhao C, Yuan G, Han CC. Stabilization, aggregation and gelation of microgels induced by thermosensitive microgels. *Macromolecules.* 2012;45(23):9468–9474.
272. Zhao C, Yuan G, Han CC. Bridging and caging in mixed suspensions of microsphere and adsorptive microgel. *Soft Matter.* 2014;10(44):8905–8917.
273. Zhao C, Yuan G, Jia D, Han CC. Macrogel induced by microgel: Bridging depletion mechanisms. *Soft Matter.* 2012;8(26):7036–7043.
274. Jia D, Cheng H, Han CC. Interplay between caging and bonding in binary concentrated colloidal suspensions. *Langmuir.* 2018;34(9):3021–3029.
275. Zong Y, Yuang G, Han CC. Asymmetrical phase separation and gelation in binary mixtures of oppositely charged colloids. *Journal of Chemical Physics.* 2016;145 (1):014904.
276. Jia D, Hollingsworth JV, Zhou Z, Cheng H, Han CC. Coupling of gelation and glass transition in a biphasic colloidal mixture-from gel-to defective gel-to glass. *Soft Matter.* 2015;11(45):8818–8826.
277. Truzzolillo D, Marzi D, Marakis J, Capone B, Camargo M, Munam A, et al. Glassy states in asymmetric mixtures of soft and hard colloids. *Physical Review Letters.* 2013;111 (20):208301.
278. Merola MC, Parisi D, Truzzolillo D, Vlassopoulos D. Asymmetric soft-hard colloidal mixtures: Osmotic effects, glassy states and rheology. *Journal of Rheology.* 2018;62 (1):63–79.
279. Cloitre M. Yielding, flow, and slip in microgel suspensions: From microstructure to macroscopic rheology. In Fernandez-Nieves A, Wyss HM, Mattsson J, Weitz DA (eds.) *Microgel Suspensions: Fundamentals and Applications.* Weinheim: Wiley; 2011; pp. 285–310.
280. Akcora P, Liu H, Kumar SK, Moll J, Li Y, Benicewicz BC, et al. Anisotropic self-assembly of spherical polymer-grafted nanoparticles. *Nature Materials.* 2009;8 (4):354–359.
281. Srivastava S, Agarwal A, Archer LA. Tethered nanoparticle-polymer composites: Phase stability and curvature. *Langmuir.* 2012;28(15):6276–6281.
282. Gohr K, Schärtl W. Dynamics of copolymer micelles in a homopolymer melt: Influence of the matrix molecular weight. *Macromolecules.* 2000;33(6):2129–2135.
283. Lindenblatt G, Schärtl W, Pakula T, Schmidt M. Structure and dynamics of hairy spherical colloids in a matrix of nonentangled linear chains. *Macromolecules.* 2001;34(6):730–736.
284. Green DL, Mewis J. Connecting the wetting and rheological behaviors of poly (dimethylsiloxane)-grafted silica spheres in poly(dimethylsiloxane) melts. *Langmuir.* 2006;22(23):9546–9553.
285. Borukhov I, Leibler L. Stabilizing grafted colloids in a polymer melt: Favorable enthalpic interactions. *Physical Review E.* 2000;62(1):R41–44.
286. Mangal R, Nath P, Tikekar M, Archer LA. Enthalpy-driven stabilization of dispersions of polymer-grafted nanoparticles in high-molecular-weight polymer melts. *Langmuir.* 2016;32(41):10621–10631.

7 Biocolloid Rheology

Surita Bhatia and Wendy Hom

The preceding chapters have provided the fundamental framework for colloidal forces and their role in the structure, dynamics, and rheology of particulate dispersions and polymeric assemblies. Here and in subsequent chapters, we describe the importance of these concepts to the design and understanding of colloidal systems in a variety of applications. We focus in this chapter on biocolloids, which we define broadly as (i) colloidal assemblies whose primary applications are biomedical in nature (e.g., block copolymers used in pharmaceutical formulations and biomaterials applications), and/or (ii) biomacromolecules that can be reasonably described with colloidal descriptions of the interparticle interactions; namely globular proteins and protein assemblies such as casein micelles. Obviously, this still encompasses a wide range of complex materials; our discussion mainly focuses on systems where concepts from colloidal interactions prove useful in interpreting the rheological behavior.

7.1 Landmark Observations and Example Applications

7.1.1 Colloid Rheology in Pharmaceutical and Biomaterials Applications

In 1992, Malmsten and Lindman [1] and Yu et al. [2] independently reported thermally-triggered gelation in aqueous dispersions of poly(ethylene oxide)-poly(propylene oxide)-poly(ethylene oxide) (PEO-PPO-PEO) triblock copolymers, whereby dispersions that behave rheologically as liquids at room temperature and gels at higher temperatures could be created. Note this behavior is counter to most systems where heating melts the solid phase. Further studies by these groups, as well as by Wanka et al. [3], Alexandridis et al. [4], Mortensen et al. [5], and others established the self-assembly of these polymers into spherical micelles, with some evidence that the observed liquid-gel transition could be connected to close-packing of spherical micelles into ordered structures. Mortensen et al. [5] first drew an analogy between the gelation transition in these block copolymer systems and crystallization of colloidal hard spheres, with gelation occurring in the triblock solutions at an effective volume fraction of ~0.53. SANS studies performed by Prud'homme et al. [6] confirmed that the gel phase comprises spherical micelles packed into a cubic structure, although solutions below the gel point were also found to contain some ordered domains. As discussed in this section, these landmark studies have led to an explosion of work on related systems with thermoreversible rheology.

This novel, inverse melting behavior has potential applications for these materials in pharmaceutical formulations and drug delivery. If the liquid–gel transition is tuned to occur near physiological temperatures (35–37°C), dispersions can be created that are low viscosity and can be easily injected at room temperatures, but show an increase in viscosity and/or storage modulus in vivo, enabling the extended release of active agents [7].

In a similar manner, these thermally-induced rheological transitions can be used as a means to trigger gelation through a relatively mild route, which can be beneficial for biomaterials applications, such as cell delivery, cell encapsulation, and seeding of cells in stiffer tissue engineering scaffolds. Early studies [8–12] demonstrated some limitations of these systems in terms of long-term stability of the gel phase in vivo and issues with copolymers impacting cell membrane permeability. However, more recent studies focusing on chemically-modified copolymers and blends of PEO-based block copolymers with polysaccharides have shown promise.

7.1.2 Colloidal Rheology of Proteins in Food and Biopharmaceutical Applications

Biomolecules are undoubtedly quite varied and complex in terms of their chemistry, solution interactions, shape, and rheology. Nevertheless, concepts from colloid suspension rheology have proved to be useful in understanding the flow properties behavior biomolecules. As early as 1921, Bogue [13] utilized the expression due to Einstein for the viscosity of two-phases systems, Eq. (1.1), modified by Hatschek to account for deformation of the dispersed phase, to describe rheology of concentrated gelatin sols. Much of the pioneering work in this area was motivated by the proteins found in food applications, in particular casein. The protein casein plays a major role in forming and maintaining the structure of dairy products such as yogurt and cheese, and the texture and consistency of these foods can be attributed in part to the rheology of casein assemblies in solution [14]. Casein usually associates into casein micelles in cow's milk and serves to solubilize calcium phosphate. In 1992, de Kruif et al. [15] observed that casein micelles exist as sterically stabilized particles having a radius of about 100 nm, and that they interact similarly to hard spheres under physiological conditions. In other words, milk can be described as a dilute colloidal dispersion of fat globules and casein micelles [16], where casein micelles are dispersed in a continuous phase containing mostly water and dissolved substances such as salts and protein. The casein micelles are quite stable, and can maintain their stability when dried, frozen, and heated; allowing for a wide range of production processes in the food industry [14].

de Kruif and others also noted that casein micelles can be destabilized by decreasing the pH or by the process of renneting in cheesemaking. This causes the micelles to attract and flocculate, resulting in a rheology and structure similar to that described in Chapter 5. The renneting process has been studied using shear rheology, and the behavior of the resulting solutions and gels has been described in terms of theories for adhesive hard spheres [17] and fractal aggregation [18,19]. Denatured whey proteins can also associate with casein micelles, increasing the micelle size and modifying the intermicellar interactions. While casein micelles do not denature when milk is heated, whey proteins do denature upon heating, and thus the interaction of whey proteins with casein micelles is important for processes involving heating of milk and milk products, such as yogurt production.

In 1999, the shear rheology of yogurt was quantitatively described by van Marle et al. [20] in terms of the intermicellar interaction strengths, inspired by work on the behavior of weakly flocculated colloidal suspensions of Potanin et al. [21], Buscall et al. [22], and Goodwin et al. [23]. More recently, this colloidal view has also found utility in describing rheological properties of concentrated dispersions of casein micelles in small amplitude oscillatory shear. Figure 7.1 shows the work of Olivares et al. [24], where the normalized high frequency storage modulus is compared to the expression derived by Buscall [25] for dense colloidal dispersions,

$$G'_{\infty} = \frac{N\phi_{\max}}{5\pi r}\frac{d^2U(r)}{dr^2}, \tag{7.1}$$

with a maximum packing fraction $\phi_{\max}$ of 0.68 for casein micelles [26], number of nearest neighbors N taken to be 7.5 in the disordered state, and interaction potentials proposed by Tuinier and de Kruif [27] and Berli and Quemada [28]. The model of Buscall provides a reasonable framework for describing G'_{∞} of these concentrated suspensions of food colloids. A further discussion of the use of Eq. (1.1) for studying soft colloids can be found in section 4.4.3.3 of *CSR* [29].

Although these landmark studies on protein rheology were motivated by food applications, recent advances in biomedicine have resulted in new therapies based on proteins and other biomacromolecules, leading to a renewed interest in understanding protein rheology for optimal processing of pharmaceutical formulations, as reviewed recently by Zhang and Liu [31]. This is complicated by the fact that such formulations are often multicomponent mixtures that contain surfactants or cosolvents

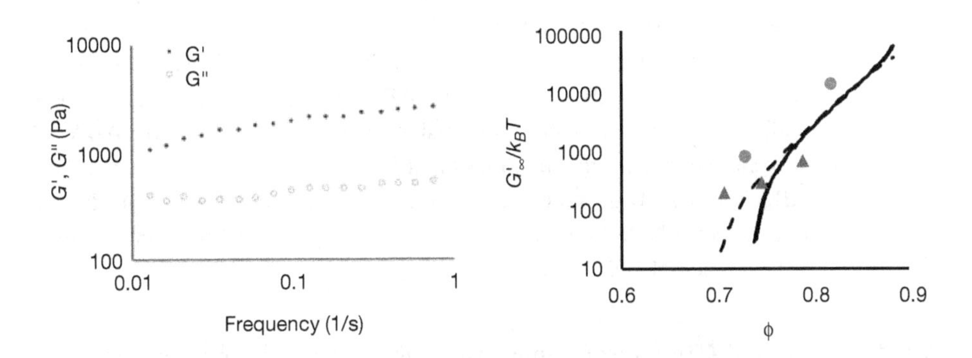

Figure 7.1 Left: Storage and loss modulus for skim milk concentrate, with a volume fraction of casein micelles of 0.723. Right: Dimensionless high frequency storage modulus of casein micelles as a function of volume fraction Circles are data from Bouchoux et al. [30], triangles are from Olivares et al. [24]. Solid and dashed lines are the predictions from Eq. (7.1) with interaction potentials proposed by Tuinier and de Kruif [27] and Berli and Quemada [28], respectively. After Olivares et al. [24]

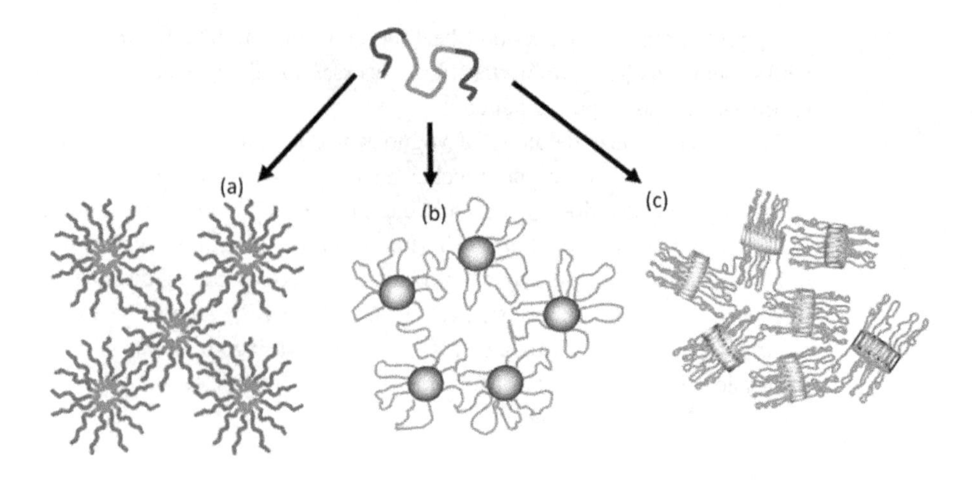

Figure 7.2 Schematic of some types of block copolymer micelles. (a) ABA triblocks in a solvent selective for the endblock, which may assemble into cubic gels; (b) ABA triblocks in a solvent selective for the midblock, which may form associative "flowerlike" micelles; (c) ABA triblocks with more specific interactions between the endblocks.

to promote the solubility of active agents, and additives and adjuvants to aid in administration and efficacy.

7.2 Self-assembled Colloids: Block Copolymer Micelles

As described in Chapter 6, block copolymer assemblies can be characterized by the parameters such as an effective volume fraction, softness of intermicellar repulsion, and strength and range of intermicellar attractions. The rheology can oftentimes be interpreted in terms of models that have been developed for colloidal systems. Here, we focus on three particular structures that have found use in several applications (Figure 7.2): (a) AB diblocks and ABA triblocks in a solvent selective for the endblock which have thermoresponsive rheological behavior, (b) ABA triblocks in a solvent selective for the midblocks that form associative "flowerlike" micelles, and (c) ABA triblocks with more specific interactions between the endblocks, such that the micelle cores have some structure and order (e.g., formation of crystalline domains or helical assemblies).

7.2.1 AB and ABA Block Copolymers: Cubic Gels and Thermoresponsive Behavior

Among the most widely studied family of block copolymers for biomedical applications are those containing PEO blocks. Landmark studies involving these systems have been alluded to above. Early work focused on PEO-PPO-PEO and PEO-PPO copolymers [1–6], often referred to as poloxamers or by the trade names Pluronic®, Lutrol®, Synperonic®, and Kolliphor®. More recent studies have investigated

PEO-PBO-PEO systems and more complex polymer architectures, such as penta-blocks with various endblocks [32–35]. Here, we focus on the behavior of aqueous formulations of AB and ABA block copolymers in which the "A" block is PEO and the "B" block is another poly(alkylene oxide), such as PPO or poly(butylene oxide) (PBO). The solution morphology of these copolymers has been well-characterized using SAXS and SANS. Most poly(alkylene oxides) exhibit a lower critical solution temperature (LCST) in water. This behavior leads to a decrease in solubility as temperature is increased, with implications for both self-assembly and rheology.

As concentration or temperature is increased, many PEO-containing block copolymers show a progression of isotropic micellar-cubic micellar-hexagonal-lamellar structure, analogous to the phase behavior seen in surfactant solutions [3,36,37]. Many biomedical applications of these materials focus on the first transition in this series, i.e., from an isotropic liquid of spherical micelles to a micellar phase with cubic order. The effect of increasing temperature on these systems is complex. The decreasing solubility of the PEO blocks in water as temperature increases can lead to a more compact micellar corona; however, increasing temperature can also lead to the development of attractive interactions between micelles, again due to the increasingly unfavorable interaction between PEO corona chains and water at higher temperatures.

The relative importance of these two effects, and the structure of the cubic phase itself, depends strongly upon the specific copolymers, block lengths, and concentrations being explored. As noted, Prud'homme et al. [6] demonstrated that the gel phase of a widely-studied PEO-PPO-PEO triblock, Pluronic® F127, has a cubic structure. This copolymer has the approximate formula (EO)100–(PO)70–(EO)100. Prud'homme et al. argue for a simple cubic structure; however, several other studies show a face-centered cubic (FCC) structure, and some a body-centered cubic (BCC) structure. These variations in the structure of the gel phase in commercial block copolymers have been attributed to batch-to-batch variation and a small population of diblocks present in some commercial copolymers.

Research by Hamley et al. [38,39] and Pople et al. [40] on PEO-PBO diblocks and PEO-PBO-PEO triblocks showed that sharp, short-range interactions, typical of hard spheres, favor a face-centered cubic (FCC) structure, while softer repulsive interactions favor a body-centered cubic (BCC) structure, as would be expected based on results from soft colloids. Details of the hard sphere and soft sphere potentials are described in Chapter 1. In this system, the FCC structure is favored at high temperatures, due to contraction of the PEO corona [41], consistent with studies on other soft sphere colloids described in Chapter 6.

Formation of this cubic phase at higher temperatures corresponds to the development of a gel from low viscosity solutions as the temperature is increased [1–3,6], although ordered domains have also been reported in the liquid phase. An example of the linear viscoelasticity of these systems, obtained by small amplitude oscillatory shear measurements, is shown in Figure 7.3. These data are the storage and loss moduli for a widely-studied PEO-PPO-PEO system, Pluronic® F127, formulated as an aqueous dispersion at a polymer concentration of 20 wt%. The system transitions from viscoelastic liquid behavior at 15°C to gel-like rheological behavior at 37°C,

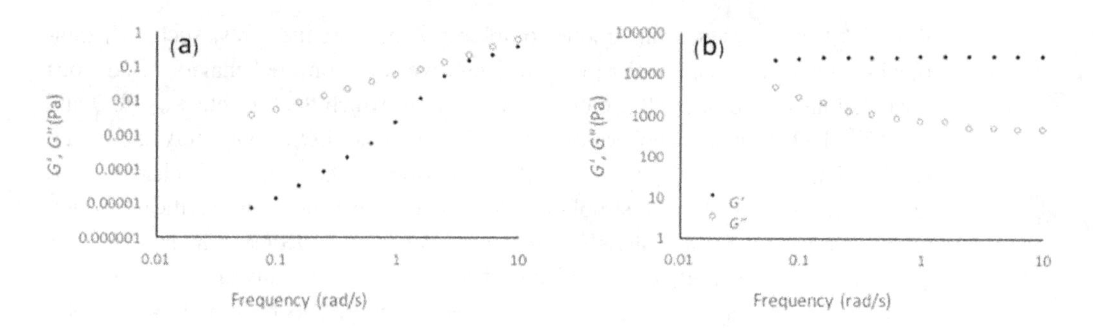

Figure 7.3 Storage and loss moduli of 20 wt% F127 in water as a function of frequency and at temperatures of (a) 15°C and (b) 37°C. Data from Quah et al. [42]

with *G'* nearly independent of frequency and *G'* > *G"* over the measurable frequency range [42].

Interesting temperature effects are also observed in the steady shear measurements conducted on systems that are in the liquid region. Early work by Prud'homme et al. [6] on aqueous solutions of Pluronic® F127 shows a transition from low viscosity Newtonian to high viscosity shear thinning behavior as temperature is increased. The region of shear thinning behavior appears to correspond to liquids that show both disordered domains and some domains with cubic order, suggesting that the presence of these larger partially-ordered domains is related to the non-Newtonian behavior. More recently, Jalaal et al. [43] have explored Pluronic® F127 solutions at higher temperatures in more detail, and report viscoplastic behavior, with systems exhibiting a yield stress as large as several hundred Pa. The data at higher temperatures appear to fit well to a Herschel–Bulkley model (see Table 1.A1).

Although many block copolymers of this type display liquid-gel transitions corresponding to formation of a micellar crystal, some show a behavior that is analogous to other types of dynamic transitions observed in colloidal dispersions. Chen et al. [44] explored dynamics in the short triblock copolymer $(EO)_{13}(PO)_{30}(EO)_{13}$ (Pluronic® L64) as a function of polymer concentration and temperature. They find state transitions that map onto the expected state diagram for adhesive hard sphere (AHS) colloids, with a liquid–glass transition as temperature is increased at $\phi_{eff} = 0.535$, and repulsive glass-attractive glass transitions as temperature is increased at $\phi_{eff} = 0.542$, 0.544, and 0.546 [44].

Rheological state diagrams have been reported for several copolymers, which show gel–liquid transitions as a function of temperature and polymer concentration. In some cases, regions of liquids, "soft gels", and "hard gels" have been reported, although the rheological distinctions between these different states is not always evident. Data for copolymers of differing molecular weights can be scaled using the critical micelle concentration and temperature; in general, increasing the PEO block length shifts the phase boundaries to lower concentrations [45]. However, for a given copolymer, such as Pluronic® F127, there are variations in the rheological state diagrams reported in the literature. As with structural studies, these differences may be due to batch-to-

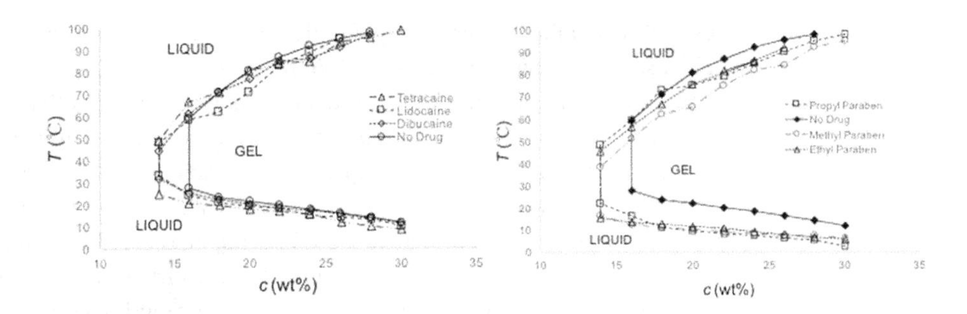

Figure 7.4 The lower and upper gelation boundaries of the ABA triblock copolymer Pluronic® F127 in the presence and absence of the (left) local anesthetic drugs dibucaine, tetracaine, and lidocaine; and (right) the pharmaceutical additives methyl paraben, ethyl paraben, and propyl paraben. Data from Sharma and Bhatia [46] and Sharma et al. [47]

batch variation in commercially-available copolymers. Additionally, as discussed further later in this section, variations in reported liquid–gel state diagrams may also arise from differences in how the gel and liquid states are defined and measured, as well as from temperature history/hysteresis effects.

As many of these polymers are used in multicomponent formulations with active agents and excipients, it is important to recognize that these other components will impact the assembly and rheology of these systems. Sharma and Bhatia [46] and Sharma et al. [47] demonstrated that hydrophobic small-molecule pharmaceuticals and additives tend to shift the gelation boundary of Pluronic® F127 to lower temperatures and concentrations; examples are shown in Figure 7.4. The extent to which the addition of small molecules shifts the liquid–gel boundary has been found to correlate with the micelle-water partition coefficient (which can be taken as an indirect measure of the hydrophobicity of the additive) and with the solubility of the additive in the copolymer solution [47]. In many cases, these changes in the phase boundary are driven by an increase in the micellar size and decrease in the aggregation number [48], both of which contribute to an increase in the effective volume fraction at fixed polymer concentration.

Similarly, many salts, buffers, additives, cosolvents, and biomacromolecules used in typical pharmaceutical, biomaterials, and personal care formulations have been found to decrease the critical micelle concentration (CMC) and/or increase the micelle sizes for these types of copolymers. This again leads to a relative increase in the effective volume fraction, thus decreasing the polymer concentration and temperature needed for gelation. Matthew et al. [11] and Sharma et al. [49] reported this type of effect for Pluronic® F127 in common formulations used for mammalian cell culture, minimum essential media (MEM), and MEM with added fetal bovine serum (MEM-FBS).

The impact of larger biomacromolecules on the assembly, intermicellar interactions, and gelation behavior of these polymers is complex, as would be expected. However, in many cases, the resulting multicomponent formulations still display thermoresponsive rheology, and the ability of the copolymers to form micellar assemblies with crystalline order is surprisingly robust. Blends of Pluronic® F127 with the glycosaminoglycan hyaluronic acid, of interest for injectable treatments for arthritis and other joint and cartilage disorders, display thermosensitive rheology [50]. Formulations of PEO-PPO-PEO copolymers with the polysaccarhide alginate have also been explored by several groups, for applications such as transdermal delivery [51] and controlled release of proteins and riboflavin [52]. These formulations can exhibit an increase in the storage modulus of up to two orders of magnitude as temperature is increased [42], and SANS studies show that the alginate does not hinder micelle formation or organization into a cubic lattice at higher temperatures [53]. The rheological transitions in these systems are thermoreversible although, as often in such cases, some hysteresis is observed (Figure 7.5), possibly due to the slower reorganization of micelles with temperature due to the presence of the alginate chains.

It is important to recognize that several of the liquid–gel boundaries reported in the literature have been determined by tube inversion, rather than by more detailed rheological measurements. Some comparisons between tube-inversion studies and quantitative rheological measurements have been performed. For example, Li et al. [54] demonstrated that the sol–gel boundaries for a PEO-PBO copolymer determined by tube inversion corresponded to the crossover point between G' and G'' measured at a frequency of 1.0 Hz. Early work by Kelarakis et al. [45] attempted to quantify studies that report "hard gel" and "soft gel" states by specifying a "hard gel" as one that does not flow upon tube inversion, and reported this to correspond to solutions with a yield stress exceeding 40 Pa. Systems with $G' > G''$ and a small but measurable yield stress were defined as "soft gels", and solutions with $G'' > G'$ that flow were defined as sols [45]. More recently, Hyun et al. [55] demonstrated that the

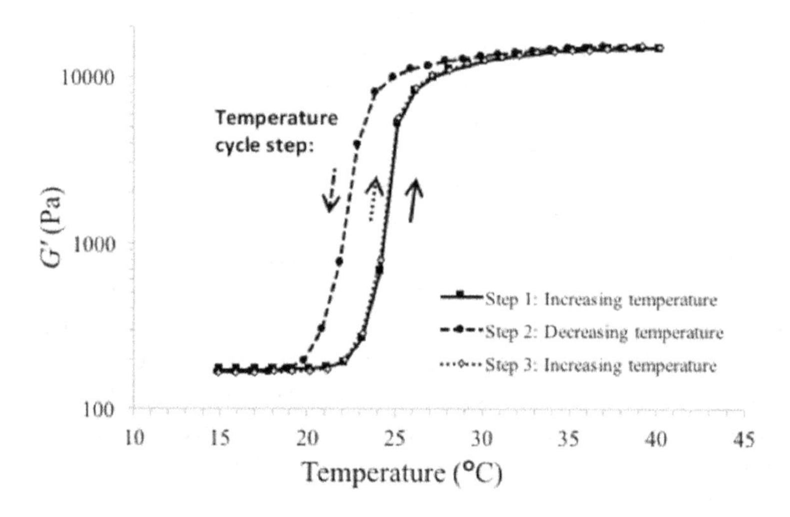

Figure 7.5 Storage modulus from oscillatory shear at fixed frequency and stress amplitude during thermal cycling of a thermoreversible hydrogel containing 0.5 wt% alginate and 20 wt% Pluronic® F127. Data from Quah et al. [42]

regions typically designated as "soft gels" and "hard gels" display different rheology under large amplitude oscillatory shear (LAOS). Systems in the "soft gel" state display a combination of both type I (strain thinning in both the viscous and elastic response) and type IV (strain overshoot in both the viscous and elastic response) LAOS behavior, with a Lissajous pattern that evolves from an elliptical shape to a tilted, rounded rectangular to a symmetrical ellipse as strain is increased. By contrast, systems in the "hard gel" regime show type III LAOS behavior (strain-thinning in the elastic response and an overshoot in the viscous response), and a Lissajous pattern that rapidly transitions from an ellipse to a rounded rectangle with increasing strain [55].

7.2.2 Associative ABA Triblock Copolymers

An alternative strategy for creating block copolymer systems with novel rheology is to utilize ABA triblocks where the "A" endblocks are hydrophobic and the "B" mid-blocks are hydrophilic. This morphology leads to formation of "flowerlike micelles" with bridging chains between micelles (Figure 7.2, center), corresponding to an increased intermicellar attraction and an increase in viscosity and elasticity as polymer concentration is increased. In this section, we describe the rheology of systems containing only associative polymer micelles. Applications of similar polymers as rheological modifiers in particulate dispersions, relevant to paints and coatings, are discussed in Chapter 9. With respect to pharmaceutical formulations, these types of polymers have also been used in multicomponent systems together with microemulsions and emulsions, for example to control the rheology of ointments, lotions, and injectables. Although assembly in these multicomponent systems can be quite complex, oftentimes the physics underlying the rheology remains the same – bridging polymer chains impart an attractive interaction between assemblies, leading to an increase in the viscosity and/or storage modulus.

Early work on the rheology of associative polymers was based on modifying the classic description of polymer networks of Green and Tobolsky [56], which describes the elasticity as being proportional to the crosslink density of the network, n. This concept was adapted to associative polymers by Annable et al. [57] and Tanaka and Edwards [58], who envisage these systems to be comprised of network junctions that are reversible, with a network junction lifetime of t. The storage and loss moduli can then be described in terms of the chain breakage rate under shear, which is determined by the process of the hydrophobic endblock binding to the micellar core [58]. With this picture, the viscoelastic moduli show one characteristic timescale, the junction lifetime t, and there is a single characteristic crossover frequency at which G' exceeds G''. The junction lifetime is expected to depend exponentially on the interaction strength χ_{as}. Experimentally, this can be realized by varying the length of the hydrophobic block. These models predict that χ_{as} should alter the dynamics in the system (e.g., the crossover frequency), but not necessarily G'_{∞}, the high frequency plateau of the storage modulus, which can be thought of as a measure of the gel stiffness.

Both experiments and simulations, however, show that the interaction strength impacts not only the dynamics of the network, but also the stiffness of the gel. Simulations of triblock copolymers performed by Nguyen-Misra et al. [59] show that a larger χ_{as} leads to both slower dynamics and an increased G'_{∞}. These simulations predict a strong dependence of G' on the length of the hydrophobic block L_{end}, but only a weak dependence on the length of the midblock L_{mid}, and polymer concentration c. Experimental work on model ABA copolymer gels by Tae et al. [60] suggests that the detailed nature of the hydrophobic interactions and self-assembly must also be taken into account when describing the viscoelasticity of these systems. Physically, these can be quantified by the micelle aggregation number N_{agg}, micelle size R_{mic}, and volume fraction of micelles ϕ.

One aspect of the physics that is missing from the above description is the impact of the looping midblock chains on the viscoelasticity. From a colloidal standpoint, these looping chains give rise to a soft intermicellar repulsion, which must be accounted for when considering the dynamics of the system. Semenov et al. [61] first accounted for this in a scaling theory that described movement of associative polymer micelles as being hindered not only by intermicellar attractions from bridging midblocks, but also by crowding from neighboring micelles. When the system is subject to a stress, in order for the micelles to rearrange and the system to relax, bridging midblocks must dissociate from neighboring micelles, and the micelle coronas formed from looping chains must deform as micelles squeeze past one another. This second mechanism only comes into play at higher volume fractions, near random close packing. Thus, the rheology can be described in terms of two characteristic timescales: (i) the junction lifetime, as described above, which depends upon the associative strength, and (ii) a timescale for micelle diffusion that accounts for the energetics of deformation. The relative importance of these timescale leads to different types of rheological behavior as ϕ is increased. The predictions of this model, as well as results from others, are summarized in Table 7.1. For some studies (e.g., Monte Carlo simulations), analytical expressions are not available, but comparisons to trends and qualitative predictions can be made.

7.2.3 Specific Interactions: Stereocomplexation and Crystallinity

Biological polymers and gels, such as cellulose and spider silk, often contain well-organized or crystalline domains within a softer matrix. This has led researchers to explore similar types of strategies with block copolymers where the "A" blocks can form structures such as crystallites, helices, or beta sheets. Such an approach has been employed for synthetic block polypeptides, as described by Deming [62] and reviewed recently by Carlsen and Lecommandoux [63], an hierarchical assembly can be observed in such systems (Figure 7.6), along with complex rheological properties. Associative block copolymers with blocks that can form crystalline or semicrystalline domains, i.e., Figure 7.2(c), such as block copolymers containing PEO and poly(lactic acid) (PLA), are more amenable to colloidal interpretation of their rheological behavior, as we can more directly predict how parameters in the models shown in Table 7.1, such as χ_{as} and Nagg, are impacted by crystalline domains.

Table 7.1 Summary of selected theoretical predictions for polymeric networks and associative polymers, showing dependence on various physical parameters

Model	Predictions for Rheological Properties
Associative polymer networks [56]	$G'_{\infty} \sim \nu k_B T$
MC simulations on associative polymers modified for topology [57]	$G'_{\infty} \sim \nu f(c) kTG'_{\infty}$ f nonlinear function of c that increases quickly at low c and plateaus at high c
Scaling theory for associative polymers [61]	$G'_{\infty} \sim N_{agg}^{9/10} R_{mic}^{-3} k_B T$, low ϕ $G'_{\infty} \sim N_{agg}^{3/2} (\phi - 1)^2 R_{mic}^{-3} k_B T$, moderate $\phi G'_{\infty} \sim N_{agg}^{3/2} \phi^{7/12} R_{mic}^{-3} k_B T$, high ϕ
MC simulations on associative triblocks [59]	G' increases as χ_{as} or L_{end} increases Weak dependence of G' on c and L_{mid} $G' \sim \omega^2$, $G'' \sim \omega$ for low L_{end} $G' \sim \omega$, $G'' \sim \omega^{0.5}$ for high L_{end}

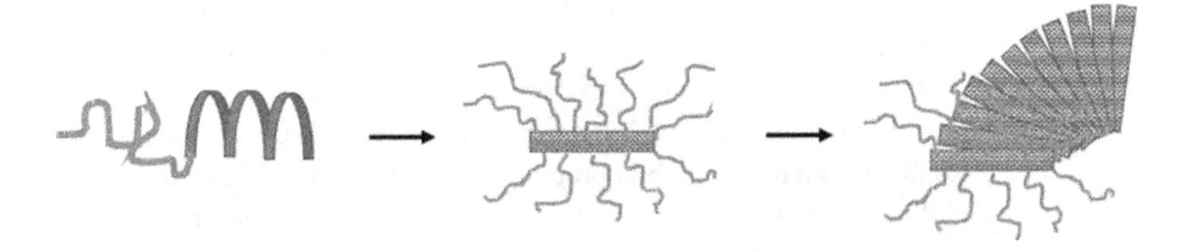

Figure 7.6 An example of specific interactions in a block polypeptide, leading to disk-shaped micelles and further assembly of micelles into fibrils. After Deming [62]

Rheology of proteins, which obviously also contain specific interactions, is discussed in Section 7.3, although there are some parallels to the systems described here.

Block copolymers of PLA and PEO in both the solid state and in aqueous media have been widely studied for biomedical applications [64–73] due to the general biocompatibility of both blocks and the biodegradability of the PLA block. PLA-PEO-PLA triblock copolymers have been shown to form micellar structures in water. They undergo a liquid–gel transition when the polymer concentration is increased, with the liquid–gel transition depending upon block length [72,74]. Gels comprising blends of PLLA and PDLA have the possibility of forming PLLA/PDLA stereocomplexes within the hydrogel [75]. Temperature-dependent, injectable hydrogels can be formed by inducing stereocomplex formation of enantiomeric micelle mixtures of PLA-PEO-PLA triblock copolymer [76]. The presence of these stereocomplexes leads to formation of very strong junction points within the physically associated hydrogel, and the resulting materials have been explored for drug release and tissue engineering scaffolds [75].

Agrawal et al. [77] have explored another route for incorporating strong junction points into associative PLA-PEO-PLA hydrogels through the formation of micelles with crystalline PLA domains, by using copolymers containing poly(L-lactic acid), PLLA-PEO-PLLA, and copolymers synthesized from a racemic mixture of L-lactide and D-lactide, PL50D50LA-PEO-PL50D50LA. Wide-angle x-ray diffraction (WAXD) data on systems with a L/D ratios of 100 show peaks at scattering angles of 17° and 19° which are characteristic of crystalline PLLA [73], and samples with L/D ratios of 50/50 show no evidence of PLA crystals. The crystallites present in the PLLA-PEO-PLLA systems persist even when the polymers are hydrated in the gel state. SANS experiments [78] yielded spectra that fit well to models of spherical micelles for systems with a L/D ratio of 50/50, and flat disk-shaped micelles for polymers with an L/D ratio of 100. This is consistent with the concept that the stereoregular polymers form crystalline domains. Moreover, previous theoretical work had suggested that the morphology we observed for the PLLA-PEO-PLLA systems should only be accessible in the "superstrong segregation" regime [79]; this work demonstrates that polymer stereochemistry alone could be used to drive formation into this unexpected morphology of disk-like or sheet-like micelles.

The difference in the nanoscale structure has important implications for the rheology of these systems. For the PLLA-PEO-PLLA systems with crystalline domains, the storage modulus, G', is very sensitive to the length of the PLA block [77,80]. SANS studies show that increasing the PLLA block length does not strongly affect the thickness of the disk-shaped micelles, but does result in a strong increase in their radius. Thus, from a colloidal perspective, the effective volume fraction of the crystalline domains, which is proportional to the radius of the disks cubed, increases rapidly with PLA block length. Moreover, in comparing the rheology of gels with the same polymer molecular weight and concentration, it is found that the presence of crystalline PLA domains results in an increase in G' of up to three orders of magnitude at low frequencies [77].

It is important to note that the effect of these types of specific interactions on the rheology may not follow from the chemical structure in a straightforward manner. For example, associative micellar gels comprising PLA-PEO-PLA triblocks with varying amounts of L- and D-lactide have been found to display an unexpected maximum in the storage modulus at an intermediate L/D ratio [81]. No stereocomplexation is observed in the gels. Rather, the unique rheology has been attributed to competition between an increase in the time for PLA endblocks to pull out of micelles as the L/D ratio is increased and PLLA crystallization occurs, and a decrease in the number of bridging chains for micelles with crystalline PLA domains, as formation of bridges may be hindered by crowded crystalline PLA domains. Thus, these types of specific interactions may have competing effects on the various parameters, listed in Table 7.1, that characterize the assembly (e.g., N_{agg} R_{mic}) and the nature of the intermicellar interactions (e.g., ν, χ_{as}).

7.3 Protein Solutions

While early work on the rheology of proteins was focused on food applications, more recent studies have been motivated by protein-based biotherapeutics, including rDNA-derived monoclonal antibody (mAb) drugs. The latter have been very effective in treating a host of diseases. The rheological behavior of protein solutions can have a profound effect on the delivery and bioactivity of pharmaceutical formations, especially if there are factors that modify the interactions that dictate protein structure. The ability to understand and tune the behavior of these multicomponent materials has a tremendous impact on applications in biotherapeutics and in commercially-available products. Using proteins in pharmaceutical formulations is challenging as their structure needs to be maintained throughout all stages of manufacturing, storage, and delivery to have optimal efficacy. Physical and/or chemical instabilities such as aggregation, precipitation, oxidation, and deamidation can arise during suboptimal conditions of protein expression and purification, drug formulation, storage, and handling [82,83].

There are four main types of proteins: soluble, which includes globular proteins; membrane; fibrous; and intrinsically disordered [84]. We will focus here on globular proteins, as their characteristics make them most amenable to drawing analogies to colloidal dispersions with respect to their solution rheology. Some examples of globular proteins include serum albumins, hemoglobin found in blood, and immunoglobulins. Bovine serum albumin (BSA) has long been used as a model system for globular proteins, and it has been found to act approximately as a hard colloidal particle as a result of long-range repulsive interactions [85,86]. Lysozyme has also been used as a model system, displaying dynamical arrest behavior and structures at intermediate and large length scales that have been interpreted in terms of analogies to colloids with a short range attraction and a long-range repulsion [87,88].

Early work on BSA solutions by Ikeda and Nishinari [86,89] showed an apparent yield stress, with a flow curve in steady shear that could be described with a Bingham plastic model, and properties in low amplitude oscillatory shear characteristic of a soft solid. However, more recent experiments [85,90,91] clearly show that this apparent yield stress and solid-like behavior is due, at least in part, to interfacial effects. In other words, formation of an adsorbed layer of proteins or "protein skin" at the air–water interfaces dominates the bulk rheology. This calls for added scrutiny of earlier measurements on protein solutions performed with conventional rotational rheometers. This effect can be eliminated by the use of microfluidic rheometers, which are based on flow through channels imprinted on a microfluid chip [85,90]. It has also been proposed to incorporate surfactants such as polysorbate 80 and sodium dodecyl sulfate (SDS) to remove proteins at the interface and eliminate concerns for errors due to interfacial effects [90]. Once interfacial effects are removed, the data can be more clearly interpreted in terms of the protein volume fraction and solution interactions; for example, viscosity data from microfluidic rheometry on BSA solutions [85] has been found to follow the familiar expression for spherical colloids [92]:

$$\eta/\eta_0 = 1 + 2.5\phi + s\phi^2, \tag{7.2}$$

with the empirical constant $s = 10$, consistent with spherical charge-stabilized colloids [85]. More on this expansion can be found in Section 1.4.

While many groups have utilized a colloidal picture to describe the rheology of various globular protein solutions, such as eye lens α-crystalline and BSA under various buffer conditions [85,93,94], there has been some doubt cast on the validity of using these types of models to describe protein rheology more generally [85,93–95]. This holds in particular for limitations of these models in considering the protein primary structure, flexibility, and conformational changes [96]. There has been some success applying expressions developed for colloidal solutions to protein solutions at lower concentrations, such as the Krieger–Dougherty equation:

$$\frac{\eta}{\eta_0} = \left(1 - \frac{\phi}{\phi_{\max}}\right)^{-[\eta]\phi_{\max}}, \tag{7.3}$$

where $[\eta]$ is the intrinsic viscosity. A derivation of this and related expressions also used for protein solutions can be found in chapter 1, appendix B of *CSR*. However, these approaches apply best for globular proteins having only electrostatic repulsions, where a model of a hard sphere core with electrostatic repulsion can be used to describe protein–protein interactions. Even for these cases, Sharma et al. [85] find that Eq. (7.3) overpredicts the viscosity in the case of repulsive ellipsoidal proteins.

Expressions developed for colloidal dispersions, modified to account for shape anisotropy and more complex interparticle interactions, have also been applied to protein solutions. For example, expressions such as the Ross–Minton equation [97], which accounts for short-range attractions and was first applied to hemoglobin solutions, has found utility:

$$\eta_r = \exp\left(\frac{[\eta]c}{1 - (k/S)[\eta]c}\right). \tag{7.4}$$

Here, c is the protein concentration, k is a parameter to describe self-crowding, and S is the Simha shape parameter. In many cases, the protein–protein interaction has both short-range attraction and long-range repulsion components, with the repulsion range sometimes comparable to the size of a protein itself. This competition of interactions have been found to alter solution viscosity [88]. For example, Lilyestrom et al. [98] used a modified Ross–Minton equation to characterize the solution viscosity of mAb1, and found that attractive intermolecular interactions directly affected solution viscosity at high protein concentrations, which supports that molecular interactions of mAb1 play a large role in dynamic behavior of bulk solutions. Godfrin et al. [99] measured the viscosity of very concentrated lysozyme in D_2O with a capillary rheometer and found that the theoretical and experimental viscosity data deviate when the volume fractions of lysozyme are over ~0.15. In addition, when the temperature decreases from 50°C to 5°C, the viscosity of solutions with high protein concentrations significantly increases due to the heightened attraction strength, which also results in formation of structures with intermediate-range order. Although a large viscosity change is observed here, the lysozyme solutions were still found to behave as Newtonian fluids at all concentrations studied.

As mentioned before, despite their chemical and structural complexity, solutions of relatively low concentrations of protein often exhibit Newtonian behavior. This has implications for use of these solutions in pharmaceutical formulations, such as mAb drugs.

Although many approved mAb drugs are formulated for the intravenous route, there has been a great need to produce pharmaceutical formulations containing high protein concentrations of greater than 50 mg/ml, especially in the case of delivering a dose of high concentrations of monoclonal antibodies in the small volumes (<1.5 ml), such as those required via the subcutaneous route [100–102]. Compared to globular protein solutions, controlling the rheology of mAB solutions presents a challenge; however, some understanding derived from studies of globular proteins may still be useful [31].

Higher protein concentration solutions have different possible protein–protein interactions than dilute protein solutions [103,104]. Specifically, distances between molecules of protein are greater at dilute concentrations, which allows long-range interactions to dominate over the short-range protein–protein attractions. With increasing protein concentration, however, the relative effect of short-range interactions increases. These competing interactions may result in reversible cluster formation, the impacts of which on the rheology have been discussed by Godfrin et al. [105], among others. Concentrated solutions are often non-Newtonian and exhibit shear thinning behavior [106,107]. Some interactions such as hydrogen bonding, ionic and hydrophobic, can influence the rheological behavior of protein solutions, which allows for further characterization [106]. Added salts can also increase the viscosity of various mAB formulations [105,106]. Without salts, Yearley et al. [107,108] found that a mAB of interest could reversibly associate into a loosely connected transient cluster composed of monomers. Upon adding 50 mM Na_2SO_4, their mAB was able to associate into long dimers over longer time periods and then form larger transient clusters at a higher concentration, and it is these clusters that result in a significant increase in the viscosity of the mAB solution. It was also observed that the viscosity increased when small clusters with extended open structures are formed, and that added salts disrupted these structures and decreased the mAB solution viscosity [108,109]. However, an interesting finding is that, regardless of the change in viscosity based on the presence/absence of salt, the structure–viscosity relationship for both mABs studied were the same [105,108].

An example is shown in Figure 7.7, where Lilyestrom et al. [98] use Eq. (7.4) written in the form known as the modified Ross–Minton equation:

$$[\ln(\eta_r)]/c = \left(\frac{k}{S}\right)[\eta][\ln(\eta_r)] + [\eta] \tag{7.5}$$

Eq. (7.5) is constructed such that suspensions dominated by excluded volume interactions should exhibit a straight line when plotted as inherent viscosity versus relative viscosity on a log–log plot, as observed for the hard spheres shown in Figure 7.7 [99]. The intercept is the intrinsic viscosity, while the slope is the crowding factor. Note that straight line fits are observed for both the hard sphere colloidal suspensions as well as mAb suspensions that are dominated by repulsive interactions. However, a mAb suspension that is thought to show associations in the form of dimerization exhibits significant curvature, consistent with increasing dimerization with increasing crowding [99]. Associations leading to nontrivial rheological behavior have been reported for a variety of mAb protein solutions [105,111].

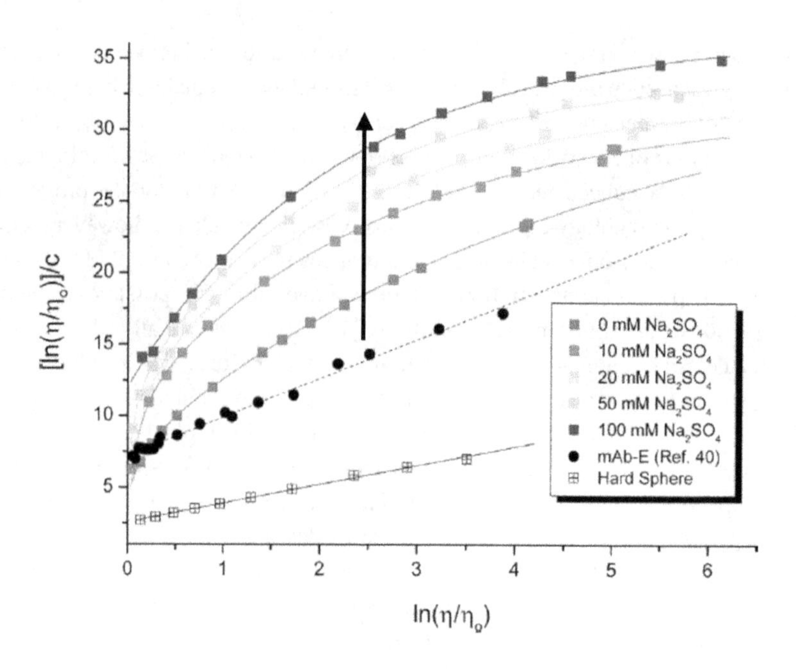

Figure 7.7 Plot for the viscosity of a variety of biocolloid suspensions according to the modified Ross–Minton equation (7.5) for a variety of suspensions as labeled. The linear behavior is characteristic of hard sphere dispersions, with the remaining plots for mAb solutions with increasing sodium sulfate concentration, as labeled and indicated by the arrow. Added electrolyte drives attractive interactions and dimerization, which leads to a deviation from hard sphere behavior. Reprinted with permission from Lilyestrom et al., Monoclonal antibody self-association, cluster formation, and rheology at high concentrations. *The Journal of Physical Chemistry B* 2013;117(21):6373–6384 [98] by the American Chemical Society

Interactions between proteins, particularly in solutions at higher concentrations, may also lead to irreversible aggregation. Impacts of the aggregation process on rheological behavior have been discussed by Castellanos et al. [90]. For both BSA and the mAb IgG1, the presence of aggregates leads to a bulk yield stress [90]. Aging of these solutions should also be considered when characterizing the rheology, as the aggregation process may be slow and can take months under storage conditions [90]. Aggregate formation may be accompanied by protein unfolding and conformational changes, with a corresponding loss of biological function. However, some proteins are able to maintain their primary structure while forming ordered super-structures [82]. These concentrated formulations also display increased viscosity due to either aggregation or other protein–protein interactions, which gives rise to a host of challenges in manufacturing and delivery. Protein solution viscosity can be affected by various variables, including pH, temperature, surface charge, ionic strength, molecular shape, shear rate, as well as solvent viscosity. Measuring rheological behavior through creep/relaxation, steady shear, and small and large amplitude oscillatory shear experiments can provide a better understanding of the complex behavior needed to develop more stable and efficacious protein formulations [31]. Once armed with an understanding of the behavior of a protein formulation, experimental conditions (e.g., co-ions, counter-ions, pH) at different stages of manufacturing, as well as conditions during storage and delivery can be tuned to see their effect on protein–protein interactions [102,105,110].

7.4 Conclusions and Outlook

Despite the chemical complexity of biocolloids, various models of colloidal interactions can provide a useful framework to understand, and ultimately tune, the rheology of these systems. These must be applied together with an understanding of polymer physics and/or protein biochemistry to develop a complete picture of how the structure, self-assembly, and solution interactions of such systems may vary with temperature, pH, chemical structure, and concentration. For block copolymer micelles utilized in biomedicine, the ability to more precisely tune the rheology through specific interactions is a promising direction. Recent advances in polymer synthesis, such as the segmer assembly polymerization (SAP) approaches of Meyer and co-workers [111–113], allow for the creation of sequence-specific regions within block copolymers. These approaches have been applied to PLA and poly(glycolic acid) copolymers [111–113] to control the stereochemistry on a finer scale than the PLA-PEO-PLA studies described; however, the effects on self-assembly and solution rheology remain to be explored.

The case for protein rheology is also challenging, where new approaches based on colloidal science and suspension rheology are needed to create low viscosity dispersions (e.g., for injection and/or ease of processing) at high protein concentrations. The use of emulsions, which were discussed in Chapter 6, including micro- and nanoemulsions, have been explored as a solution, but maintenance of biological activity is of concern, as changes in chemical and physical properties of the protein's environment will inevitably affect its therapeutic ability. Experimental rheological measurements of protein solutions benefit from careful understanding of the possible role of protein aggregation at hydrophobic interfaces, such as the air–water interface, and the associated contribution from surface rheology.

Acknowledgments

We thank Madani Khan for assistance in preparing graphics for this chapter.

Chapter Notation

c polymer or protein concentration [g/cm^3]
k self-crowding parameter [–]
L_{end} length of hydrophobic endblock [m]

L_{mid} length of midblock [m]
N number of nearest neighbors [–]
N_{agg} aggregation number [–]
R_{mic} micelle radius [m]
S Simha shape factor [–]

Greek Symbols

χ_{as} strength of association [J]
τ junction lifetime [s]
ν junction number density [m^{-3}]

Abbreviations

mAb monoclonal antibody
PLA poly(lactic acid)
PLLA poly(L-lactide)
PDLA poly(D-lactide)
PEO poly(ethylene oxide)
PPO poly (propylene oxide)
PBO poly(butylene oxide)

References

1. Malmsten M, Lindman B. Self-assembly in aqueous block copolymer solutions. *Macromolecules*. 1992;25(20):5440–5445.
2. Yu G-E, Deng Y, Dalton S, Wang Q-G, Attwood D, Price C, et al. Micellisation and gelation of triblock copoly (oxyethylene/oxypropylene/oxyethylene), F127. *Journal of the Chemical Society, Faraday Transactions*. 1992;88(17):2537–2544.
3. Wanka G, Hoffmann H, Ulbricht W. Phase diagrams and aggregation behavior of poly (oxyethylene)-poly (oxypropylene)-poly (oxyethylene) triblock copolymers in aqueous solutions. *Macromolecules*. 1994;27(15):4145–4159.
4. Alexandridis P, Holzwarth JF, Hatton TA. Micellization of poly (ethylene oxide)-poly (propylene oxide)-poly (ethylene oxide) triblock copolymers in aqueous solutions: Thermodynamics of copolymer association. *Macromolecules*. 1994;27(9):2414–2425.
5. Mortensen K, Brown W, Joergensen E. Phase behavior of poly (propylene oxide)-poly (ethylene oxide)-poly (propylene oxide) triblock copolymer melt and aqueous solutions. *Macromolecules*. 1994;27(20):5654–5666.
6. Prud'homme RK, Wu G, Schneider DK. Structure and rheology studies of poly (oxyethylene–oxypropylene–oxyethylene) aqueous solution. *Langmuir*. 1996;12(20):4651–4659.

7. Anderson BC, Pandit NK, Mallapragada SK. Understanding drug release from poly (ethylene oxide)-b-poly (propylene oxide)-b-poly (ethylene oxide) gels. *Journal of Controlled Release*. 2001;70(1):157–167.
8. Arevalo-Silva CA, Eavey RD, Cao Y, Vacanti M, Weng Y, Vacanti CA. Internal support of tissue-engineered cartilage. *Archives of Otolaryngology – Head and Neck Surgery*. 2000;126(12):1448–1452.
9. Cao YL, Rodriguez A, Vacanti M, Ibarra C, Arevalo C, Vacanti CA. Comparitive study of the use of poly(glycolic acid), calcium alginate and pluronics in the engineering of autologous porcine cartilage. *Journal of Biomaterials Science – Polymer Edition*. 1998;9(5):475–487.
10. Matthew J, Bhatia S, Roberts S. Pluronic F127 gels as materials for mammalian cell encapsulation. *Polymer Preprints*. 2002;43(2):769–770.
11. Matthew J, Nazario Y, Roberts S, Bhatia S. Effect of mammalian cell culture medium on the gelation properties of Pluronic F127. *Biomaterials*. 2002;23(23):4615–4619.
12. Khattak SF, Bhatia SR, Roberts SC. Pluronic® F127 as a cell encapsulation material: Utilization of membrane stabilizing agents. *Tissue Engineering*. 2005;11(5–6):974–983.
13. Bogue RH. The viscosity of gelatin sols. *Journal of the American Chemical Society*. 1921;43(8):1764–1773.
14. de Kruif C. Chapter VI caseins. In Aalbersberg WY, Hamer RJ, Jasperse P, de Jongh HHJ, de Kruif CG, Walstra P, de Wolf FA (eds.) *Progress in Biotechnology*, vol. 23. Amsterdam: Elsevier; 2003. pp. 219–269.
15. de Kruif C, Jeurnink TJ, Zoon P. The viscosity of milk during the initial stages of renneting. *Netherlands Milk and Dairy Journal (Netherlands)*. 1992;46(2):123–137.
16. Walstra P, Geurts TJ, Walstra P, Wouters JT. *Dairy Science and Technology*. Boca Raton, FL: CRC press; 2005.
17. de Kruif CG, Jeurnink TJM, Zoon P. The viscosity of milk during the initial-stages of renneting. *Netherlands Milk and Dairy Journal*. 1992;46(2):123–137.
18. Bremer LGB, Bijsterbosch BH, Schrijvers R, Vanvliet T, Walstra P. On the fractal nature of the structure of acid casein gels. *Colloids and Surfaces*. 1990;51:159–170.
19. Dickinson E, Bergenstahl B. *Food Colloids: Proteins, Lipids and Polysaccharides*. Cambridge: Woodhead Publishing Limited; 1997.
20. van Marle ME, van den Ende D, de Kruif CG, Mellema J. Steady-shear viscosity of stirred yogurts with varying ropiness. *Journal of Rheology*. 1999;43(6):1643–1662.
21. Potanin AA, Derooij R, Vandenende D, Mellema J. Microrheological modeling of weakly aggregated dispersions. *Journal of Chemical Physics*. 1995;102(14):5845–5853.
22. Buscall R, McGowan JI, Mortonjones AJ. The rheology of concentrated dispersions of weakly attracting colloidal particles with and without wall slip. *Journal of Rheology*. 1993;37(4):621–641.
23. Goodwin JW, Hughes RW, Partridge SJ, Zukoski CF. The elasticity of weakly flocculated suspensions *Journal of Chemical Physics*. 1986;85(1):559–566.
24. Olivares ML, Berli CLA, Zorrilla S. Connection between dynamic rheometry and pair interactions of casein micelles in concentrated skim milk. *Food Hydrocolloids*. 2018;74:104–107.
25. Buscall R. Effect of long-range repulsive forces on the viscosity of concentrated latices: Comparison of experimental data with an effective hard-sphere model. *Journal of the Chemical Society, Faraday Transactions*. 1991;87(9):1365–1370.
26. Nöbel S, Weidendorfer K, Hinrichs J. Apparent voluminosity of casein micelles determined by rheometry. *Journal of Colloid and Interface Science*. 2012;386(1):174–180.

27. Tuinier R, De Kruif C. Stability of casein micelles in milk. *The Journal of Chemical Physics*. 2002;117(3):1290–1295.
28. Berli CL, Quemada D. Rheological modeling of microgel suspensions involving solid–liquid transition. *Langmuir*. 2000;16(21):7968–7974.
29. Mewis J, Wagner NJ. *Colloidal Suspension Rheology*. Cambridge: Cambridge University Press; 2012. 393 p.
30. Bouchoux A, Debbou B, Gesan-Guiziou G, Famelart MH, Doublier JL, Cabane B. Rheology and phase behavior of dense casein micelle dispersions. *Journal of Chemical Physics*. 2009;131(16):165106.
31. Zhang Z, Liu Y. Recent progresses of understanding the viscosity of concentrated protein solutions. *Current Opinion in Chemical Engineering*. 2017;16:48–55.
32. Anderson BC, Cox SM, Bloom PD, Sheares VV, Mallapragada SK. Synthesis and characterization of diblock and gel-forming pentablock copolymers of tertiary amine methacrylates, poly (ethylene glycol), and poly (propylene glycol). *Macromolecules*. 2003;36(5):1670–1676.
33. Determan MD, Guo L, Lo C-T, Thiyagarajan P, Mallapragada SK. pH-and temperature-dependent phase behavior of a PEO-PPO-PEO-based pentablock copolymer in aqueous media. *Physical Review E*. 2008;78(2):021802.
34. Cohn D, Lando G, Sosnik A, Garty S, Levi A. PEO–PPO–PEO-based poly (ether ester urethane) as degradable reverse thermo-responsive multiblock copolymers. *Biomaterials*. 2006;27(9):1718–1727.
35. Jiang J, Malal R, Li C, Lin MY, Colby RH, Gersappe D, et al. Rheology of thermoreversible hydrogels from multiblock associating copolymers. *Macromolecules*. 2008;41(10):3646–3652.
36. Alexandridis P, Olsson U, Lindman B. Structural polymorphism of amphiphilic copolymers: Six lyotropic liquid crystalline and two solution phases in a poly (oxybutylene)-b-poly (oxyethylene)–water–xylene system. *Langmuir*. 1997;13(1):23–34.
37. Hamley I. Amphiphilic diblock copolymer gels: The relationship between structure and rheology. *Philosophical Transactions of the Royal Society of London A: Mathematical, Physical and Engineering Sciences*. 2001;359(1782):1017–1044.
38. Hamley IW, Daniel C, Mingvanish W, Mai S-M, Booth C, Messe L, et al. From hard spheres to soft spheres: The effect of copolymer composition on the structure of micellar cubic phases formed by diblock copolymers in aqueous solution. *Langmuir*. 2000;16(6):2508–2514.
39. Hamley I, Mortensen K, Yu G-E, Booth C. Mesoscopic crystallography: A small-angle neutron scattering study of the body-centered cubic micellar structure formed in a block copolymer gel. *Macromolecules*. 1998;31(20):6958–6963.
40. Pople J, Hamley I, Fairclough J, Ryan A, Komanschek B, Gleeson A, et al. Ordered phases in aqueous solutions of diblock oxyethylene/oxybutylene copolymers investigated by simultaneous small-angle X-ray scattering and rheology. *Macromolecules*. 1997;30(19):5721–5728.
41. Hamley I, Pople J, Diat O. A thermally induced transition from a body-centred to a face-centred cubic lattice in a diblock copolymer gel. *Colloid & Polymer Science*. 1998;276(5):446–450.
42. Quah SP, Smith AJ, Preston AN, Laughlin ST, Bhatia SR. Large-area alginate/PEO-PPO-PEO hydrogels with thermoreversible rheology at physiological temperatures. *Polymer*. 2018;135:171–177.

43. Jalaal M, Cottrell G, Balmforth N, Stoeber B. On the rheology of pluronic F127 aqueous solutions. *Journal of Rheology*. 2017;61(1):139–146.
44. Chen S-H, Chen WR, Mallamace F. The glass-to-glass transition and its end point in a copolymer micellar system. *Science*. 2003;300(5619):619–622.
45. Kelarakis A, Mingvanish W, Daniel C, Li H, Havredaki V, Booth C, et al. Rheology and structures of aqueous gels of diblock (oxyethylene–oxybutylene) copolymers with lengthy oxyethylene blocks. *Physical Chemistry Chemical Physics*. 2000;2(12):2755–2763.
46. Sharma PK, Bhatia SR. Effect of anti-inflammatories on pluronic (R) F127: micellar assembly, gelation and partitioning. *International Journal of Pharmaceutics*. 2004;278 (2):361–377.
47. Sharma PK, Reilly MJ., Bhatia SK, Sakhitab N, Archambault JD, Bhatia SR. Effect of pharmaceuticals on thermoreversible gelation of PEO-PPO-PEO copolymers. *Colloids and Surfaces B–Biointerfaces*. 2008;63(2):229–235.
48. Sharma PK, Reilly MJ, Jones DN, Robinson PM, Bhatia SR. The effect of pharmaceuticals on the nanoscale structure of PEO-PPO-PEO micelles. *Colloids and Surfaces B–Biointerfaces*. 2008;61(1):53–60.
49. Sharma PK, Matthew JE, Bhatia SR. Structure and assembly of PEO-PPO-PEO copolymers in mammalian cell-culture media. *Journal of Biomaterials Science–Polymer Edition*. 2005;16(9):1139–1151.
50. Lee Y, Chung HJ, Yeo S, Ahn C-H, Lee H, Messersmith PB, et al. Thermo-sensitive, injectable, and tissue adhesive sol–gel transition hyaluronic acid/pluronic composite hydrogels prepared from bio-inspired catechol-thiol reaction. *Soft Matter*. 2010;6(5):977–983.
51. Chen CC, Fang, CL, Al-Suwayeh, SA, Leu, YL, Fang, JY. Transdermal delivery of selegiline from alginate-Pluronic composite thermogels. *International Journal of Pharmaceutics*. 2011;415(1–2):119–128.
52. Stoppel WL, White, JC, Horava, SD, Bhatia, SR, Roberts, SC. Transport of biological molecules in surfactant-alginate composite hydrogels. *Acta Biomaterialia*. 2011 (7):3988–3998.
53. White JC, Saffer, EM, Bhatia, SR. Alginate/PEO-PPO-PEO composite hydrogels with thermally-active plasticity. *Biomacromolecules*. 2013;14(12):4456–4464.
54. Li H, Yu G-E, Price C, Booth C, Hecht E, Hoffmann H. Concentrated aqueous micellar solutions of diblock copoly (oxyethylene/oxybutylene) E41B8: A study of phase behavior. *Macromolecules*. 1997;30(5):1347–1354.
55. Hyun K, Nam JG, Wilhellm M, Ahn KH, Lee SJ. Large amplitude oscillatory shear behavior of PEO-PPO-PEO triblock copolymer solutions. *Rheologica Acta*. 2006;45 (3):239–249.
56. Green M, Tobolsky A. A new approach to the theory of relaxing polymeric media. *The Journal of Chemical Physics*. 1946;14(2):80–92.
57. Annable T, Buscall R, Ettelaie R, Whittlestone D. The rheology of solutions of associating polymers: Comparison of experimental behavior with transient network theory. *Journal of Rheology*. 1993;37(4):695–726.
58. Tanaka F, Edwards S. Viscoelastic properties of physically crosslinked networks. 1. Transient network theory. *Macromolecules*. 1992;25(5):1516–1523.
59. Nguyen-Misra M, Misra S, Mattice WL. Bridging by end-adsorbed triblock copolymers. *Macromolecules*. 1996;29(5):1407–1415.
60. Tae G, Kornfield JA, Hubbell JA, Lal J. Ordering transitions of fluoroalkyl-ended poly (ethylene glycol): Rheology and SANS. *Macromolecules*. 2002;35(11):4448–4457.

61. Semenov A, Joanny J-F, Khokhlov A. Associating polymers: Equilibrium and linear viscoelasticity. *Macromolecules*. 1995;28(4):1066–1075.
62. Deming TJ. Polypeptide hydrogels via a unique assembly mechanism. *Soft Matter*. 2005;1 (1):28–35.
63. Carlsen A, Lecommandoux S. Self-assembly of polypeptide-based block copolymer amphiphiles. *Current Opinion in Colloid & Interface Science*. 2009;14(5):329–339.
64. Tew GN, Sanabria-DeLong N, Agrawal SK, Bhatia SR. New properties from PLA-PEO-PLA hydrogels. *Soft Matter*. 2005;1(4):253–258.
65. Agrawal SK, Sanabria-DeLong N, Coburn JM, Tew GN, Bhatia SR. Novel drug release profiles from micellar solutions of PLA-PEO-PLA triblock copolymers. *Journal of Controlled Release*. 2006;112(1):64–71.
66. Agrawal SK, Sanabria-DeLong N, Jemian PR, Tew GN, Bhatia SR. Micro- to nanoscale structure of biocompatible PLA-PEO-PLA hydrogels. *Langmuir*. 2007;23(9):5039–5044.
67. Sanabria-DeLong N, Agrawal SK, Bhatia SR, Tew GN. Controlling hydrogel properties by crystallization of hydrophobic domains. *Macromolecules*. 2006;39(4):1308–1310.
68. Metters AT, Anseth KS, Bowman CN. A statistical kinetic model for the bulk degradation of PLA-b-PEG-b-PLA hydrogel networks: Incorporating network non-idealities. *Journal of Physical Chemistry B*. 2001;105(34):8069–8076.
69. Anseth KS, Metters AT, Bryant SJ, Martens PJ, Elisseeff JH, Bowman CN. In situ forming degradable networks and their application in tissue engineering and drug delivery. *Journal of Controlled Release*. 2002;78(1–3):199–209.
70. Garric X, Garreau H, Vert M, Moles JP. Behaviors of keratinocytes and fibroblasts on films of PLA(50)-PEO-PLA(50) triblock copolymers with various PLA segment lengths. *Journal of Materials Science: Materials in Medicine*. 2008;19(4):1645–1651.
71. Molina I, Li SM, Martinez MB, Vert M. Protein release from physically crosslinked hydrogels of the PLA/PEO/PLA triblock copolymer-type. *Biomaterials*. 2001;22 (4):363–369.
72. Lee HT, Lee DS. Thermoresponsive phase transitions of PLA-block-PEO-block-PLA triblock stereo-copolymers in aqueous solution. *Macromolecular Research*. 2002;10 (6):359–364.
73. Li SM, Rashkov I, Espartero JL, Manolova N, Vert M. Synthesis, characterization, and hydrolytic degradation of PLA/PEO/PLA triblock copolymers with long poly(L-lactic acid) blocks. *Macromolecules*. 1996;29(1):57–62.
74. Li F, Li SM, Vert M. Synthesis and rheological properties of polylactide/poly(ethylene glycol) multiblock copolymers. *Macromolecular Bioscience*. 2005;5(11):1125–1131.
75. Jing Y, Quan C, Liu B, Jiang Q, Zhang C. A mini review on the functional biomaterials based on poly (lactic acid) stereocomplex. *Polymer Reviews*. 2016;56(2):262–286.
76. Fujiwara T, Mukose T, Yamaoka T, Yamane H, Sakurai S, Kimura Y. Novel thermo-responsive formation of a hydrogel by stereo-complexation between PLLA-PEG-PLLA and PDLA-PEG-PDLA block copolymers. *Macromolecular Bioscience*. 2001;1 (5):204–208.
77. Agrawal SK, Sanabria-DeLong N, Tew GN, Bhatia SR. Rheological characterization of biocompatible associative polymer hydrogels with crystalline and amorphous endblocks. *Journal of Materials Research*. 2006;21(8):2118–2125.
78. Agrawal SK, Sanabria-DeLong N, Tew GN, Bhatia SR. Structural characterization of PLA-PEO-PLA solutions and hydrogels: Crystalline vs amorphous PLA domains. *Macromolecules*. 2008;41(5):1774–1784.

79. Semenov A, Nyrkova I, Khokhlov A. Polymers with strongly interacting groups: Theory for nonspherical multiplets. *Macromolecules*. 1995;28(22):7491–7500.
80. Aamer KA, Sardinha H, Bhatia SR, Tew GN. Rheological studies of PLLA-PEO-PLLA triblock copolymer hydrogels. *Biomaterials*. 2004;25(6):1087–1093.
81. Yin X, Hewitt DR, Quah SP, Zheng B, Mattei GS, Khalifah PG, et al. Impact of stereochemistry on rheology and nanostructure of PLA–PEO–PLA triblocks: Stiff gels at intermediate l/d-lactide ratios. *Soft Matter*. 2018;14(35):7255–7263.
82. Jezek J, Rides M, Derham B, Moore J, Cerasoli E, Simler R., et al. Viscosity of concentrated therapeutic protein compositions. *Advanced Drug Delivery Reviews*. 2011;63 (13):1107–1117.
83. Perez-Ramirez B, Guziewicz N, Simler R. Preformulation research: Assessing protein solution behavior during early development. In Jameel F, Hershenson S (eds.) *Formulation and Process Development Strategies for Manufacturing Biopharmaceuticals*. Hoboken: Wiley; 2010; pp. 119–146.
84. Andreeva A, Howorth D, Chothia C, Kulesha E, Murzin AG. SCOP2 prototype: A new approach to protein structure mining. *Nucleic Acids Research*. 2013;42(D1):D310–D314.
85. Sharma V, Jaishankar A, Wang Y-C, McKinley GH. Rheology of globular proteins: Apparent yield stress, high shear rate viscosity and interfacial viscoelasticity of bovine serum albumin solutions. *Soft Matter*. 2011;7(11):5150–5160.
86. Ikeda S, Nishinari K. On solid-like rheological behaviors of globular protein solutions. *Food Hydrocolloids*. 2001;15(4–6):401–406.
87. Cardinaux F, Zaccarelli E, Stradner A, Bucciarelli S, Farago B, Egelhaaf SU, et al. Cluster-driven dynamical arrest in concentrated lysozyme solutions. *The Journal of Physical Chemistry B*. 2011;115(22):7227–7237.
88. Liu Y, Porcar L, Chen J, Chen W-R, Falus P, Faraone A, et al. Lysozyme protein solution with an intermediate range order structure. *The Journal of Physical Chemistry B*. 2010;115(22):7238–7247.
89. Ikeda S, Nishinari K. Intermolecular forces in bovine serum albumin solutions exhibiting solidlike mechanical behaviors. *Biomacromolecules*. 2000;1(4):757–763.
90. Castellanos MM, Pathak JA, Colby RH. Both protein adsorption and aggregation contribute to shear yielding and viscosity increase in protein solutions. *Soft Matter*. 2014;10 (1):122–131.
91. Tein YS, Zhang Z, Wagner NJ. Competitive surface activity of monoclonal antibodies and nonionic surfactants at the air–water interface determined by interfacial rheology and neutron reflectometry. *Langmuir*. 2020;36(27):7814–7823.
92. Russel WB, Saville DA, Schowalter WR. *Colloidal Dispersions*. Cambridge: Cambridge University Press; 1991.
93. Foffi G, Savin G, Bucciarelli S, Dorsaz N, Thurston GM, Stradner A, et al. Hard sphere-like glass transition in eye lens α-crystallin solutions. *Proceedings of the National Academy of Sciences*. 2014;111(47):16748–16753.
94. Heinen M, Zanini F, Roosen-Runge F, Fedunová D, Zhang F, Hennig M, et al. Viscosity and diffusion: Crowding and salt effects in protein solutions. *Soft Matter*. 2012;8 (5):1404–1419.
95. Castellanos MM, Clark NJ, Watson MC, Krueger S, McAuley A, Curtis JE. Role of molecular flexibility and colloidal descriptions of proteins in crowded environments from small-angle scattering. *The Journal of Physical Chemistry B*. 2016;120 (49):12511–12518.

96. Sarangapani PS, Hudson SD, Jones RL, Douglas JF, Pathak JA. Critical examination of the colloidal particle model of globular proteins. *Biophysical journal*. 2015;108 (3):724–737.
97. Ross PD, Minton AP. Hard quasi-spherical model for viscosity of hemoglobin solutions. *Biochemical and Biophysical Research Communications*. 1977;76(4):971–976.
98. Lilyestrom WG, Yadav S, Shire SJ, Scherer TM. Monoclonal antibody self-association, cluster formation, and rheology at high concentrations. *The Journal of Physical Chemistry B*. 2013;117(21):6373–6384.
99. Godfrin PD, Hudson SD, Hong K, Porcar L, Falus P, Wagner NJ, et al. Short-time glassy dynamics in viscous protein solutions with competing interactions. *Physical Review Letters*. 2015;115(22):228302.
100. Shire SJ, Shahrokh Z, Liu J. Challenges in the development of high protein concentration formulations. *Journal of Pharmaceutical Sciences*. 2004;93(6):1390–1402.
101. Rader RA. *Biopharmaceutical Products in the US and European Markets*. Rockville, MD: BioPlan Associates, Incorporated; 2007.
102. Liu J, Nguyen MD, Andya JD, Shire SJ. Reversible self-association increases the viscosity of a concentrated monoclonal antibody in aqueous solution. *Journal of Pharmaceutical Sciences*. 2005;94(9):1928–1940.
103. Saluja A, Badkar AV, Zeng DL, Kalonia DS. Ultrasonic rheology of a monoclonal antibody (IgG2) solution: Implications for physical stability of proteins in high concentration formulations. *Journal of Pharmaceutical Sciences*. 2007;96(12):3181–3195.
104. Saluja A, Badkar AV, Zeng DL, Nema S, Kalonia DS. Application of high-frequency rheology measurements for analyzing protein–protein interactions in high protein concentration solutions using a model monoclonal antibody (IgG(2)). *Journal of Pharmaceutical Sciences*. 2006;95(9):1967–1983.
105. Godfrin PD, Zarraga IE, Zarzar J, Porcar L, Falus P, Wagner NJ, et al. Effect of hierarchical cluster formation on the viscosity of concentrated monoclonal antibody formulations studied by neutron scattering. *The Journal of Physical Chemistry B*. 2016;120(2):278–291.
106. Saluja A, Kalonia DS. Nature and consequences of protein–protein interactions in high protein concentration solutions. *International Journal of Pharmaceutics*. 2008;358(1–2):1–15.
107. Patel AR, Kerwin BA, Kanapuram SR. Viscoelastic characterization of high concentration antibody formulations using quartz crystal microbalance with dissipation monitoring. *Journal of Pharmaceutical Sciences*. 2009;98(9):3108–3116.
108. Yearley EJ, Godfrin PD, Perevozchikova T, Zhang H, Falus P, Porcar L, et al. Observation of small cluster formation in concentrated monoclonal antibody solutions and its implications to solution viscosity. *Biophysical Journal*. 2014;106(8):1763–1770.
109. Yearley EJ, Zarraga IE, Shire SJ, Scherer TM, Gokarn Y, Wagner NJ, et al. Small-angle neutron scattering characterization of monoclonal antibody conformations and interactions at high concentrations. *Biophysical Journal*. 2013;105(3):720–731.
110. Dear BJ, Hung JJ, Truskett TM, Johnston KP. Contrasting the influence of cationic amino acids on the viscosity and stability of a highly concentrated monoclonal antibody. *Pharmaceutical Research*. 2017;34(1):193–207.
111. Weiss RM, Short AL, Meyer TY. Sequence-controlled copolymers prepared via entropy-driven ring-opening metathesis polymerization. *ACS Macro Letters*. 2015;4 (9):1039–1043.

112. Weiss RM, Li J, Liu HH, Washington MA, Giesen JA, Grayson SM, et al. Determining sequence fidelity in repeating sequence poly(lactic-co-glycolic acid)s. *Macromolecules*. 2017;50(2):550–560.
113. Washington MA, Swiner DJ, Bell KR, Fedorchak MV, Little SR, Meyer TY. The impact of monomer sequence and stereochemistry on the swelling and erosion of biodegradable poly(lactic-co-glycolic acid) matrices. *Biomaterials*. 2017;117:66–76.

8 Hemorheology

Antony N. Beris

8.1 Introduction

Cardiovascular diseases represent the leading cause of death worldwide, being responsible for roughly one out of three deaths [1]. According to the latest statistics, in the USA about 92.1 million adults are living with some form of cardiovascular disease, with a total cost for healthcare and lost productivity estimated to be 329.7 billion dollars annually [2]. This explains the enormous interest and the many research undertakings dealing with blood flow, testimony to the many books, also recent ones, which deal with this subject [3–20].

Blood is a concentrated suspension of cells (about 46% on average) within plasma, which is an aqueous solution (approximately 90–92% water) of various proteins, electrolytes, gases, nutrients, and waste products [21]. The majority of the cells (more than 99%) are red blood cells, RBC (erythrocytes). When undeformed they have a disk-doughnut shape of about $7.7 \pm 0.7\,\mu\text{m}$ diameter and about $1.4 \pm 0.5\ \mu\text{m}$ central and $2.8 \pm 0.5\ \mu\text{m}$ peripheral thickness [15]. The rest of the cells are white blood cells (leukocytes), which are roughly spherical in shape with a 7–22 μm diameter, and platelets of rounded or oval shape and dimensions from 2 to 4 μm [14]. Roughly, for every white cell there are about 35 platelets and 700 red blood cells [21], with the volume fraction of the RBCs constituting the hematocrit (approximately 0.45 for healthy men and 0.42 for healthy women). A key physical characteristic of the red blood cells is their elasticity and deformability. Aided by their nonspherical shape, it allows them to pass through capillaries as narrow as 2.9 μm in diameter and to be compacted through centrifugation to almost 100% (in contrast to a maximum of 60% for hardened red blood cells) [22]. In the presence of suitable biomacromolecules, typically fibrinogen, the RBCs are known to form reversible "rouleaux" aggregate structures at equilibrium and under low shear rate conditions, at least above a certain critical hematocrit value [22]. In turn, all the colloidal characteristics mentioned above are relevant for the rheology of blood: the fact that blood consists primarily of a concentrated suspension of RBCs that they deform under flow and/or high concentration, and that they form extended aggregates that actually are able to sustain stress without flow, are responsible for a complex, non-Newtonian, thixotropic, and viscoelastic blood rheology. Although we know quite a lot about it today, as discussed in the books referenced above and as also elaborated in the rest of this chapter, this know-how took a long time to develop – the following story is about the early history of blood, blood flow, and hemorheology.

Story 8.1 Early History of the Study of Blood, Blood Flow, and Hemorheology

From the medical and physiological point of view, interest in blood can be traced all the way back to Ancient Egyptian medicine or Mesopotamia [23]. However, it was with the ancient Greeks that its role was properly appreciated. Empedocles of Agrigentum (490–430 BC) characterized blood as the carrier of soul and health [23]. In Hippocrates's treatise "ΠΕΡΙ ΦΥΣΙΟΣ ΑΝΘΡΩΠΟΥ" ("On the nature of man") blood is described as one of the four constituents of the human body: "Τὸ δὲ σῶμα τοῦ ἀνθρώπου ἔχει ἐν ἑωυτῷ αἷμα καὶ φλέγμα καὶ χολὴν ξανθὴν καὶ μέλαιναν . . ." ("The human body contains blood, phlegm, yellow bile, and black bile. . .") [24,25]. Of interest is that these four ingredients, or humors, illustrated in Figure 8.1, that can be correlated to different states of blood: "Blood" to the normal, oxygenated, blood, "phlegma" to the mucus (buffy coat) of a clot consisting of leucocytes, platelets, and fibrin, "yellow bile" to the serum that separates from the blood clot, and "black bile" to the red blood cells (RBC) that failed to be oxygenated – those were analyzed by allowing blood to settle and separate without an anticoagulant. However, we had to wait until 1658 for the true science of blood to emerge [23]. That was the year when Jan Swammerdam unequivocally saw in the blood of a frog, through one of his more than 400 microscopes that he developed, red blood cells. A few years later, in 1674, Anton van Leeuwenhoek,

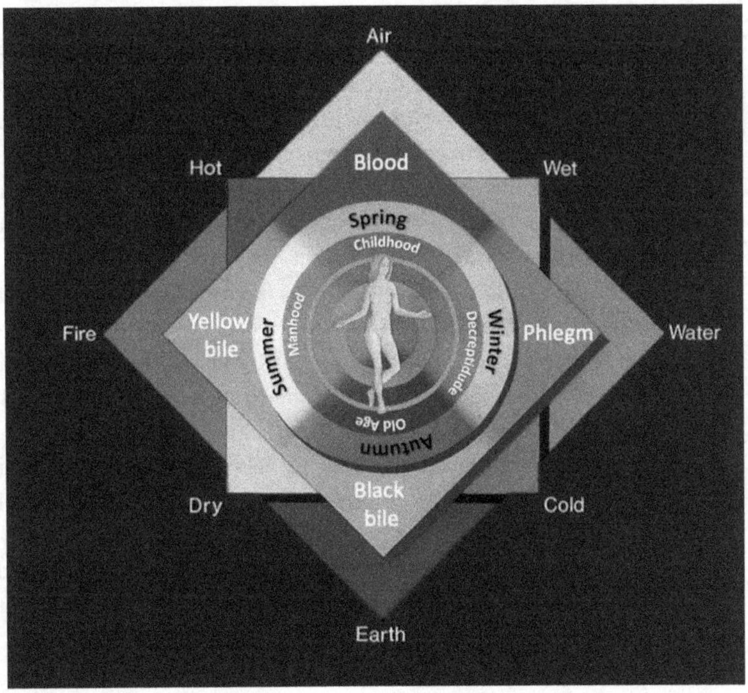

Figure 8.1 The four elements, four qualities, four humors, four seasons, and four ages of man. Adapted from airbrush by Lois Hague, 1991. Credit: Wellcome Collection. Attribution-NonCommercial 4.0 International (CC BY-NC 4.0)

Story 8.1 *(cont.)*

a leading proponent of microscopy, made the first observations of RBC in human blood that he characterized as "very minute globules" [23]. It then took several more years, in 1735, until Boerheave recognized the role of RBC in capillary circulation, and many more years to elucidate the RBC production, morphology, functions in the blood, life span (120 days), and most importantly their role in immunology [23]. Indeed, it was as recently as in 1900 that Karl Landsteiner first described the ABO blood types and recognized their incompatibility and even more recently in the 1940s when the rhesus (Rh) groups and their immunological role were properly identified [23].

Most importantly for rheology, the mixing of incompatible types of blood is the primary reason leading to uncontrollable blood cells aggregation, or "agglutination," with catastrophic consequences for the blood viscosity and health. On the other hand, the reversible aggregation of red blood cells into the well-known "rouleaux" structures, although probably observed from Hippocrates time (as those aggregates do play a role in red blood settling and clotting), had to wait for their first scientific observations until the work of Fåhræus in 1921 [26–29]. Robin Fåhræus (1888–1968) was a pathologist from the University of Upsala in Sweden interested in the stability of blood, the topic of his doctoral thesis [26]. He also reported observations on the aggregates of both healthy and diseased blood and showed the latter to be more "solid" in appearance, noting for the first time aging effects [26]. He was also the first to note the role of fibrinogen as the principal protein involved in sedimentation [26,28] and the formation of the rouleaux structures [27]. Additional contributions of Fåhræus come in his pioneering observations of the Fåhræus [28] and Fåhræus-Lindqvist [29] effects, that have to do with the observation of a cell-free layer near the wall, and the concomitant decrease in the hematocrit [28] and the apparent viscosity [29] in blood flows through narrow capillaries with a diameter less than 0.3 mm. Of further interest to rheology is the further observation by Fåhraeus on the role that the rouleaux structures have not only in enhancing sedimentation but also causing the phenomenon of syneresis, where the mass of RBC contracts on itself as they are also pulled towards the center in very slow capillary pipe flows [27].

Around 450 BC, Hippocrates came up with the theory (humoralism) that all human ailments are caused by an imbalance of the "four humors": blood, yellow bile, black bile, and phlegm, corresponding to four basic personality types: sanguine (courageous, hopeful, amorous); choleric (easily angered, bad tempered), melancholic (despondent, sleepless, irritable), and phlegmatic (calm, unemotional). Galen expanded this a bit, but since then almost nothing changed until it was disregarded in 1858 AD when Rudolf Virchow discovered the cell. In humoral medicine, disease and insanity was believed to be caused by a humoral imbalance of the four humors. The theory behind the treatments was that removing one of the humors would restore the imbalance. Strangely, blood was seen as a poison to be removed, and this is the origin of blood-letting [24].

Story 8.1 *(cont.)*

From a continuum mechanics point of view, the modern history of blood flow studies starts with William Harvey (1578–1657) and his treatise on "Exertitato Anatomica de Motu Cordis et Sanguinis in Annimalibus" ("Anatomical Essay on the Motion of the Heart and Blood in Animals"), on which we can pinpoint the beginning of modern cardiovascular physiology with a first analysis of blood circulation [18]. The next development was by Reverend Stephen Hales (1677–1761) who in his treatise "Haemastaticks" (1733) described the first measurement of arterial pressure [18,30], leading eventually (after the general fluid flow developments by Newton, Euler, and Bernoulli) to Poiseuille (1799–1869), an engineer turned into a physician, most well-known for obtaining the linear pressure–flowrate relationship for viscous flow down a cylindrical pipe [18,30]. What is not as widely known is the fact that Poiseuille, between 1828 and 1848, was the first to use a mercury sphygmomanometer to measure arterial pressure, and he was the first to describe the preferential around the axis flow of red blood cells in microcirculation as well as the first to correctly measure and interpret the changes in central venous pressure during respiration [30].

In 1815 Poiseuille studied engineering in the Ecole Polytechnique of Paris, France, while afterwards he studied medicine with a specialization in experimental physiology. His doctoral dissertation that he presented on August 8, 1828 to the faculty of medicine to become a doctor in medicine was entitled "Recherches sur la force de cœur aortique" ("Research on the arterial force in the heart") and includes the first instance that a mercury sphygmomanometer was used to measure arterial pressure [30]. He developed the "Poiseuille law" for the flow in cylindrical vessels during winter 1840–1841, which he presented to the French Academy of Sciences, creating such a stir that some of the members of the academy were obliged to repeat Poiseuille's experiments of flow within glass vessels, eventually confirming his results and recognizing fully Poiseuille's contribution by 1843 [30].

From a rheology viewpoint though, the first contribution was that of Hess in 1915 [31], further referenced in Baskurt et al. [26] and Goldsmith et al. [27], that noted the shear rate dependence of the apparent steady state blood viscosity correcting the Poiseuille solution for the analysis of blood flow in a pipe, which provides only a first order approximation and only for high shear rates. He believed that the primary reason for the non-Newtonian shear thinning behavior of blood was the elastic deformation of RBC while he also considered plausible the role of the rouleaux structures there as well as in blood's viscoelasticity [27].

But the most important contribution came without doubt from Casson and his development of the Casson fluid or Casson equation. He developed that in "Rheology of Disperse Systems" (C. C. Mills, ed.) [32], aiming to describe the rheology of mixtures of pigments and oils [33]. The most important characteristic of the Casson model is its prediction of a critical stress (yield stress) below which no flow is possible. Scott Blair was the first to note that Casson's model can be

Story 8.1 *(cont.)*

used to describe the blood rheology not only from humans, but also cows and pigs using all available capillary data [34]. They notice a linear relationship between the square root of the diameter and the square root of the wall shear stress with a nonzero x-axis intercept, for capillary diameters larger than 150 μm [34]. This is exactly what the application of the Casson model for capillary flow predicts [35], with the nonzero intercept being a significant correction to Poiseuille's zero result and a signature of a nonzero yield stress for blood rheology. A few years later Merrill et al. [36] systematically used the Casson equation in describing human blood steady-state rheology, employing a variety of viscometric data, and were the first to note the fibrinogen concentration effect on the rheology. Based on these and other results, the same researchers [37] were the first ones to insist that human blood possesses a nonzero yield stress and the first to provide the first parametric expressions for it, in terms of the hematocrit values, above a minimum that is donor-dependent [38]. Later, they also showed a fibrinogen-dependence [39] whereas Chien et al. [40] demonstrated most clearly the role of aggregates to nonzero yield stress by showing the disappearance of the yield stress under conditions under which the RBC aggregation into rouleaux structures is inhibited.

The next important contributions to the analysis and evaluation of non-Newtonian effects in blood rheology were those associated with time-dependent history effects. Thurston [41] was the first one to observe and study the viscoelasticity of blood in small and high amplitude oscillatory flow in a pipe, whereas Sun and De Kée [42] were the first to propose a thixotropy model for blood. Finally, in this short history review we should also mention the first detailed microscopic numerical simulations for RBC solutions of Fedosov, Karniadakis and coworkers, based on a detailed mechanical description of RBCs and their interactions under flow. Among other findings, they demonstrated both shear thinning effects in planar Couette flow simulations [43] as well as the Fåhræus and Fåhræus–Lindqvist effects in thin capillary flow simulations [44]. Many more researchers have continued this effort to understand and describe blood rheology and blood flow behavior to our days.

The non-Newtonian character of the blood flow rheology is one of the reasons why the exact local fluid dynamics of the cardiovascular system is not well understood, others being the complexity of the geometry, the viscoelastic properties of the vascular walls, and the boundary conditions to name a few. Thus, it is still difficult to use simulation methods to reproduce the *in vivo* measured flow profiles or even effectively represent the arterial flow in high risk regions, one of the main applications of computational fluid dynamics in blood flow circulation. Nevertheless, three-dimensional (3D) computational fluid dynamic simulations have been performed with increasing frequency lately, on various vascular geometries, ranging from the carotid [45–56], the coronary [57–68], the abdominal aorta [69,70], as well as the cerebral [71–76], and pulmonary [77] arteries to name a few.

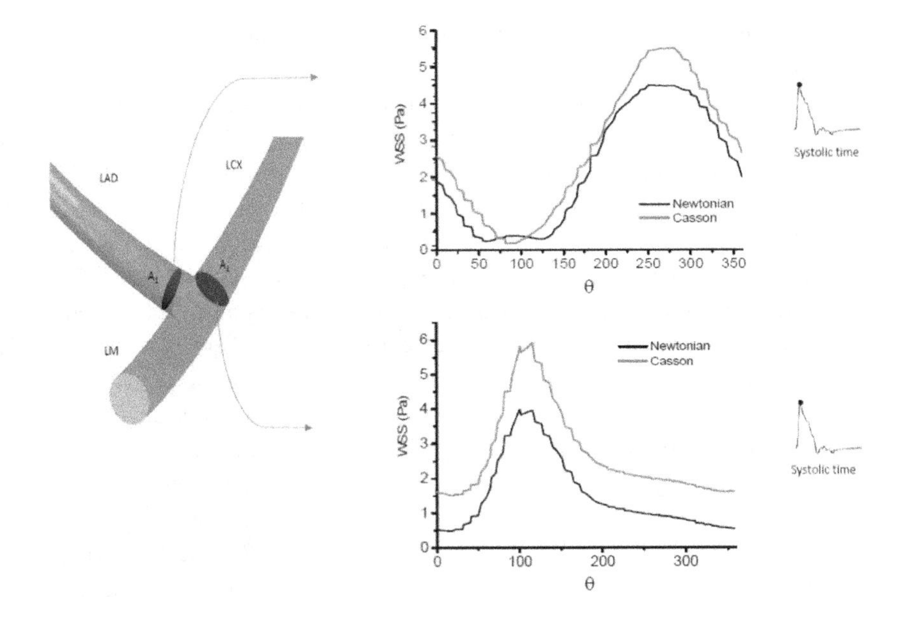

Figure 8.2 Wall Shear Stress (WSS) azimuthal profiles (in Pascals) along the circumference of two cross-sections of the first bifurcation of the left main coronary (LM) artery into the left anterior descending (LAD) and the left circumflex (LCX) arteries at the peak of the systolic time as simulated using FLUENT® and based on the Newtonian (black) and the Casson (gray) fluid models. Reprinted with permission from Apostolidis et al., Non-Newtonian effects in realistic blood flow simulations of coronary arterial flow. *Journal of Non-Newtonian Fluid Mechanics* 2016;233:155–165 [68]. By Elsevier, DOI 10.1016/j.jnnfm.2016.03.008

As an illustrative example we show in Figure 8.2 the wall shear stress (WSS) predictions in two cross-sections in the first bifurcation in the left main coronary artery (LM) to the left anterior descending (LAD) coronary and the left circumflex (LCX) arteries. The values are calculated at the peak of the systolic time as obtained in a full 3D and time-dependent FLUENT® simulation [68]. The numerical simulations show a considerable difference between the predictions obtained assuming a Newtonian and non-Newtonian, viscoplastic (Casson model) rheology. The wall stress is thus shown to be a sensitive function of the non-Newtonian blood rheology. Vessel geometry, topology and surface boundary conditions also affect the results, as discussed in recent reviews of non-Newtonian CFD applications, such as those written on cerebral aneurisms [78], global circulation models [79], and microcirculation [80].

The wall shear stress is an important quantity since it is shown to influence the state of the endothelium cells (see Figure 8.2) with low shear stress values (e.g., less than 0.5 *Pa*) leading to inflammation that triggers platelet activation, aggregation, and atherosclerosis [81–84]. As Figure 8.2 shows, the wall shear stress is significantly affected by the non-Newtonian nature of the blood, especially concerning the region of low stress levels in the LAD entrance. The latter is the location where the most dangerous stenoses are seen in medical cases, thus explaining the attribution "widowmaker" [85]. It is therefore clearly important to capture correctly the non-Newtonian blood rheology effects.

Not only is the non-Newtonian nature of the blood rheology an issue in simulations. The inhomogeneous red blood cell concentration [86,87], the appropriate conditions describing the interactions between blood vessel walls and flow [88], and their tethering within the human body [3] are often neglected or at best severely simplified. Traditionally, non-Newtonian effects have been approximated through generalized Newtonian models [45,50,55,68,89–94]. As we know from theory [95] though, such a description is only general enough to cover any physically encountered rheological behavior for steady shear flows. Theoretically, for any Lagrangian unsteady flows the concentrated colloidal structure, cellular and multicomponent, of blood, along with accompanying red blood cell aggregation and deformation, is anticipated to lead to a complex, history-dependent, rheology [43–96]. Experimentally, this dependence of the blood rheology on the deformation history is well documented, both in the early literature [97] as well as in the more recent [98]. This history dependence cannot be captured within a generalized Newtonian description. Still, such a description can and should be used as a constraint to more general time and structure-dependent rheological descriptions [99].

At first, generic, viscoelastic model-based descriptions were proposed to capture the history dependence of blood rheology [100]. Those efforts were followed by models that attempted to link the macroscopic rheological behavior of blood to its microstructure. Two main avenues have been followed to achieve that goal. In the first, the blood rheology is described by making the parameters of the viscoelastic model dependent on the microstructure of blood [96,101–111]. In the second, the stress is described through a thixotropic model as the result of an elastic and viscous contribution [42,99]. Introducing an elastic contribution to the stress in thixotropy can produce a more comprehensive *elastoviscoplastic* description [98]. These non-Newtonian developments, closely following those established for other concentrated colloidal systems, as described in *CSR* and in this book, will be discussed in Section 8.3 in more detail. We start first with a presentation of the essential microscopic characteristics of blood, in Section 8.2.

8.2 Structural Overview – Mesoscopic Micromechanical Effects and Models

Blood is a complex fluid. It is essentially a concentrated suspension of red blood cells in plasma. However, as mentioned, there are many other ingredients, ranging from white blood cells and platelets to a variety of proteins [112]. With the influence of certain proteins (of which fibrinogen is a key one) red blood cells tend to form cylindrical aggregates called "rouleaux" [26,33], as shown in Figure 8.3.

Under flow, the rouleaux structures tend to break and then reform. This explains not only the non-Newtonian blood flow characteristics but also the presence of a yield stress and the time-dependency, i.e., the presence of thixotropy. Several efforts have been made to reconstruct the non-Newtonian characteristic of blood from

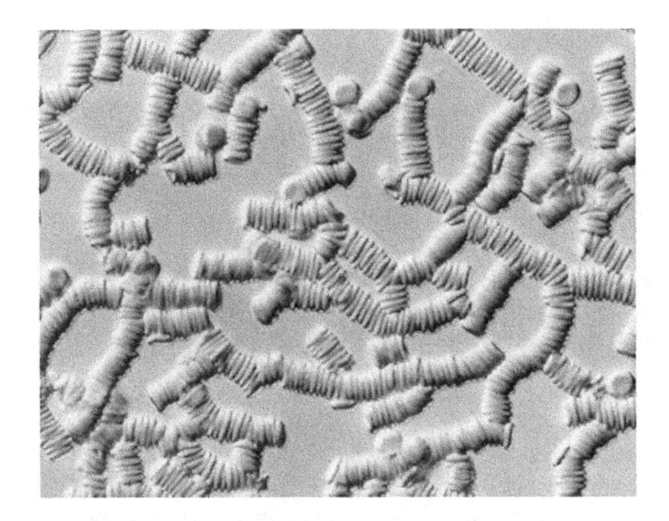

Figure 8.3 Characteristic rouleaux structures in human blood at rest. Reproduced with permission from Baskurt et al., *Red Blood Cell Aggregation*, Boca Raton, FL: CRC Press, 2012 [26]

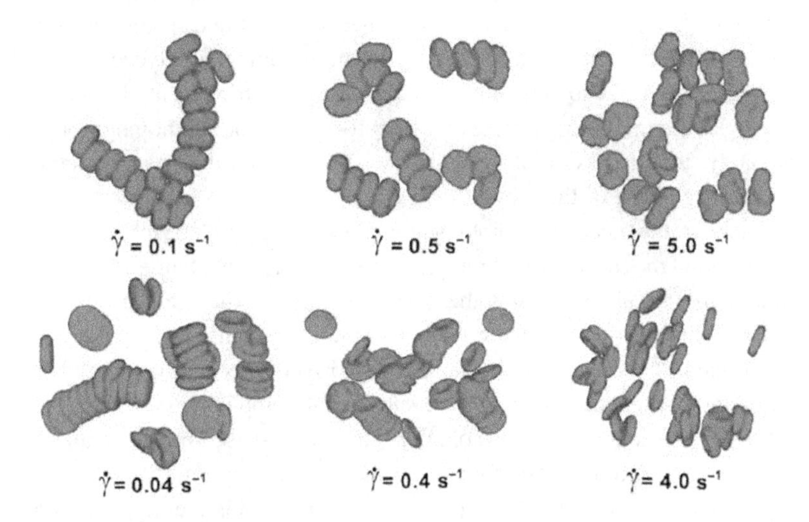

Figure 8.4 Characteristic rouleaux structures obtained in microscopic simulations at different shear rates. Reproduced with permission from Fedosov et al., Predicting human blood viscosity in silico, *Proceedings of the National Academy of Sciences of the United States of America* 2011;108:11772–11777 [43], DOI 10.1073/pnas.1101210108

micromechanical models based on first principles [43,113,114] – some representative microstructure simulation results are shown in Figure 8.4, which can be compared with the image presented in Figure 8.3.

Due to poorly understood biological effects between its ingredients, a full a priori construction of blood rheology remains elusive. Adjustable parameters to describe the red blood cell behavior are still necessary in microscopic as well as mesoscopic and

Figure 8.5 Schematic of the wall depletion layer developing during blood flow at the adjacency of a wall due to the exclusion volume of the red blood cells. Redrawn from Wikipedia.org, https://en.wikipedia.org/wiki/F%C3%A5hr%C3%A6us_effect/

macroscopic modeling approaches. Where microscopic simulations have been particularly useful is in resolving wall effects, due to the presence of a depletion layer next to the wall surface [113], illustrated in Figure 8.5. These depletion layers are important in flows through small vessels, with diameter, $D \lesssim 300$ μm, becoming especially prominent when $D \lesssim 31$ μm, leading to the well-known Fåhræus [28] and Fåhræus–Lindqvist [29] effects. They refer to local reductions in the local hematocrit values (Fåhræus [28]) and in the local viscosity near the wall (Fåhræus–Lindqvist [29]) and they have been further analyzed and quantified in the literature [86]. Together with the full analysis of the endothelial glycocalyx, and the wall elasticity, such phenomena are essential when one tries to model blood flow in the microvasculature [115]. However, as we focus here on the homogeneous bulk blood rheology, the discussion of those inhomogeneous and flow-wall structure effects are beyond the scope of the present review.

Lacking the necessary biological understanding to deduce the non-Newtonian blood flow rheology from first principles the use of semi-empirical macroscopic constitutive relations remains the best alternative at the moment [33]. Microscopic, colloidal systems-based, hydrodynamic simulations still require an empirical modeling of the in general unknown biologically driven effects of cell-interactions and cell properties. In addition, they require substantial computational resources. In contrast, macroscopic models with physiological parameters-based model parameters allow for a direct connection between rheology and physiology and for more complex flow applications. As the first step for any of these models is the description of the steady state blood rheology, we start first in Section 8.3 with a description of the steady state blood rheology before addressing more involved models able to represent transient effects in Section 8.4.

8.3 Steady State Shear Blood Rheology Models

Starting with the simplifying assumptions that blood is a simple fluid, i.e., that nonlocal effects are not present [95], and that it remains homogeneous and of constant properties as should hold in reasonably large scale flow geometries, as compared to the microscopic length scale characterizing the RBC: 7–8 μm, then the most general equation for the shear stress, σ, in steady-state shear flow is given by the generalized Newtonian model:

$$\sigma = \eta(|\dot{\gamma}|)\dot{\gamma}, \tag{8.1}$$

where the (apparent) viscosity, η, depends only on the magnitude of the local shear rate, $\dot{\gamma}$.

Based on the above discussion, a plethora of generalized Newtonian models have been proposed for steady state blood flow predictions in the literature – see, for example, Yilmaz and Gundogdu [91] and Marcinkowska-Gapińska et al. [116] and references therein. However, of interest here are the following questions:

(a) Is the simple fluid hypothesis valid for blood? That is, can one reliably describe its rheological behavior using only local functions that are also independent of the flow?
(b) If the answer to the previous question is yes, what is the expression for the apparent viscosity that describes most of the data?
(c) As no two blood samples are ever the same, can one proceed beyond the fitting of the curve and propose physically motivated correlations between the model parameters and key physiological descriptors, such as hematocrit. This would allow us to arrive at quantitative predictions of the steady state shear blood rheology.

The first work that systematically tried to address all these important questions was that of Apostolidis and Beris [94]. By using the most reliable available data from the literature at that time, obtained on whole blood and in both Couette and capillary flows, the authors were able to answer affirmatively the first question (to within 10% accuracy, for flows in macroscopic flow geometries and shear rates > ~0.7 s^{-1}). They also established that, among a number of possible constitutive models for the viscosity, the Casson equation [32] is by far the best, in agreement with the work of Merrill et al. [117]. These authors were the first to systematically use the Casson model to describe steady state blood rheology [117]. Apostolidis and Beris [94] also obtained, for physiological blood samples, reliable parametric expressions connecting the Casson model parameters to two key physiological parameters: the hematocrit and the fibrinogen concentrations. These relationships, derived from a large number of available data from the literature, represented a significant improvement over earlier relations [36,38,39,118,119].

Apostolidis and Beris [94] developed a parametric steady state model based on their available shear data. The steady state shear stress, σ, following the Casson model, see Chapter 1, is given for positive shear rates $\dot{\gamma} > 0$ as [32]:

$$\sqrt{\sigma} = \sqrt{\sigma_y} + \sqrt{\eta^C}\sqrt{\dot{\gamma}}; \quad \sigma \geq \sigma_y, \tag{8.2}$$

where σ_y represents the yield stress and η^C is the model viscosity. First, the yield stress parametric form is given as:

$$\sigma_y = \begin{cases} (Hct - Hct_c)^2 (0.5084 c_f + 0.4517)^2 & Hct > Hct_c \\ 0 & Hct \leq Hct_c \end{cases}, \tag{8.3}$$

where the yield stress, σ_y, is in dyne/cm^2, the fibrinogen concentration, c_f in g/dl, Hct represents the hematocrit fraction, and Hct_c the critical hematocrit value for yield stress given as:

$$Hct_c = \begin{cases} 0.3126c_f^2 - 0.468c_f + 0.1764 & c_f < 0.75 \\ 0.0012 & c_f \geq 0.75 \end{cases}. \tag{8.4}$$

The dependence of the yield stress on the fibrinogen concentration has been advocated in the pioneering work of Merrill and coworkers [36,39,118,119]. It was motivated by the previously proven role of fibrinogen in facilitating the formation of rouleaux aggregates [120] which were deemed to be of critical importance to the development of yield stress [117]. This has been first demonstrated qualitatively [36] and subsequently proven quantitatively [39,118,119]. However, Merrill's models addressed only the direct dependence of the yield stress on the fibrinogen concentration for a fixed hematocrit value [39,118,119], not the equally important dependence of the critical hematocrit on the fibrinogen concentration. Yet, the existence of a critical hematocrit for a yield stress has been recognized from the very beginning, with the critical value reported to be between 0.02 and 0.08 [39,119] (note that the values obtained from Eq. (8.4) for the physiological limiting values of fibrinogen concentration of 0.1 and 0.4 g/dl [119] are 0.04 and 0.13). To determine the dependence of the critical hematocrit value on the fibrinogen content, the much more complete blood yield stress data reported by Morris et al. [121] were utilized [94]. Note that a parametric dependence of the yield stress on the fibrinogen has also been presented in Morris et al. [121]; however, it combined the effects of the fibrinogen and hematocrit and did not make use of the critical hematocrit concept.

The only drawback in using the more complete, and physically relevant, model for the yield stress presented in Eqs. (8.3) and (8.4) is that it only involves the hematocrit and the fibrinogen values. Additional factors, such as the type and rhesus factor of the blood, sex and age, cholesterol and triglyceride levels, etc., being not available, had not been taken into account. Nevertheless, even under physiological conditions, these factors can vary and are known to cause measurable differences to blood rheology, and in particular, the yield stress. For example, Merrill et al. [118] have reported differences in the yield stress of as much as 25% from the average depending on the type of the RBCs. Similarly, the effect on blood rheology of total cholesterol, high and low density lipoprotein and triglycerides, has also been documented [122]. Taking these factors into account led to significant corrections to the above-proposed model equations [123]. Still, at present, only very fragmented data, restricted to limited values of hematocrit and fibrinogen, are available. They do not therefore provide the necessary basis for proposing a more complete expression for the yield stress involving the additional factors.

The exact mechanism through which this critical dependence of the yield stress on the fibrinogen concentration is obtained is still a matter of debate. Two theories in particular are prevalent. The first assumes that an increased fibrinogen concentration facilitates the formation of the rouleaux aggregates because of the development of depletion-generated attractive forces between adjacent red blood cells (generically generated with polymers in solution in colloidal suspensions – see, for example, Asakura and Oosawa [124] and Chapter 1). In the second one, aggregation is attributed to the formation of fibrinogen bridges between neighboring red blood cells, thus connecting them together into rouleaux structures. In fact, it is this second theory that has been used by Merrill et al. [118] to establish a model for the yield stress that involves the square of the number of links between the cells, leading to a dependence on the fibrinogen concentration squared. This result is not far from Eq. (8.3). Also, note that the expression for the yield stress above the critical hematocrit value involves a square dependence on the difference between the actual and critical hematocrit values. This is to be contrasted with the third power proposed by Merrill and coworkers [36,38,39,119]. The quadratic dependence, in addition of emerging naturally from the fitting of data [94], is also consistent with the ideas linking the yield stress to the interparticle attraction. This is the route followed by Merrill himself in proposing the second power dependence on the fibrinogen concentration mentioned above [118]), and which is also discussed in section 6.6.1 of *CSR*. It is also compatible with the critical character of the emergence of a nonzero yield stress value upon increasing values of the hematocrit above a critical value [94].

Predicting a nonzero yield stress is of particular interest in most of the physiological situations in which the blood hematrocrit, *Hct*, is above the critical hematocrit value, Hct_c, as defined in Eq. (8.4). Nonzero yield stresses characterize viscoplastic behavior, as commonly seen in concentrated, aggregated, suspensions, see also section 6.6.1 of *CSR*. The model described above has also been rather successfully compared with available data in the literature, presented in Figure 8.6, by considering the expected dependence of the yield stress on additional physiological parameters and the limitations of the available data. The latter were based on only partially characterized blood samples, often just with the hematocrit value.

The parametric form for the model viscosity developed in Apostolidis and Beris [94] is given as:

$$\eta^C = \eta_{p0}\left(1 + 2.0703 \times Hct + 3.7222 \times Hct^2\right) \times \exp\left(-7.0276(1 - T_0/T)\right), \quad (8.5)$$

where T_0 is the reference temperature of 273.16 + 23 = 296.16°K, at which the plasma viscosity $\eta_{p0} = 1.67 \times 10^{-3} Pa\ s$ is measured. An issue of interest to observe here is that the intrinsic viscosity, represented by the coefficient of the linear term in Eq. (8.5), is 2.07, a value that is lower than the Einstein result of 2.5, as presented in Chapter 1. This deviation is however expected as Einstein's relation is applicable for noninteracting rigid spheres while the RBCs are deformable and interacting. Under these conditions one expects an intrinsic viscosity below 2.5 [129], which is consistent with the measured value of 2.07, albeit a direct a priori evaluation is not possible due to the unknown interactions between the RBCs. The quality of the parametrization offered by Eqs. (8.3)–(8.5) has been evaluated in Apostolidis and Beris [94] through a comparison between predictions and additional data. Two typical examples are shown in Figure 8.7, which are Casson plots.

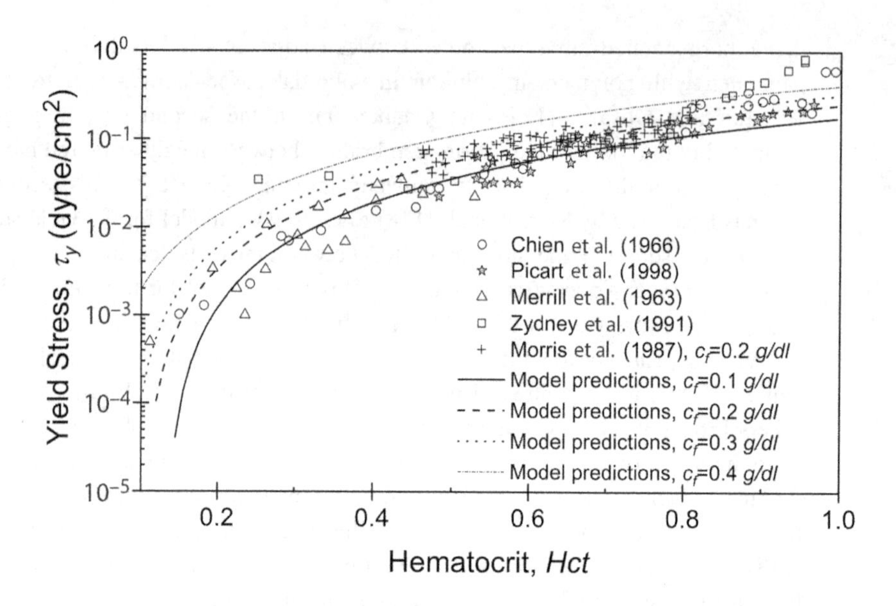

Figure 8.6 Yield stress vs. hematocrit. Continuous lines: model predictions for different values of fibrinogen concentration as shown in the legend. Symbols: directly measured yield stresses (Picart et al. 1998 [125]), derived from chamber sedimentation data (Morris et al. 1987 [126]) and extrapolated from low shear data (Chien et al. 1966 [127]; Merrill et al. 1963 [36,38]; Zydney et al. 1991 [128]). After Apostolidis and Beris [94]

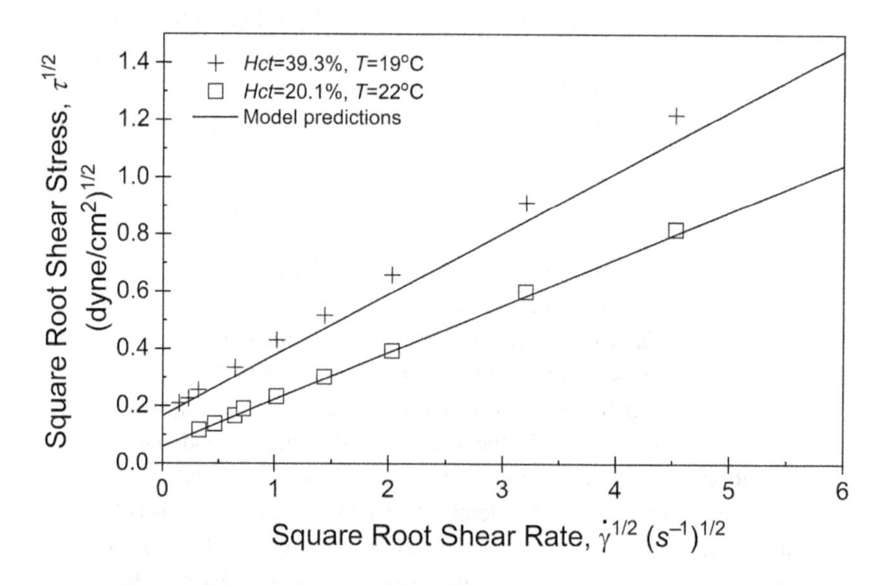

Figure 8.7 Casson plot for Couette viscometer data of two blood samples for conditions shown. Solid lines: model predictions; symbols: experimental data [130]. After Apostolidis and Beris [94]

Variations of the Casson model have also been developed, especially recently, as more detailed experimental data and/or more detailed models have become available. However, given the close connection of those variations to the transient shear flows they are discussed in Section 8.4.

8.4 Models for Transient Shear Flow

As already mentioned, blood, as a complex colloidal fluid, exhibits in addition to a yield stress a flow behavior that depends on time and flow history, i.e., thixotropy and/or viscoelasticity. This has been confirmed early on with experimental observations on the linear viscoelasticity of prepared suspensions of red blood cells in plasma at different hematocrits [41,131] and then in transient shear experiments [97,132,133]. Additional data have subsequently been obtained on the linear viscoelasticity of full human blood [134,135] and on nonlinear transient shear blood flows [136]. The latter exploited recent experimental developments in connection with probing complex time-dependent rheological behavior by systematically varying both the characteristic time scale and the amplitude of the deformation through Large Amplitude Oscillatory Shear (LAOS)(See also section 9.4.5 of *CSR*). Of particular relevance to blood rheology, considering the pulsatile characteristics of vascular blood flow, are recent data obtained in Uni-Directional Large Amplitude Oscillatory Shear Flow (UD-LAOS) [98]. UD-LAOS is realized by superimposing on the LAOS a steady shear in such a manner that the oscillating shear flow always remains in the same direction. An example is discussed in the following, along with the predictions of various history-dependent continuum models. This wealth of time-dependent flow data has provided ample evidence on the complexity of the blood rheology. In addition, they provided a testbed for models. In general, any acceptable constitutive model for blood rheology has to describe all the above-mentioned time–history effects. In addition it obviously has to reproduce the previously mentioned shear thinning and often viscoplastic effects under steady state shear flow conditions.

Initially, two parallel approaches have been followed in order to describe macroscopically the transient shear rheology of blood at the continuum level. In the first, a more or less standard viscoelastic constitutive equation is adapted to fit the special transient and steady state rheological characteristics of blood. This led to a number of variants of a "White–Metzner"-type modification of the Oldroyd-B or even the Johnson–Segalman [137] viscoelastic model [138–140] adapted for blood rheology [42,96,100,103,104,110,141]. In the second one, the starting point is a structural model for thixotropy [142] (see also Chapter 1) with the stress provided as a superposition of elastic and a viscous contributions [42,99,143]. In both of these approaches the model parameters may be functions of a structural parameter describing the red blood cells aggregation that is assumed to obey a kinetic relaxation equation [42,99,101–103,143,144]. More recently, hybrid models that combine and extend both approaches have also been developed.

Some are based on either multimode extensions of the generalized viscoelastic models taking also into account the effect of rouleaux aggregation and/or inhomogeneous stress-induced migration [106–111]. In still others, viscoelastic contributions are added to thixotropic models [98]. In the following subsections representative examples of each of the three types of models will be discussed and their predictions compared to recently obtained steady and transient shear data [98]. This section will be concluded with a short discussion on the latest multimode viscoelastic blood flow models.

8.4.1 Simple Viscoelastic Models: The Anand–Kwack–Masud (AKM) Model

The AKM model [110] has been proposed to eliminate discontinuities and convergence difficulties encountered in the numerical solution of the original generalized Oldroyd-B model [104]. The latter is based on a general nonequilibrium thermodynamics/continuum mechanics framework to describe rate-type models [145]. The AKM model can be considered as typical for the first class of models (homogeneous viscoelastic) where one tries to describe the time-dependent behavior of blood through a suitably modified viscoelastic constitutive model. It follows an extended White–Metzner modeling approach, using a conformation tensor [139,140]. The latter, $\mathbf{B}$, is used to describe the fluid structure in the AKM model and is assumed to have a unit determinant: $\det(\mathbf{B}) = 1$. The governing equations contain a viscoelastic evolution equation for $\mathbf{B}$ along with an equation that expresses the extra stress tensor $\mathbf{S}$ in terms of the conformation tensor $\mathbf{B}$ and the symmetric component of the velocity gradient $\mathbf{D} \equiv \frac{1}{2}(\nabla\mathbf{v} + \nabla\mathbf{v}^T)$ [110]:

$$\begin{aligned} \overset{\nabla}{\mathbf{B}} &= -2\frac{G_{ve}}{\eta_{ve}}(\mathbf{B} - \beta\mathbf{I}) \\ \beta &= \frac{3}{\mathrm{tr}\left(\mathbf{B}^{-1}\right)} \qquad , \\ \mathbf{S} &= G_{ve}\mathbf{B} + \mu_m\mathbf{D} \end{aligned} \tag{8.6}$$

where the superscript ∇ denotes the upper-convected time derivative [95],

$$\eta_{ve} = \mu_{ve}(\mathrm{tr}(\mathbf{B}))^{-m}, \tag{8.7}$$

and $G_{ve}, \mu_{ve}, m, \eta_m$ are positive constant model parameters, physically representing a viscoelastic modulus, viscoelastic viscosity, power law exponent, and "medium" (background) viscosity, respectively. Note that some of the model parameters are written here in a slightly different form than in Bird et al. [95] to facilitate their physical interpretation and their comparison against [98].

For a one-dimensional simple shear flow at a shear rate of $\dot{\gamma}$ these equations reduce to the following [98,110]:

$$\frac{d}{dt}B_{12} = -\frac{2G_{ve}}{\eta_{ve}}B_{12} + \dot{\gamma}B_{22}, \tag{8.8}$$

$$\frac{d}{dt}B_{11} = -\frac{2G_{ve}}{\eta_{ve}}(B_{11} - \beta) + 2\dot{\gamma}B_{12}, \tag{8.9}$$

$$\frac{d}{dt}B_{22} = -\frac{2G_{ve}}{\eta_{ve}}(B_{22} - \beta), \tag{8.10}$$

where

$$\eta_{ve} = \mu_{ve}(B_{11} + 2B_{22})^{-m}, \tag{8.11}$$

and

$$\beta = \frac{3B_{22}\left(B_{11}B_{22} - B_{12}^2\right)}{B_{22}^2 + 2B_{11}B_{11}}22 - B_{12}^2\mu_{ve}. \tag{8.12}$$

In this case, the shear stress, σ_{12}, is

$$\sigma_{12} = G_{ve}B_{12} + \eta_m\dot{\gamma} \tag{8.13}$$

In total, the AKM model contains four· parameters $G_{ve}, \mu_{ve}, m, \mu_m$ that must be determined by fitting to experimental data. Anand et al. [110] propose fitting these parameters to the steady shear data where the steady shear stress is given implicitly by:

$$\sigma_{12} = \left(\frac{1}{2}\eta_{ve}\beta + \mu_m\right)\dot{\gamma}, \tag{8.14}$$

$$\eta_{ve} = \mu_{ve}\left(3 + \frac{1}{2}\left(\frac{\eta_{ve}}{G_{ve}}\dot{\gamma}\right)^2\right)^{-m}\left(1 + \frac{1}{4}\left(\frac{\eta_{ve}}{G_{ve}}\dot{\gamma}\right)^2\right)^{m/3} \tag{8.15}$$

$$\beta = \left(1 + \frac{1}{4}\left(\frac{\eta_{ve}}{G_{ve}}\dot{\gamma}\right)^2\right)^{-1/3}. \tag{8.16}$$

The AKM model offers a good representation of the viscoelasticity of blood at high shear rates. However, the model was found to deviate from the experimental data at low shear rates due to the fundamentally different elastic response, the absence of a yield stress, and a failure to account for thixotropy in addition to viscoelasticity [99]. Its behavior in transient data is discussed in Section 8.5.

8.4.2 Structural Thixotropic Models: The Apostolidis–Armstrong–Beris (AAB) Model

The AAB model was developed by Apostolidis et al. [99] following standard practices for the development of a scalar structural parameter-based thixotropic model, as discussed in chapter 7 of *CSR*. In particular, Apostolidis et al. [99] adapted the structural thixotropic model of Armstrong et al. [146] so that it reduces to the Casson model under steady state shear conditions, at least for low shear rates.

This led to a model structural thixotropic model described by the following set of governing equations.

The basic equation expresses the shear stress, σ, as the superposition of an elastic contribution, proportional to the elastic strain, γ_e, and a viscous one, proportional to the plastic strain rate, $\dot{\gamma}_p$:

$$\sigma = G\gamma_e + \eta_m \dot{\gamma}_p, \tag{8.17}$$

where G represents an elastic modulus. The elastic and plastic strains are then defined through two equations. First, the total strain is decomposed into the sum of elastic and plastic contributions:

$$\gamma = \gamma_e + \gamma_p \quad \Leftrightarrow \quad \dot{\gamma} = \dot{\gamma}_e + \dot{\gamma}_p. \tag{8.18}$$

Second, the rate of change of the elastic strain is determined, following a straightforward extension of the kinematic hardening model [147] as:

$$\dot{\gamma}_e = \dot{\gamma}_p - \frac{\gamma_e}{\gamma_{\max}} |\dot{\gamma}_p|, \tag{8.19}$$

where $\gamma_{\max}$ represents the maximum allowed elastic strain. For the blood flow model the following relationship was proposed for $\gamma_{\max}$:

$$\gamma_{\max} = \min\left(\frac{\gamma_0}{\lambda^2}, \gamma_\infty\right), \tag{8.20}$$

where γ_0, γ_∞ are two dimensionless parameters representing, respectively, the limiting values at zero and infinite shear rate limit for the maximum elastic strain supported within the material. The values for $\gamma_0 < \gamma_\infty$ are expected to range between order 0.01 and 1, roughly.

Finally, two kinetic equations are proposed for the time evolution of the scalar structural parameter λ and the elastic modulus G[99], respectively. For the elastic modulus G this is provided by a relaxation equation of the form:

$$\frac{d}{dt}G = \frac{\sigma_y}{\eta_m} k_G \lambda (G_e - G), \tag{8.21}$$

where $\sigma_y$is the zero shear rate yield stress given by:

$$\sigma_y \equiv G_0 \gamma_0, \tag{8.22}$$

G_e represents an equilibrium modulus value assumed to depend proportionally on the structural parameter λ and on its zero shear rate value G_0:

$$G_e = \lambda G_0, \tag{8.23}$$

and k_G is a dimensionless kinetic coefficient of order 1.

The kinetic equation for the scalar structural parameter is:

$$\frac{d}{dt}\lambda = \frac{\sigma_y}{\mu} k_\lambda \left((1-\lambda) - \lambda \sqrt{\frac{4\eta_m \dot{\gamma}_p}{\sigma_y}}\right), \tag{8.24}$$

with k_λ another dimensionless kinetic parameter of order 1 characterizing the structural rebuild rate. Note that in Eq. (8.24) the square root dependence on the plastic shear rate, as well as the ratio of the kinetic coefficients for breakdown over rebuild, the latter represented by the term $\sqrt{4\eta_m/\sigma_y}$, result in the following solution for λ in steady state simple shear flow:

$$\lambda = \frac{1}{\sqrt{\frac{4\mu\dot{\gamma}}{\sigma_0}} + 1}. \tag{8.25}$$

When this solution is introduced into the constitutive relations for G_e and $\gamma_{\max}$, Eqs. (8.23) and (8.20), respectively, and those in turn are substituted into the constitutive equation for the shear stress, Eq. (8.17), the Casson model expression, Eq. (8.2), arises naturally, provided that $\gamma_0/\lambda^2 \leq \gamma_\infty \Leftrightarrow \sqrt{4\eta_m/\sigma_0} \leq \sqrt{\gamma_\infty/\gamma_0} - 1$. Note that for higher shear rates a correction to the Casson model is naturally introduced which reduces the effective model viscosity asymptotically to η_m [99]. Although such a transition from Casson to Newtonian has also been observed experimentally [148], this is not a generic feature of blood rheology [98].

Figure 8.8 is a comparison of the model fitting to transient data from hysteresis experiments [99]. The data from Bureau et al. [97] are obtained by imposing a linearly increasing shear rate from zero to a maximum, followed by decreasing it linearly back to zero. The two graphs in Figure 8.8 represent the two regimes that have been investigated: the one on the left corresponds to a maximum shear rate of $0.12 s^{-1}$, while the one on the right corresponds to a maximum shear rate of $1\ s^{-1}$.

To evaluate the AAB model predictions, the three parameters added to the steady state Casson model parametrization, as shown in Eqs. (8.17)–(8.24), are calculated using the hematocrit value reported by Bureau et al. [97]. Because the fibrinogen concentration was not reported for the blood sample used, a value has been obtained by fitting the low shear rate data reported on the left graph in Figure 8.8; the

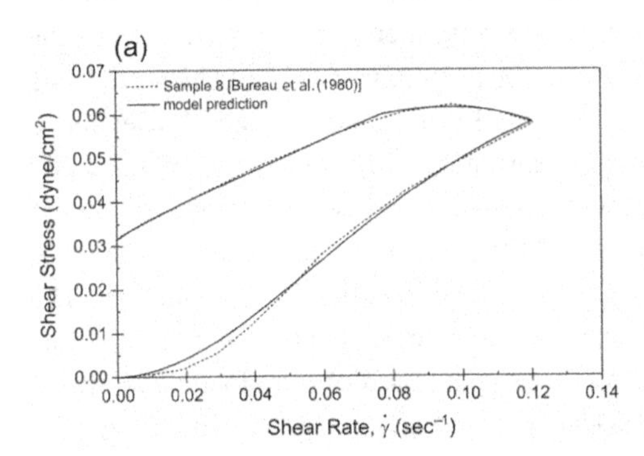

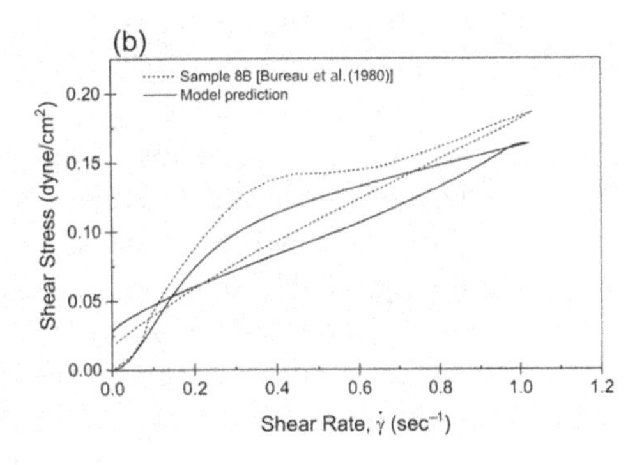

Figure 8.8 Comparison of model predictions (shear stress vs. shear rate) against the data of Sample 8 of Bureau et al. [97]. Model parameters used: $\gamma_0 = 0.039$; $c_f = 0.125\ g/\mathrm{dl}$; $k_\lambda = 1.214$; $k_G = 0.595$, from Apostolidis et al. [998]

comparison on the right being therefore a comparison of data against pure model predictions. These graphs show both the complexity encountered in transient blood flow behavior (Newtonian results would have been represented as just a straight line instead of a hystersis loop) as well as the power of the structural thixotropic model in capturing well the main experimental data features. Similar good, semi-quantitative, agreement with the experimental data has also been obtained for the transient start-up and shear flow cessation data of Bureau et al. [132] as well as the LAOS data reported by Sousa et al. [136]; see Apostolidis et al. [99] for more details.

8.4.3 Hybrid Thixotropic–Viscoelastic Models: The Horner–Armstrong–Wagner–Beris (HAWB) Model

More recently, the blood collection and experimental analysis protocol, previously developed by Baskurt et al. [149], has been carefully validated [150]. Further, a careful study of the effects of the time of storage on both steady and transient blood rheology measurements has been conducted [150]. The main conclusions from that study were that the biological origin of blood gives it a very special status as a colloidal fluid: Even in the presence of anticoagulants, even when employing standard high shear rate preconditioning of the samples to be tested before each measurement, the sample rheological measurements showed systematic changes over time (aging), statistically significant, even within the recommended allotted 4 hours time window. Furthermore, these changes were donor-dependent, indicating their origin to arise from an underlying biology. This study points out the need for a better understanding of those biological changes and the role that blood physiology plays, an issue that is probably common to all biofluids.

The same work generated high quality experimental data from several healthy donors under a variety of conditions, both steady as well as transient shear [98]. The new data made possible several refinements of the previously discussed (AAB) model [98]. First, they showed small but statistically significant deviations from the Casson model predictions at steady state. Henceforth, there was no need to force the transient model to comply with Casson behavior at steady state shear. That led to a simpler and simultaneously richer structure of the kinetic model equations. Second, a viscoelastic contribution to the stress was introduced in order to more faithfully represent the stress developed due to the deformation of the red blood cells under flow. These two developments led to the HAWB model [98]. This is representative of the hybrid thixotropic/viscoelastic models introduced earlier in this section. The overall extra stress of the system, $\boldsymbol{\sigma}$, consists of two components: a thixotropic one, $\boldsymbol{\sigma}_R$ that accounts for the presence of the rouleaux aggregates of red blood cells, which form at low shear rates, and a viscoelastic component, $\boldsymbol{\sigma}_C$ described through an extended White–Metzner approach:

$$\boldsymbol{\sigma} = \boldsymbol{\sigma}_R + \boldsymbol{\sigma}_C. \tag{8.26}$$

The governing equations of the HAWB model are described in more detail next.

In the HAWB model, as in the AAB one, thixotropy is modeled using a scalar structural parameter, λ. It represents, in dimensionless form, the level of rouleaux formation and varies between 0 and 1. A value of 1 represents the maximum structure that can be established for any given set of physiological conditions at static equilibrium. The minimum value, 0, is asymptotically reached at high shear rates. Following previous developments, as for example discussed in Chapter 1, chapter 7 of *CSR*, and more recently in Armstrong et al. [146] and Mwasame et al. [151], the time evolution of the structure parameter, λ, is assumed to be described by a general kinetic equation. It contains a Brownian aggregation term, a shear induced aggregation term, and a shear break-up term [98]:

$$\frac{d}{dt}\lambda = \frac{1}{\tau_\lambda}\left((1-\lambda) + (1-\lambda)\tau_a|\dot{\gamma}_p| - \lambda\left(\tau_b|\dot{\gamma}_p|\right)^2\right), \tag{8.27}$$

where, τ_λ, τ_a, and τ_b are time constants governing the rate of structure formation due to Brownian motion, the relative rate of shear-induced structure formation to that due to Brownian motion, and the relative rate of shear induced structure breakdown, respectively. The thixotropy contribution to the shear stress due to the rouleaux formation, σ_R, has been modeled as the sum of elastic and viscous components [98]:

$$\sigma_R = \lambda G_R \gamma_e + \lambda^3 \eta_R \dot{\gamma}_p, \tag{8.28}$$

where G_R is an elastic modulus, η_R is a thixotropic viscosity, and γ_e and $\dot{\gamma}_p$ are the red blood cell elastic strain and plastic rate of shear, respectively. The λ^3 proportionality to the thixotropic viscosity has been selected to better represent the link of the structure parameter to the average length of the rouleaux [98]. In contrast, the elastic stress is assumed to depend linearly on the structure parameter, λ, to better represent a network contribution that naturally disappears in the absence of structure.

In a similar fashion to the AAB model, the elastic modulus G_R is assumed to also obey a relaxation equation [98]:

$$\frac{d}{dt}G_R = \frac{1}{\tau_G}\left(\frac{G_{0,R}}{\lambda} - G_R\right), \tag{8.29}$$

where τ_G is the corresponding relaxation time, and where the equilibrium value for the elastic modulus, established after long times, is taken to be inversely proportional to the structure parameter, reflective of a strain stiffening system. Of course, an upper bound for the elastic modulus could also be prescribed following Eq. (8.20). This is, however, typically sufficiently large to not affect the implementation of the model at low and intermediate shear rates where the thixotropy contribution to the stress is nonzero. Finally, the fully structured (zero shear rate) elastic modulus, $G_{0,R}$, is defined as:

$$G_{0,R} = \sigma_y/\gamma_{0,R}, \tag{8.30}$$

where σ_y represents the dynamic yield stress and $\gamma_{0,R}$ is the maximum elastic strain at asymptotically zero shear rate.

The rest of the thixotropy model follows closely the analysis offered in connection to the AAB model, namely the decomposition of the total strain into an elastic and a plastic component, as described in Eq. (8.17), with one difference. The elastic strain rate of Eq. (8.19) is now given by:

$$\dot{\gamma}_e = \begin{cases} \dot{\gamma}_p - \dfrac{\gamma_e}{\gamma_{\max}} \left|\dot{\gamma}_p\right| & \dfrac{d}{dt}\gamma_{\max} \geq 0 \\ \dot{\gamma}_p - \dfrac{\gamma_e}{\gamma_{\max}} \left|\dot{\gamma}_p\right| + \dfrac{\gamma_e}{\gamma_{\max}} \dfrac{d}{dt}\gamma_{\max} & \dfrac{d}{dt}\gamma_{\max} < 0 \end{cases}, \tag{8.31}$$

in which the maximum elastic strain sustained by the rouleaux, $\gamma_{\max}$, is defined as:

$$\gamma_{\max} = \gamma_{0,R}\lambda. \tag{8.32}$$

The deviations from the AAB model as far as the thixotropic contributions are concerned have been proposed to account for the changes in the maximum elastic strain due to structure break-up.

The second and major difference in the HAWB from the AAB model is now that there is an additional contribution to the shear stress that arises from the deformation of the individual red blood cells due to the flow. The modeling of this contribution is patterned after experimental results involving red blood cells suspended in an albumin-Ringer solution, conditions under which it is assumed that red blood cells aggregation into rouleaux formation is prevented [22]. This is shown by the dotted line in Figure 8.9. As shown, under these conditions the apparent viscosity of the red blood cell suspension is much lower than when they are suspended in plasma. As noted, in the latter case

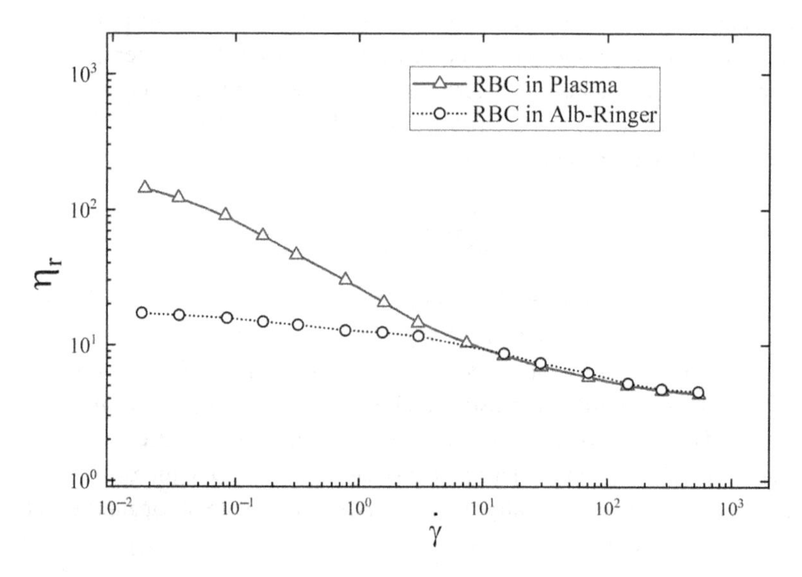

Figure 8.9 Dependence of the relative viscosity η_r on the apparent shear rate $\dot{\gamma}$ in steady shear flow for a 45% hematocrit suspension of red blood cells in (a) plasma (solid line) and in (b) an albumin-Ringer solution under conditions under which rouleaux aggregates are not observed. After Chien [22]

viscoplasticity with a nonzero yield stress occurs as well. It is this difference that in the HAWB model is modeled with the thixotropic contribution described by Eqs. (8.27)–(8.32). The contribution of the individual deformed cells corresponding to the dotted line still needs to be added to it to complete the blood flow model. It is of interest to note here that, despite the absence of rouleaux formation, the free red blood cells continue to demonstrate a non-Newtonian shear thinning behavior that is characterized at steady state by a shear rate dependent viscosity $\eta_{Css}(\dot{\gamma})$ changing from a zero shear rate viscosity, η_{C0}, to an infinite shear rate viscosity, $\eta_{C\infty}$. This steady shear stress, σ_{Css}, is represented in the HAWB model by a Cross model with the exponential term in the Cross model set to 1:

$$\sigma_{Css} = \eta_{Css}(\dot{\gamma})\dot{\gamma} \equiv \left(\frac{\eta_{C0} - \eta_{C\infty}}{1 + \tau_C|\dot{\gamma}|} + \eta_{C\infty}\right)\dot{\gamma}, \tag{8.33}$$

where τ_C is a time constant governing the dependence of the steady state viscosity on the shear rate.

Additional experimental work has shown that free red blood cells not only induce non-Newtonian behavior but also viscoelastic effects [22]. This is reflected by contributions to both the storage and loss modulus in the linear viscoelastic behavior [22]. In the HAWB model the (generally tensorial and time-dependent) free cell stress contribution σ_C is modeled through a full tensorial viscoelastic constitutive model. In particular, the extended White–Metzner approach [139,140] is adopted. This offers adaptability to fit the steady state viscosity represented by Eq. (8.33) while allowing for model stability [140] and a full viscoelastic behavior (including normal stresses). It can be conveniently represented by [98]:

$$\boldsymbol{\sigma}_C + \left(\frac{\eta_C(\mathrm{I}_{\sigma_C})}{G_C}\right)\overset{\nabla}{\boldsymbol{\sigma}}_C = \eta_C(\mathbf{I}_{\sigma_C})\dot{\gamma}, \tag{8.34}$$

where I_{σ_C} is the first invariant of the stress tensor and $\dot{\gamma} \equiv \nabla\mathbf{v} + \nabla\mathbf{v}^T$ is the rate of strain tensor defined in Chapter 1, G_C represents the elastic modulus of the suspension of free red blood cells, and η_C is the corresponding viscosity, now expressed though as a function of the first invariant, I_{σ_C}. As shown in the supplemental material of Horner et al. [98], to derive from Eq. (8.34) the steady state behavior given by Eq. (8.33) requires first a choice of the invariant(s) in terms of which to describe the dependence of η_C. If we choose just the first invariant, I_{σ_c}, then the expression for $\eta_C = \eta_C(\mathrm{I}_{\sigma_C})$ which reduces to $\eta_{Css}(\dot{\gamma})$ for steady shear flow is [98]:

$$\eta_C = \frac{-b + \sqrt{b^2 + 4c}}{2}, \tag{8.35}$$

where

$$b \equiv \tau_C\sqrt{\frac{G_C\mathrm{I}_{\sigma_C}}{2}} - \eta_{0,C}; \quad c \equiv \eta_{\infty,C}\tau_C\sqrt{\frac{G_C\mathrm{I}_{\sigma_C}}{2}}. \tag{8.36}$$

A comparison between the predictions obtained with the various models and recent data on human blood in steady shear and specific LAOS flows follows in Section 8.5.

8.4.4 Multimode Viscoelastic Models

Alternate, more sophisticated, multimode viscoelastic blood flow models were developed initially by Owens and coworkers [96,105–109]. In the first of these models an attempt was made to take into account, through a set of viscoelastic/viscoplastic phenomenological equations extracted through a polymer network theory analog, the aggregation and disaggregation of the erythrocytes [96]. This model was subsequently applied to simple shear flows [96] as well as steady, oscillatory, and pulsatile flow in rigid vessels [105]. Later, Moyers-Gonzalez et al. [107] developed a further refinement of that model to take into account inhomogeneous erythrocyte concentrations due to stress-induced migration following the modeling described in earlier work of Beris and Mavrantzas [152]. This allowed for the incorporation of the Fåhræus and Fåhræus–Lindqvist effects, such that the local hematocrit and the apparent blood viscosity decreases with tube diameter for sufficiently small vessels [16]. The steady state analysis in Moyers-Gonzalez et al. [107] was followed by an asymptotic analysis of the wall boundary layers at high Péclet numbers [106], and a numerical [108] as well as multiple scale analysis at high frequencies [109] for oscillatory flow in straight rigid tubes.

Most recently, Tsimouri et al. [111] considerably improved upon the Owens red blood cell aggregation model [107] by redeveloping the governing multimode viscoelastic equations within the context of nonequilibrium thermodynamics [140,153] following recent developments from the modeling of stress-influenced breakage rates in rod-like micellar solutions [154]. They thus avoided a number of arbitrary parameters that have appeared in the Owens model while maintaining very similar predictive capabilities for the red blood cell aggregates in the results. Clearly, this work has shown the benefit for accommodating a genuine time-dependent non-Newtonian description for the blood flow through rigorous, thermodynamically consistent modeling, allowing for a more detailed description of the involved microstructural changes, especially the red blood cell aggregation process. However, there is still more work needed in order to extend this analysis to conditions more appropriate to those encountered in vivo in the arterial circulation (also allowing for example, for fluid–structure interactions, as well as multiple vessels, geometric effects, etc.). Most importantly, what is lacking is a systematic development of the model parameters as a function of the physiological characterization of the blood and the ensuing biological processes. As the authors noted [111] there are several issues that need to be addressed in future work before this type of model can be ready to be applied in practice – clearly it shows potential, but still, as also noted in Section 8.2 in connection with genuine microscopic models, it remains to be further developed.

8.5 Comparison of Model Predictions to Steady Shear and UD-LAOS Experimental Data

The availability of several new series of detailed experimental data from eight healthy donors allowed Horner et al. [98] to pursue a systematic evaluation of the three main models described in Section 8.4: AKM, AAB, and HAWB. Model parameters were fitted using the same parallel-tempering stochastic optimization approach [155] for all models to compare with the experimental data while avoiding any model bias. Some indicative results from that work are given in Sections 8.5.1 and 8.5.2. They allow us to appreciate the details of the newly available data and simultaneously best assess the quality of the continuum model predictions. Both steady state and transient results are shown. For additional results the reader is referred to the original publication and its supplemental material [98].

8.5.1 Steady State Blood Rheology Data and Model Fits

The experimental data and the results of model fitting to the initial steady shear experiments conducted on blood from donors 1 and 2 of Horner et al. [98] are shown in Figure 8.10. From that figure, and especially the Casson plots shown in Figure 8.10(b), it is evident that there is a small but consistent departure from the more commonly seen Casson behavior of blood flow in the new data. As such, the HAWB model is the only one able to capture this detail. The AAB model is restricted by construction to only show a Casson behavior, whereas the AKM model fails to account for the yield stress behavior that is significant at low shear rates.

8.5.2 UD-LAOS Time-Dependent Blood Rheology Data and Model Fits

For the AKM model all parameters can be obtained by fitting steady state data. To complete the fit of the model, parameters for the AAB and HAWB models through four separate UD-LAOS experiments were used at conditions bracketing physiological flows. These tests involved sets at a high and a low frequency, and at a high and a low amplitude. The elastic projections of these fits/predictions for donor 1 are shown in Figure 8.11. Similar results were obtained for donor 2 [98]. The viscous projections (Supplemental Material of Horner et al. [98]) show a similar quality of the fits/predictions.

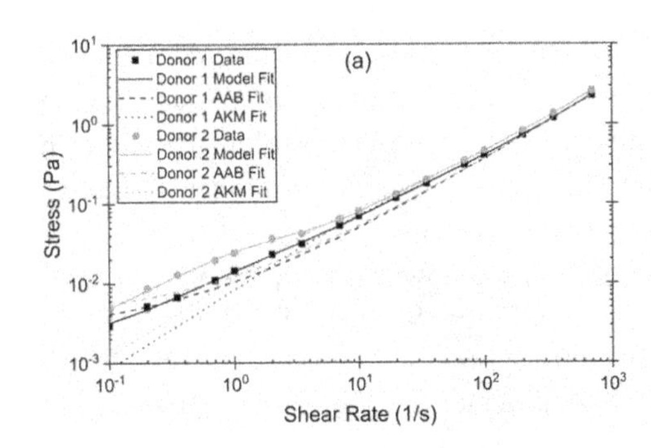

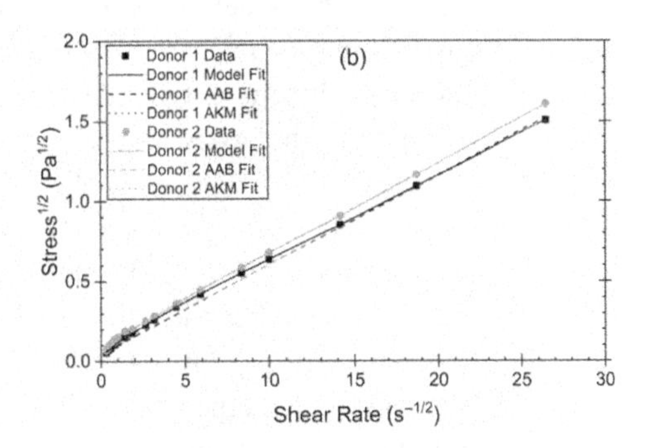

Figure 8.10 Steady shear data (symbols) taken on blood from two donors (a) in standard coordinates and (b) Casson plots, compared with the AAB, AKM, and HAWB (noted simply as "Model") models. After Horner et al. [98]

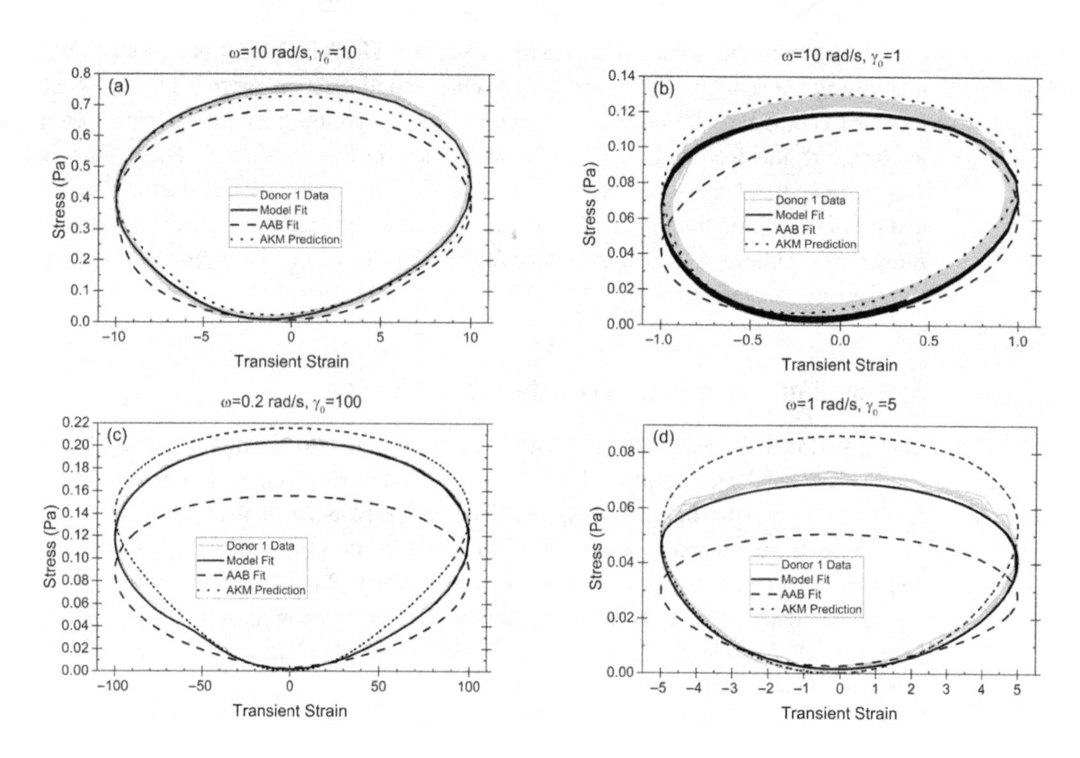

Figure 8.11 Elastic projection (for the HAWB = model, AAB, and AKM) model fits/predictions to UD-LAOS experimental data on blood from Donor 1 at (a) a frequency of 10 rad/s and a strain amplitude of 10, (b) a frequency of 10 rad/s and a strain amplitude of 1, (c) a frequency of 0.2 rad/s and a strain amplitude of 100, and (d) a frequency of 1 rad/s and a strain amplitude of 5.
After Horner et al. [98]

The experimental results display a primarily thixotropic behavior. This is particularly evident in the top–bottom asymmetry for tests shown in Figure. 8.11(c and d). Also, some viscoelastic effects are present, as shown by the left–right asymmetry, e.g., in Figure 8.11(a). The HAWB model is able to capture the transient behavior exhibited by these UD-LAOS tests very well across the range of frequencies, in contrast to the AAB model. The best model fit slightly underpredicts the stress in Figure 8.11(b). However, the deviations are relatively small and the qualitative behavior is still captured well. The AKM model works well at high shear rates and frequencies but deviates from the experimental data at lower shear rates, e.g., the poor qualitative prediction of the UD-LAOS curve in Figure 8.11(d). This shortcoming of the AKM model illustrates the fundamentally different elastic responses that blood demonstrates at high and low shear rates, thixotropy effects dominate the latter and are poorly captured within the purely viscoelastic AKM setting.

The data and model fits/predictions for Donor 2 (see Horner et al. [98], not shown here) demonstrate qualitatively similar behavior to that of Donor 1. However, subtle differences between the data sets exist, as could already be seen in the steady state data of Figure 8.10. This is a result of the unique physiological profiles that characterize each blood sample, associated with different blood count and cholesterol/triglyceride characteristics. The differences observed between the various data sets suggest the possibility to relate the bulk rheology of the samples to the physiology. Knowing this connection could unlock the potential of personalized modeling, or, inversely, using rheology as a quick and inexpensive tool to identify pathological issues with a blood sample. The idea of using blood rheology as a diagnostic tool is not new; it was introduced more than 40 years ago – see, for example, Dintenfass [156]. What is new now is the capability through substantially more elaborate experiments and models to make that connection much more quantitative and systematic. It further makes the case for a closer connection between rheology modeling and the underlying biology.

8.6 Summary and Outlook

In this review the study of the rheology of human blood is approached from the perspective of that of a complex, colloidal, fluid. Given the tremendous practical interest of the subject, and therefore the considerable amount of research work along various avenues, this treatment cannot be comprehensive. Lacking space, many issues have only be touched upon briefly or even not mentioned at all. Most notably among the latter are those associated with inhomogeneities in the flow field and fluid–wall surface interactions as well as those considering microscopic effects and nonphysiological, i.e., pathological, conditions. Given the tremendous complexity of the problem and the fact that it is still a very active area of research, providing an all-encompassing comprehensive review at this stage might be of limited value in time. Instead, we tried to present the main issues underscoring the macroscopic non-Newtonian rheological behaviour of blood, especially as associated with thixotropy, viscoplasticity, and viscoelasticity. These are the properties that most directly reflect blood to be a colloidal, concentrated suspension in which aggregate forming and breaking processes occur.

By focusing on macroscopic continuum models we wanted to offer tools that can potentially be used in computational fluid mechanics simulations. In the process the close connection to the underlying rouleaux structure has also become obvious, leaving many challenges and opportunities for future progress. For example, recent particle-level population balance modeling shows promise to mechanistically connect red blood cell properties and blood physiology to rouleaux formation and breakdown and the consequent thixotropy [151,157]. Not only the role of the red blood cells, their deformation and interactions leading to their rouleaux aggregation should be further considered, but also the effect of platelets with their potential activation and aggregation, along with margination and flow-induced migration effects. The connection to biology and the underlying interactions of the red and white blood cells and platelets with several proteins also emerges as a critical issue to further understand and better model blood flow. We believe that such an understanding is needed if we ever aspire to arrive at a state where hemorheology can be properly described with many potential applications to the health sector and medicine.

Acknowledgments

Financial support of the National Science Foundation through Award No. CBET 1510837 is gratefully acknowledged, along with the invaluable help provided by the co-PIs (Profs. Norman J. Wagner and Donna Woulfe) and the graduate student, Jeff S. Horner.

Chapter Notation

$\mathbf{B}$ configuration tensor [–]
c_f fibrogen concentration [g/dl]
$\mathbf{D}$ rate-of-strain tensor [s^{-1}]
D diameter blood vessel [μm]
G_e equilibrium modulus [Pa]
k_G kinetic coefficient (Eq. 8.21) [–]
k_λ kinetic coefficient (Eq. 8.24) [–]
T_0 reference temperature [K]

Greek Symbols

β model parameter of the AKM model (Eqs. 8.14 and 8.16) [–]
σ_y yield stress [Pa]
η_C free cell model viscosity [Pa s]
η_m viscosity of the suspending medium [Pa s]
η_{p0} plasma viscosity at reference temperature T_0 [Pa s]
μ model parameter viscosity [Pa s]
λ structural parameter [–]
σ extra stress tensor [Pa]
τ_a time constant governing the rate of shear-induced structure formation relative to that due to Brownian motion [–]
τ_b time constant governing the rate of shear-induced structure breakdown [–]
τ_C time constant for the viscosity in the HAWB model [s]
τ_G relaxation time associated with G_R [t]
τ_λ time constant governing the rate of structure formation due to Brownian motion [s]

Subscripts

C free cell contribution in the HAWB model
e elastic
p plastic

R thixotropic contribution due to rouleaux formation
r relative
ss steady state
ve viscoelastic part in the AKM model

References

1. GBD 2016, Causes of Death Collaborators, regional, and national age-sex specific mortality for 264 causes of death, 1980–2016: A systematic analysis for the Global Burden of Disease Study 2016. *Lancet.* 2017;390(10100):1151–1210.
2. American Heart Association Council on Epidemiology and Prevention Statistics Committee and Stroke Statistics Subcommittee, Heart Disease and Stroke Statistics – 2018 Update: A Report from the American Heart Association. *Circulation.* 2018;137(12): e64–e492.
3. Pedley T. *The Fluid Mechanics of Large Blood Vessels.* Cambridge: Cambridge University Press; 1980.
4. Cheer AY, van Dam CP. (eds.) *Fluid Dynamics in Biology. Contemporary Mathematics,* vol. 141, Providence, RI: American Mathematical Society; 1993.
5. Fung YC. *Biomechanics. Circulation,* 2nd ed. New York: Springer-Verlag; 1997.
6. Drzewiecki GM, Li JK-J. *Analysis and Assessment of Cardiovascular Function.* New York: Springer-Verlag; 1998.
7. Li JK-J. *The Arterial Circulation. Physical Principles and Clinical Applications.* Totowa, NJ: Humana Press; 2000.
8. Li JK-J. *Dynamics of the Vascular System (Series on Bioengineering and Biomedical Engineering – Vol. 1).* Singapore: World Scientific; 2004.
9. Zamir M. *The Physics of Pulsatile Flow.* New York: AIP Press, Springer-Verlag; 2000.
10. Zamir M. *The Physics of Coronary Blood Flow.* New York: AIP Press, Springer-Verlag; 2005.
11. Kowalewski TA. (ed.) *Blood Flow Modelling and Diagnostics.* Warsaw, Poland: Institute of Fundamental Technological Research; 2005.
12. Waite L, Fine J. *Applied Biofluid Mechanics.* New York: McGraw Hill; 2007.
13. Batzel JJ, Kappel F, Schneditz T, Tran HT. *Cardiovascular and Respiratory Systems. Modeling, Analysis, and Control.* Philadelphia, PA: SIAM; 2007.
14. Galdi GP, Rannacher R, Robertson AM, Turek S. (eds.) *Hemodynamical Flows. Modeling, Analysis and Simulation.* Basel, Switzerland: Birkhäuser; 2008.
15. Thiriet M. *Biology and Mechanics of Blood Flows,* 2 vols. New York: Springer; 2008.
16. Truskey GA, Yuan F, Katz DF. *Transport Phenomena in Biological Systems,* 2nd ed. Upper Saddle River, NJ: Prentice Hall; 2009.
17. Formaggia L, Quarteroni A, Veneziani A. (eds.) *Cardiovascular Mechanics, Vol.1, Modeling and Simulation of the Circulatory System.* Milano, Italy: Springer-Verlag; 2009.
18. Nichols WW, O'Rourke MF, Vlachopoulos C. *McDonald's Blood Flow in Arteries. Theoretical, Experimental and Clinical Principles,* 6th ed. London: Hodder Arnold; 2011.
19. Chandran KB, Rittgers SE, Yoganathan AP. *Biofluid Mechanics. The Human Circulation,* 2nd ed. Boca Raton, FL: CRC Press; 2012.
20. Peattie RA, Fisher RJ, Bronzino JD, Peterson DR. *Transport Phenomena in Biomedical Engineering. Principles and Practices.* Boca Raton, FL: CRC Press; 2013.

21. Vander AJ, Sherman JH, Luciano DS. *Human Physiology. The Mechanisms of Body Function*, 6th ed. New York: McGraw-Hill; 1994.
22. Chien S. Biophysical behavior of red cells in suspensions. In Surgenor DM. (ed.) *The Red Blood Cell*, 2nd ed., vol. 2. New York: Academic Press; 1975; pp. 1031–1133.
23. Ness PM, Stengle JM. Historical introduction. In Surgenor DM. (ed.) *The Red Blood Cell*, 2nd ed., vol. 1. New York: Academic Press 1974; pp. 1–50.
24. Humorism – Wikipedia article: https://en.wikipedia.org/wiki/Humorism
25. Loeb Classical Library: www.loebclassics.com/view/hippocrates_cos-nature_man/1931/pb_LCL150.11.xml
26. Baskurt OK, Neu B, Meiselman HJ. *Red Blood Cell Aggregation*. Boca Raton, FL: CRC Press; 2012.
27. Goldsmith HL, Cokelet GR, Gaehtgens P. Fåhræus R. Evolution of his concepts in cardiovascular physiology. *American Journal of Physiology*. 1989;257(3):H1005–H1015.
28. Fåhræus R. The suspension stability of blood. *Physiological Reviews*. 1929;9(2):241–274.
29. Fåhræus R, Lindqvist T. The viscosity of blood in narrow capillary tubes. *American Journal of Physiology*. 1931;96(3):562–568.
30. Stavridis IK. Contributions of Jean Leonard Marie Poiseuille to microcirculation research (in Greek). In I. K. Stavridis, ed. *Graduate Lessons of the Physiology of Microcirculation*. Athens, Greece: Medical Publications of P.Ch. Paschalides; 1993, Ch. 1, pp. 9–17.
31. Hess WR. Gehört das Blut dem allgemeinen Strömungsgesetz der Flüssigkeiten? *Pflügers Archiv*. 1915;162(5):187–224.
32. Casson N. Rheology of disperse systems. In Mill CC (ed.) *Flow Equation for Pigment Oil Suspensions of the Printing Ink Type. Rheology of Disperse Systems*. London: Pergamon Press; 1959; pp. 84–102.
33. Robertson AM, Sequeira A, Kameneva MV. Hemorheology. In Caldi GP, Rannacher R, Robertson AM, Turek S. (eds.) *Hemodynamical Flows. Modelling, Analysis and Simulation, Oberwolfach Seminars*, Vol. 37, Basel, Switzerland: Birkhäuser Verlag; 2008, pp. 63–120.
34. Scott Blair GW. An equation for the flow of blood, plasma and serum through glass capillaries. *Nature*. 1959;183(4661):613–614.
35. Reiner M, Scott Blair GM. The flow of blood through narrow tubes. *Nature*. 1959;184 (4683):354.
36. Merrill EW, Cokelet GC, Britten A, Wells RE Jr. Non-Newtonian rheology of human-blood-effect of fibrinogen deduced by "subtraction". *Circulation Research*. 1963;13 (1):48–55.
37. Cokelet GC, Merrill EW, Gilliland ER, Shin H. The rheology of human blood – Measurement near and at zero shear rate. *Transactions of the Society of Rheology*. 1963;7(1):303–317.
38. Merrill EW, Gilliland ER, Cokelet GC, Shin H, Britten A, Wells Jr. RE. Rheology of human blood, near and at zero flow: Effects of temperature and hematocrit level. *Biophysical Journal*. 1963;3(3):199–213.
39. Merrill EW, Gilliland ER, Lee TS, Salzman EW. Blood rheology: Effect of fibrinogen deduced by addition. *Circulation Research*. 1966;18(4):437–446.
40. Chien S, Usami S, Dellenback RJ, Gregersen MI. Shear dependence of effective cell volume as a determinant of blood viscosity. *Science*. 1970;168(3934):977–979.
41. Thurston GB. Viscoelasticity of human blood. *Biophysical Journal*. 1972;12(9):1205–1217.
42. Sun N, De Kée D. Simple shear, hysteresis and yield stress in biofluids. *Canadian Journal of Chemical Engineering*. 2001;79(1):36–41.

43. Fedosov DA, Pan WX, Caswell B, Gompper G, Karniadakisb GE. Predicting human blood viscosity in silico. *Proceedings of the National Academy of Sciences of the USA*. 2011;108(29):11772–11777.
44. Fedosov DA, Dao M, Karniadakis GE, Suresh S. Computational biorheology of human blood flow in health and disease. *Annals of Biomedical Engineering*. 2014;42 (2):368–387.
45. Rindt CCM, van Steehoven AA. Unsteady flow in a rigid 3-D model of the carotid artery bifurcation. *Journal of Biomechanical Engineering*. 1996;118(1):149–159.
46. Milner JS, Moore JA, Rutt BK, Steinman DA. Hemodynamics of human carotid artery bifurcations: Computational studies with models reconstructed from magnetic resonance imaging of normal subjects. *Journal of Vascular Surgery*. 1998;28(1):143–156.
47. Gijsen FJH, van de Vosse FN, Janssen JD. The influence of the non-Newtonian properties of blood on the flow in large arteries: Steady flow in a carotid artery bifurcation model. *Journal of Biomechanics*. 1999;32(6):601–608.
48. Steinman DA, Thomas JB, Ladak HM, Milner JS, Rutt BK, Spence JD. Reconstruction of carotid bifurcation hemodynamics and wall thickness using computational fluid dynamics and MRI. *Magnetic Resonance in Medicine*. 2002;47(1):149–159.
49. Kato M, Dote K, Habara S, Takemoto H, Goto K, Nakaoka K. Clinical implications of carotid artery remodeling in acute coronary syndrome. Ultrasonographic assessment of positive remodeling. *Journal of American College of Cardiology*. 2003;42(6):1026–1032.
50. Valencia A, Zarate A, Galvez M, Badilla L. Non-Newtonian blood flow dynamics in a right internal carotid artery with a saccular aneurism. *International Journal for Numerical Methods in Fluids*. 2006;50(6):751–764.
51. Nguyen KT, Clark CD, Chancellor TJ, Papavassiliou DV. Carotid geometry effects on blood flow and on risk for vascular disease. *Journal of Biomechanics*. 2008;41 (1):11–19.
52. Gay M, Zhang L. Numerical studies of blood flow in healthy, stenosed, and stented carotid arteries. *International Journal for Numerical Methods in Fluids*. 2009;61(4):453–472.
53. Wake AK, Oshinski JN, Tannenbaum AR, Giddens DP. Choice of in vivo versus idealized velocity boundary conditions influences physiologically relevant flow patterns in a subject-specific simulation of flow in the human carotid bifurcation. *Journal of Biomechanical Engineering*. 2009;131(2):021013.
54. Bevan RTL, Nithiarasu P, Van Loon R, Savonov I, Luckraz H, Garnham A. Application of a locally conservative Galerkin (LCG) method for modelling blood flow through a patient-specific carotid bifurcation. *International Journal for Numerical Methods in Fluids*. 2010;64(10–12):1274–1295.
55. Morbiducci U, Gallo D, Massai D, Ponzini R, Deriu MA, Antiga L, et al. On the importance of blood rheology for bulk flow in hemodynamic models of the carotid bifurcation. *Journal of Biomechanics*. 2011;44(13):2427–2438.
56. Kamenskiy AV, MacTaggart JN, Pipinos II, Bikhchandani J, Dzenis YA. Three-dimensional geometry of the human carotid artery. *Journal of Biomechanical Engineering – Transactions ASME*. 2012;134(6):064502.
57. Hutchins GM, Miner MM, Boitnott JK. Vessel caliber and branch-angle of human coronary artery branch-points. *Circulation Research*. 1976;38(6):572–576.
58. Nissen SE, Gurley JC, Grines CL, Booth DC, McClure R, Berk M, et al. Intravascular ultrasound assessment of lumen size and wall morphology in normal subjects and patients with coronary artery disease. *Circulation*. 1991;84(3):1087–1099.

59. Brinkman AM, Baker PB, Newman WP, Vigorito R, Friedman MH. Variability of human coronary artery geometry: An angiographic study of the left anterior descending arteries of 30 autopsy hearts. *Annals of Biomedical Engineering*. 1994;22(1):34–44.
60. Friedman MH, Baker PB, Ding Z, Kuban BD. Relationship between the geometry and qualitative morphology of the left anterior descending coronary artery. *Arteriosclerosis*. 1996;125(12):183–192.
61. Perktold K, Hofer M, Rappitsch G, Loew M, Kuban BD, Friedman MH. Validated computation of physiologic flow in a realistic coronary artery branch. *Journal of Biomechanics*. 1998;31(3):217–228.
62. Changizi MA, Cherniak C. Modeling the large-scale geometry of human coronary arteries. *Canadian Journal of Physiology and Pharmacology*. 2000;78(8):603–611.
63. Seron FJ, Garcia E, del Pico J. MOTRICO project – geometric construction and mesh generation of blood vessels by means of the fusion of angiograms and IVUS. In: Perales FJ, Campilho AJC, de la Blanca NP, Sanfeliu A. (eds.) *Pattern Recognition and Image Analysis. IbPRIA 2003. Lecture Notes in Computer Science*, vol 2652. Berlin: Springer; 2003. https://doi.org/10.1007/978-3-540-44871-6_110.
64. Ramaswamy SD, Vigmostad SC, Wahle A, Lai YG, Olszewski ME, Braddy KC, et al. Fluid dynamic analysis in a human left anterior descending coronary artery with arterial motion. *Annals of Biomedical Engineering*. 2004;32(12):1628–1641.
65. Frauenfelder T, Boutsianis E, Schertler TL, Husmann L, Leschka S, Poulikakos D, et al. In-vivo flow simulation in coronary arteries based on computed tomography datasets: Feasibility and initial results. *European Radiology*. 2007;17(5):1291–1300.
66. Johnson DA, Naik UP, Beris AN. Efficient implementation of the proper outlet flow conditions in blood flow simulations through asymmetric arterial bifurcations. *International Journal for Numerical Methods in Fluids*. 2011;66(11):1383–1408.
67. Apostolidis AJ, Beris AN, Dhurjati PS. Introducing CFD through a cardiovascular application in a fluid mechanics course. *Chemical Engineering Education*. 2014;48(3):175–184.
68. Apostolidis AJ, Moyer AP, Beris AN. Non-Newtonian effects in simulations of coronary arterial blood flow. *Journal of Non-Newtonian Fluid Mechanics*. 2016;233:155–165.
69. Taylor CA, Hughes TJR, Zarins CK. Finite element modeling of three-dimensional pulsatile flow in the abdominal aorta: Relevance to atherosclerosis. *Annals of Biomedical Engineering*. 1998;26(6):975–987.
70. Gohil TB, McGregor RHP, Szczerba D, Burckhardt K, Muralidhar K, Székely G. Simulation of oscillatory flow in an aortic bifurcation using FVM and FEM: A comparative study of implementation strategies. *International Journal for Numerical Methods in Fluids*. 2011;66(8):1037–1067.
71. Moore S, David T, Chase JG, Arnold J, Fink J. 3D models of blood flow in the cerebral vasculature. *Journal of Biomechanics*. 2006;39(8):1454–1463.
72. Alastruey J, Parker KH, Peiró J, Byrd SM, Sherwin SJ. Modeling the circle of Willis to assess the effects of anatomical variations and occlusions on cerebral flows. *Journal of Biomechanics*. 2007;40(8):1794–1805.
73. Alastruey J, Moore SM, Parker KH, David T, Peiró J, Sherwin SJ. Reduced modeling of blood flow in the cerebral circulation: Coupling 1-D, 0-D and cerebral autoregulation models. *International Journal for Numerical Methods in Fluids*. 2008;56(8):1061–1067.
74. Passerini T, de Luca MR, Formaggia L, Quarteroni A, Veneziani A. A 3D/1D geometrical multiscale model of cerebral vasculature. *Journal of Engineering Mathematics*. 2009;64 (4):319–330.

75. Reymond P, Perren F, Lazeyras F, Stergiopulos N. Patient-specific mean pressure drop in the systemic arterial tree, a comparison between 1-D and 3-D models. *Journal of Biomechanics*. 2012;45(15):2499–2505.
76. Fahy, P, Delassus P, McCarthy P, Sultan S, Hynes N, Morris L. An in vitro assessment of the cerebral hemodynamics through three patient specific circle of Willis geometries. *Journal of Biomechanical Engineering -Transactions of the ASME*. 2014;136(1):011007.
77. Spilker RL, Feinstein JA, Parker DW, Reddy VM, Taylor CA. Morphometry-based impedance boundary conditions for patient-specific modeling of blood flow in pulmonary arteries. *Annals of Biomedical Engineering*. 2007;35(4):546–559.
78. Campo-Deaño L, Oliveira MSN, Pinho FT. A review of computational hematodynamics in middle celebral aneurysms and rheological models for blood flow. *Applied Mechanics Reviews*. 2015;67(3):030801–16.
79. Bessonov N, Sequeira A, Simakov S, Vassilevskii Yu, Volpert V. Methods of blood flow modelling. *Mathematical Modelling of Natural Phenomena*. 2016;11(1):1–25.
80. Secomb TW. Blood flow in the microcirculation. *Annual Review of Fluid Mechanics*. 2017;49(1):443–461.
81. Malek AM, Alper SM, Izumo S. Hemodynamic shear stress and its role in arteriosclerosis. *Journal of the American Medical Association*. 1999;282(21):2035–2042.
82. Cecchi E, Giglioli C, Valente S, Lazzeri C, Gensini GF, Abbate R, et al. Role of hemodynamic shear stress in cardiovascular disease. *Atherosclerosis*. 2011;214(2):249–256.
83. Chaichana T, Sun ZH, Jewkes J. Computational fluid dynamics analysis of the effect of plaques in the left coronary artery. *Computational and Mathematical Methods in Medicine*. 2012;(4i):504367.
84. Wentzel JJ, Chatzizisis YS, Gijsen FJH, Giannoglou GD, Feldman CL, Stone PH. Endothelial shear stress in the evolution of coronary atherosclerotic plaque and vascular remodelling: Current understanding and remaining questions. *Cardiovascular Research*. 2012;96(2):234–243.
85. Topol EJ, Califf RM, Prystowsky EN, Thomas JD, Thompson PD. *Textbook of Cardiovascular Medicine*. Philadelphia: Lippincott Williams & Wilkins; 2007; 283 p.
86. Pries AR, Secomb TW, Gaehtgens P, Gross JF. Blood flow in microvascular networks. *Experiments and Simulation. Circulation Research*. 1990;67(4):826–834.
87. Turitto VT. Blood viscosity, mass transport, and thrombogenesis. *Progress in Hemostasis and Thrombosis*. 1982;6:139–177.
88. Olufsen MS. Modeling flow and pressure in the systemic arteries. In Ottesen JT, Olufsen MS, Larsen JK (eds.) *Applied Mathematical Models in Human Physiology*. Philadelphia: SIAM; 2004; pp. 91–136.
89. Gijsen FJH, Allanic E, van de Vosse FN, Janssen JD. The influence of the non-Newtonian properties of blood on the flow in large arteries: Unsteady flow in a 90 degrees curved tube. *Journal of Biomechanic*. 1999;32(7):705–713.
90. Lee SW, Steinman DA. On the relative importance of rheology for image-based CFD models of the carotid bifurcation. *Journal of Biomechanical Engineering-Transactions ASME*. 2008;129(2):273–278.
91. Yilmaz F, Gundogdu MY. A critical review on blood flow in large arteries; relevance to blood rheology, viscosity models, and physiologic conditions. *Korea-Australia Rheology Journal*. 2008;20(4):197–211.
92. Wang SZ, Chen JL, Ding GH, Lu G, Zhang XL. Non-Newtonian computational hemodynamics in two patient-specific cerebral aneurysms with daughter saccules. *Journal of Hydrodynamics*. 2010;22(5):639–646.

93. Seo T. Numerical simulations of blood flow in arterial bifurcation models. *Korea-Australia Rheology Journal*. 2013;25(3):153–161.
94. Apostolidis AJ, Beris AN. Modeling of the blood rheology in steady-state shear flows. *Journal of Rheology*. 2014;58(3):607–633.
95. Bird RB, Armstrong RC, Hassager O. *Dynamics of Polymeric Fluids*, vol. 1, 2nd ed. New York: Wiley-Interscience; 1987.
96. Owens RG. A new microstructure-based constitutive model for human blood. *Journal of Non-Newtonian Fluid Mechanics*. 2006;140(1):57–70.
97. Bureau M, Healy JC, Bourgoin D, Joly M. Rheological properties of blood at low shear rate. *Biorheology*. 1980;17(1–2):191–203.
98. Horner JS, Armstrong MJ, Wagner NJ, Beris AN. Investigation of blood rheology under steady and unidirectional large amplitude oscillatory shear. *Journal of Rheology*. 2018;62 (2):577–591.
99. Apostolidis AJ, Armstrong MJ, Beris AN. Modeling of human blood rheology in transient shear flows. *Journal of Rheology*. 2015;59(1):275–298.
100. Thurston GB. Rheological parameters for the viscosity, viscoelasticity and thixotropy of blood. *Biorheology*. 1979;16(3):149–162.
101. Quemada D. A rheological model for studying the hematocrit dependence of red-cell red-cell and red-cell protein interactions in blood. *Biorheology*. 1981;18(3–6):501–516.
102. Quemada D. Blood rheology and its implication in flow of blood. In Rodkiewicz CM (ed.) Int. Centre for Mechanical Sciences Courses and Lectures No 270. Wien: Springer-Verlag; 1983; pp. 1–127.
103. Quemada D. A nonlinear Maxwell model of biofluids – Application to normal blood. *Biorheology*. 1993;30(3–4):253–265.
104. Anand M, Rajagopal KR. A shear-thinning viscoelastic fluid model for describing the flow of blood. *International Journal of Cardiovascular Medicine and Science*. 2004;4 (2):59–68.
105. Fang JN, Owens RG. Numerical simulations of pulsatile blodd flow using a new constitutive model. *Biorheology*. 2006;43(5):637–660.
106. Moyers-Gonzalez M, Owens RG. A non-homogeneous constitutive model for human blood. Part II. Asymptotic solution at large Péclet numbers. *Journal of Non-Newtonian Fluid Mechanics*. 2008;155(3):146–160.
107. Moyers-Gonzalez M, Owens RG, Fang J. A non-homogeneous constitutive model for human blood. Part 1. Model derivation and steady flow. *Journal of Fluid Mechanics*. 2008;617:324–354.
108. Moyers-Gonzalez M, Owens RG, Fang J. A non-homogeneous constitutive model for human blood. Part III. Oscillatory flow. *Journal of Non-Newtonian Fluid Mechanics*. 2008;155(3):161–173.
109. Moyers-Gonzalez M, Owens RG, Fang J. On the high frequency oscillatory tube flow of healthy human blood. *Journal of Non-Newtonian Fluid Mechanics*. 2009;163(1–3):45–61.
110. Anand M, Kwack J, Masud A. A new generalized Oldroyd-B model for blood flow in complex geometries. *International Journal of Engineering Science*. 2013;72:78–88.
111. Tsimouri ICh, Stephanou PS, Mavrantzas VG. A constitutive rheological model for agglomerating blood derived from nonequilibrium thermodynamics. *Physics of Fluids*. 2018;30(3):030710.
112. Keener J, Sneyd J. *Mathematical Physiology*. New York: Springer; 1998.

113. Fedosov DA, Caswell B, Popel AS, Karniadakis GE. Blood flow and cell-free layer in microvessels. *Microcirculation*. 2010;17(8):615–628.
114. Li XJ, Peng ZL, Lei, H, Dao D, Karniadakis GE. Probing red blood cell mechanics, rheology and dynamics with a two-component multi-scale model. *Philosophical Transactions of the Royal Society A – Mathematics, Physics and Engineering Science*. 2014;372:20130389.
115. Lipowsky HH. Microvascular rheology and hemodynamics. *Microcirculation*. 2005;12 (1):5–15.
116. Marcinkowska-Gapińska A, Gapiński J, Elikowski W, Jaroszyk F, Kubisz L. Comparison of three rheological models of shear flow behavior studied on blood samples from post-infarction patients. *Medical & Biological Engineering & Computation*. 2007;45(9):837–844.
117. Merrill EW, Gilliland ER, Cokelet GC, Shin H, Britten A, Wells Jr RE. Rheology of blood and flow in the microcirculation. *Journal of Applied Physiology*. 1963;18(2):1–3.
118. Merrill EW, Cheng CS, Pelletier GE. Yield stress of normal blood as a function of endogenous fibrinogen. *Journal of Applied Physiology*. 1969;26(1):1–3.
119. Merrill EW. Rheology of blood. *Physiological Reviews*. 1969;49(4):863–888.
120. Fåhræus R. Influence of the rouleau formation of erythrocytes on the rheology of the blood. *Acta Medica Scandinavica*. 1958;161(2):151–165.
121. Morris CL, Rucknagel DL, Shukla R, Gruppo RA, Smith CM, Blackshear P. Evaluation of the yield stress of normal blood as a function of fibrinogen concentration and hematocrit. *Microvascular Research*. 1989;37(3):323–338.
122. Moreno L, Calderas F, Sanchez-Olivares G, Medina-Torres L, Sanchez-Solis A, Manero O. Effect of cholesterol and triglycerides levels on the rheological behavior of human blood. *Korea-Australia Rheology Journal*. 2015;27(1):1–10.
123. Apostolidis AJ, Beris AN. The effect of cholesterol and triglycerides on the steady state shear rheology of blood. *Rheologica Acta*. 2016;55(6):497–509.
124. Asakura S, Oosawa F. On interaction between two bodies immersed in a solution of macromolecules. *Journal of Chemical Physics*. 1954;22(7):1255–1256.
125. Picart C, Piau JM, Galliard H. Human blood shear yield stress and its hematocrit dependence. *Journal of Rheology*. 1998;42(2):1–12.
126. Morris CL, Smith II CM, Blackshear Jr. PL. A new method for measuring the yield stress in thin layers of sedimenting blood. *Biophysical Journal*. 1987;52(2):229–240.
127. Chien S, Usami S, Taylor HM, Lundberg JL, Gregersen MI. Effects of hematocrit and plasma proteins on human blood rheology at low shear rates. *Journal of Applied Physiology*. 1966;21(1):81–87.
128. Zydney AL, Oliver JD, Colton CK. A constitutive equation for the viscosity of stored red cell suspensions: Effect of hematocrit, shear rate, and suspending phase. *Journal of Rheology*. 1991;35(8):1639–1679.
129. Happel J, Brenner H. *Low Reynolds Number Hydrodynamics*. The Hague, the Netherlands: Martinus Nijhoff; 1983.
130. Merrill EW, Benis AM, Gilliland ER, Sherwood TK, Salzman EW. Pressure-flow relations of human blood in hollow fibers at low flow rates. *Journal of Applied Physiology*. 1965;20(5):954–967.
131. Chien S, King RG, Skalak R, Usami S, Copley AL. Viscoelastic properties of human blood and red cell suspensions. *Biorheology*. 1975;12(6):341–346.
132. Bureau M, Healy JC, Bourgoin D, Joly M. Etude rhéologique en régime transitoire de quelques échantillons de sangs humains artificiellement modifiés. *Rheologica Acta*. 1979;18(6):756–768.

133. McMillan DE, Strigberger J, Utterback NG. Rapidly recovered transient flow resistance: A newly discovered property of blood. *American Journal of Physiology*. 1987;253(4): H919–926.
134. Alves MM, Rocha C, Gonçalves MP. Study of the rheological behavior of human blood using a controlled stress rheometer. *Clinical Hemorheology and Microcirculation*. 2013;53(4):369–386.
135. Campo-Deaño L, Dullens RPA, Aarts DGAL, Pinho FT, Oliveira MSN. Viscoelasticity of blood and viscoelastic blood analogues for use in polydimethylsiloxane in vitro models of the circulatory system. *Biomicrofluidics*. 2013;7(3):034102.
136. Sousa PC, Carneiro J, Vaz R, Cerejo A, Pinho FT, Alves MA, et al. Shear viscosity and nonlinear behavior of whole blood under large amplitude oscillatory shear. *Biorheology*. 2013;50(5–6):269–282.
137. Johnson MW, Segalman D. A model for viscoelastic fluid behavior which allows non-affine deformation. *Journal of Non-Newtonian Fluid Mechanics*. 1977;2(3):255–270.
138. White JL, Metzner AB. Development of constitutive equations for polymeric melts and solutions. *Journal of Applied Polymer Science*. 1963;7(5):1867–1889.
139. Souvaliotis A, Beris AN. An extended White–Metzner viscoelastic fluid model based on an internal structural parameter. *Journal of Rheology*. 1992;36(2):241–271.
140. Beris AN, Edwards BJ. *Thermodynamics of Flowing Systems with Internal Microstructure*. New York: Oxford University Press; 1994.
141. Anand M, Rajagopal KR. A short review of advances in the modelling of blood rheology and clot formation. *Fluids*. 2017;2(3):35.
142. Doraiswamy D, Mujumdar AN, Tsao I, Beris AN, Danforth SC, Metzner AB. The Cox-Merz rule extended: A rheological model for concentrated suspensions and other materials with a yield stress. *Journal of Rheology*. 1991;35(4):647–685.
143. Clarion M, Deegan M, Helton T, Hudgins J, Monteferrante N, Ousley E, et al. Contemporary modeling and analysis of steady state and transient human blood rheology. *Rheologica Acta*. 2018:57(2):141–168.
144. Sun N, De Kée D. Linear viscoelasticity and stress growth in blood. *Canadian Journal of Chemical Engineering*. 2002;80(3):495–498.
145. Rajagopal KR, Srinivasa AR. Modeling anisotropic fluids within the framework of bodies with multiple natural configurations. *Journal of Non-Newtonian Fluid Mechanics*. 2001;99(2–3):109–124.
146. Armstrong MJ, Beris AN, Rogers SA, Wagner NJ. Dynamic shear rheology of a thixotropic suspension: Comparison of an improved structure-based model with large amplitude oscillatory shear experiments. *Journal of Rheology*. 2016;60(3):433–450.
147. Dimitriou CJ, Ewoldt RH, McKinley GH. Describing and prescribing the Large Amplitude Oscillatory Shear Stress (LAOStress). *Journal of Rheology*. 2013;57(1):27–70.
148. Merrill EW, Pelletier GA. Viscosity of human blood: Transition from Newtonian to non-Newtonian. *Journal of Applied Physiology*. 1967;23(2):178–182.
149. Baskurt OK, Boynard M, Cokelet GC, Connes P, Cooke BM, Forconi S, et al. New guidelines for hemorheological laboratory techniques. *Clinical Hemorheology and Microcirculation*. 2009;42(2):75–97.
150. Horner JS, Beris AN, Woulfe DS, Wagner NJ. Effects of ex vivo aging and storage temperature on blood viscosity. *Clinical Hemorheology and Microcirculation*. 2018;70 (2):155–172.

151. Mwasame PM, Beris AN, Diemer RB, Wagner NJ. A constitutive equation for thixotropic suspensions with yield stress by coarse-graining a population balance model. *A.I.Ch.E. Journal*. 2017;63(2):517–531.
152. Beris AN, Mavrantzas VG. On the compatibility between various macroscopic formalisms for the concentration and flow of dilute polymer solutions. *Journal of Rheology*. 1994;38 (5):1235–1250.
153. Öttinger HC. *Beyond Equilibrium Thermodynamics*. Hoboken, NJ: Wiley-Interscience; 2005.
154. Germann N, Cook LP, Beris AN. Nonequilibrium thermodynamic modeling of the structure and rheology of concentrated wormlike micellar solutions. *Journal of Non-Newtonian Fluid Mechanics*. 2013;196:51–57.
155. Armstrong MJ, Beris AN, Wagner NJ. An adaptive parallel tempering method for the dynamic data-driven parameter estimation of nonlinear models. *A.I.Ch.E. Journal*.2017;63(6):1937–1958.
156. Dintenfass L. Blood rheology as diagnostic and predictive tool in cardiovascular diseases. *Angiology*. 1974;25(6):365–372.
157. Jariwala S, Horner JS, Wagner NJ, Beris AN. Application of population balance-based thixotropic model to human blood. *Journal of Non-Newtonian Fluid Mechanics*. 2020;281:104294.

9 Applications

9.1 Introduction

Jan Mewis and Norman J. Wagner

During the various stages of their lifetime, industrial colloidal dispersions are subjected to different kinds of flow. Often their rheological behavior is critical for ensuring optimal processing as well as faultless application, and requires fine-tuning within narrow limits. Failure to do so might cause various production or application problems and even make it impossible to use a particular product for a given application.

Printing illustrates some of the issues involved. A wide selection of processes is available to apply thin layers of material in a given pattern onto a substrate. Each process requires inks that have specific rheological properties under the widely varying flow conditions before, during, and after the actual printing. An example is shown in the vignette in this chapter about early women in industrial rheology, where hysteresis loops were used to characterize the significant thixotropy often observed for printing inks. Some of the many possible problems that may be caused by unsuitable ink flow, e.g., in the case of a rotary press, are lack of ink pick-up by the rollers, ejecting ink droplets from the nip between rotating rollers, and insufficient ink transfer to the substrate or to an already printed area. Print quality problems also include uneven ink layer thickness and the inability to generate sufficiently fine patterns because of distortion and expansion of printed dots. In extreme cases, even the (paper) substrate can be damaged. As in many other industrial processes a wide range of shear rates are to be considered, as well as transient flows and mixtures of shear and extensional flow. Inkjet printing methods have substantially extended the range of printing applications, even into 3D. Also, here suspensions of particles, sometimes metals or polymers, have to be formulated for a specific, shear, and time-dependent rheology in order to be successfully printed yet support the weight of the next layers. Suspensions of conductive silver nanoparticles are also screen printed to form connective wires on solar voltaic panels as well as other electronics applications. In the formulation of all these systems colloidal phenomena play an important role in setting the rheology and yielding properties. Unsuitable flow properties of inks can result for instance in a rejected production batch, but the consequences can be much more dramatic in other industries, as illustrated by the failure of dams containing mine tailings (see Section 9.5).

Efficient formulation of industrial dispersions requires some ability to predict the rheology of a given formulation as well as knowledge of how changes in formulation will affect rheology. Although quantitative predictions can be made for well-defined model systems, as described in *CSR* [1] and in the earlier chapters of this book, predicting the flow of an industrial system becomes much more challenging. On the one hand one has to take into account much more complex flow situations in industry than commonly generated on rheometers. The large number of components in typical formulations with complex and often unknown characteristics and associations further complicates the prediction of rheological properties. In addition, the degrees of freedom in adjusting the formulation are normally limited as the main components are to a large extent dictated by the physical and chemical requirements of the final application. Hence, the basic rules of colloid suspension rheology rarely provide direct quantitative guidance in industrial formulation problems.

The challenges that arise are illustrated in this chapter with examples that cover diverse materials. Paints are a case in point, as their rheological properties have to be controlled over a wide range of shear rates and for various types of flow, and time-dependency of these properties is critical. Sometimes apparently contradictory requirements have to be satisfied during and after application onto a substrate (Section 9.2). Section 9.3 deals with applications for carbon blacks, the presence of which readily leads to a complex rheological behavior and therefore, their application in high performance systems remains challenging. Asphalts (Section 9.4) constitute a critically important construction material used on a vast scale worldwide, where a colloidal viewpoint provides valuable guidance even while other structural levels have to be considered as well. Manmade or natural mineral products often also provide complex suspensions containing particles with vast differences in particle size. Cements and mortars are among the most used materials by humanity and colloidal suspension rheology plays a most central role in their properties and processing, as is covered in Section 9.5. Finally, a broader overview of the problems encountered with processing operations of slurries is discussed in Section 9.6.

9.2 Paints

Alex Routh

A tin of paint, purchased from the hardware store, is a highly engineered colloidal dispersion. It represents a broad class of products that, upon physical or chemical drying, form a continuous polymer layer on a substrate. The dried film must satisfy various chemical and physical requirements. For example, it must provide protection from water and oxygen, be sufficiently mechanically strong, have suitable optical properties such as a defined color, gloss, or opacity, and produce an esthetically pleasing appearance. It can be necessary to superimpose layers of different composition to achieve the necessary properties. The rheology of the liquid coating determines whether the paint can be applied properly, either by brush, roller, spraying, or

Table 9.1 Typical formulations for white emulsion paint.

Component	Function	wt % in a Gloss Paint [2]	wt% in a Matte paint
Polymer particles (latex)	Binds film together	26.5	15
TiO_2	Creates opacity and white appearance	23.0	10
Defoamer	Prevents bubbles and foams	0.15	3
Thickener	Controls rheology	12.0	
Pigment dispersant	Stabilizes the pigment in dispersion	1.0	
Coalescent	Lowers the temperature for coherent film formation	2.35	
Extender	Filler – typically $CaCO_3$		32
Water	Solvent/dispersing medium	35.0	40

any other method. In addition, during the drying process itself flow occurs, which, under certain circumstances, can result in film defects and even holes. These can also be controlled through the dispersion rheology, which often requires the use of high-tech additives. Each formulation is proprietary and specific to the manufacturer and is optimized for a given application. It is, however, possible to give some typical compositions, as illustrated for emulsion paints in Table 9.1. Here, the film-forming polymer is not dissolved in a solvent but is present as a dispersed phase via small particles. In this waterborne system, drying occurs by evaporating water and subsequent coalescence of the latex particles.

For matte paint, a mirror-like surface causing specular reflection of light should be avoided. The surface can be made rougher by adding suitable particles such as calcium carbonate. For gloss paint the opposite is true: it is necessary to have a very smooth surface after drying, and hence large filler particles are avoided.

9.2.1 Relevant Shear Rates

All coatings experience a vast range of shear rates during manufacture and application. When stored, paint will experience sedimentation of nonbuoyant dispersed particles – which involves shear of very low magnitude. When applied, for example by brush, spray gun, or roller, the shear rates can be as high as 10^3–10^5 s^{-1}. Once applied the coating will experience flows due to leveling of any surface unevenness or to sagging if on a nonhorizontal surface. The relevant shear rates of these flows are again small. The desired rheological properties of the coating will need to be engineered across this vast range of applied shear conditions. Figure 9.1 provides an

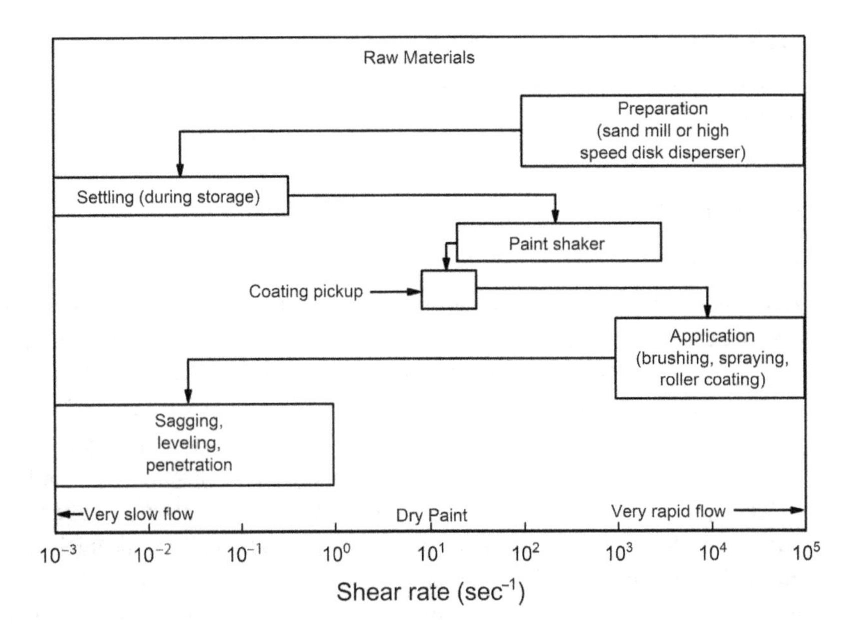

Figure 9.1 Different shear rates experienced by a coating from manufacture, through storage, application, and drying. Adapted from Patton [3]

overview, as suggested by Patton [3], of the shear rates experienced by an emulsion coating as it is manufactured, through storage, application, and drying.

In the following sections, we discuss various examples of how the rheology of the colloidal dispersion controls the final properties. Firstly, we examine cases for which the low shear viscosity of the formulation is crucial. The first example concerns the case of film leveling, which requires a sufficiently low viscosity value. We then consider film sagging and sedimentation, two cases where a high viscosity value is desirable. As an example of another feature required from a wet paint, we examine the recession of water from the edge of drying films. This is related to the wet edge time, which measures the time over which the edge of a film is re-dispersible. Another related measure is the open time, which is the time the edge of a film may be touched-up and reworked. To achieve a large wet edge time a large value of the low shear viscosity is preferable.

Secondly, we look at high shear flows. In these cases, such as brush application and spraying, it is desirable to have a low value of the viscosity, for ease of application, but this can make it impossible to obtain a sufficiently thick layer. We also consider some of the possible defects that can occur within coatings and the role that rheology can have in solving these cosmetic issues. Finally, we discuss the desired global rheology of coatings and examine how the formulation can be tuned to achieve these properties. This includes the use of associative thickeners – additives that give a solidification of the film at low shear to prevent drips, although under shear the viscosity reduces considerably.

As well as the limits at high and low shear, the viscosity at intermediate shear rates is also relevant [3]. To allow paint to be picked up by a brush a sufficient viscosity is required. Typical shear rates during brush transfer are estimated to be around 15–30 s^{-1} with a viscosity above 1.5 Pa s being sufficient for good paint transfer [3].

9.2.2 Low Shear Flows

Film Leveling

Leveling involves surface irregularities such as ripples and brush marks, in a paint film vanishing under the action of surface tension. If the viscosity is too high then surface leveling will not occur before the film dries and a textured surface will be observed. As shown in Figure 9.1, typical shear rates during leveling are between 0.001 and 1 s^{-1}. To understand film leveling, consider a thin film of Newtonian fluid with a sinusoidal surface profile, as shown in Figure 9.2. The driving force for flow is the surface tension induced pressure, determined by Γ/R, with Γ the surface tension and R the radius of curvature. The latter depends on the time dependent amplitude of the thickness variation $a(t)$ and the wavelength of oscillation, λ. The hydrodynamic resistance is proportional to the viscosity and inversely proportional to the film thickness h, as the latter determines the shear rate in the film. Assuming h to be constant and small compared to the wavelength, the amplitude $\alpha(t)$ is then given by [4]:

$$\alpha(t) = \alpha(0) \exp\left[\frac{-16\pi^4 h^3 \Gamma}{3\lambda^4 \eta} t\right]. \tag{9.1}$$

Eq. (9.1) shows the strong effect of film thickness and wavelength on the height of the surface pattern $\alpha(t)$. The question now is whether the surface profile will flatten out before drying of the film makes further leveling impossible.

An alternative, but entirely equivalent, way to consider leveling is to look at the timescales for flow and evaporation. Here we consider evaporation as the primary drying mechanism although, in other cases this could be a chemical reaction or

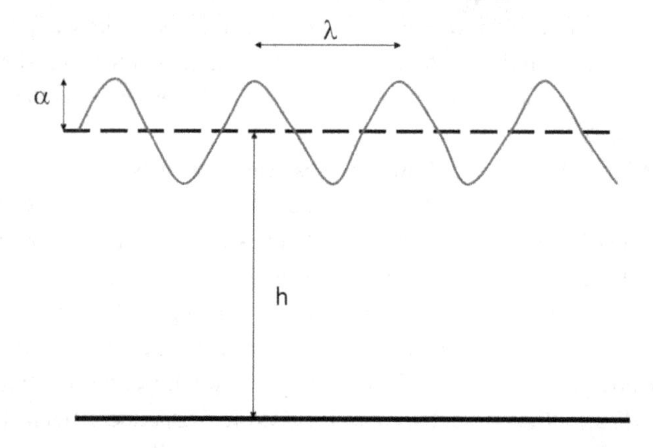

Figure 9.2 Fluctuations in a film surface will be reduced by flow driven by surface tension.

another structure formation mechanism. In emulsion paint, evaporation will increase the viscosity of the drying film, thus reducing the effect of surface tension and eventually locking in the film topology. The viscosity increase will initially be gradual but will accelerate as the particle volume fraction approaches close packing. The relative timescale of the two processes determines a "victor" and whether the film will be flat or not. This simplistic argument incorporates a surface tension flow timescale $\eta_0\lambda^4/h^3\Gamma$ and an evaporative timescale $h/\dot{E}$, where $\dot{E}$ is the rate of evaporation expressed in volume per unit surface area and per unit time. The resulting condition for obtaining a smooth surface is then: $\eta_0\dot{E}\lambda^4/h^4\Gamma < 1$. Thus, the same result can be obtained by inserting the evaporation time, $h/\dot{E}$ into Eq. (9.1) and ignoring the constants in the expression.

The typical value of the evaporation rate, $\dot{E}$ is 0.3 cm/day for waterborne systems. For a surface tension, Γ, of order 10^{-2} N/m and a low shear viscosity, η_0 of say 1 Pa s, the leveling of coatings is seemingly assured for initial wavelengths of perturbations, λ, up to 25 times the initial film thickness. The weakness of this scaling argument is that the viscosity is assumed to the constant and the value of 1 Pa s used here may be too low. Irrespective, the argument shows the importance of the low shear viscosity in film leveling.

For simple Newtonian fluids that cannot display any surface tension gradients, there is good experimental evidence in favor of this surface tension driven flow mechanism [5]. For paints, it was however also observed that after surface ripples decayed away as expected, they reappeared at later times [5,6] but with the surface troughs becoming peaks and vice versa. This phase shift was explained by the presence of surface tension gradients during evaporation, leading to a shear stress at the film–air interface. The change in boundary condition then alters the predicted film leveling. Linear stability analysis shows that the presence of surface tension gradients will induce flow instabilities in drying coatings [7,8]. Physical causes such as the presence of surfactants [9] or a mixture of two solvents with different volatility [10] have been considered.

Sagging

When a paint film is applied to a vertical substrate, such as a wall, gravity will cause the film to flow downward. This is called "sagging" and in extreme cases it can lead to tear shaped patterns or even sliding down of sheets of paint termed *slumping*. At a distance x from the free surface the shear forces should balance the weight of the paint layer between x and the free surface. Hence the local shear stresses are proportional to the distance to the outer surface where the stresses become zero. The stress distribution holds for all kind of fluids. For a Newtonian fluid, it causes a linear change of shear rate with depth and hence a quadratic velocity profile

$$\upsilon = \frac{\rho g}{2\eta}\left(h^2 - x^2\right), \tag{9.2}$$

which is shown in Figure 9.3(a). If the paint has a yield stress and this stress is reached at a depth x, the layer between x and the surface will slide down as a solid, termed

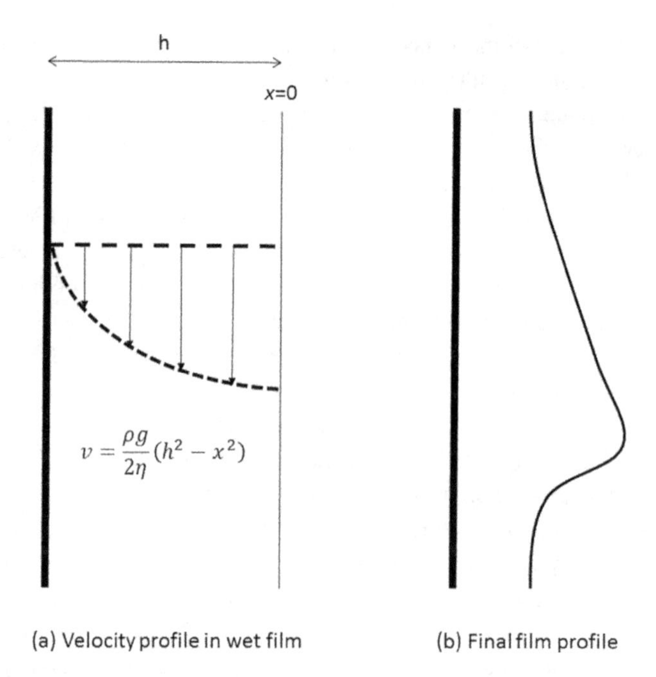

Figure 9.3 (a) Velocity profile for a Newtonian fluid flowing under the influence of gravity on a vertical wall. (b) The yielding of the film at a given point can lead to a nonuniform film profile, a process called sagging.

curtaining. To characterize sagging we define a *sag length* as the total distance moved by the surface [11]. For Newtonian fluids:

$$l_{\text{sag}} = \frac{\rho g}{2} \int_0^{t_f} \frac{h^2}{\eta} dt, \tag{9.3}$$

where the upper limit of the integral, t_f, is either taken as either 10 minutes [11], or an infinite drying time if the final film appearance is desired. In Eq. (9.3) a time evolution of the viscosity can also be taken into account. A film profile, typically observed in sagging systems, is sketched in Figure 9.3(b).

Sagging is measured by applying horizontal lines of paint with varying thickness on a tilted or vertical substrate and noting the minimum thickness for sagging [11,12]. A more direct method consists of applying a horizontal line of dye across the paint film and tracking its downward motion [13,14]. In this manner, a sagging distance can be measured as a function of time. A length greater than 0.05 cm will result in noticeable depressions around dust particles and for a value greater than 0.1 cm the coating will not be acceptable [13].

Whilst film leveling requires as low a viscosity as possible, as can be seen in Eq. (9.3), the issue of sagging necessitates a sufficiently high, or sufficiently rapidly increasing viscosity. The rheological properties of the coating are complex and time dependent so it is difficult to correlate an optimal leveling/sagging balance with the rheology of the various types of coatings [12,15].

Sedimentation

Coatings may be stored in a tin for over a year before use. Consequently, different components may well settle, resulting in a nonuniform composition. As shown in Figure 9.1, for sedimentation a typical shear rate is less than 0.1 s^{-1}. Although it is common practice to stir a paint formulation prior to application, it is still preferable to avoid sedimentation. At the least a dense layer of particles, near the maximum random packing, should be avoided as it would be shear thickening during stirring and thus difficult to homogenize. The Stokes settling velocity of a single isolated particle is given by

$$U_s = \frac{2\Delta\rho g a^2}{9\eta}. \tag{9.4}$$

Different coatings display a large range of possible parameter values. For an emulsion paint the binder particle size, a, is around 100 nm and the density difference, $\Delta\rho$, is about 50 kg/m^3. Consequently, in water, with a viscosity of 10^{-3} Pa s, the particles will sediment about 3.5 cm in one year. There are several caveats to the simple calculation above: The calculated sedimentation velocity is for a single isolated particle in an infinite fluid, while a swarm of particles has a hindered settling which reduces the sedimentation velocity. Additionally, the viscosity of the suspension medium is higher than the assumed value of water. However, it should be noted that pigment particles are larger than the binder and the density is also greater. Hence components such as TiO_2 will sediment faster than calculated above. Increasing the low shear viscosity will delay the settling of the particles, although this will have an adverse effect on the film leveling. An alternative approach is to hold the particles within a thixotropic, gelled structure. If the gravitational stress imposed by the particles is below the yield stress of the formulation, then the particles will not sediment at all. Stirring the paint will break the gelled structure and restore a lower, low shear viscosity.

Wet Edge Time

When a coating is applied to a substrate there is an amount of time during which the coating can be touched up or reapplied. This is called the *open time*. It is related to the *wet edge time*, which is the time that the edge of the film remains re-dispersible. It is widely reported that waterborne coatings have inferior wet edge characteristics to solvent-borne coatings [2]. The current drive towards waterborne coatings, to reduce atmospheric emissions, therefore necessitates an understanding of the properties which determine the wet edge and open times.

When coatings are dried, the edge is seen to consolidate first, because of the reduced film height. Evaporation continues from this region of packed particles and a horizontal flux of solvent towards the edge results, which propagates a front of consolidated particles laterally across the drying film. These fronts have been termed horizontal drying fronts. The flow of solvent through the solidified region results in a pressure drop and when the edge pressure reaches its largest (negative) supportable value, the solvent is seen to recede from the film edge. This type of edge drying has been termed coffee-ring flow [16].

The pressure, and hence recession, of solvent in drying particulate coatings is determined by a single dimensionless group, called P_{cap}, which is the maximum pressure that can be supported by a particulate array compared with the magnitude of the pressure which can be generated by flow through an array of packed particles. The dimensionless group is determined by [17]:

$$P_{\mathrm{cap}} = \frac{20}{75}\left(\frac{3\Gamma\eta_0}{\dot{E}}\right)^{\frac{1}{2}} \frac{a(1-\phi_{\mathrm{max}})^2}{\mu\phi_{\mathrm{max}}^2 h}, \tag{9.5}$$

where ϕ_{max} is the volume fraction at close packing.

Salamanca et al. [18] used 2D MRI imaging to track the position of water in drying coatings. The magnetic resonance detects mobile protons, in this case water, and it is possible to obtain a precise measure of the water location. Sample images are shown in Figure 9.4 and one can see the easy detection of the time when water leaves the film edge. The results for a range of experiments are shown in Figure 9.5 and demonstrate the desire to have as large a value of P_{cap} as possible, to obtain as long an open time as possible. The parameters involved in P_{cap} are mostly fixed and not open to control, although it is noticeable that as large a low shear viscosity as possible will increase the open time.

9.2.3 High Shear Flows during Application of the Coating

Brushing

When applied by brush, the shear rate experienced by paint is approximately 10^4 s^{-1} [3]. When applied by roller the shear rate can be even higher, approaching 10^5 s^{-1}. If the corresponding high shear viscosity is too high, then the coating will provide

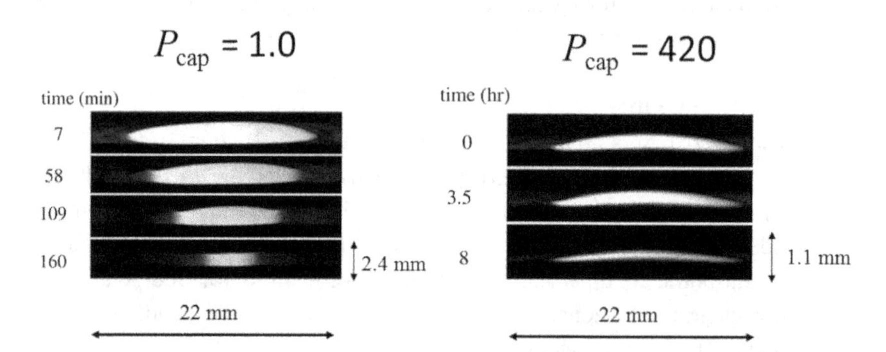

Figure 9.4 2D MRI images of a drying aqueous based coating at various times for two different values of P_{cap}. Reprinted with permission from Salamanca et al., Lateral drying in thick films of waterborne colloidal particles. *Langmuir* 2001;17(11):3202 [18] Copyright (2001) American Chemical Society

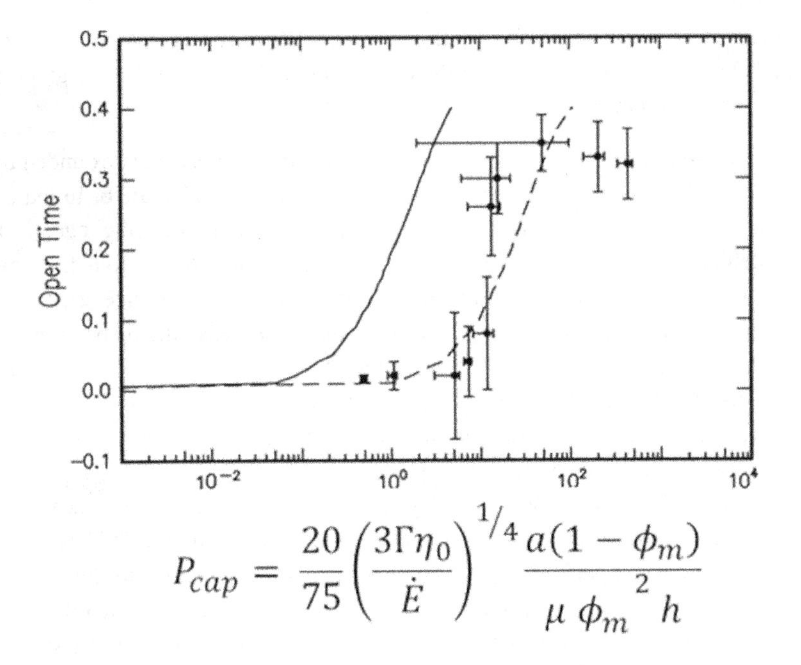

Figure 9.5 The time that water was observed to leave the edge of drying coatings as a function of a single dimensionless group. Reprinted with permission from Salamanca et al., Lateral drying in thick films of waterborne colloidal particles. *Langmuir* 2001;17(11):3202 [18] Copyright (2001) American Chemical Society

excessive drag and prove very difficult to spread uniformly on the substrate. With too low a high shear viscosity it becomes impossible to apply a sufficiently thick layer in a single application.

Spraying

When a coating is applied by spraying it will experience high shear rates within the nozzle of up to 10^5 s^{-1} [3]. If the high shear viscosity is too high, then the pressure drop through the gun will be too high. Equally, on spraying it is desired for individual droplets to readily flatten onto the surface and coalesce into a uniform film. This also necessitates a low value for the high shear viscosity. It is also possible to coalesce the constituent particles during the spraying process, as commonly used in powder coatings. Interestingly, the composition of the paint at the substrate differs from that in the sprayer as solvent flashes off during spraying. Spraying often involves dilution and commercial paint sprayers are often accompanied by a simple dip-stick viscometer for guiding viscosity adjustment by dilution at the customer.

9.2.4 Coating Defects

As well as the issues at low and high shear, a common occurrence in coatings is the appearance of surface defects. These will be most noticeable in high gloss applications and consequently automotive coatings manufacturers go to great lengths to understand and avoid them. A discussion of the many causes of surface defects is given by Schoff [19] and a summary is provided here:

Crater Formation: Craters are observed as a central depression surrounded by a raised ridge. The cause can be contamination either on the substrate or in the coating, or an excess of surfactant or another surface tension reducing agent, such as a defoamer. In the case of surfactant accumulation, a Marangoni flow away from such a region results in a film height depression. The immediate remedy is to avoid the contamination or, if this is not possible, to increase the dispersion viscosity to counteract the flow. To hinder crater formation, it may counter intuitively be sensible to take the opposite approach and have a low viscosity, allowing the Marangoni flow to occur, and then to use the leveling properties of the coating to subsequently remedy the defect [19].

Orange Peel: When a coating displays an uneven surface, it often appears as a regular mottle pattern like the skin of an orange. There are many possible causes to what is termed orange peel. One possibility is incomplete leveling, so that lack of flow is responsible. Another possible cause is the presence of contaminants, such as silicones, excess defoamer, or incompatible solvent mixtures. In such cases the orange peel will appear from an initially smooth surface, indicating that surface tension driven flow instabilities are likely to be responsible [19].

Bénard Cells: These often-observed patterns are caused by temperature gradients across the coating leading to a circulating flow. The hotter and hence lighter fluid is drawn upwards and the cooler heavier fluid settles. A complication arises in that the temperature variation can also lead to surface tension gradients and consequently Marangoni flows. The pattern formation is due to an instability in the thin film fluid mechanics determined by a single dimensionless group which balances surface tension with the dispersion viscosity, as discussed by Pearson [20]. Unsurprisingly, if the dispersion viscosity is high enough, flow ceases and no Bénard cells are observed.

Pin-holes: Small holes in the coating, called pin-holes, are commonly encountered in protective coatings and their existence negates the barrier properties. These could become a major source of corrosion, for example in the case of automobiles. There are many possible causes of pin-holing. As discussed above a foreign object on the substrate or an accumulation of surfactant could lead to localized crater formation. In these cases, a low viscosity will hopefully enable film leveling to restrict the crater formation. Additional causes of pin-hole formation center around air or dissolved gases on the substrate being expelled through the formed film. The air can come about through rapid application of the coating, or even excess surfactant, causing a foam to form as the coating is applied [21]. An example concerns glove dipping, where a thin film is cast on a hand shaped former [22]. Strict quality control measures can mean the existence of pin-holes leading to entire glove batches being discarded.

9.2.5 Viscosity Curve and the Design of Desired Rheology

The scenarios discussed above present different desired rheological properties, such that the formulation engineer will need to strike a balance to satisfy all design requirements. For example, Overdiep [5] discusses how painting of a sharp corner necessitates a coating rheology that is challenging to achieve, with a required high value of the low shear viscosity meaning that leveling is often not achieved.

The desired material properties can be split into the shear regimes experienced by the coating. This is shown visually in Figure 9.6 for the case of emulsion paints [3]. In the low shear regime, one requires an extended open time and good leveling of the coating. In addition, the sagging must not be excessive. This trade off results in a desired range of viscosity and yield stress. For the high shear flows it is necessary to have a consistency which allows ease of application without excessive drag or run off during spraying. The solid lines in Figure 9.6 relate to assuming a constitutive equation, in this case the Casson equation, and determining the parameter values to give desired viscosity values in the two relevant regimes.

As discussed above, to obtain good leveling properties it is desired to achieve as low a value for the low shear viscosity as possible. To obtain good wet edge time properties as well as sagging and sedimentation properties it is desirable to have as high a low shear viscosity as can be achieved. This contradiction leads to compromises having to be made, depending on the application of the coating. For example, matte coatings will not require as good a leveling as a high gloss coating. One possible way to satisfy the various design constraints is through thixotropic structure formation within the coating. Additives, such as clays and high molecular weight polymers such as acrylates and polyalkylene ethers as well as co-crystals of hexamethylene

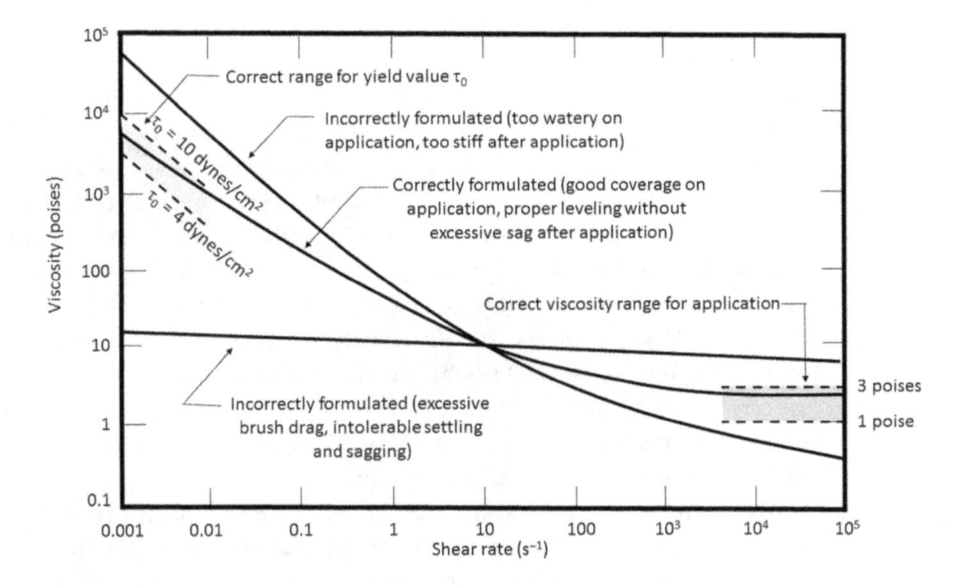

Figure 9.6 Desired universal flow curve for a paint formulation. Adapted from Patton [3]

diisocyanate and benzylamine, are called sag control agents. These are components that induce flocculation and thus generate a space-filling network, allowing the coating to display a yield stress [15]. This hinders sag and sedimentation as well as wet edge issues. However, upon application of a sufficient stress the network breaks and allows flow during application. The resulting low shear viscosity should be low enough to allow leveling to occur. After a suitable timescale the structure should then reform, either because of thixotropy or drying, to avoid sagging or wet edge issues. Thixotropy is discussed in Chapter 1. A yield stress of even 0.1 Pa is sufficient to impair leveling, so it is necessary to ensure the recovery in viscosity is slow enough to allow leveling to occur. A discussion of how the rheology of high performance coatings can be tuned using different additives highlights how a designer needs to consider the time for structure formation relative to the different flows which need to be controlled within the coating [23].

Dispersions of polymer lattices, at concentrations used in waterborne coatings, are nearly Newtonian and have a low viscosity value of around 10 mPa s. They will display some shear thinning because of their Brownian nature, but the magnitude of thinning is small. *Rheological modifiers* or *thickeners* are added to coatings formulations to generate the required flow behavior. Adding water-soluble polymers that do not adsorb on the latex particles, such as modified cellulosics, increases the viscosity of the suspending medium and induces shear thinning. With these additives it remains, however, difficult to combine the low shear requirements of adequate leveling and sagging with those of good film formation during application. In addition, complications can arise because of the extensional flow behavior of these polymers, e.g., formation of strings with possibly ejecting droplets during roller application [24]. This inability to generate the desired dispersion rheology across all shear rates has led to the development of associative polymers, as discussed next.

Associative Polymers

Associative polymers (AP) can provide a suitable alternative method to achieve shear thinning in dispersions [25]. They are amphiphilic, containing hydrophilic chain elements in the backbone to which two or more hydrophobic groups are attached. The hydrophilic backbone ensures compatibility with water, whereas hydrophobic groups induce structure. The latter makes it possible to reach high viscosity levels without high molar mass. Common examples are the *Hydrophobically modified Ethoxylated URethanes* (HEUR) and *Hydrophobically modified Alkali Soluble Emulsion* (HASE) polymers. The HEUR polymers consist of a hydrophilic backbone, such as poly(ethylene-oxide) (PEO), modified with hydrophobic groups via urethane spacers. Often a *telechelic* structure is used, i.e., with hydrophobic groups situated at each end of the backbone chain. A HASE thickener is an emulsion at low pH but develops an associative structure at high pH.

When APs are dissolved in water, above a critical concentration their hydrophobic groups tend to cluster together in micellar structures. The different hydrophobic groups on a polymer molecule can be attached to the same micellar aggregate or they may bridge between micelles, as sketched in Figure 9.7.

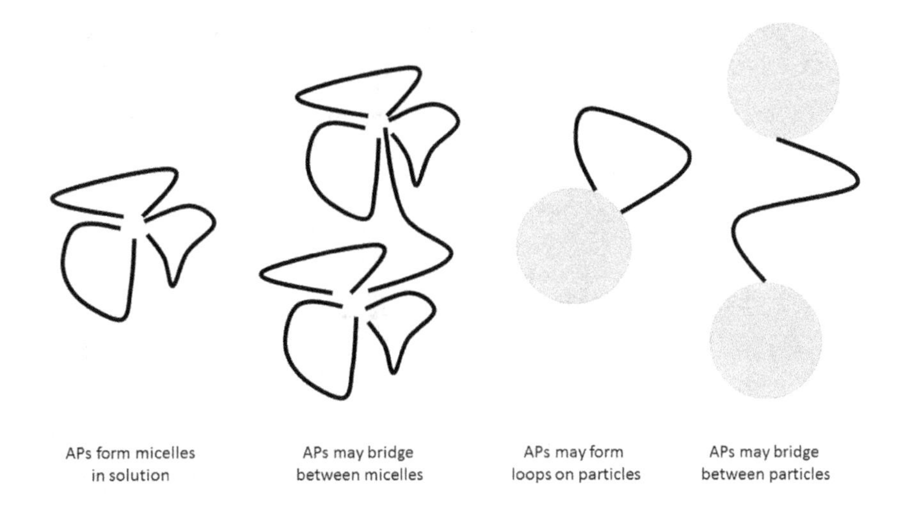

Figure 9.7 Associative polymers in solution may form micelles. Individual chains can also bridge between different micelles. Upon addition of latex particles, the associative polymer may form loops on the surface or bridge between different particles.

Because the attachment/detachment process of the hydrophobes is dynamic, as for surfactant micelles, the links and structures are of a transient nature. The kinetics of these processes and therefore also the rheology is controlled, at least to some extent, by the length of the hydrophobic groups. Other dissolved components, such as surfactants, can participate or interfere in the formation of the micellar aggregates, which is the topic of the rest of this section.

Aqueous solutions of APs, especially the telechelics, demonstrate a specific rheological behavior. While the hydrophobic chain ends cluster together, the hydrophilic backbone chains form loops on the outside, giving flower-like structures. The presence of micelles increases the viscosity, especially at higher concentrations when bridge interconnections become more likely. Because of the transient nature of the links the low shear behavior can remain Newtonian. Shear affects the micellar interactions, resulting in substantial shear thinning. Over some intermediate concentration range shear thickening can occur [26,27] In oscillatory flow the telechelics display nearly Maxwellian behavior with a single dominating relaxation time, controlled by the association/dissociation kinetics of the hydrophobes [27]. The detailed rheological behavior depends on the molecular architecture of the APs, especially the length of the hydrophobic segments and, for the nontelechelic ones, the number and distribution of hydrophobic groups along the backbone.

Adding other components to an AP solution can change the rheological behavior. For example, if surfactants or dispersants are present, their hydrophobic groups can participate in the formation of micelles and thus increase viscosities and moduli. The thickening effect peaks at a given surfactant concentration [28]. At that stage actual surfactant micelles are present, which can capture single hydrophobes from the associative polymer and, thus, reduce the connectivity of the AP network. Water-soluble

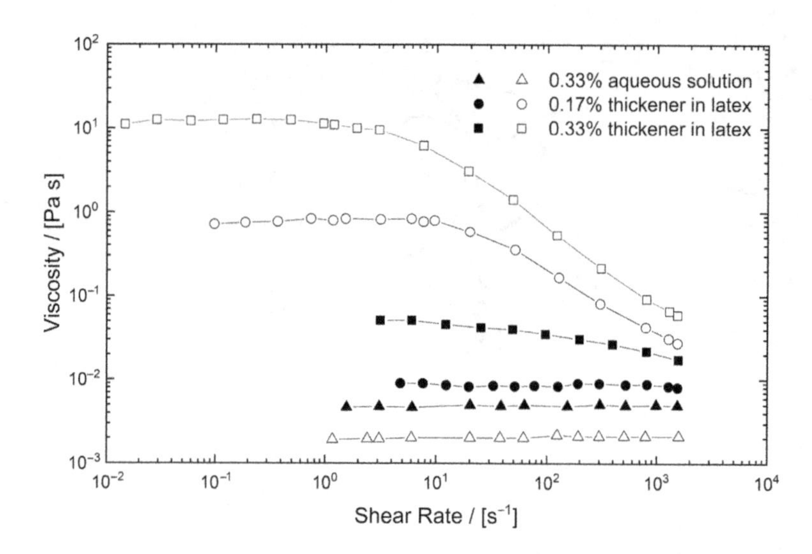

Figure 9.8 Effect of associative thickener (open symbols) compared with that of a cellulosic thickener (closed symbols) on the flow curves for latex. Adapted from Huldén [33]

cosolvents, such as propylene glycol, also modify the rheological behavior of an AP solution, with an effect similar to surfactants. In the case of HASE polymers the electrolyte content will also be important [29].

Mixing APs with a latex dispersion causes a drastic thickening effect. Typical flow-curves for a latex dispersion with and without APs are shown in Figure 9.8. The hydrophobic groups of the AP adsorb on the surface of the latex particles as well as on other particles such as pigments. For this they compete with the surfactants and dispersants present in the latex system. With limited amounts of associative polymer present, as in many formulations, the total amount might be adsorbed [30]. The adsorbed layer increases the effective volume of the dispersed phase and hence the viscosity. Low shear viscosities, however, turn out to be much higher than predicted on the basis of the effective volume, which makes associate polymers such efficient thickeners [31]. The stronger viscous effect is attributed to the potential of AP polymers to form direct but transient bridges between latex particles, as sketched in Figure 9.7. With increasing shear rate the adsorbed layer becomes thinner and the association between latex particles is reduced, although a residual structure of latex particles remains, as well as separate flocs of pigment [32]. The net result is a strong shear thinning. The hydrodynamic effect of the effective volume provides a sufficiently high viscosity at high shear rates to ensure sufficient film building during application. When the shear rate suddenly drops to that of leveling and sagging, most of the low shear viscosity is recovered, almost instantaneously, with some subsequent thixotropic recovery [12]. If properly formulated this ensures suitable leveling as well as acceptable sagging. The elongational viscosities with APs present are lower than those when using normal polymers, which reduces problems such as spattering and misting in roller applications and ensures proper droplet formation when spraying [24].

Not surprisingly, considering the complex rheology and flow during and after application of coatings, it is difficult to establish accurate correlations between rheological properties and leveling/sagging behavior for these systems [12]. The competition between APs, surfactants, and dispersants for adsorption on particle surfaces can affect the final properties of the dry coating because of changes in colloidal stability of the particles in the coating. This requires a suitable balance between the components [29].

9.3 Carbon Blacks

Jeffrey Richards

9.3.1 Introduction

Due to its ubiquity as a structural reinforcing agent and pigment, the dispersion and stabilization of carbon black powders in polar and apolar fluids have concerned formulation scientists for decades [34–37]. The emergence of advanced carbon blacks as electrically conductive additives has further motivated attention to the mechanical and electrical properties of composites containing them [38]. An area of active investigation is their incorporation into electrochemical devices, such as super capacitors, lithium ion batteries, and slurry-based electrochemical flow-cells [39,40]. In particular in electrochemical flow cells, carbon black is suspended in a salt-containing polar solvent and pumped through a channel to provide scalable electrochemical storage for load-leveling, uninterruptable power-supplies, power firming, stand-alone power applications, and capacitive water purification for deionization and desalinization [41]. The role of carbon black in these applications is to impart electrical conductivity to the working fluid. A key design challenge associated with carbon black is that increases in electrical conductivity occur concomitantly with undesirable rheological performance, such as large viscosities, which leads to unacceptably large pumping losses. This section reviews the factors that must be considered to understand the physical foundation for both the rheological and electrical properties of filled carbon fluids within the framework of colloidal suspension rheology.

9.3.2 Background: Structural Hierarchy in Carbon Black Suspensions

As discussed and demonstrated throughout this book, the rheology of suspensions of colloidal particles are most commonly characterized by the effective volume fraction and the strength and range of interparticle interactions [1]. These factors influence the physical state of the suspension, be it a fluid, gel, or glass, and thus its physical characteristics. Therefore, the chemical and physical properties of the conductive carbon black and suspending fluids used in various applications are important to understand to achieve formulation properties. Unlike, however, spherical particle

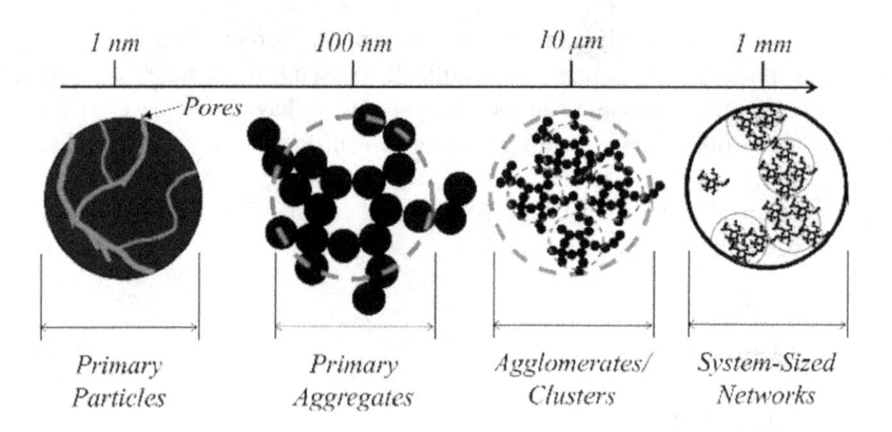

Figure 9.9 Schematic diagram of the structural hierarchy of highly structured carbon blacks.

where the volume fraction and interactions strength are well-defined, these properties are convoluted in carbon black suspensions due to a complex structural hierarchy.

Conductive carbon blacks are synthesized by oxygen-free thermal decomposition of a carbon feedstock or through the thermal-oxidative decomposition of hydrocarbons [38]. The process conditions during decomposition determine both the chemical and physical nature of the carbon black including its conductivity, level of structure, and surface chemistry. These characteristics, in turn, encode a complex structural hierarchy into suspensions of carbon. This structural hierarchy can be divided into two categories: (1) the intrinsic properties that arise due the physical structure of the carbon black aggregate and (2) the extrinsic properties that arise due to interactions between carbon black aggregates in suspensions.

At the smallest length scales, carbon black aggregates consist of primary particles that contain mesoscale pores, as shown in Figure 9.9. Solvents capable of wetting the carbon surface fill these pores and raise the effective volume fraction. The porosity is indicated by the Brunauer–Emmett–Teller (BET) surface area and oil-adsorption number (OAN). As the BET surface area and the OAN increase, so does the amount of porosity. High structured carbon blacks possess extreme porosity (e.g., OAN exceeding 170 mL/100 g), such that they can achieve desired performance at extremely low weight loadings. These porous primary particles, typically 10–50 nm in diameter, are fused chemically to form primary aggregates, typically 100–500 nm in size. These aggregates are themselves characterized by random connections of primary particles and their density can be quantified using a mass fractal dimension. The fractal structure of the primary aggregates implies that fluid flow within the hydrodynamic volume of the primary aggregate is hindered. Therefore, the primary aggregate structure influences the volume of entrained solvent that flows along with the particles, with higher fractal dimension corresponding to less effective porosity. These physical traits (i.e., porosity of the primary particle and morphology of the primary aggregate) are distinct from those that determine the interactions between the carbon structural elements constituting the suspension.

Once a type of carbon black is selected for an application, formulation scientists must therefore seek to influence the strength and range of interparticle interactions of the primary aggregates to influence the suspension properties. Due to strong van der Waals interactions that arise from the carbon chemistry, carbon black particles experience relatively strong, short-range interactions in both polar and nonpolar solvents [36]. These strong interactions can drive a phase instability at very small effective volume fractions [42]. Phase separation in dilute suspensions leads to agglomeration and clustering. The distinction between carbon black agglomerates and their primary aggregates is often made by the energy required to break bonds between primary carbon particles. Whereas enormous stresses must be supplied to break the bonds that create primary aggregates, much lower stresses are required to break the bonds that hold agglomerates together. Therefore, it is the number distribution of agglomerates, their size, and phase stability that are determined by the solution conditions [43]. Generally agglomeration leads to further solvent entrainment in the carbon black structural hierarchy and further increase of the effective volume fraction. It is also the reorganization of the agglomerate microstructure under typical mixing, shear, and extensional flows found in most processing operations which require consideration for the ultimate properties of a carbon black formulation.

9.3.3 Rheological Characterization

With appropriate consideration of the carbon black structural hierarchy, it is in principle possible to categorize carbon black rheology in a similar fashion as in the case of spherical particles. However, rheological measurements on carbon black suspensions are plagued by significant experimental challenges. Nonuniform flow fields caused by slip, sedimentation, vorticity banding, and dynamic shear banding are all possible depending on the polarity, viscosity, and density of the suspending fluid [44,45]. Additionally, after preparation, syneresis is sometimes observed, which leads to aging of the sample in the vial and likely contributes to the sometimes confusing and widely varying rheological properties measured in the literature [44,46,47]. Putting aside these confounding experimental challenges, it is instructive to highlight several important findings related to the steady and transient shear measurements of carbon black where care has been taken to ensure uniform and homogenous flow. The steady shear rheology of concentrated carbon black suspensions is generally characterized by the presence of an apparent yield stress and strong shear thinning. The time-dependent rheology has also frequently been shown to be thixotropic [48,49].

Due to the reorganization of the carbon black agglomerates under shear, sample preparation and shear protocol are key factors to successful rheological characterization. As carbon blacks are frequently provided as a dry powder, the first task is to prepare the suspension in the desired solvent at a known weight fraction. This is preferably done using a homogenizer, such as a three-roll mixer or high shear vortex mixer. As highlighted by Dullaert and Mewis [48], high shear mixing is necessary to obtain a reversible, thixotropic fluid. Such processing ensures that the shear rates experienced during preparation far exceed those that the sample will experience in a measurement. Once the appropriate sample preparation is identified, a commonly selected protocol for carbon black suspensions is a preshear step protocol. The goal of this protocol is to rejuvenate the suspension by restoring the agglomerates to a consistent structure such that the history of previous tests is erased. This ensures a consistent starting point for any subsequent rheological test.

As an example of successful implementation of this protocol, a steady state flow curve from carbon black (FW2 from Degussa) dispersed in a nonrefined naphthenic oil (Shell S6141) is shown in Figure 9.10(a) [50]. This flow curve includes the measured elastic and viscous contributions to the overall shear stress [48]. This sample exhibits a yield stress of ~11 Pa and shear thinning above 1 s^{-1}. Time dependent rate jumps are shown in Figure 9.10(b). These tests are an important component to the characterization of any thixotropic material [50], as also discussed in Chapter 1. Rate jumps from high shear rate to low shear rates lead to an increase in viscosity with time and rate jumps in the opposite direction to a decrease in viscosity with time. The microstructural assumption underlying the transient behavior in Figure 9.10(b) is that fractal agglomerates breakdown and build-up self-similarly in shear flow. In this way as the rate is increased the average size of agglomerates is reduced thereby decreasing the effective volume fraction and the viscosity. In a similar fashion, as the rate is decreased, the average size of the agglomerates increases until the agglomerates fill space leading to a solid-like response, i.e., a yield stress.

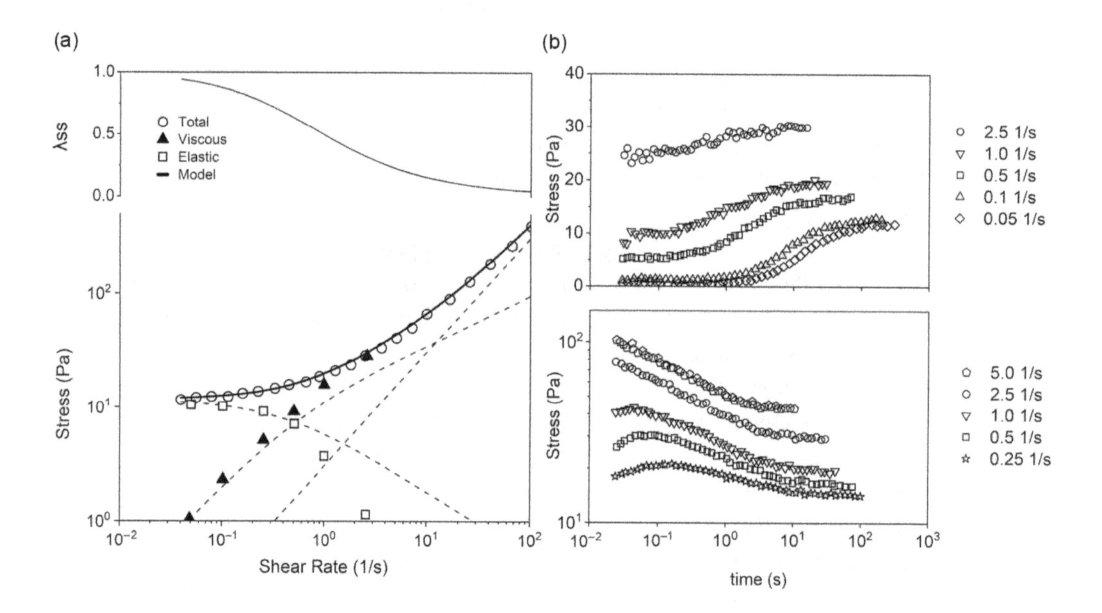

Figure 9.10 (a) Steady state flow curve for FW2 Degussa Carbon Black suspensions in napthenic oil with total stress, viscous, and elastic contributions, (b) Rate step-down experiments from 5 s^{-1} to the value specified in the legend (top) and rate step-up experiments from an initial shear rate of 0.1 s^{-1} to that specified in the legend (bottom). Adapted from Dullaert and Mewis [48]

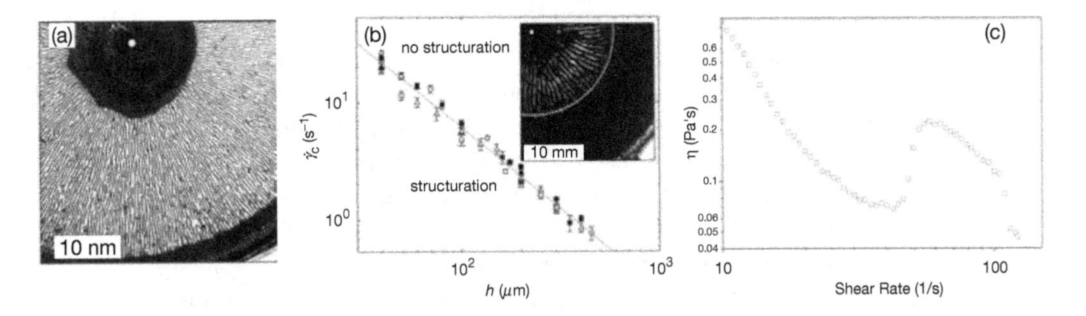

Figure 9.11 (a) Vorticity aligned log-rolling structures imaged in a parallel plate rheometer under steady shear conditions (h = 250 μm and $\dot{\gamma} = 0.5s^{-1}$); (b) stability window for structuration characterized by the critical shear rate, $\dot{\gamma}c$, and gap spacing, h. (a) and (b) Reprinted with permission from Grenard et al., Shear-induced structuration of confined carbon black gels: Steady-state features of vorticity-aligned flocs. Soft Matter 2011;7(8):3920–3928 [53] by the American Chemical Society. (c) Steady shear rheology of carbon black suspension at a single 8 wt% measured in a cone-plate rheometer. Adapted from Osuji et al. [47]

From these rheological measurements, quantitative relationships between the instantaneous structure and shear stress can be established by assuming an appropriate constitutive model. Traditional approaches achieve this through the application of a aggregation/disaggregation kinetic equation via a structural parameter that is related to the bond number [48,50]. Dullaert and Mewis proposed a constitutive model for the evolution of structure in the sample whose fit to the steady state flow curve is shown in Figure 9.10(a). The model describes the elastic and viscous contributions, but also the time-dependent changes in shear stress adequately. The elastic, viscous, and hydrodynamic contributions to the stress are plotted along with the structural parameter lambda in Figure 9.10(a). While the kinetic equation approach provides a means to predict the rheological behavior from a set of simple basis experiments, the connection to the underlying physical microstructure and lambda is challenging to verify.

An alternative approach is the use of population balance modeling, that describes the aggregation/disaggregate processes [51]. Unlike traditional kinetic thixotropic models, this population balance approach formulates aggregation and breakage kernels that explicitly incorporate course-grained agglomerate distributions. The structural parameter is then related to the aggregation number, which is more closely connected to what can be measured experimentally. While these methods hold promise, there remain challenges associated with incorporating structural anisotropy, which is necessary to reach agreement with recent experiments [52]. There are also significant underlying uncertainties regarding the appropriate functional form to describe the elastic stress contribution [50].

A curious effect has been observed in the steady shear rheological measurements of carbon black. Below a critical shear rate, log-rolling structures are observed in carbon black suspensions sheared in confinement [53]. These structures consist of dense flocs that align in the vorticity direction and are persistent under steady flow conditions, as shown in Figure 9.11(a). The critical shear rate, $\dot{\gamma}_c$, at which log-rolling structures are observed is a power-law function of the extent of confinement quantified by the geometry gap, h, is shown for carbon black suspended in mineral oil in Figure 9.11(b). The persistence of these log-rolling structures is attributed to a purely hydrodynamic coupling between the walls of the rheological tooling and the dense flocs in suspension [54]. Depending on how the test has been performed, the onset of this hydrodynamic instability precedes an apparent shear thickening transition at high shear rates, as shown in Figure 9.11(c) [47]. Shear thickening appears similar in nature to that observed in hard sphere colloidal suspensions whose microstructural origin are the formation of hydroclusters [55]. However, in this case, the shear thickening transition is associated with the break-up and resuspension of dense-flocs that lead to an increase in viscosity with time and shear rate.

A closely related observation in many carbon black suspensions is that under certain conditions the transient shear stress behavior is opposite of the thixotropic case (this phenomenon is often referred to as rheopexy or anti-thixotropy) [47,56,57]. This is characterized macroscopically by a decrease in apparent viscosity as a function of time in rate controlled experiments when the shear rate is stepped from a high shear rate to one whose shear stress is near or below the yield stress of the suspension. Microscopically, this phenomenon has been proposed to be associated with loose, open agglomerates becoming compacted under shear flow. This differs from the structural build-up assumed in traditional thixotropy, which implies that aggregation occurs in a self-similar fashion such that the fractal dimension is preserved as agglomerates grow. Instead, this compaction resembles the syneresis effect leading to the expulsion of solvent and a lower effective volume fraction despite the growth of particle size.

Rheopectic kinetic models for such behavior have been described to capture the underlying kinetics of the restructuring of agglomerates in the suspension, and there is some simulation work available that suggests the ability of shear to systematically modify the fractal dimension of agglomerates [56,58]. The presence of such structural evolution can account for several experimental observations present in the literature, namely viscosity bifurcation, apparent shear thickening, and preshear-dependent elasticity and yield stress [47]. Perhaps the most important consequence is that the mechanical properties of quiescent gels exhibit a yield stress and elastic modulus that depends sensitively on the time, rate, and preshear stress before cessation of flow [44,47,49,59]. This is shown in Figure 9.12(a). A carbon black gel is presheared at shear stresses near or below the yield stress of the material, leading to restructuring and a subsequent systematic relationship between the preshear stress applied and the ultimate yield stress measured after cessation. The fact that the flow does stop at the yield stress of the apparent yield stress of the material has been observed previously and is often called dynamic yielding. This behavior is also apparent in the steady flow curve (inset Figure 9.12) as viscosity bifurcation at low shear rates [49]. An important consequence of this behavior is that the preshear protocol can systematically tune both the elasticity and conductivity of a material, as shown by Helal et al. [44]. This scaling

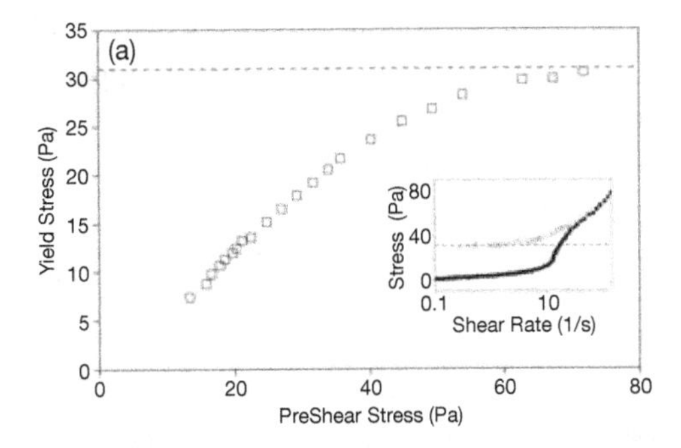

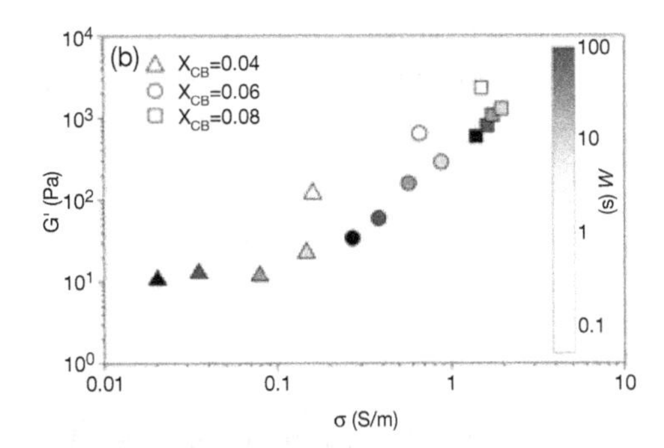

Figure 9.12 (a) Yield stress (Pa) vs preshear stress for a carbon black suspension in mineral oil (weigth fraction xCB = 0.06 (Vulcan XC72R), inset: grey points – steady flow curve measured by taking 96 stress steps from 80.9 Pa to 0.6 Pa with 15 minutes measured at each stress, black points – flow curve determined from MRI measurements of the actual fluid velocity;
(b) Observed scaling between elastic modulus, G' (Pa) and dc-conductivity, σ (S/m), using stress steps from 100 Pa to 0 Pa with the markers indexed by the time interval at each step, Δt (s). Adapted from Ovarlez et al. [49]

can be understood within the context of microstructural rearrangement and compaction leading to weaker, less electrically connected gels [44,59].

9.3.4 Electrical Characterization

Electrical measurements of carbon black suspensions also have a long history motivated by the formulation of polymer composites for electrostatic dissipative coatings and electrorheological fluids [38,60]. In both cases, it is recognized that carbon black facilitates the activated hopping of electrons within the insulating matrix. This implies that a critical effective volume fraction of carbon is required to achieve dc-like transport of charge carriers throughout the material. This critical volume fraction, ϕ_C, marks the boundary between the suspension electrical properties being dominated by the fluid and by the hopping paths present between the carbon particles. Materials filled with carbon black span a wide range of conductivities from 10–12 S/m to 1 S/m, whereas, the conductivity of compressed carbon powders ranges from 10 S/m to 104 S/m. That the conductivity of the pure carbon is rarely achieved in suspensions of carbon black implies that carbon particles, even at very high volume fractions, do not form intimate contact, but maintain a thin solvated layer between their surfaces [61]. Therefore, the probability of hopping is determined solely by the distribution of carbon–carbon surface separation, and therefore it is anticipated that the suspension microstructure will play an important role in determining the conductivity.

Equally important to the microstructure, however, is the ionic character of the carrying fluid. Double-layer charging induced by electrical polarization along the conducting pathways in carbon filled fluids leads to large capacitances, time-dependent currents, and nonlinear field dependent electrical phenomena. The electrical properties of carbon black suspensions are frequently measured using analytical electrochemical impedance spectroscopy (EIS) where a sinusoidal voltage is applied, and the current amplitude and phase shift of the response are measured.

$$Z^* = \frac{Z_e^* Z_i^*}{Z_e^* + Z_i^*}\delta + \frac{Z_e^* Z_i^*}{Z_e^* + Z_i^*}\frac{2\Delta}{\sin h\left(\frac{\delta}{\lambda}\right)} + \frac{Z_e^{*2} + Z_i^{*2}}{Z_e^* + Z_i^*}\frac{\Delta}{\tan h\left(\frac{\delta}{\lambda}\right)}$$
$$\Delta = \left(\frac{Z_\zeta^*}{Z_e^* + Z_i^*}\right)^n \quad (9.6)$$

The frequency dependence of this complex impedance, Z^*, Eq. (9.6), encodes the dynamics of charge carriers and dipolar relaxations within the sample. An EIS model that describes a carbon black suspension is that of a transmission line of finite length [62]. The total impedance is therefore, in the absence of faradaic current (i.e., electrochemical reaction), described by a constant phase element, Δ, with the exponent n taking a value between 0 and 1 ($n = 1/2$ for Warburg element) in series with an electrical and ionic resistance. The capacitance of the double layer is given as $Z_\zeta^* \sim (j \cdot 2\pi f \cdot C_{DL})^{-1}$. Z_e^* and Z_i^* are the electrical impedance and the ionic impedance, respectively, with the electrical impedance containing the frequency dependent response of the mobile charge carriers in the system (i.e., electrons, holes, or polarons). In the limit of high frequency, $f \to \infty$, the transmission line equation reduces to $Z^* = Z_e^* Z_i^* \delta / Z_e^* + Z_i^*$, which is simply the equation of two parallel resistances, whose conductivities can simply be summed $\sigma_{ac} = \sigma_{e,ac} + \sigma_{i,ac}$. Such is the case when carbon black is suspended in water or other polar solvents such as propylene carbonate, as shown in Figure 9.13. At low concentrations of carbon black, the spacing between particles on average is very large so the conductivity is dominated by the ionic strength of the solvent and independent of the concentration of carbon. At some critical weight fraction, $x_{CB} = 0.001$, the electrical properties become significantly affected by the presence of the carbon in the suspension. These fluid samples of carbon black are percolated electrically though they exhibit no

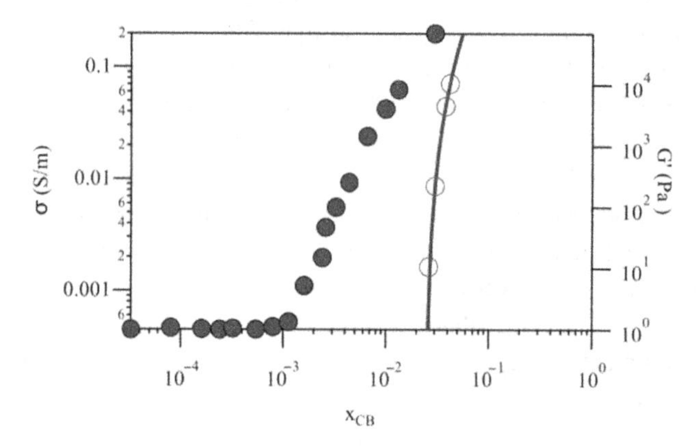

Figure 9.13 ac conductivity, σ_{ac}, measured at 20 kHz (closed symbols) and elastic modulus, G' (Pa), (open symbols) vs weight fraction of carbon black, x_{CB} (Vulcan XC72 R) suspended in propylene carbonate. Data from Richards et al. [61]

measurable elasticity. As the concentration increases, clusters of carbon black form and those clusters jam, leading to a solid-like response. Critically, the electrical properties below the liquid–solid transition resemble those above it only with more carbon leading to larger conductivities.

High frequency measurements (f~kHz) of carbon black suspensions can therefore be used to decouple the electrical and ionic contributions. In the absence of double-layer charging, the electrical component is the only response observed. The electrical component, describing the motion of electrons along the percolated path, is often observed to exhibit a Debye-like response and power-law response at high frequencies [61,63]. The Debye-like response is associated with mobile charge carriers in the conduction band, where the power-law response at high frequencies encodes the temporal distribution of trapping/hopping processes active within the suspension [64,65]. This high frequency power-law behavior is directly related to a low frequency dc-like response and the dc-conductivity that would be measured from a traditional dc-measurement. Suspensions above the electrical percolation threshold will maintain dc-like electrical transport characteristics even when subjected to shear. This is because electrical percolation does not correspond to stress percolation, i.e., the presence of stress bearing bonds [61]. However, shearing can affect the electrical conductivity as the microstructure of carbon black is coupled to the rheological properties and flow history [44,66].

9.3.5 Conclusions

The literature on rheo-dielectric measurements is replete with examples of correlations between the elastic moduli, yield stress, and the dc conductivity at various volume fractions and preshear conditions. It is assumed that this implies a fundamental connection between these macroscopic properties and the microstructure of the suspension. Elucidating these relationships is the subject of ongoing investigation. What is clear is that to gain a full appreciation for the electrical and mechanical properties of suspensions of carbon black for electrochemical flow applications, attention must be paid both to the physical characteristics of the high structure carbon black and to their interactions in specific solutions as these control the evolution of agglomerate size distributions and heterogeneity under processing conditions.

9.4 Bitumen/Asphalt

Norman J. Wagner

Ancients used bitumen (natural pitch) as a binder, and its use as a road construction material can be traced back to Babylonian times [67]. Today, bitumen, or asphalt binder, is a major component of highway asphalt concrete, which is produced on the scale of ~100 million tonnes per year. Consequently, the thermo-rheological properties of asphalt binder are a major source of concern and research, accordingly, given the

Table 9.2 Major components in asphalt binder

Source	Saturates (%)	Aromatics (%)	Resins (%)	Asphaltenes (%)	η 140°F (P)
Lloydminster, AAA-1	10.6	31.8	37.3	19.6	86.4
WY Sour, AAB-1	8.6	33.4	38.3	19.3	102.9
CA Coast, AAD-1	8.6	25.1	41.3	23.9	105.5
CA Valley, AAG-1	8.5	32.5	51.2	8.3	186.2

enormous costs associated with road paving and repair. As a natural product comprised of higher hydrocarbons, some containing heteroatoms such as oxygen, nitrogen, sulfur, and various metals, bitumen is typically classified by sequential dissolution in solvents of varying polarity into fractions termed "saturates, aromatics, resins, and asphaltenes", i.e., SARA. More details about the detailed composition, and the methods of determination can be found in review articles, such as Lesueur [68], which is extensively referenced throughout this section. Compositions of bitumen in SARA fractions are shown in Table 9.2, along with a measurement of viscosity at 140°F [69]. These fractions can be further divided along polarity and solubility lines, with average chemical analyses reported and characteristic molecular structures proposed.

Given the complexity of bitumen chemistry, understanding bitumen rheology has been advanced by a "colloidal model" of bitumen, where the solid fraction of asphaltenes is considered to be a colloidal dispersion of asphaltene micelles (~100 nm in size) in the oily maltenes, which comprise the liquid fraction of aromatics, resins, and saturates. The asphaltene micelles are considered to be coated by a resin core and dispersed to varying degrees in the remaining liquid maltenes. This is illustrated in Figure 9.14, which represents a greatly simplified model for bitumen, but one that enables an understanding of bitumen rheology within the context of colloidal suspension rheology. A quick inspection of the data in Table 9.2 suggests that the viscosity does not trivially scale with the asphaltene fraction alone, such that the operative volume fraction for the solid phase must include a solvation shell of resin. Indeed, the conversion factor from asphaltene mass fraction to effective volume fraction ranges from a factor of 3 to 8 with varying bitumens, with a typical value of ~5.5 that is weakly temperature dependent (see Lesueur [68] for tables of values). It is important to note that this viewpoint is not universally accepted in the field [70]. However, scattering experiments [71,72] and atomic force microscopy (AFM) [73] can be interpreted via a colloidal-like structure in bitumen.

Bitumen is performance graded by limiting temperatures for specific rheological properties. The high temperature limit is defined as where the inverse viscous compliance of the binder drops below 1 kPa when measured at a frequency of

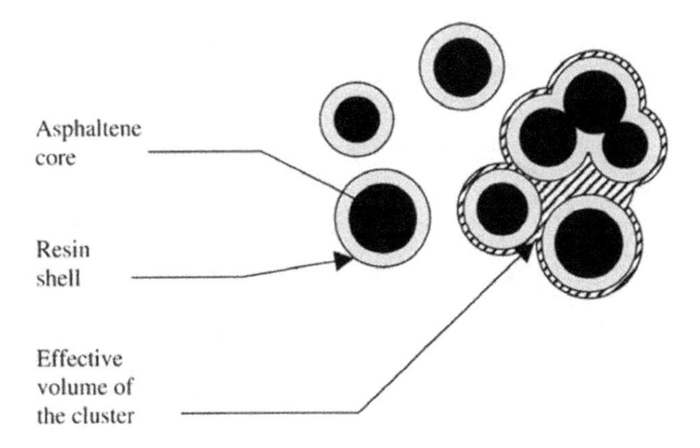

Figure 9.14 Proposed colloidal structure of bitumen as asphaltene micelles with a resin solvation layer dispersed in an oily dispersion medium comprised of the remaining maltenes.
Reprinted with permission from Lesueur, The colloidal structure of bitumen: Consequences on the rheology and on the mechanisms of bitumen modification. Advances in Colloid and Interface Science 2009;145(1):42–82 [68] by Elsevier

10 rad/s (AASHTO T315), where use at higher temperatures risks pavement "rutting". This behavior is associated with the softening of the asphaltene solid phase, leading to a reduction in the rheology, as discussed more generally in Chapter 6. The limiting low temperature corresponds to where the flexural creep modulus of the binder becomes greater than 300 MPa when measured at a loading time of 60 s (AASHTO T313), which risks pavement cracking. This is associated with a glass transition of the maltene phase, which is distinct from the glass formation discussed in Chapter 5. Performance Grades (PG) indicate a high–low range, such that a binder with PG 58–28 means its limiting high temperature is 58°C while its limiting low temperature is 28°C.

The thermal rheology in the linear viscoelastic regime is typically viewed within the context of time–temperature superposition (TTS), and representative master curves for bitumen before and after aging are shown in Figure 9.15. These materials are only approximately thermorheologically simple and deviations from TTS can be observed [68]. This linear viscoelasticity resembles the data shown in Chapter 5 for colloidal gels, and bitumens are considered physical gels of the asphaltene phase in the maltenes. Williams Landel Ferry (WLF) shifting is often used and a variety of models have been proposed to describe these properties, as presented in table 4 of Lesueur [68].

Bitumen is subject to chemical degradation, which yields irreversible aging, as already shown in Figure 9.15, as well as in more detail in Figure 9.16. The increase in viscosity is attributed to the oxidative attack of the asphaltene itself, leading to an increase in the solids fraction. The noticeable difference in the temperature dependence of the oxidation kinetics for sample AAD-1 in contrast to the insensitivity for sample AAG-1 is attributed to aggregation of the asphaltene micelles in AAD-1 at 60°C, restricting access to the sites susceptible to oxidation. These experiments

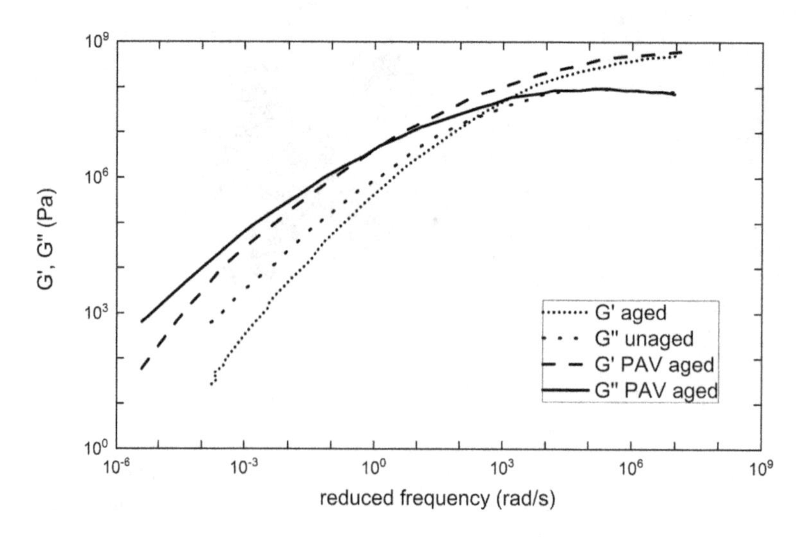

Figure 9.15 Time–temperature superposition for bitumen for a sample before and after aging, showing master curves and the shift to longer relaxation times and lower moduli upon aging. Adapted from Lesueur [68]

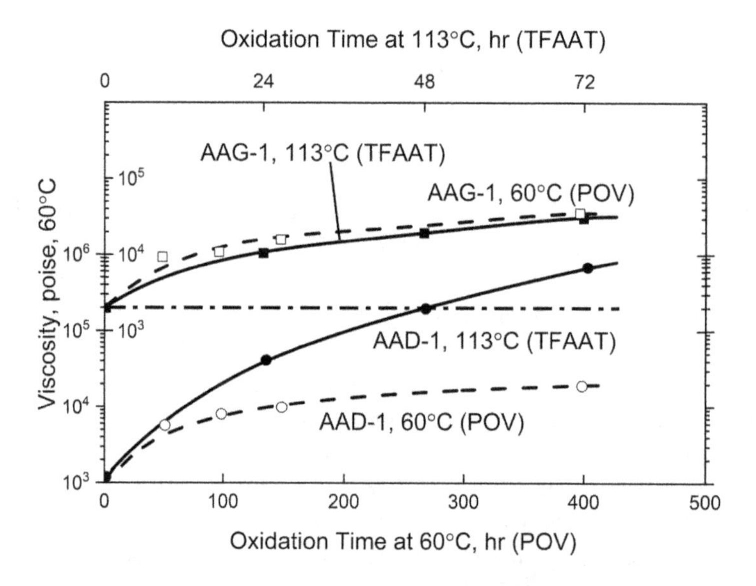

Figure 9.16 Effects of oxidation on bitumen rheology for two bitumens (see Table 9.2), showing different effects of temperature on aging. The rise in viscosity is attributed to oxidation of the asphaltene, which is limited in part by aggregation in sample AAD-1 at low temperature. Adapted from Lesueur [68]

highlight the rheological complexity of bitumen binders and the insights that colloidal models can provide.

A simple model attributed to Roscoe, which is along the lines of the phenomenological correlations for the volume fraction dependence of the viscosity already

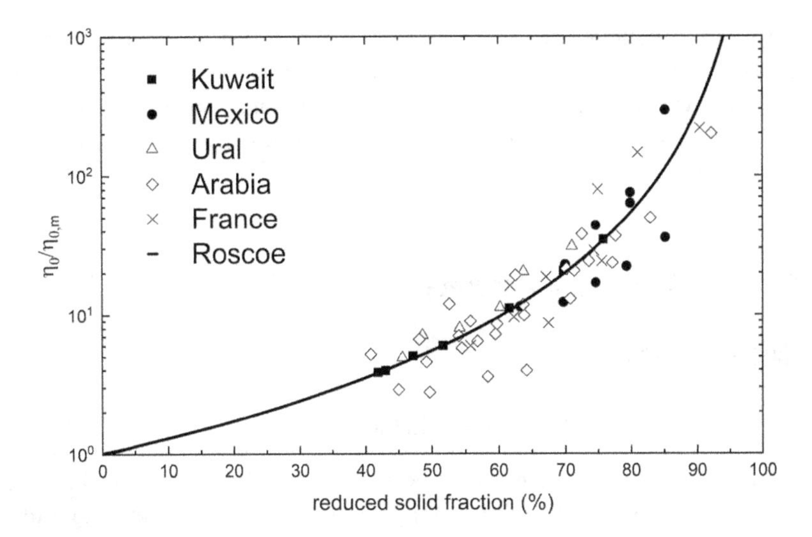

Figure 9.17 Roscoe model for the relative, zero shear viscosity of suspensions compared to measurements on various bitumen samples with dilutions, showing the colloidal-like viscosity of bitumen viewed as resin solvated asphaltene in maltene. Adapted from Lesueur [68]

presented in Chapter 1, is used to translate the colloidal model of bitumen to understand the effective zero shear viscosity.

$$\frac{\eta_0}{\eta_{0,m}} = \left(1 - \phi_{\text{eff}} \middle/ \phi_{\text{max}}\right)^{-2.5}. \tag{9.7}$$

The medium viscosity $\eta_{0,m}$ is taken to be that of the maltenes, while the zero shear suspension viscosity is related to the effective volume fraction comprised of the asphaltene micelles plus the solvating resin. A plot of various measurements against the model of Roscoe in Figure 9.17 shows reasonable agreements and trends that, within the uncertainty, justify the general use of a colloidal model to describe bitumen as a suspension of solvated asphaltene in a maltene oily matrix.

The success of the colloidal model of bitumen to rationalize the long relaxation times, oxidative aging, effects of polymer additives, and blending [68] suggest that further advances in our understanding and control of bitumen will benefit from more clearly linking the colloidal structure of bitumen to its thermorheological properties. This very brief review has necessarily limited the discussion to a few properties of neat bitumen, but the use of additives to modify bitumen is also a critically important area of research. For example, adding a small amount of styrene-butadiene-styrene copolymer can significantly increase the high temperature stability of a bitumen binder. This effect can be attributed to the most soluble fractions of the maltene being absorbed in the block-copolymer micelles, thus increasing the concentration and softening temperature of the hard asphaltene phase [72], and by the colloidal model, the modulus of the binder. Thus, the colloidal model is a fruitful approach to modern asphalt research and is it noteworthy that advanced rheological techniques being advanced for colloidal dispersions, such as LAOS, are being applied to understand the effects of various modifiers on bitumen performance under nonlinear deformations [74]. We envision that many of the properties and approaches to understand the rheology of soft particles, e.g., those presented in Chapter 6, will be of significant value in bitumen rheology research.

9.5 Cement, Mortar, and Concrete

Nicolas Roussel

9.5.1 Background: "Fresh" Cement-Based Products

Cement-based products are economically viable and readily available materials that have the property of being fluid during the first hours of their industrial life and then turn, through a chemical hydraulic reaction, into porous solids with mechanical properties roughly similar to the ones of some natural stones.

Cement-based products result from the mixing of water with inert and reactive mineral particles. Cement is one of these reactive particles. It is produced from the calcination of a mixture of limestone and clay and mostly contains silica, calcium, and aluminum. It is commercially available under the form of a powder with an average particle size around 10 micrometers and a density around 3 g/cm^3.

Cement-based products can be used under the form of, e.g., grouts, adhesives, coatings, render mortars, or even structural concrete. There exist some standards around the world to precisely define the boundaries between these industrial products but, as we are here mainly interested in the underlying physics, we will keep in mind, at this stage, that grouts and pastes only contain particles smaller than 100 micrometers while for sand-based mortars the maximum particle size is below a few millimeters. Finally, concrete may contain particles up to a couple of centimeters. The particles finer than 100 micrometers are often considered as belonging to the so-called cementitious "matrix" while larger particles are designated under the general term "aggregates." Cement-based products also differ in terms of their water-to-cement mass ratio. This ratio dictates the chemical reaction between cement and water, which leads to the "setting" of the product and the filling of its initial porosity by the resulting hydrates. This ratio also dictates the final porosity and hence the long-term properties of the product in terms of mechanical strength and durability. Finally, cement-based products can also differ in their proportions of so-called "admixtures," which are either mineral or organic components able to tailor a given property of the product even when introduced at extremely low dosages in the system. After mixing, the family of cement-based products therefore covers a very wide range of properties both in the so-called "fresh" state before setting and in the hardened state and service life.

We will obviously focus here on fresh cement-based products being "pasty" mineral suspensions. The idea behind this section is, first, to present their macroscopic behavior along with the typical shaping processes that do dictate the rheological requirements for fresh cement-based products. We will then discuss the various interactions at the origin of their macroscopic behavior. Finally, we will give a short overview on the technological levers, allowing for the tuning of both these interactions and the resulting behavior to the process.

9.5.2 Typical Rheological Behavior and Typical Shaping Processes

Fresh cement-based materials, as many materials in industry or nature, behave as yield stress fluids [75–78]. For stresses higher than the yield stress, they behave as visco-plastic materials. In the case of cement pastes or grouts (i.e., those containing fine particles), the yield stress can be measured using conventional rheological tools [79,80]. In the case of mortars or concretes containing large particles, specific rheometers had to be developed [81–84]. The so-called slump test (ASTM standard C143/C143M-20) remains, however, the most common industrial test for yield stress assessment of fresh cement-based materials (Figure 9.18). Typically, in a slump test, a conical mold is filled with the material to be tested. The mold is lifted and flow occurs. As inertia effects can be neglected, it is generally accepted that flow stops when gravity-induced shear stresses at any point in the tested sample become equal to or smaller than the yield stress [85,86]. Consequently, the final shape of the sample is directly linked to the material yield stress. From a practical point of view, in the construction industry two geometrical quantities may be measured: "slump" and "spread." The slump is the difference between the height of the mold and the thickness of the sample at the end of the test. The spread is the final diameter of the collapsed sample. In the case of cement pastes or grouts, the yield stress of which is low compared to that of mortars or concrete (see Section 3), a smaller cone may be used, namely the ASTM C230–90 mini-cone or the Kantro Mini cone [87]. Several attempts to relate slump to yield stress can be found in the literature

Figure 9.18 Slump flow measurement for yield stress assessment of a fluid concrete.

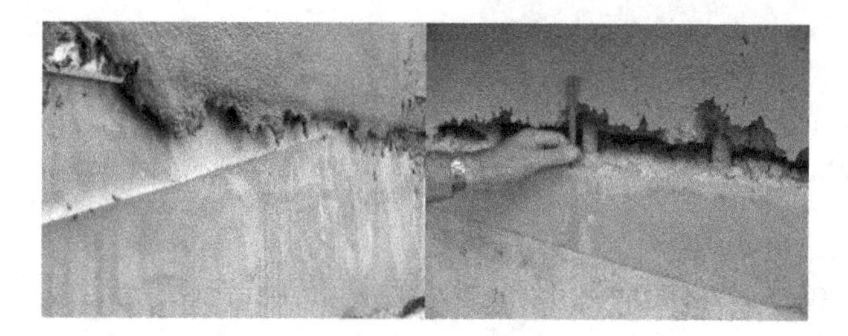

Figure 9.19 Typical shaping processes: (left) coating of a render mortar; (right) mold filling of structural concrete (failure).

[85,86], which allows for a computation of the material yield stress from one of the above geometrical measurements.

Most cement-based materials are shaped under the sole effect of gravity. Exceptions can be found in some specific extrusion or injection processes; however, for most processes of interest, the final shape only results from the competition between gravity-induced stresses and yield stress. Moreover, the large dimensions of most objects shaped using cement-based products make any contribution of the surface tension to the final shape negligible. Within this framework, we are therefore facing two main types of rheological requirements for shape control of cement-based products. Either the material shall display a minimum yield stress to be able to, for instance, adhere to a substrate as illustrated in Figure 9.19 (left) in the case of a render mortar coating, or the material shall display a maximum value of its yield stress in order to guarantee the proper filling of a mold under the sole effect of gravity, as illustrated in Figure 9.19 (right) in the case of structural concrete. The role of the viscosity of this kind of materials is often limited to ensuring, on the one hand, that the power required for mixing or pumping shall stay within reasonable technological boundaries and, on the other hand, that the shaping or filling rates under the sole effect of gravity are compatible with industrial production rates.

In the case of many cementitious materials, the knowledge of the yield stress at the end of the product mixing phase alone is not sufficient to guarantee a proper shaping. An evolution of the material rheological behavior is indeed often noted during the first hours after mixing and through processing. At rest, the material apparent yield stress increases; when sheared the material yield stress decreases.

It has been shown that this evolution can be correlated to the flow history of the material using simple thixotropic models [80,88]. It has to be noted that, although all cement-based materials are thixotropic, in practice only a few of them are described as such. It seems that, in industrial practice, a cement-based material is considered as thixotropic only if it displays a rather short flocculation characteristic time (typically several minutes) and a de-flocculation characteristic time of several tens of seconds in the shear rate range of industrial interest [88]. In the case of cement-based materials, the above reversible evolution of the properties goes along with an overall (slower) nonreversible evolution of the system.

The consequences of thixotropy on industrial shaping processes are many but, in the last decades, research and development have mostly focused on two major industrial aspects: formwork or mold pressure and multi-layers casting. During molding, given the high fluidity of modern cement-based materials, it is expected that the pressure on the mold is hydrostatic. With an average density for concrete around 2,400 kg/m^3, such pressures can be at the origin of either mold failures or high mold reinforcement costs. However, in most cases, far lower mold pressures are often reported. It was recently concluded that, during slow placing or at rest, thixotropy progressively gives the material the ability to withstand the load from the material above without increasing the lateral stress against the mold [88,89]. Moreover, during placing, the yield stress of a layer of a thixotropic cement-based material increases during the short time interval before a second layer of material is cast above it. If the apparent yield stress increases above a critical value, then the two layers do not intermix and a weak interface between these layers may threaten the mechanical integrity of the final object [88,90].

9.5.3 Physical Origin of the Rheological Behavior: Upscaling between the Cement Matrix and Mortar or Concrete

It is accepted in the literature that most of the complexity of the rheological behavior of cement-based materials lies in the cementitious matrix. Adding particles such as sand and/or coarse aggregates to this matrix to produce mortar or concrete can conceptually be considered as adding rigid inert non-Brownian noncolloidal particles into a yield stress thixotropic fluid as long as the typical size of the particles (i.e., sand or coarse aggregates of a respective average size of 1 and 10 mm) is much larger than that of the constitutive elements of the matrix. The rheological properties of the resulting suspension were shown to depend only on the rheological properties of the suspending fluid on the one hand, and on the particle volume fraction, shape, and size distribution on the other (see chapter 2 of *CSR*). The above studies emphasize the fundamental role played by the packing properties of the aggregates added to the cement matrix. Most standard mortars and concretes are mix-designed with aggregate volume fraction between random loose packing and random close packing (see Figure 9.20) [91]. Random loose packing generally refers to the most loose, mechanically stable packing state of the particles, while random close packing refers to the densest state particles can achieve when randomly packed [92]. There thus exists in these materials a percolated network of rigid particles in contact. In this range of aggregate concentrations, the amplification of the rheological behavior of the matrix by the aggregates is dictated by the amount of direct frictional interparticle contacts in the system and the yield stress of the material can easily be more than 1000 times that of its constitutive matrix. Extremely fluid cement-based products can, however, be designed below the random loose packing fraction of the aggregates. In this regime, although particles are able to slightly amplify the rheological behavior of the constitutive matrix, the absence of a percolated direct frictional contact network in the system allows for the mix design of so-called self-compacting or self-leveling mortars

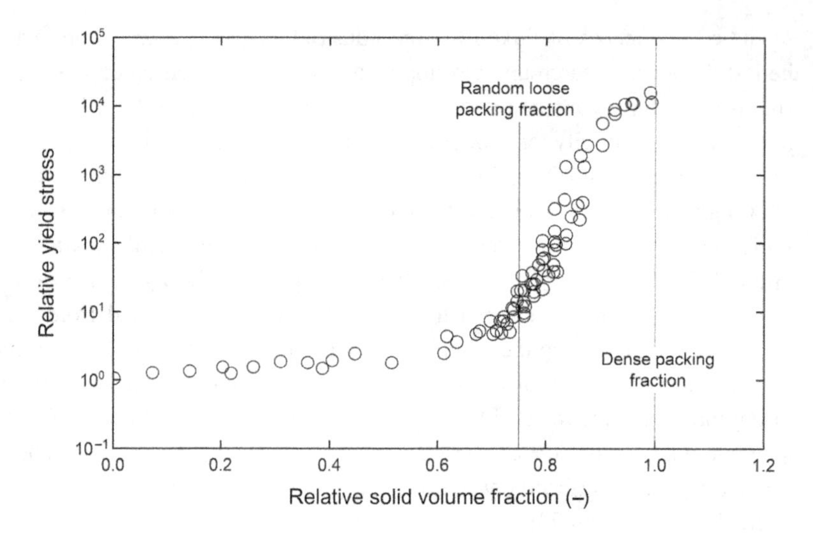

Figure 9.20 Relative yield stress of various mortars or concretes (i.e., ratio between the yield stress of the material and the yield tress of its constitutive cement paste) as a function of relative solid volume fraction (i.e., ratio between aggregates volume fraction and random close packing fraction). Adapted from Roussel et al. [93]

or concretes. Nevertheless, in practice, as the economic cost of the matrix is far higher than that of the aggregates, this fluidity comes at a price.

9.5.4 Physical Origin of the Rheological Behavior: Interactions within the Cement Matrix

From the literature, the interactions of interest in a fresh cement paste are the following: van der Waals attractive forces, electrostatic repulsive forces, gravity forces, Brownian effects, hydrodynamic forces (including laminar viscous dissipation and kinetic energy) and more or less direct contacts (potentially both friction (i.e., long duration contacts) and collision (i.e., short duration contacts)) [93]. These forces have been introduced in Chapter 1. Note that, once mixing is over and the interstitial fluid is forming a continuous phase, we shall neglect any capillary forces [93].

Forces Competing at Rest

The first type of competition in the system occurs at rest. It results in one of the following regimes [94]:

- If colloidal attractive forces dominate Brownian motion, which is the case for most cementitious materials in industry, a percolated network of interacting particles appears within the suspension. After flocculation, the interparticle distance at the pseudo contact points between cement particles results from an equilibrium between the van der Waals attractive forces and the short-range electrostatic repulsive forces. It was estimated to be, in the absence of any adsorbed organic molecules, of the order of a few nanometers [95]. This network of interacting cement particles is able to withstand an external stress up to the yield stress. This yield stress was shown to scale with the inverse square of the interparticle separating distance. Therefore, it strongly depends on the presence of adsorbed polymers at pseudo-contact points between particles. Moreover, it was shown that the length scale that dictates the magnitude of these attractive forces is not the size of the cement particles themselves (around 10 μm) but that of their characteristic surface roughness (i.e., around several hundreds of nanometers) [95].

- If Brownian motion dominates colloidal attractive forces (in the case of dilute or highly deflocculated systems), the network of attractive colloidal particles is destroyed by Brownian agitation and is not able to withstand any external stress. Moreover, because of the large size and density of cement particles, gravity forces do dominate Brownian motion and such cement-based suspensions shall always display high levels of sedimentation.
- If colloidal attractive forces dominate gravity forces, particles are trapped in the colloidal interactions network and are therefore not able to rearrange their relative positions under the effect of gravity. The cement suspension is stable and there is no sedimentation.
- If gravity forces dominate colloidal attractive forces and Brownian motion, particles may settle. This induces a relative displacement of the cement particles within the interstitial fluid (sedimentation) or a relative displacement of the interstitial fluid between the cement particles (consolidation). These phenomena are time driven and are influenced by the viscosity of the interstitial fluid and the permeability of the porous medium formed by the interacting cement grains. Within very specific conditions, sedimentation or consolidation may be adequately limited by mix design. It is in this regime only that fluid cement pastes allowing for the production of fluid mortars or concretes can be mix-designed.

Forces Competing during Flow

The second type of competition starts as soon as the material starts to flow. Van der Waals colloidal attractive forces for cement-based suspensions dominate hydrodynamic forces in the low strain rates regime of industrial interest and give rise to a shear thinning macroscopic behavior, as shown in Figure 9.21 [93]. Hydrodynamic forces dominate at intermediate strain rates. In such concentrated systems these forces involve lubrication forces potentially amplified by interparticle friction [96]. It is therefore expected that the interparticle region, in which most of the shear is concentrated, shall play a major role. Recent results [97] suggest indeed that hydrodynamic dissipation in cement pastes scales with both the viscosity of the interstitial fluid and the inverse surface-to-surface interparticle distance. Finally, it can be noted that the inertia of the cement particles may dominate the high strain rates response and was suggested to be at the origin of the shear thickening behavior observed in the right part of Figure 9.21 [88]. The transitions between the flow regimes mentioned above are governed by critical shear rates that depend on average particle size, fluid viscosity, cement density, Hamaker constant, particle roughness, and surface-to-surface interparticle distance [88].

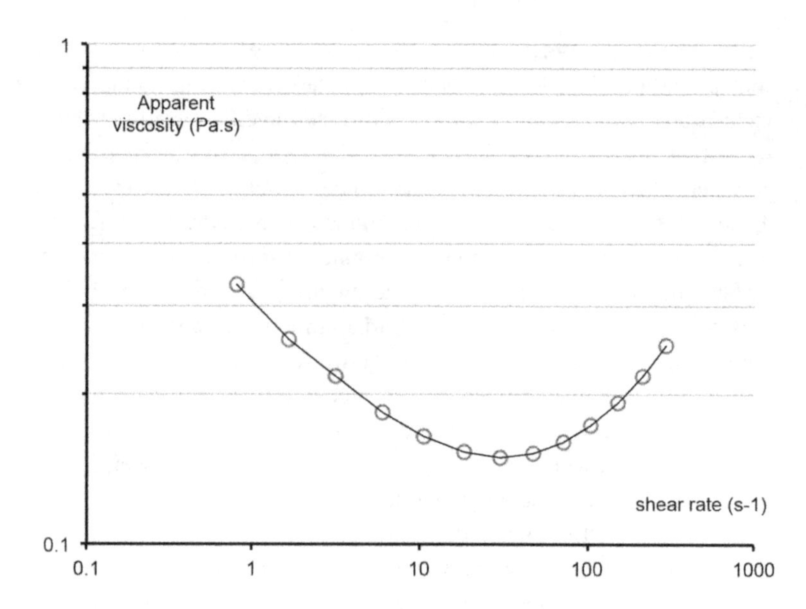

Figure 9.21 Typical flow curve for cement paste. Experimental results for a W/C ratio equal to 0.35 on a Portland cement CEM I. The flow curve was obtained using a Couette viscometer with serrated cylinders. Adapted from Roussel et al. [93]

Contribution of the Hydration Reaction

Finally, from a microstructural point of view, the ability of a typical cementitious material to build up a structure at rest and display therefore a thixotropic behavior was shown to originate both from a reversible flocculation process typical of colloidal attractive particles and from early hydrates nucleation at the pseudo-contact points between cement grains in the developing flocculated structure. The following framework describing the two origins of the time dependent behavior was shown to apply to standard cement-based products [98]:

- At the end of the mixing phase, cement particles are dispersed (see Figure 9.22(a)).
- Because of colloidal attractive forces, cement particles flocculate and form a network of interacting particles displaying an initial elastic modulus and an initial yield stress (see Figure 9.22(b)). This progressive flocculation is controlled by competing forces acting on the cement particles: the colloidal interparticle attraction and the viscous squeezing forces in the interstitial water between the cement grains. Yield stress and elastic behavior develop in the cementitious suspension in a matter of a few seconds after the flow stops [98].
- Simultaneously, at the pseudo-contact points between particles within the network (i.e., at a separation distance of the order of a few nanometers), nucleation of hydrates occurs (see Figure 9.22(c)). This nucleation turns locally the soft colloidal interactions between cement particles into higher energy interactions, which can be roughly seen as solid bridges. As a consequence, at the macroscopic scale, both elastic modulus and apparent yield stress increase. After

a few hundred seconds, a percolation path of cement particles interacting purely through hydrate bridges appears.

- Further increase of the macroscopic elastic modulus and yield stress is the result of an increase in size or number of hydrate bridges between percolated cement particles (see Figure 9.22(d)). This phase is often referred to as structuration in cement literature. At the microscopic level, it is a nonreversible chemical reaction that creates the hydrate bonds between particles. These bonds may, however, be weak enough to be broken by shear and/or remixing, whereas new bonds may spontaneously appear again at rest as long as there is a sufficient reservoir present of chemical species. The formation of these bonds is hence not incompatible with a macroscopic reversible evolution as expected from a thixotropic behavior. From a practical point of view, hydration may therefore have reversible macroscopic consequences as long as the available remixing power is sufficient to break the hydrate bridges between cement particles. On long timescales, hydration and

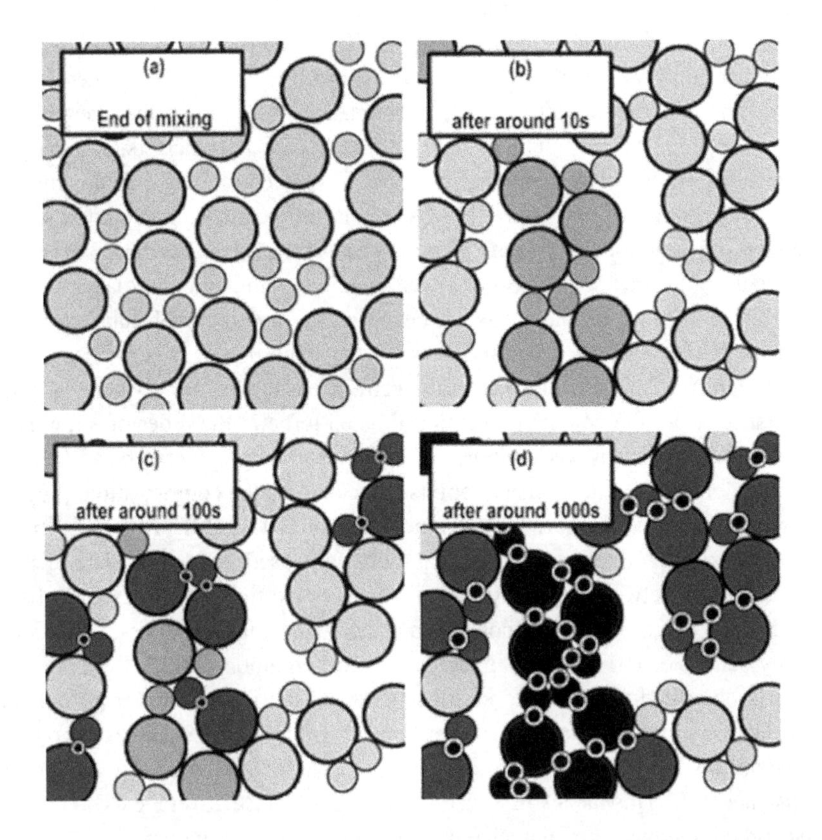

Figure 9.22 Network(s) of interacting cement particles in the dormant period. All times are given as orders of magnitude. Small black dots between cement grains are hydrates. Darker grey cement particles belong to a colloidal forces percolation path. Black large particles belong to a path of percolating hydrate bridges. Reprinted with permission from Roussel et al., The origins of thixotropy of fresh cement pastes. *Cement and Concrete Research* 2012;42(1):148–157 [98] by Elsevier

nucleation are however at the origin of a nonreversible aging of the system as soon as the available remixing power becomes insufficient to break these interparticle connections [98].

9.5.5 Tuning the Rheological Behavior of Fresh Cement-Based Products

First, it has to be noted that we do not consider here the specific situation in which high molar mass adsorbing polymers are added to the system that causes bridging forces between cement grains [99,100]. We moreover neglect any potential depletion forces due to nonadsorbed polymers as most standard admixture industrial dosages result in concentrations of nonadsorbed polymer that are too low to generate any depletion forces that could compete with the other forces in the system [99]. When ignoring these forces, all changes in the rheology of a cement paste upon addition of rheology-modifying polymers can be attributed to changes in both the surface-to-surface particle separation distance and the viscosity of the interstitial fluid.

Within this framework, adding polymers to a cement paste simply consists of adding polymer coils, which, if adsorbed, may modify the surface-to-surface separation distance and which, if not adsorbed, may increase the viscosity of the interstitial fluid (see Figure 9.23). The first family contains all the so-called plasticizers or super-plasticizers depending on the magnitude of their effect. The second family contains most so-called viscosity modifying admixtures (VMA) or viscosity enhancing admixtures (VEA). The effects of both these families of macromolecules are expected to increase with the size of the coils (adsorbed or nonadsorbed) and therefore with the molar mass of these organic products. We neglect here any influence of the specific molecular structure of the polymers (for instance, comb or linear, co- or homo-polymers, see Chapter 6).

When there are no polymers in the cement paste, the surface-to-surface separation distance at pseudo-contact points is estimated to be of the order of a few nanometers [101]. When there is steric hindrance from adsorbed polymers, the surface-to-surface separation distance at "contact" points is dictated by the conformation of the adsorbed polymer coils at the surface of cement grains [102]. Orders of magnitude for the adsorbed layer thickness of typical polymers used to deflocculate cement-based material are between 5 and 10 nm. It can be noted that recent advances have shown that, even in a system as complex as a cement paste, it seems possible to correlate the polymer molecular structure to its surface conformation [102].

As described above, yield stress scales with the inverse square of the surface-to-surface separating distance between cement grains while recent results suggest that viscous dissipation scales with the inverse power of the surface-to-surface separating distance [97]. This means first that, at full grain surface coverage (when two layers of adsorbed polymers are intercalated between two interacting grains), yield stress can be decreased by up to two decades whereas viscosity can be decreased by one decade in the typical range of dosages and molar mass used in industry.

The presence of nonadsorbed high molar mass polymers will increase the viscosity of the interstitial fluid between cement grains, increasing proportionally the macroscopic viscosity of the paste. This is the case, for instance, for nonadsorbed cellulose

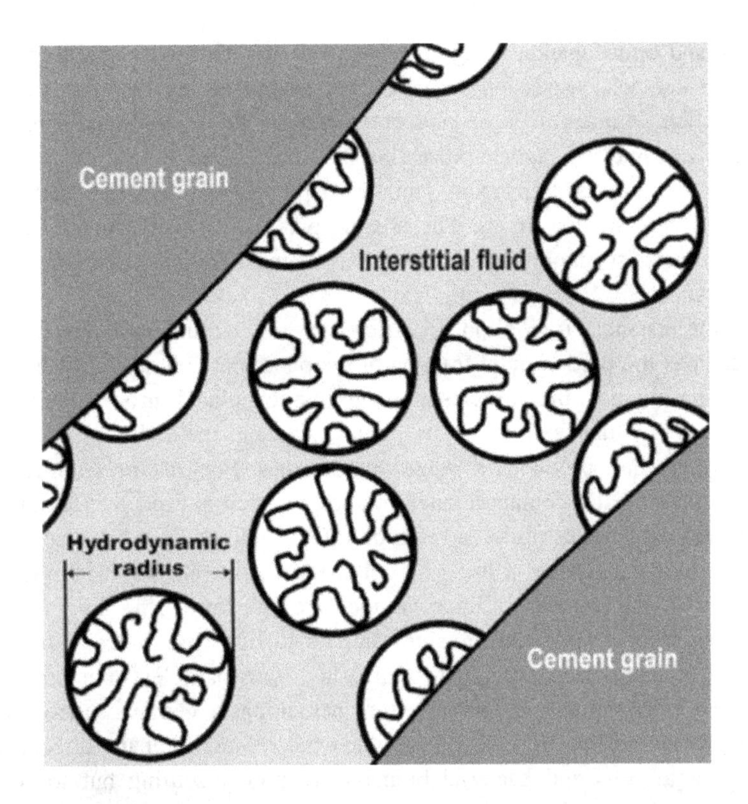

Figure 9.23 Conceptual localization of polymer coils in a cement paste. The macromolecules can either be adsorbed or stay in the interstitial fluid depending on their affinity with the surface of the cement grains. Adapted from Brumaud [106]

ethers [103–105]. This increase in the viscosity of the interstitial fluid, although detrimental for the macroscopic viscosity of the cement paste, may however slow down the rate of sedimentation of an unstable cement paste and allow for its meta-stabilization before chemical setting "freezes" the system.

9.6 Large Scale Processing

Peter Scales

Colloidal and noncolloidal particulate suspensions are common to a wide range of industries. In some instances, the process application is representative of millions of tonnes on a daily basis. The particles in these suspensions are almost never mono-sized and the carrier fluid is usually water. The particles are sometimes dispersed such that particle–particle interactions are repulsive but the nature of large scale industrial suspension processing is such that coagulation or flocculation is required to produce larger aggregates so as to enhance the sedimentation and suspension network permeability in solid–liquid separation processes. These processes, with sedimentation/thickening, filtration and centrifugation being examples, look to either recover water for reuse or create a solid rich suspension so as to reduce volumes for disposal and/or reduce transportation volumes. It is perhaps not surprising that solid–liquid separation is a feature of nearly every particle processing flowsheet, whereby low solids suspensions are taken on a phase separation journey to produce particle rich and water rich (nominally zero solids) phases. The mixing, pumping, and transport of the particle rich phase are also important unit operations of most particle suspension processing flowsheets.

In some instances (i.e., alum or ferric rich water treatment sludges), the coagulant becomes the predominant fraction of the particulate phase. In other cases (i.e., waste water sludges) the particles are naturally coagulated through the presence of salts or extra-cellular exudates. The particle size and methodology of coagulation varies widely and produces a range of outcomes. For instance, minerals tailing suspensions typically contain a small fraction of micron-sized particles but show a distribution of particles up to an order of 100s of micrometers top-size. The use of very high molecular weight bridging polymers ensures all particles in the suspension can be flocculated to improve the rate of dewatering in a solid–liquid separation process. In contrast, alum and ferric rich sludges contain aggregates made up of nano-sized precipitates enmeshed with particles of micrometer size. Algal cell rich sludges, as with wastewater sludges, contain micrometer sized particulates enmeshed in an extra-cellular exudate such that the almost neutrally buoyant algal cells and bacterial biomass are poorly settling but form a strong interconnected network.

When increasing the solids content, all coagulated and flocculated particulate systems form a space filling network of aggregates, a gel, at a unique concentration: the gel transition concentration GTC, as discussed in detail in Chapter 5. This concentration should not be confused with the gel point of polymer systems or crosslinking in gelled glasses. At the GTC the flocculated suspension becomes self-supporting and, as such, does not sediment. This is an important transition point since gravity alone will not cause further consolidation or settling, this now requires an external force to be applied that exceeds the network strength. In a particle packing sense, the GTC can be thought of as an analog of the random close packing phase transition in hard sphere systems [107,108] except that the network is now a collection of aggregates, the internal particle volume fraction of which may be very low [109]. In the case of particulate suspensions, where the network is dominated by nano-sized precipitates or crosslinked by a molecular gel (algal cell or waste water biomass sludge), the GTC may be as low as 1–2 volume percent solids [110]. For mineral suspensions, it may vary from a few to of order 40 volume percent solids, depending on the particle size distribution and depth of the attractive well between particles. The GTC has been shown to be a scaling factor for both of these parameters, as demonstrated through work on (strongly aggregated) model suspensions [111–115].

In a rheological sense, aggregated particulate suspensions at concentrations above their GTC all have a measureable network strength, or shear yield stress, that increases in an exponential fashion with increasing solids concentration [116]. Using a simple rheometer configuration such as a vane (or bob) in a cup, the strength is manifest as a peak in the torque as a function of time upon rotation of the vane at a constant rate [117,118], a transition from constant compliance behavior at long times to flow behavior across a narrow range of increasing applied stresses in a creep compliance test [119], or a peak stress as a function of increasing strain in a constant strain rate test [120,121]. The detailed measurement of these shear-based parameters is given in chapter 9 of *CSR*. An alternative method that does not involve a rotational device and that is used extensively in the minerals industry as a quality control test is simply to measure the residual height in a cylinder slump test, as discussed in Section 9.4. This is a useful in field test requiring limited equipment and expertise whereby the residual height is equated to a network strength parameter [122,123]. The shear yield stress is a useful engineering parameter in that it is the minimum stress required to cause the suspension to flow and an easy reference point for the design of applications that require startup of flow or the movement of the suspension when the torque in mixing is limited. This is the case for all process flow sheets as all processes are rheologically limited in some manner.

The scientific issue with the designation of a shear yield stress to an aggregated particulate suspension is that the shear yield stress value is not unique across the many modes of measurement. The detailed measurement is discussed in chapter 9 of *CSR*, but the question here is which value to use for the design and operation of large scale processes. The measured shear yield stress shows a strong strain rate dependence, although there exists a strain rate condition wherein the peak stress at flow is a minimum [117,121]. This allows one to measure and equate the parameter in a consistent fashion for operational quality control and for design that ensures pumps, mixers, and other equipment are not undersized. The reasons for this strain rate dependence are complex and will not be exalted here, suffice to say that the failure of the network exhibits elastic deformation at very low strains followed by strain hardening, strain softening, and then flow with an increase in strain [120]. The behavior is consistent with a cage melting mechanism where the strain at failure varies depending on the volume fraction and the Péclet number [121]. From the point of view of applications, using a single rheological parameter such as the yield stress is useful in both design and operations, suffice to say, those that choose to use a single value need to be cognizant of the limitations of the approach.

From an engineering applications perspective, the shear yield stress is used in models of pipeline flow to provide an indicator of the maximum pressure drop in pipeline transport of mineral suspensions over long distances [124] and in limiting the flow of suspensions in open channels [125]. It is also used in models to predict the maximum torque on a rake in thickening [126], the maximum torque of a mixer, and the predicted “dry stacking” slope in a tailings storage facility [127,128]. In the latter example, being able to control tailings rheology to a fixed yield stress behavior is a key determinant of the slope and in turn, the type of containment wall. Data shows the slopes of up to 10 percent gradient are possible at yield stress values of order 100 Pa. Those companies that elect for either historic or cost reasons not to dewater and then control the rheology of their tailings to give a measureable slope on their storage facility require dam style containment and show a far greater risk of tailings dam failure, the consequences of which can be devastating [129].

In the pipeline transport of mineral backfill, product or tailings materials where the particle size distribution ranges from micrometers to many millimeters (sometimes centimeters), the optimum point of transport to avoid particle segregation and pipe blockage [130] is just above the laminar to turbulent transition. The calculation of the appropriate Reynolds number for design is at issue here since the key rheological properties of the suspension are governed by the smaller size fraction. The common approach to understanding this suspension is of a mixture of large particles (up to centimeters in size) in a carrier fluid. The rheology of the carrier fluid is determined by the colloidal particle fraction. If the concentration of the colloidal fraction (nominally less than 30 μm) is below the GTC, the description of the viscosity term in the Reynolds number is well described in work developed by Wasp et al. [131] using the original descriptions of Metzner and Reed [132]. However, for suspensions where the carrier fluid is above the GTC the fluid is now nonNewtonian (remembering that for some suspensions, this concentration may be only a few volume percent solids) and the suspension will exhibit a shear yield stress. A viscosity descriptor incorporating a Herschel–Bulkley (HB) type model (see Chapter 1) is a necessary addition to the Metzner and Reed descriptions to achieve an adequate prediction of the actual behavior of suspensions in pipeline and open channel flow [133,134]. For laminar flow, however, where segregation of the larger particles in the colloidal carrier fluid is likely, more complex descriptions are required [124]. Incorporation of the strain rate dependence of the failure of the particulate network into models is an important next step in this evolving field.

In the case of suspension mixing or the transport of suspensions in thickeners with rakes, the role of the shear yield stress is also critical to both design and operations. Rudman et al. [126] showed the maximum torque on the rake in a thickener is a strong function of the shear yield stress. They stated that "over a range of rake speeds, the measured torque was an almost linear function of yield stress." One would argue that this is hardly surprising in that the rake turning in a thickener is a very close analog of a large rotational rheometer. Despite this correlation, the behavior of the suspension after failure and before flow is fully established is perhaps the least understood attribute of the rheology of particulate suspensions. In the case of mixing, suspensions with a yield stress are known to "rat-hole" or show stagnant zones in the mixing vessel. Computational fluid dynamics (CFD) models are popular in modeling these systems as the three-dimensional and not just two-dimensional behaviors are important here. CFD allows both separation and coupling of the hydrodynamic and carrier fluid rheology elements of the flow from the momentum and sedimentation behavior of the particulate phase. Despite the sophistication here, the prediction of the interface between flow and stagnant zones (in the limit of no flow) is still poor. This is perhaps not surprising as many of the rheological descriptions of the carrier fluid are simple Herschel–Bulkley style models that assume a single value for the yield stress [135].

Another aspect of the rheology of particulate suspensions is their behavior in compression. Whilst uniaxial compression is clearly the simplest to consider, combined shear and compressional effects are important and commonplace in the application space. Despite its importance, a unified theory of combined shear and compression for aggregated particulate suspensions is still an area of intense study. At the heart of the issue is the nature of the description of the so-called elastic–plastic transition that traditionally finds a home in powder mechanics using the likes of a powder shear cell as a route to parameterization of the behavior of a material [136,137]. The nature of the failure of aggregated particle networks in shear is such that total strain energy rather than a yield stress is likely to be the material property parameter that is constant for a given concentration of the suspension.

The predominant difference between compression and shear is due to the incompressibility of both the fluid and particulate phases. As a consequence, under compression movement of the suspension or phase separation are not possible without solid-liquid separation. The GTC marks a point above which, for the strongly flocculated suspensions typical of the minerals and water and waste water industries, the network strength manifests as a shear yield stress. This same resistance of the network to failure manifests as a pressure dependence of the extent of dewatering in compression whereby the applied pressure rises from zero at the GTC to nominally infinity at close packing. The applied pressure at equilibrium reflects a condition whereby the network dewaters and can resist further change, without creep. In modern dewatering theory [138], this condition describes the so-called compressive yield stress, $P_y(\phi)$ of the suspension. It mimics the solids concentration (ϕ) response of the suspension in shear failure, albeit the stress at failure is at least an order of magnitude higher in compression across the majority of the particle volume fraction range [115,139]. It is not clear that this condition holds in the limit of the GTC. An example of the relationship between the shear and compressive yield stress of a suspension (as a function of volume fraction) for a series of 0.2–0.75 μm ceramic alumina colloidal particles is shown in Figure 9.24 [115]. Here the yield stress, in either compression or shear, is scaled by the particle diameter (d) squared.

Differences in the particulate network, determined by particle size and the depth of the interparticle interaction, mean that some suspensions show a strong dependence on the compressive yield stress as a function of concentration and others a very weak dependence. This is illustrated by wastewater treatment sludges, which are a mixture of cellular fragments flocculated by an extra cellular polymer crosslinked matrix. The GTC of this system is of order 1–2 volume percent solids, but the equilibrium extent of dewatering in high pressure filtration, a technique commonly used for this purpose, is of order 50 volume percent. In contrast, mineral suspensions with a coarse particle distribution with a d_{80} of say 120 μm and a limited colloidal (<20 μm) fraction, are likely to have a GTC of order 40–50 volume percent and an

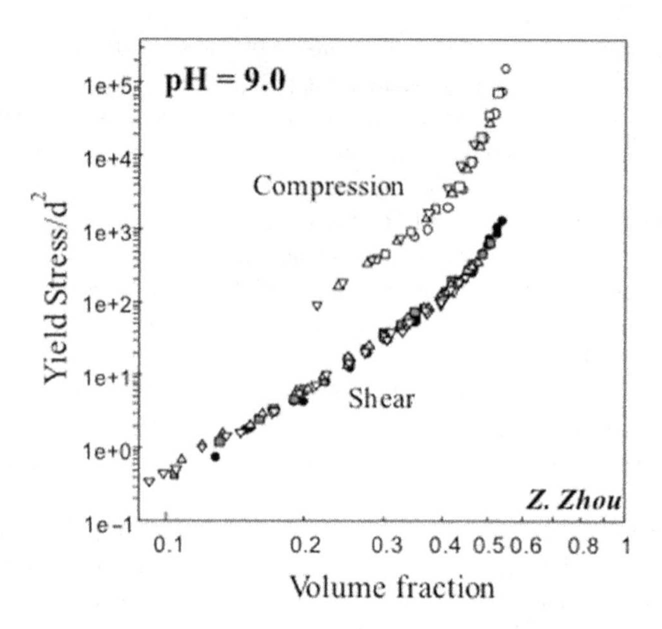

Figure 9.24 Comparison of the yield stress in compression versus that in shear for suspensions of aluminium oxide particles at pH 9.0 with particles sizes in the range 0.2–0.75 μm. Adapted from the work of Zhou et al. [115]

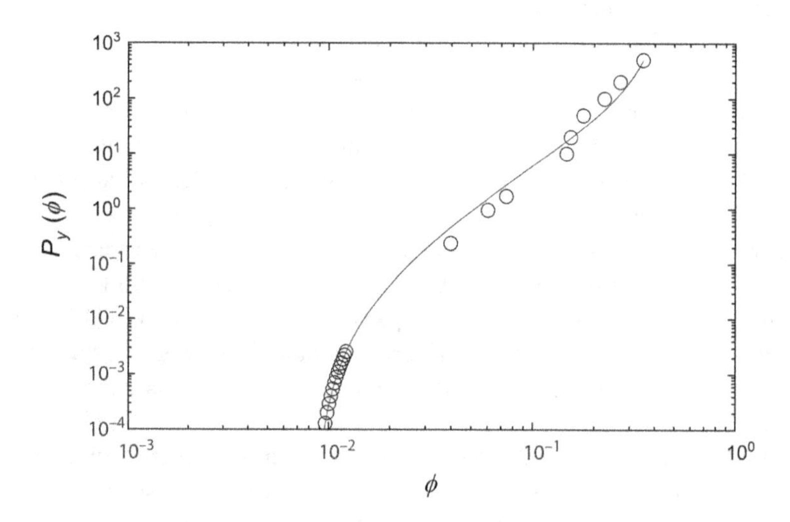

Figure 9.25 Compressive yield stress of a wastewater treatment suspension as a function of the volume fraction of solids.

extent of dewatering that is not significantly higher. These suspensions are often referred to as noncompressional in that they show very little deformation as a function of pressure. An example of the measured compressive yield stress behavior of a wastewater sludge is shown in Figure 9.25.

A governing equation for dewatering in one dimension is derived from conservation of mass and momentum for the solid and liquid phases [140]:

$$\frac{\partial \phi}{\partial t} = \frac{\partial}{\partial z}\left[D(\phi)\frac{\partial \phi}{\partial z} + \phi q(t) + \phi \Delta \rho g h_0 \frac{(1-\phi)^2}{R(\phi)}\right], \quad (9.8)$$

where the solids diffusivity (not the particle diffusivity described earlier in Chapter 1), $D(\phi)$, is given by:

$$D(\phi) = \frac{dP_y(\phi)}{d\phi}\frac{(1-\phi)^2}{R(\phi)}, \quad (9.9)$$

and t is the time, z is the vertical direction coordinate with h_0 the initial sediment height and the solids concentration profile, $\phi(z)$, is dependent on the gravitational constant, g, and the difference in density between the solid and liquid phase, $\Delta\rho$. The bulk flow, $q(t)$, is zero for the case of gravity batch settling (in a solid based container) and dV/dt in the case of constant pressure dead-end filtration (where V is the specific volume of filtrate). This governing equation assumes that compression is irreversible and the rate of the rearrangement of interparticle bonds is much quicker than the rate of drainage, so the solids pressure $p_s \approx P_y(\phi)$ [138].

Eq. (9.8) can be used as the basis for extracting the material properties from dewatering experiments. This is an example of an inverse problem. Conversely, the forward problem of solving Eq. (9.8) for known material properties with appropriate initial and boundary conditions allows the prediction of dewatering behavior. Predictions can be made for operation of full scale dewatering devices including continuous thickening [141–143], batch settling [144–146], and pressure filtration [147–149], if a model for these devices exists.

The rate of network collapse is assumed to be limited only by the rate of escape of the fluid from the network, described through a drag or hindered settling function, $R(\phi)$, in this approach. The form of the hindered settling function for a wastewater treatment sludge is shown in Figure 9.26. The laboratory measurements required to attain the data in Figures 9.25 and 9.26 will not be presented here but the methodology is well described in a series of publications [144,150–153]. Skinner et al. [110] presented an internally consistent approach for highly compressible suspensions, including materials that show either quadratic or nonquadratic filtration behavior [154]. Approaches for noncompressible suspensions are available for a range of solid–liquid separation devices and a range of models to predict the performance, to optimize operations as well as to aid in the design of devices, are presented in the book of Tarleton and Wakeman [155].

In summary, the key descriptors in compressive rheology of aggregated suspensions are:

- The compressive yield stress function, $P_y(\phi)$, is a quantification of the strength of a networked suspension under compression and determines the maximum solids volume fraction reached for a given applied pressure. It is also known as the compressive strength.

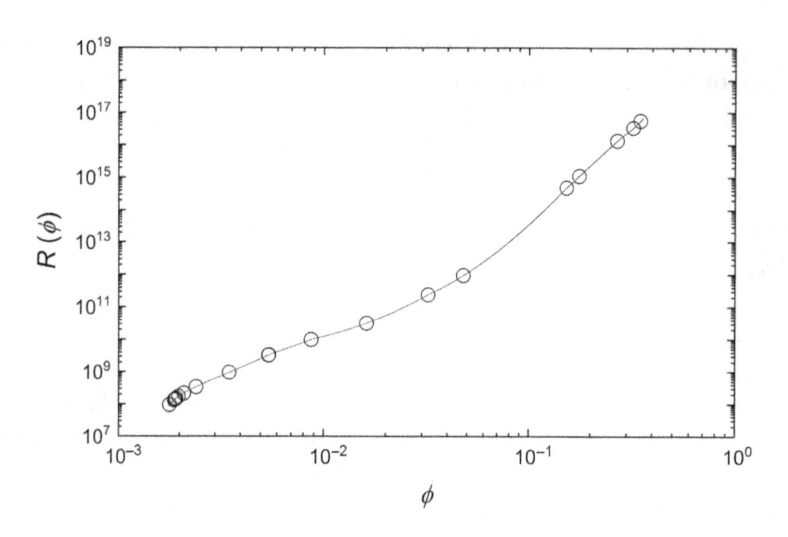

Figure 9.26 Hindered settling function of a wastewater treatment suspension as a function of the volume fraction of solids.

- The hindered settling function, $R(\phi)$, is a quantification of the hydrodynamic drag forces and is proportional to the inverse of the settling velocity during sedimentation or Darcian permeability during filtration.
- The solids diffusivity, $D(\phi)$, is a quantification of the rate of propagation of a concentration gradient after a network has been formed. It is a ratio of the rate of strengthening of the network, $dP_y(\phi)/d\phi$, with the increased resistance to flow, $R(\phi)$, as a network is compressed.

The characterization of dewatering parameters for aggregated particle suspensions in compression allows for incorporation into semiempirical models of devices. This can then be used for operations optimization and, ultimately, the design of new devices. An example of the modeling and optimization of the plate and frame filtration of an alum rich water treatment sludge is shown in the work of Stickland et al. [147,148]. Correlation with full scale data allowed validation of the approach and provides confidence that processes such as plate and frame filtration [156], which is a good mimic of uniaxial compression, are well described by the continuum model approach.

Difficulties arise in matching theory, laboratory practice, and application when one moves to thickening, filtration, and centrifugation processes. It is implicit in these applications that there is a significant shear component overlaid on the uniaxial compression that drives the solid–liquid separation process. In the case of thickening, low shear processes are induced by rake transport of the solids to the outlet, flow against sloping walls, and thermal movements that cause aggregate collision. These collisions all lead to changes in the aggregate structure that are independent of compressional processes. Typically this involves aggregate densification resulting in a change from fractal to nonfractal aggregate structure [157–159] through compression of the outer layers of the aggregate. This process effectively changes the compressive yield stress and the hindered settling curve for a material in time as the solids move through the thickener device [142,160]. This effect is enhanced in the presence of compression [161]. The result is a vast improvement (factor of 5 to 20 times) in the rate of dewatering over that predicted using models that assume only a uniaxial compressive contribution to dewatering. The consequence in practice is a significant reduction in the size of devices required to produce a constant output. This has seen a major reduction in the diameter of thickeners in the past 20 years to achieve an equivalent solids flux [162]. Despite the success in describing the mechanics of changes induced through combined shear and compression through empirical approaches, a unified theory is still missing. A further example is provided by belt press filtration, which combines compression and shear through differential movement of a suspension between two belts moving around a set of rollers. The device produces a solids output that is in excess of that predicted with uniaxial compression alone.

A large number of industrial filtration processes, particularly those involving highly compressible materials, do not reach the equilibrium solids content within the time-scale of the industrial solid–liquid separation process. In the case of thickening, the downward flux of solids in the device is a combination of the sedimentation rate and the withdrawal of solids from the device. Once the latter exceeds the hindered settling rate, consolidation reaches a limiting value, despite the fact that the compressive yield stress at the solids pressure in the system would suggest a far higher concentration was possible. Approaches that increase network permeability are thus important to achieve a greater extent of dewatering, as many of the processes are limited by permeability rather than compression. In many instances it is, however, not clear whether the improvement of dewatering is the result of an increase in the equilibrium extent of dewatering or just an improvement in permeability. The role of flocculants and coagulants comes into question here since all processes that increase the level of attractive particle–particle interaction in a suspension have a negative effect on the theoretical extent of dewatering at a fixed solids concentration. Overuse of coagulants and flocculants, whilst improving permeability, limits the final possible solids concentration and also provides the aggregate with a higher resistance to deformation in shear. The goal for solid–liquid separation processes that combine shear and compression is thus to always minimize the dose of coagulants and flocculants to achieve a clarity and settling rate in the system that exceeds a minimum requirement associated with device design but does not start to significantly limit the top end of dewatering.

Chapter Notation

$D(\phi)$ solids diffusivity, Eq. (9.8) [m^2/s]
d particle diameter [m] (in Figure 9.24)
$\dot{E}$ evaporation rate [m/s]
h film thickness [m]
l_{sag} sagging length [m]

$P_y(\phi)$ Compressive yield stress [kPa]
P_{cap} capillary pressure of evaporating solvent in particle containing coating, Eq. (9.5) [–]
$R(\phi)$ Hindered settling function [kg/s-m^3]
t_f sagging time [t]
V_s settling velocity [m/s]
x distance from the surface [m]
Z^* complex impedance [Ω]
Z^*_ζ impedance of the double layer [Ω]

Greek symbols

Γ surface tension [N/m]
$\eta_{0,m}$ zero shear viscosity of the maltenes [Pa s]
λ wave length of the surface pattern [m]
$\dot{\gamma}_c$ critical shear rate for log-rolling [s^{-1}]
σ_{ac} ac-conductivity [S/m]
ϕ_C critical volume fraction for electrical properties [–]

Subscripts

e due to mobile charge carriers (electrons, holes, and polarons)
i ionic

References

1. Mewis J, Wagner NJ. *Colloidal Suspension Rheology*. Cambridge, UK: Cambridge University Press; 2012. 393 p.
2. Keddie JL, Routh AF. *Fundamentals of Latex Film Formation: Processes and Properties*. Dordrecht: Springer; 2010. 308 p.
3. Patton TC. *Paint Flow and Pigment Dispersion: A Rheological Approach to Coating and Ink Technology*. 2nd ed. New York: John Wiley and Sons; 1979.
4. Orchard SE. On surface leveling in viscous liquids and gels. *Applied Scientific Research A*. 1963;11(4):451–464.
5. Overdiep WS. The effect of a reduced solvent content of solvent-bourne solution paints on film formation *Progress in Organic Coatings*. 1986;14(1):1–21.
6. Overdiep WS. The leveling of paints. *Progress in Organic Coatings*. 1986;14 (2):159–175.
7. Wilson SK. The leveling of paint films. *IMA Journal of Applied Mathematics*. 1993;50 (2):149–166.
8. Howison SD, Moriarty JA, Ockendon JR, Terrill EL, Wilson SK. A mathematical model for drying paint layers. *Journal of Engineering Mathematics*. 1997;32(4):377–394.

9. Evans PL, Schwartz LW, Roy RV. A mathematical model for crater defect formation in a drying paint layer. *Journal of Colloid and Interface Science*. 2000;227(1):191–205.
10. Eales AD, Dartnell N, Goddard S, Routh AF. Thin, binary liquid droplets, containing polymer: An investigation of the parameters controlling film shape. *Journal of Fluid Mechanics*. 2016;794:200–232.
11. Lade RK, Song JO, Musliner AD, Williams BA, Kumar S, Macosko CW, et al. Sag in drying coatings: Prediction and real time measurement with particle tracking. *Progress in Organic Coatings*. 2015;86:49–58.
12. Bhavsar R, Shreepathi S. Evolving empirical rheological limits to predict flow-levelling and sag resistance of waterborne architectural paints. *Progress in Organic Coatings*. 2016;101:15–23.
13. Wu S. Rheology of high solid coatings. 1. Analysis of sagging and slumping. *Journal of Applied Polymer Science*. 1978;22(10):2769–2782.
14. Wu S. Rheology of high solid coatings. 2. Analysis of combined sagging and leveling. *Journal of Applied Polymer Science*. 1978;22(10):2783–2791.
15. Bosma M, Brinkhuis R, Coopmans J, Reuvers B. The role of sag control agents in optimizing the sag/leveling balance and a new powerful tool to study this. *Progress in Organic Coatings*. 2006;55(2):97–104.
16. Deegan RD, Bakajin O, Dupont TF, Huber G, Nagel SR, Witten TA. Capillary flow as the cause of ring stains from dried liquid drops. *Nature*. 1997;389(6653):827–829.
17. Routh AF, Russel WB. Horizontal drying fronts during solvent evaporation from latex films. *AIChE Journal*. 1998;44(9):2088–2098.
18. Salamanca JM, Ciampi E, Faux DA, Glover PM, McDonald PJ, Routh AF, et al. Lateral drying in thick films of waterborne colloidal particles. *Langmuir*. 2001;17(11):3202–3207.
19. Schoff CK. Surface defects: Diagnosis and cure. *Journal of Coatings Technology*. 1999;71(888):56–73.
20. Pearson JRA. On convection cells induced by surface tension. *Journal of Fluid Mechanics*. 1958;4(5):489–500.
21. Bader HF. How to stop pin holes in exam gloves. *Rubber Asia*. 1996;Sept–Oct:85.
22. Groves R, Routh AF. Film deposition and consolidation during thin glove coagulant dipping. *Journal of Polymer Science Part B: Polymer Physics*. 2017;55(22):1633–1648.
23. Deka A, Dey N. Rheological studies of two component high build epoxy and polyurethane based high performance coatings. *Journal of Coatings Technology Research*. 2013;10(3):305–315.
24. Fernando RH, Xing LL, Glass JE. Rheology parameters controlling spray atomization and roll misting behavior of waterborne coatings. *Progress in Organic Coatings*. 2000;40(1–4):35–38.
25. Chassenieux C, Nicolai T, Benyahia L. Rheology of associative polymer solutions. *Current Opinion in Colloid & Interface Science*. 2011;16(1):18–26.
26. Suzuki S, Uneyama T, Inoue T, Watanabe H. Nonlinear rheology of telechelic associative polymer networks: Shear thickening and thinning behavior of hydrophobically modified ethoxylated urethane (HEUR) in aqueous solution. *Macromolecules*. 2012;45(2):888–898.
27. Ianniruberto G, Marrucci G. New interpretation of shear thickening in telechelic associating polymers. *Macromolecules*. 2015;48(15):5439–5449.
28. Annable T, Buscall R, Ettelaie R, Shepherd P, Whittlestone D. Influence of surfactants on the rheology of associating polymers in solution. *Langmuir*. 1994;10(4):1060–1070.

29. Kostansek E. Using dispersion/flocculation phase diagrams to visualize interactions of associative polymers, latexes, and surfactants. *Journal of Coatings Technology*. 2003;75 (940):27–34.
30. Beshah K, Izmitli A, Van Dyk AK, Rabasco JJ, Bohling J, Fitzwater SJ. Diffusion-weighted PFGNMR study of molecular level interactions of loops and direct bridges of HEURs on latex particles. *Macromolecules*. 2013;46(6):2216–2227.
31. Chatterjee T, Nakatani AI, Van Dyk AK. Shear-dependent interactions in hydrophobically modified ethylene oxide urethane (HEUR) based rheology modifier-latex suspensions: Part 1. *Molecular Microstructure. Macromolecules*. 2014;47(3):1155–1174.
32. Van Dyk AK, Chatterjee T, Ginzburg VV, Nakatani AI. Shear-dependent interactions in hydrophobically modified ethylene oxide urethane (HEUR) based coatings: Mesoscale structure and viscosity. *Macromolecules*. 2015;48(6):1866–1882.
33. Huldén M. Hydrophobically modified urethane–ethoxylate (HEUR) associative thickeners 2. Interaction with latex. *Colloids and Surfaces A: Physicochemical and Engineering Aspects*. 1994;88(2):207–221.
34. van der Waarden M. Stabilization of carbon black dispersions in hydrocarbons. *Journal of Colloid Science*. 1950;5(3):317–325.
35. Hartley PA, Parfitt GD. Dispersion of powders in liquids 1. Contributions of the van der Waals force to the cohesiveness of carbon black powders. *Langmuir*. 1985;1(6):651–657.
36. Dagastine RR, Prieve DC, White LR. Calculations of van der Waals forces in 2-dimensionally anisotropic materials and its application to carbon black. *Journal of Colloid and Interface Science*. 2002;249(1):78–83.
37. Hartley PA, Parfitt GD, Pollack LB. The role of the van der Waals force in the agglomeration of powders containing submicron particles. *Powder Technology*. 1985;42(1):35–46.
38. Spahr ME, Gilardi R, Bonacchi D. Carbon black for electrically conductive polymer applications. In Palsule S. (ed.) *Encyclopedia of Polymers and Composites*. Berlin: Springer; 2013. pp. 1–20. https://doi.org/10.1007/978-3-642-37179-0_32-1.
39. Hatzell KB, Beidaghi M, Campos JW, Dennison CR, Kumbur EC, Gogotsi Y. A high performance pseudocapacitive suspension electrode for the electrochemical flow capacitor. *Electrochimica Acta*. 2013;111:888–897.
40. Duduta M, Ho B, Wood VC, Limthongkul P, Brunini VE, Carter WC, et al. Semi-solid lithium rechargeable flow battery. *Advanced Energy Materials*. 2011;1(4):511–516.
41. Choo KY, Yoo CY, Han MH, Kim DK. Electrochemical analysis of slurry electrodes for flow-electrode capacitive deionization. *Journal of Electroanalytical Chemistry*. 2017;806:50–60.
42. Kroy K, Cates ME, Poon WCK. Cluster mode–coupling approach to weak gelation in attractive colloids. *Physical Review Letters*. 2004;92(14):148302.
43. Eggersdorfer ML, Kadau D, Herrmann HJ, Pratsinis SE. Fragmentation and restructuring of soft-agglomerates under shear. *Journal of Colloid and Interface Science*. 2010;342 (2):261–268.
44. Helal A, Divoux T, McKinley GH. Simultaneous rheoelectric measurements of strongly conductive complex fluids. *Physical Review Applied*. 2016;6(6):064004.
45. Jamali S, McKinley GH, Armstrong RC. Microstructural rearrangements and their rheological implications in a model thixotropic elastoviscoplastic fluid. *Physical Review Letters*. 2017;118(4):048003.
46. Collins IR, Taylor SE. The microstructural properties of coagulated dispersions of carbon black. *Journal of Colloids and Interfacial Science*. 1993;155(2):471–481.

47. Osuji CO, Kim C, Weitz DA. Shear thickening and scaling of the elastic modulus in a fractal colloidal system with attractive interactions. *Physical Review E: Statistical, Nonlinear, and Soft Matter Physics*. 2008;77(6 Pt 1):060402.
48. Dullaert K, Mewis J. A structural kinetics model for thixotropy. *Journal of Non-Newtonian Fluid Mechanics*. 2006;139(1–2):21–30.
49. Ovarlez G, Tocquer L, Bertrand F, Coussot P. Rheopexy and tunable yield stress of carbon black suspensions. *Soft Matter*. 2013;9(23):5540–5549.
50. Wei Y, Solomon MJ, Larson RG. Quantitative nonlinear thixotropic model with stretched exponential response in transient shear flows. *Journal of Rheology*. 2016;60 (6):1301–1315.
51. Mwasame PM, Beris AM, Diemer RB, Wagner NJ. A constitutive equation for thixotropic suspensions with yield stress by coarse-graining a population balance model. *AIChE Journal*. 2017;63(2):517–531.
52. Colombo G, Kim S, Schweizer T, Schroyen B, Clasen C, Mewis J, et al. Superposition rheology and anisotropy in rheological properties of sheared colloidal gels. *Journal of Rheology*. 2017;61(5):1035–1048.
53. Grenard V, Taberlet N, Manneville S. Shear-induced structuration of confined carbon black gels: steady-state features of vorticity-aligned flocs. *Soft Matter*. 2011;7 (8):3920–3928.
54. Varga Z, Swan JW. Large scale anisotropies in sheared colloidal gels. *Journal of Rheology*. 2018;62(2):405–418.
55. Wagner NJ, Brady JF. Shear thickening in colloidal dispersions. *Physics Today*. 2009;62 (10):27–32.
56. Zaccone A, Gentili D, Wu H, Morbidelli M, Del Gado E. Shear-driven solidification of dilute colloidal suspensions. *Physical Review Letters*. 2011;106(13):138301.
57. Brown E, Forman NA, Orellana CS, Zhang H, Maynor BW, Betts DE, et al. Generality of shear thickening in dense suspensions. *Nature Materials*. 2010;9(3):220–224.
58. Harshe YM, Lattuada M. Breakage rate of colloidal aggregates in shear flow through stokesian dynamics. *Langmuir*. 2012;28(1):283–292.
59. Narayanan A, Mugele F, Duits MHG. Mechanical history dependence in carbon black suspensions for flow batteries: A rheo-impedance study. *Langmuir*. 2017;33 (7):1629–1638.
60. Negita K, Misono Y, Yamaguchi T, Shinagawa J. Dielectric and electrical properties of electrorheological carbon suspensions. *Journal of Colloid and Interface Science*. 2008;321(2):452–458.
61. Richards JJ, Hipp JB, Riley JK, Wagner NJ, Butler PD. Clustering and percolation in suspensions of carbon black. *Langmuir*. 2017;33(43):12260–12266.
62. Petek TJ, Hoyt NC, Savinell RF, Wainright JS. Characterizing slurry electrodes using electrochemical impedance spectroscopy. *Journal of the Electrochemical Society*. 2016;163(1):A5001–A5009.
63. Mewis J, de Groot LM, Helsen JA. Dielectric behavior of flowing thixotropic suspensions. *Colloids and Surfaces*. 1987;22(2–4):271–289.
64. Dissado LA, Hill RM. Anomalous low-frequency dispersion. Near direct current conductivity in disordered low-dimensional materials. *Journal of Chemistry Society, Faraday Transactions*. 1984;2(80):291–319.
65. Niklasson GA. Comparison of dielectric response functions for conducting materials. *Journal of Applied Physics*. 1989;66(9):4350–4359.

66. Hipp JB, Richards JJ, Wagner NJ. Structure-property relationships of sheared carbon black suspensions determined by simultaneous rheological and neutron scattering measurements. *Journal of Rheology*. 2019;63(3):423–436.
67. Abraham H. *Asphalts and Allied Substances: Their Occurrence, Modes of Production, Uses in the Arts, and Methods of Testing*, 6th ed. Princeton, NJ: Van Nostrand; 1960.
68. Lesueur D. The colloidal structure of bitumen: Consequences on the rheology and on the mechanisms of bitumen modification. *Advances in Colloid and Interface Science*. 2009;145(1):42–82.
69. Jones IV DR. *SHRP Materials Reference Library: Asphalt Cements: A Concise Data Compilation*. Washington, DC: Strategic Highway Research Program, National Research Council, Washington, DC; 1993.
70. Redelius PG. The structure of asphaltenes in bitumen. *Road Materials and Pavement Design*. 2006;7(sup1):143–162.
71. Storm DA, Sheu EY, DeTar MM. Macrostructure of asphaltenes in vacuum residue by small-angle X-ray scattering. *Fuel*. 1993;72(7):977–981.
72. Adedeji A, Grünfelder T, Bates FS, Macosko CW, Stroup-Gardiner M, Newcomb DE. Asphalt modified by SBS triblock copolymer: Structures and properties. *Polymer Engineering & Science*. 1996;36(12):1707–1723.
73. Pauli AT, Grimes RW, Beemer AG, Turner TF, Branthaver JF. Morphology of asphalts, asphalt fractions and model wax-doped asphalts studied by atomic force microscopy. *International Journal of Pavement Engineering*. 2011;12(4):291–309.
74. González E, Costa LMB, Silva HMRD, Hilliou L. Rheological characterization of EVA and HDPE polymer modified bitumens under large deformation at 20°C. *Construction and Building Materials*. 2016;112:756–764.
75. Tattersall GH, Banfill PFG. *The Rheology of Fresh Concrete*. Boston, MA: Pitman Advanced Pub. Program; 1983. xii, 356 p.
76. de Larrard F. *Concrete Mixture Proportioning: A Scientific Approach*. London: E & FN Spon, an imprint of Rotledge; 1999. xvii, 421 p.
77. Roussel N. Rheology of fresh concrete: From measurements to predictions of casting processes. *Materials and Structures*. 2007;40(10):1001–1012.
78. Hu C, de Larrard F. The rheology of fresh high-performance concrete. *Cement and Concrete Research*. 1996;26(2):283–294.
79. Banfill PFG, Saunders DC. On the viscometric examination of cement pastes. *Cement and Concrete Research*. 1981;11(3):363–370.
80. Roussel N. Steady and transient flow behaviour of fresh cement pastes. *Cement and Concrete Research*. 2005;35(9):1656–1664.
81. Tattersall GH, Bloomer SJ. Further development of the two-point test for workability and extension of its range. *Magazine of Concrete Research*. 1979;31(109):202–210.
82. Hu C, de Larrard F, Sedran T, Boulay C, Bosc F, Deflorenne F. Validation of BTRHEOM, the new rheometer for soft-to-fluid concrete. *Materials and Structures*. 1996;29(194):620–631.
83. Ferraris CF, Brower LE. Comparison of concrete rheometers: International tests at LCPC (Nantes, France) in October 2000. National Institute of Standards and Technology Interagency Report (NISTIR) 6819; 2001.
84. Ferraris CF, Brower LE. Comparison of concrete rheometers: International tests at MB (Cleveland OH, USA) in May 2003. National Institute of Standards and Technology Interagency Report (NISTIR) 7154; 2004.

85. Schowalter WR, Christensen G. Toward a rationalization of the slump test for fresh concrete: Comparisons of calculations and experiments. *Journal of Rheology*. 1998;42 (4):865–870.
86. Roussel N, Coussot P. "Fifty-cent rheometer" for yield stress measurements: From slump to spreading flow. *Journal of Rheology*. 2005;49(3):705–718.
87. Roussel N, Stefani C, Leroy R. From mini-cone test to Abrams cone test: Measurement of cement-based materials yield stress using slump tests. *Cement and Concrete Research*. 2005;35(5):817–822.
88. Roussel N. A thixotropy model for fresh fluid concretes: Theory, validation and applications. *Cement and Concrete Research*. 2006;36(10):1797–1806.
89. Ovarlez G, Roussel N. A physical model for the prediction of lateral stress exerted by self-compacting concrete on formwork. *Materials and Structures*. 2006;39(2):269–279.
90. Roussel N, Cussigh F. Distinct-layer casting of SCC: The mechanical consequences of thixotropy. *Cement and Concrete Research*. 2008;38(5):624–632.
91. Yammine J, Chaouche M, Guerinet M, Moranville M, Roussel N. From ordinary rhelogy concrete to self compacting concrete: A transition between frictional and hydrodynamic interactions. *Cement and Concrete Research*. 2008;38(7):890–896.
92. Dong KJ, Yang RY, Zou RP, Yu AB. Role of interparticle forces in the formation of random loose packing. *Physical Review Letters*. 2006;96(14):145505.
93. Roussel N, Lemaitre A, Flatt RJ, Coussot P. Steady state flow of cement suspensions: A micromechanical state of the art. *Cement and Concrete Research*. 2010;40(1):77–84.
94. Perrot A, Lecompte T, Khelifi H, Brumaud C, Hot J, Roussel N. Yield stress and bleeding of fresh cement pastes. *Cement and Concrete Research*. 2012;42(7):937–944.
95. Flatt RJ. Towards a prediction of superplasticized concrete rheology. *Materials and Structures*. 2004;37(269):289–300.
96. Ovarlez G, Bertrand F, Rodts S. Local determination of the constitutive law of a dense suspension of noncolloidal particles through magnetic resonance imaging. *Journal of Rheology*. 2006;50(3):259–292.
97. Hot J, Bessaies-Bey H, Brumaud C, Duc M, Castella C, Roussel N. Adsorbing polymers and viscosity of cement pastes. *Cement and Concrete Research*. 2014;63:12–19.
98. Roussel N, Ovarlez G, Garrault S, Brumaud C. The origins of thixotropy of fresh cement pastes. *Cement and Concrete Research*. 2012;42(1):148–157.
99. Brumaud C, Baumann R, Schmitz M, Radler M, Roussel N. Cellulose ethers and yield stress of cement pastes. *Cement and Concrete Research*. 2014;55:14–21.
100. Bessaies-Bey H, Baumann R, Schmitz M, Radler M, Roussel N. Effect of polyacrylamide on rheology of fresh cement pastes. *Cement and Concrete Research*. 2015;76:98–106.
101. Flatt RJ, Bowen P. Electrostatic repulsion between particles in cement suspensions: Domain of validity of linearized Poisson-Boltzmann equation for nonideal electrolytes. *Cement and Concrete Research*. 2003;33(6):781–791.
102. Flatt RJ, Schober I, Raphael E, Plassard C, Lesniewska E. Conformation of adsorbed comb copolymer dispersants. *Langmuir*. 2009;25(2):845–855.
103. Marliere C, Mabrouk E, Lamblet M, Coussot P. How water retention in porous media with cellulose ethers works. *Cement and Concrete Research*. 2012;42(11):1501–1512.
104. Bülichen D, Kainz J, Plank J. Working mechanism of methyl hydroxyethyl cellulose (MHEC) as water retention agent. *Cement and Concrete Research*. 2012;42(7):953–959.
105. Brumaud C, Bessaies-Bey H, Mohler C, Baumann R, Schmitz M, Radler M, et al. Cellulose ethers and water retention. *Cement and Concrete Research*. 2013;53:176–184.

106. Brumaud C. Origines microscopiques des conséquences rhéologiques de l'ajout d'éthers de cellulose dans une suspension cimentaire (in French) [PhD]. Paris: East Paris University; 2011.
107. Soppe W. Computer simulation of random packings of hard spheres. *Powder Technology.* 1990;62(2):189–196.
108. Dirksen JA, Ring TA. Fundamentals of crystallization: Kinetic effects on particle size distributions and morphology. *Chemical Engineering Science.* 1991;46(10):2389–2427.
109. Franks GV, Zhou Y. Relationship between aggregate and sediment bed properties: Influence of inter-particle adhesion. *Advanced Powder Technology.* 2010;21(4):362–373.
110. Skinner SJ, Studer LJ, Dixon DR, *et al.* Quantification of wastewater sludge dewatering. *Water Research.* 2015;82:2–13.
111. Leong YK, Scales PJ, Healy TW, Boger DV. Effect of particle size on colloidal zirconia rheology at the iso-electric point. *Journal of the American Ceramic Society.* 1995;78 (8):2209–2212.
112. Scales PJ, Johnson SB, Healy TW, Kapur PC. Shear yield stress of partially flocculated colloidal suspensions. *AIChE Journal.* 1998;44(3):538–544.
113. Franks GV, Johnson SB, Scales PJ, Boger DV, Healy TW. Ion-specific strength of attractive particle networks. *Langmuir.* 1999;15(13):4411–4420.
114. Johnson SB, Franks GV, Scales PJ, Boger DV, Healy TW. Surface chemistry-rheology relationships in concentrated mineral suspensions. *International Journal of Mineral Processing.* 2000;58(1–4):267–304.
115. Zhou ZW, Scales PJ, Boger DV. Chemical and physical control of the rheology of concentrated metal oxide suspensions. *Chemical Engineering Science.* 2001;56(9):2901–2920.
116. Kapur PC, Scales PJ, Boger DV, Healy TW. Yield stress of suspensions loaded with size distributed particles. *AIChE Journal.* 1997;43(5):1171–1179.
117. Nguyen QD, Boger DV. Yield stress measurement for concentrated suspensions. *Journal of Rheology.* 1983;27(4):321–349.
118. Nguyen QD, Boger DV. Direct yield stress measurement with the vane technique. *Journal of Rheology.* 1985;29(3):335–347.
119. Uhlherr PHT, Guo J, Tiu C, Zhang XM, Zhou JZQ, Fang TN. The shear-induced solid–liquid transition in yield stress materials with chemically different structures. *Journal of Non-Newtonian Fluid Mechanics.* 2005;125(2–3):101–119.
120. Pham K, Petekidis G, Vlassopoulos D, Egelhaaf S, Poon W, Pusey P. Yielding behavior of repulsion-and attraction-dominated colloidal glasses. *Journal of Rheology.* 2008;52 (2):649–676.
121. Buscall R, Scales PJ, Stickland AD, Teo H-E, Lester DR. Dynamic and rate-dependent yielding in model cohesive suspensions. *Journal of Non-Newtonian Fluid Mechanics.* 2015;221:40–54.
122. Pashias N, Boger DV, Summers J, Glenister DJ. A fifty cent rheometer for yield stress measurement. *Journal of Rheology.* 1996;40(6):1179–1189.
123. Clayton SA, Grice TG, Boger DV. Analysis of the slump test for on-site yield stress measurement of mineral suspensions. *International Journal of Mineral Processing.* 2003;70(1–4):3–21.
124. Pullum L, Graham L, Rudman M, Hamilton R. High concentration suspension pumping. *Minerals Engineering.* 2006;19(5):471–477.
125. Guang R, Rudman M, Chryss A, Slatter P, Bhattacharya S. A DNS investigation of the effect of yield stress for turbulent non-Newtonian suspension flow in open channels. *Particulate Science and Technology.* 2011;29(3):209–228.

126. Rudman M, Simic K, Paterson DA, Strode P, Brent A, Sutalo ID. Raking in gravity thickeners. *International Journal of Mineral Processing*. 2008;86(1–4):114–130.
127. Sofra F, Boger DV. Exploiting the rheology of mine tailings for dry disposal. Proceedings of the 2000 International Conference on Tailings and Mine Waste, Fort Collins, Colorado; 2000; pp. 169–180.
128. Sofra F, Boger DV. Environmental rheology for waste minimisation in the minerals industry. *Chemical Engineering Journal*. 2002;86(3):319–330.
129. Rico M, Benito G, Salgueiro AR, Díez-Herrero A, Pereira HG. Reported tailings dam failures. *Journal of Hazardous Materials*. 2008;152(2):846–852.
130. Graham LJW, Pullum L. An investigation of complex hybrid suspension flows by magnetic resonance imaging. *The Canadian Journal of Chemical Engineering*. 2002;80 (2):200–207.
131. Wasp EJ, Kenny JP, Gandhi RL. *Solid–Liquid Flow: Slurry Pipeline Transportation*. Clausthal: Trans Tech Publications; 1977.
132. Metzner AB, Reed JC. Flow of non-Newtonian fluids – Correlation of the laminar, transition, and turbulent-flow regions. *AIChE Journal*. 1955;1(4):434–440.
133. Founargiotakis K, Kelessidis VC, Maglione R. Laminar, transitional and turbulent flow of Herschel–Bulkley fluids in concentric annulus. *The Canadian Journal of Chemical Engineering*. 2008;86(4):676–683.
134. Slatter PT, Haldenwang R, Chhabra RP (eds.). The laminar/turbulent transition for paste sheet flow. Paste 2011 – 14th International Seminar on Paste and Thickened Tailings; Perth, Australia; April 5–7, 2011.
135. Smith LD, Rudman M, Lester DR, Metcalfe G. Mixing of discontinuously deforming media. *Chaos*. 2016;26(2):023113.
136. von Mises R. Mechanics of the ductile form changes of crystals. *Zeitschrift für Angewandte Mathematik und Mechanik*. 1928;8(3):161–185.
137. Wang Y, Koynov S, Glasser BJ, Muzzio FJ. A method to analyze shear cell data of powders measured under different initial consolidation stresses. *Powder Technology*. 2016;294:105–112.
138. Buscall R, White LR. The consolidation of concentrated suspensions. Part 1. The theory of sedimentation. *Journal of the Chemical Society, Faraday Transactions 1: Physical Chemistry in Condensed Phases*. 1987;83(3):873–891.
139. Channell GM, Zukoski CF. Shear and compressive rheology of aggregated alumina suspensions. *AIChE Journal*. 1997;43(7):1700.
140. Landman KA, White LR. Solid/liquid separation of flocculated suspensions. *Advances in Colloid and Interface Science*. 1994;51:175–246.
141. Usher SP, Scales PJ. Steady state thickener modelling from the compressive yield stress and hindered settling function. *Chemical Engineering Journal*. 2005;111(2–3):253–261.
142. Usher SP, Spehar R, Scales PJ. Theoretical analysis of aggregate densification; impact on thickener performance. *Chemical Engineering Journal*. 2009;151(1–3):202–208.
143. Betancourt F, Burger R, Diehl S, Mejias C. Advanced methods of flux identification for clarifier-thickener simulation models. *Minerals Engineering*. 2014;63:2–15.
144. Lester DR, Usher SP, Scales PJ. Estimation of the hindered settling function R(phi) from batch-settling tests. *AIChE Journal*. 2005;51(4):1158–1168.
145. Burger R, Wedland WL, Concha F. Model equations for gravitational sedimentation-consolidation processes. *Zeitschrift für Angewandte Mathematik und Mechanik*. 2000;80:79–92.

146. Burger R, Evje S, Karlsen KH, Lie K-A. Numerical methods for the simulation of the settling of flocculated suspensions. *Chemical Engineering Journal*. 2000;80(1–3):91–104.
147. Stickland AD, de Kretser RG, Usher SP, Hillis P, Tillotson MR, Scales PJ. Numerical modelling of fixed-cavity plate-and-frame filtration: Formulation, validation and optimisation. *Chemical Engineering Science*. 2006;61(12):3818–3829.
148. Stickland AD, de Kretser RG, Kilcullen AR, Scales PJ, Hillis P, Tillotson MR. Numerical modeling of flexible-membrane plate-and-frame filtration. *AIChE Journal*. 2008;54 (2):464–474.
149. Bürger R, Concha F, Karlsen KH. Phenomenological model of filtration processes: 1. Cake formation and expression. *Chemical Engineering Science*. 2001;56(15):4537–4553.
150. Green MD, Landman KA, De Kretser R, Boger DV. Pressure filtration technique for complete characterisation of consolidating suspensions. *Industrial & Engineering Chemistry Research*. 1998;37(10):4152–4156.
151. de Kretser RG, Usher SP, Scales PJ, Boger DV, Landman KA. Rapid filtration measurement of dewatering design and optimization parameters. *AIChE Journal*. 2001;47 (8):1758–1769.
152. Usher SP, de Kretser RG, Scales PJ. Validation of a new filtration technique for dewaterability characterization. *AIChE Journal*. 2001;47(7):1561–1570.
153. Usher SP, Studer LJ, Wall RC, Scales PJ. Characterisation of dewaterability from equilibrium and transient centrifugation test data. *Chemical Engineering Science*. 2013;93:277–291.
154. Stickland AD, de Kretser RG, Scales PJ. Nontraditional constant pressure filtration behavior. *AIChE Journal*. 2005;51(9):2481–2488.
155. Tarleton ES, Wakeman RJ. *Solid/Liquid Separation: Equipment Selection and Process Design*. Oxford: Butterworth-Heinemann; 2007.
156. de Kretser RG, Saha H, Biscombe C, Scales PJ. Plate and frame pressure filtration optimisation using plant load cell data: Advantages, challenges and outcomes. *Filtration*. 2010;10(2):130–135.
157. Farrow J, Fawell P, Johnston R, Nguyen T, Rudman M, Simic K, et al. Recent developments in techniques and methodologies for improving thickener performance. *Chemical Engineering Journal*. 2000;80(1–3):149–155.
158. Gladman B, de Kretser RG, Rudman M, Scales PJ. Effect of shear on particulate suspension dewatering. *Chemical Engineering Research & Design*. 2005;83 (A7):933–936.
159. van Deventer BBG, Usher SP, Kumar A, Rudman M, Scales PJ. Aggregate densification and batch settling. *Chemical Engineering Journal*. 2011;171(1):141–151.
160. Grassia P, Zhang Y, Martin AD, Usher SP, Scales PJ, Crust A, Spehar R. Effects of aggregate densification upon thickening of Kynchian suspensions. *Chemical Engineering Science*. 2014;111:56–72.
161. Spehar R, Kiviti-Manor A, Fawell P, Usher SP, Rudman M, Scales PJ. Aggregate densification in the thickening of flocculated suspensions in an un-networked bed. *Chemical Engineering Science*. 2015;122:585–595.
162. Fawell PD, Farrow JB, Heath AR, Nguyen TV, Owen AT, Paterson D, et al. 20 years of AMIRA P266 "Improving Thickener Technology" – How has it changed the understanding of thickener performance? In: Jewell R, Fourie AB, Barrera S, Wiertz J. (eds.) *Proceedings of the 12th International Seminar on Paste and Thickened Tailings*. Perth: Australian Centre for Geomechanics; 2009. pp. 59–68. https://doi.org/10.36487/ACG_repo/963_7.

Index

CPSIA information can be obtained
at www.ICGtesting.com
Printed in the USA
LVHW060852030821
694401LV00007B/449